Springer Series in Materials Science

Volume 322

The Springer Series in Materials Science covers the complete spectrum of materials research and technology, including fundamental principles, physical properties, materials theory and design. Recognizing the increasing importance of materials science in future device technologies, the book titles in this series reflect the state-of-the-art in understanding and controlling the structure and properties of all important classes of materials.

More information about this series at https://link.springer.com/bookseries/856

Zoila Barandiarán · Jonas Joos · Luis Seijo

Luminescent Materials

A Quantum Chemical Approach for Computer-Aided Discovery and Design

Springer

Zoila Barandiarán
Departamento de Química
Universidad Autónoma de Madrid
Madrid, Spain

Jonas Joos
Department of Solid State Sciences
Ghent University
Gent, Belgium

Luis Seijo
Departamento de Química
Universidad Autónoma de Madrid
Madrid, Spain

ISSN 0933-033X ISSN 2196-2812 (electronic)
Springer Series in Materials Science
ISBN 978-3-030-94983-9 ISBN 978-3-030-94984-6 (eBook)
https://doi.org/10.1007/978-3-030-94984-6

This Springer imprint is published by the registered company Springer Nature Switzerland AG
The registered company address is: Gewerbestrasse 11, 6330 Cham, Switzerland

To Sigeru Huzinaga and to the memory of Björn O. Roos and George Blasse.

Professor Huzinaga led the way from gaussian orbitals to quantum chemistry, which brought complex molecules within the reach of ab initio calculations. Our quantum chemical background grew by his side and under his guidance, together with our views and attitudes towards science.

Björn Roos led the development of solid, practical multiconfigurational methods that realized the dream of handling complex manifolds of excited states of molecules. He educated about them and about optimistic hard work to a community of researchers to which we belong.

George Blasse was a leader in the field of luminescent materials able to look decades ahead of his time. He saw hope in his Pandora's box of luminescence. And the hope he saw had the shape and colors of ab initio quantum chemistry.

To Irene. To Lotte and Sofie. To the people we love. To the memory of our beloved ones.

Foreword

Luminescent materials have a long history and are fascinating as our eyes can observe colourful emission of light. Early reports on luminescence date back to observations of mysterious afterglow from paintings in ancient China and later the Bologna stone glowing in a dark room. Science solves mysteries and in the past centuries scientists have increased our understanding of luminescence and light emission to a level that we can now accurately calculate and predict emission and absorption spectra for atoms and simple molecules. A notoriously difficult class of luminescent species are however the lanthanides. Around 1906 Becquerel observed a rich but unexplained structure of many sharp absorption lines for lanthanide doped crystals. It took until 1937 to understand the origin when Van Vleck wrote his famous paper 'The puzzle of rare earth spectra in solids'. He revealed the origin of the sharp lines: 'forbidden' transitions within the inner $4f^n$ shell of lanthanide ions.

In the decades that followed the understanding increased rapidly with important early contributions from Bethe, Racah, Hellwege, Dieke, Wybourne, Judd, Ofelt, Crosswhite (2x) and Carnall who laid the foundation for unravelling the rich energy level structure and provided a basic theory to calculate energy levels and line strengths for lanthanides. The parameter-based models were refined by Richardson, Reid, Burdick and many other clever theoreticians closely interacting with spectroscopists. Good agreement with experiments showed the strength of the models but the outcome lacked insight in many aspects for which exact knowledge of wavefunctions of ground and excited states is essential. For example, explaining the origin and width of broad band emissions, Stokes shifts, non-radiative relaxation and energy transfer rates is only possible with ab initio type calculations. This was already realized by my thesis advisor and mentor George Blasse who, 25 years ago in his famous 'Pandora's Box' address, identified these problems but also expressed hope, as beautifully quoted by the authors in the preface of this book. It is wonderful to see the book dedicated to George, one year after he passed away. It is also wonderful to see how much progress has been made in the problems he identified (and also many other problems!) through the method he saw as hope in Pandora's Box: ab initio calculations.

It is not trivial that ab initio methods have been the next step towards a better understanding of luminescent materials. Reliable calculations require an as complete as possible set of wavefunctions for an as large as possible number of atoms in a cluster embedded in an as accurate as possible potential of the surrounding host. This will eventually give all the answers. We hope. However, as with all theoretical calculations, computer power and time are crucial. This poses a problem for lanthanides. The complexity is daunting and that is why only few scientists have dared to embark on the journey to use ab initio methods to understand lanthanide luminescence. Reading the book is the best way to grasp how complex the calculations are but it is instructive to give a flavour of some of the contributing factors. Clearly, lanthanides are rather heavy atoms with many electrons and thus many wavefunctions that need to be included. The partly filled 4f shell gives rise to many different electron configurations (microstates) that have to be considered. For the $4f^6$ configuration of Eu^{3+} there are 3003 microstates. The $4f^65d$ configuration of Eu^{2+} involves 30030 microstates. Interactions and mixing of states has to be taken into account and also $4f^{n-1}6s$ states cannot be ignored. In addition, because lanthanides are heavy elements, spin-orbit coupling and relativistic effects play an important role.

The rapidly increasing computer power has been instrumental for the progress and success of ab initio methods in the field of lanthanide luminescence. Even more important has been the input of the authors of this book: Zoila Barandiarán and Luis Seijo have dedicated most of their scientific careers to make it possible to carry out these highly complex calculations. This requires deep understanding and ingenuity, applying clever tricks to make calculations more efficient and, also very important, close interaction with experiment to verify the outcome to further improve the calculations. Experimental results (when measured and interpreted correctly!) are always the correct solution of the puzzle of nature. This is something that Luis and Zoila realize, as is also evident in this book. I greatly enjoyed interacting with them during sabbatical periods they spent in our experimental spectroscopy group. With their retirement from science, the luminescence community loses two giants in the field and this will hamper further progress. However, we do not lose hope. The third author of the book, Jonas Joos, is an extremely talented scientist, combining theory and experiment. He and a few (but not enough) other young scientists have the skills and determination to continue improving the methodology and provide answers to unanswered and emerging questions in the field of luminescence.

Fortunately for the future of the field of luminescence of lanthanides there is now also this book. I am so pleased that the authors took the time and effort to write and excellent and complete book, carefully and clearly describing the background theory, calculation methodology and a plethora of examples and applications of their life's work to illustrate the power of their methods. It is a precious legacy that will serve the community to realize what has been achieved in the past decades and how our understanding of lanthanide (and as a bonus some other heavy ions) luminescence has truly increased. At the same time this book is a solid foundation for the future of the field of ab initio calculations for lanthanides. All the knowledge, theory, skills and tricks stored in the minds of the authors are now laid out in a comprehensive manner that will educate and help the field to move forward. It cannot be overestimated how important it is that this information is not lost. I hope more young scientists will be attracted to join this field—although it is certainly not an area of research for fast and easy high profile publications—and explore and solve new problems taking advantage of the ever increasing computer power.

In their preface the authors give a very insightful overview of the structure, philosophy and content of the book and how different sections can be read and used by different types of researchers in the field of luminescent materials. It is not necessary to repeat the scope and contents here. I will share my impression after a fast first reading of the book. I was impressed and thoroughly enjoyed the book. It has a very logical structure building on the foundations of quantum chemistry, giving credit to the pioneers and placing the field and their work in a historic context. The detailed description of the many theories and methods in part I also gives insight in how important experience and maybe even intuition are for successful calculations through subtle sentences like *become feasible by wise, educated, and systematic truncations that ultimately ensure systematic accuracies*. The deep familiarity with the field is also evident from interesting side comments as in the section on group theory *are known as Abelian groups, after the tragically short-lived Norwegian mathematician Niels Henrik Abel* which prompts a quick Google search to find out more. The theory is impossible to fully understand for an experimentalist like me but it is a pleasure to read and learn from the theory section to get a better sense of the theory and methodology seeing it all described in a systematic, complete and well-structured manner.

Part II of the book is called Tutorial and is somewhat unusual as it mostly really is a tutorial. It guides the reader step-by-step through calculations showing in detail the commands to use in the programs and methods, shows how input and output looks like and warns for common mistakes. For example, the *Navigation notes made during the calculation of the* $4f^1$ *and* $5d^1$ *states of* Ce^{3+} *in* BaF_2 and instructions how to *Prepare SEWARD input* provide detailed descriptions not often found in books but are extremely valuable. This type of instructions, tips and tricks are important to pass on to the next generation but are often forgotten and rarely shared as clearly as in this book. Of course, scientists not performing calculations may quickly page through this part and go to parts III to V which are for the majority of readers the most impressive and useful sections. For all lanthanides calculations are shown for $4f^n$, $4f^{n-1}5d$ and charge transfer states, an in depth comparison is made with experimental studies and implications for commercial phosphors in light emitting devices are outlined, linking purely fundamental studies to real life applications. The sections present a complete overview of the body of work by the authors and it is great to see all the information from decades of research collected and discussed in a systematic way, highlighting everything we now understand thanks to ab initio calculations and also identifying challenges. Everyone will have the opportunity to read up on their favourite topics. My first interests were topics I have been involved in such as contraction in excited $4f^{n-1}5d$ states, calculating the Stokes shift of Ce^{3+} emission in YAG:Ce, energy level structure and excited state relaxation of Tm^{2+}, Yb^{2+} and Bi^{2+} and many more topics. A comprehensive overview of the highly original recent work on intervalence charge transfer transitions concludes the book.

The book will find its way in our luminescence community and will become a must-read for everyone in the field. It provides readers with a deep understanding and the latest insights in luminescence processes of lanthanide ions. Also, it gives hope and direction to how in the future ab initio calculations, with more advanced methods and more powerful computers, can solve the remaining mysteries. The foundation is laid by the authors of this book and their names can be added to the list of great scientists in the field of lanthanide luminescence mentioned in the beginning of this foreword. I hope you will enjoy reading this book and wish you learn from it as much as I did.

Utrecht, The Netherlands Andries Meijerink
January 2022

Preface

Seminal author George Blasse concluded his lecture *Classical Phosphors: A Pandora's Box* delivered at the International Conference on Luminescence 1996 held in Prague with the following words [J. Lumin. 72–74, 129-134 (1997)]: *From what has escaped from my Pandora's box it may be concluded that broad-band luminescence in inorganic systems lacks a quantitative understanding, and in many cases even a qualitative one. After opening it, Pandora closed her box immediately, so that at least the hope remained. This seems also to apply to the present situation. We can have some hope that the information available from ab initio molecular-orbital calculations of the ground and excited states is becoming so detailed, that it might be possible to solve the type of problems sketched. However, it should be realized that such type of calculations are by no means easy and will require a years-taking effort.* This book is about Professor Blasse's hope.

In effect, the efforts of Quantum Chemistry to provide sound answers to problems on luminescence of inorganic materials have been numerous and intense and have taken decades to a large community of researchers. And the story still continues.

According to its philosophy, Quantum Chemistry has developed *ab initio* methods that, based on the principles of Quantum Mechanics alone, are appropriate to calculate properties of different types of electronic systems with high quality, ranging from ground and excites sates of atoms and small molecules, to chemical reactions of middle size and relatively large molecules, to structures of giant biomolecules involving thousands of atoms. The different types of complexities inherent to each type of system inspired specific approximations to the Schrödinger equation.

The complexities of luminescent materials, and in particular of those activated with lanthanide ions, derive from the many possible processes, relevant to luminescence, occurring within rich and crowded manifolds of local excited states of solids doped with heavy elements.

For all these features to be tackled together, the *ab initio* methods should be able to calculate electronic states localized around the optical activator under the effects of the solid host; to consider spin-independent and spin-dependent relativistic effects, which are central for lanthanide and other heavy-element activators; to deal with the multiconfigurational nature of the excited manifolds of lanthanide ions with several unfilled shells of different natures, like atomic-like $4f^N$, $4f^{N-1}5d$ and $4f^{N-1}6s$ configurations, impurity-trapped excitons, and ligand-to-metal, metal-to-metal, and other charge transfer configurations, by handling the inherent static electron correlation effects; to treat the dynamic correlation as well and to do it when the Hamiltonian includes explicit spin-orbit coupling terms; and to calculate hundreds and thousands of excited states of the luminescent material. Moreover, all this must be done with an equivalent level of high accuracy. The good news is that, nowadays, Douglas-Kroll-Hess relativistic multiconfigurational embedded cluster *ab initio* calculations can do the job and the suite of programs MOLCAS is able to perform them efficiently. This book describes it.

The authors have been active within the Quantum Chemistry community in order to solve several of the mentioned problems and bring the excited manifolds of lanthanide activated luminescent materials within reach. During the exciting period of development of relativistic quantum chemical methods, they authored a relativistic effective-core-potential method that includes spin-orbit coupling interactions. They proposed and developed the embedded cluster method used throughout the book to consider classical and quantum-mechanical host-embedding effects on the active centers. And they authored the method to transfer the dynamic correlation effects calculated with high accuracy using a spin-independent Hamiltonian to the eigenstates of many-electron Hamiltonian with explicit spin-orbit coupling terms. They were also constantly active at establishing standards for computational choices (like one-electron basis sets, truncation of multiconfigurational expansions, embedded cluster sizes, etc.) of minimal and optimal qualities, by performing many systematic multiconfigurational *ab initio* calculations and validating them against experiments.

The book is organized in five parts. Part I is a methodological one, dedicated to the methods themselves and to the particular computational choices that make the calculations feasible and sufficiently accurate. Part II is a tutorial that can be followed in detail in order to perform the individual steps that lead to the whole multiconfigurational *ab initio* calculation of the excited manifolds of an active center. In Part III, the focus is put on the different types of excited manifolds that are handled, from impurity states of atomic-like $4f^N$, $4f^{N-1}5d$ and $4f^{N-1}6s$ configurations, and impurity-trapped exciton character, to states of ligand-to-metal, inter-valence, and metal-to-metal charge transfer nature. Part IV illustrates the kind of fundamental studies on luminescence that can be carried out with multiconfigurational *ab initio* methods, where several of the above excited manifolds are found in each luminescent material. The book ends with Part V, where insights into the high complexity of luminescent materials are given; insights that have been achieved by the use of multiconfigurational *ab initio* calculations and their continuous contrast with experiments. This Part shows how the calculations are necessary ingredients to get the insights, how they complement the experiments, and how they are not aimed at being ahead or behind the experimental work, but together with it.

The book is addressed to researchers in the field of inorganic luminescent materials, both experimentalists and theoreticians, experts with an accumulated experience and students, researchers willing to initiate themselves in performing multiconfigurational *ab initio* calculations on luminescent materials and researchers willing to have a glance on concepts and procedures that facilitates their communication with those performing the calculations. Also to those that, out of curiosity, wish to be aware of the contributions of Quantum Chemistry to the field of luminescent materials, both the achievements already realized and, more importantly, the potential ones.

The book is structured so as to allow different types of reading. For instance, a student aiming at performing calculations would focus on the theoretical details of Part I and the tutorial of Part II and will, perhaps, take the descriptions of the different excited manifolds of Part III as a reference. However, an established researcher aiming at acquiring a basic theoretical background on multiconfigurational *ab initio* calculations and evaluating how these calculations can help the research in the field, will perhaps read Chap. 1 and skip the details in Chaps. 2 and 3 of the theoretical Part I, skip also the tutorial in Part II and, eventually, the excited manifolds of Part III, and focus on the fundamental studies of Part IV and the insights derived from multiconfigurational *ab initio* calculations of Part V. Also, researchers specially interested in one particular issue should also find their way. For example, those interested in $4f^{N-1}5d$ states could read Sect. 6.2 and Chaps. 8–10; those focused on charge transfer states could read Sect. 6.3 and Chap. 11; and so on.

Finally, the book only draws a narrow window of what Quantum Chemistry can do for inorganic luminescent materials. One of the goals of the book is to illustrate that the potential is huge. But for it to be fully exploited, the interplay between experiments and calculations has to find a continuity and their exchange has to become a routine in research. Reducing the gap between multiconfigurational *ab initio* calculations and experiments is, perhaps, one necessary condition for this in the long term. It requires improving accuracies, extending their applicability to still more complex problems, and enlarge the list of properties that are routinely calculated. New quantum chemical developments on dynamic correlation and embedding will be of benefit for the accuracies. Active centers involving several lanthanide ions with unfilled shells are a major challenge and new developments on perturbation-theory methods based on general active spaces are very promising. Calculations on more complex active centers will yield intensities of inter-valence and metal-to-metal charge transfer transitions. Combination of the complex electronic structure calculations with multi-state nuclear dynamics will meet the challenge of calculating vibronic couplings and their associated transition intensities, non-radiative decay rates and energy transfer rates in general.

Madrid, Spain — Zoila Barandiarán
Gent, Belgium — Jonas Joos
Madrid, Spain — Luis Seijo

Contents

Acronyms

AD	Antisite Defect
AIMP	Ab Initio Model Potential
AIMP/SM	Ab Initio Model Potential/Shell-Model
AMFI	Atomic Mean-Field Integrals
ANO	Atomic Natural Orbital
ANO-RCC	Atomic Natural Orbital—Relativistic and Correlation Corrected basis set
CAS	Complete-Active-Space
CASSCF	Complete-Active-Space Self-Consistent-Field
CBM	Conduction Band Minimum
CC	Coupled-Cluster
CCSD(T)	Coupled-Cluster with Single and Double excitations and corrections for Triples
CFT	Crystal Field Theory
CI	Configuration Interaction
CSF	Configuration State Function
CSOV	Constrained Space Orbital Variation
CT	Charge Transfer
DFT	Density Functional Theory
EPR	Electron Paramagnetic Resonance
ESA	Excited State Absorption
ESE	Excited State Excitation
EXAFS	Extended X-ray Absorption Fine Structure
FWHM	Full Width at Half-Maximum
GAS	Generalized-Active-Space
GASSCF	Generalized-Active-Space Self-Consistent Field
IPEA	Ionization Potential - Electron Affinity shift parameter
ITE	Impurity-Trapped Exciton
IVCT	Inter-Valence Charge Transfer
LF	Ligand-Field
LFDFT	Ligand-Field Density Functional Theory

LIWE	Laser Induced White Emission
LMCT	Ligand-to-Metal Charge Transfer
MBPT	Many-Body Perturbation Theory
MCSCF	Multi-Configurational Self-Consistent Field
MMCT	Metal-to-Metal Charge Transfer
MR	Multi-Reference
MRCI	Multi-Reference Configuration Interaction
MR-SDCI	Multi-Reference Single and Double excitations Configuration Interaction
MS	Multi-State
MS-CASPT2	Multi-State Complete-Active-Space 2nd order Perturbation Theory
MS-RASPT2	Multi-State Restricted-Active-Space 2nd order Perturbation Theory
NNN	Next-Nearest Neighbor
PBC	Periodic Boundary Conditions
PES	Potential Energy Surface
QDPT	Quasi-Degenerate Perturbation Theory
QM/MM	Quantum-Mechanics/Molecular-Mechanics
RAS	Restricted-Active-Space
RASSCF	Restricted-Active-Space self-Consistent-Field
RASSI	Restricted-Active-Space State-Interaction
RASSI-SO	Restricted-Active-Space State-Interaction with Spin-Orbit coupling
SA	State-Average
SA-CASSCF	State-Average Complete-Active-Space Self-Consistent-Field
SA-RASSCF	State-Average Restricted-Active-Space Self-Consistent-Field
SCEI	Self-Consistent Embedded Ions
SCF	Self-Consistent Field
SD	Single and Double excitations
SDCI	Single and Double excitations Configuration Interaction
SDTQ	Single, Double, Triple, and Quadruple excitations
SS-CASPT2	Single-State Complete-Active-Space 2nd order Perturbation Theory
SS-RASPT2	Single-State Restricted-Active-Space 2nd order Perturbation Theory
UV	Ultra-Violet
VB	Valence Bond
VBM	Valence Band Maximum
WB-AIMP	Wood-Boring relativistic Ab Initio Model Potential
WFT	Wave Function Theory
XANES	X-ray Absorption Near-Edge Structure
ZPL	Zero-Phonon Line

Part I
Multiconfigurational Ab Initio Embedded-Cluster Methods for Luminescent Materials

This Part covers the theoretical aspects of the book, which are handled in three Chapters. The first of them is dedicated to the quantum chemical theoretical methods and concepts that are used in multiconfigurational *ab initio* embedded-cluster calculations on inorganic luminescent materials. The second one is focused on the computational choices that have been found to achieve a balance between feasibility and accuracy and the procedures followed to decide upon them. The third Chapter describes the properties of the luminescent materials that are calculated and how.

Chapter 1
Quantum Chemistry Methods

Abstract The methods of Quantum Chemistry used in this book to study crowded manifolds of local excited states of luminescent materials are introduced in this chapter. The embedded-cluster approximation for local states is discussed first, which allows to treat them with the same wave function theory tools used in quantum chemistry for isolated molecules. Then, the chapter goes through the two-component Douglas-Kroll-Hess relativistic Hamiltonian derived from the four-component Dirac Hamiltonian and analyzes its scalar and spin-orbit coupling components. The last part is dedicated to the discussion of the very important but difficult to handle electron correlation and its forms: The static correlation responsible for the configurational multiplets and its handling with CASSCF and RASSCF multiconfigurational variational methods. The dynamic correlation necessary for quantitative accuracy and its handling with MS-CASPT2 and MS-RASPT2 multi-reference perturbational methods. And the simultaneous consideration of electron correlation and spin-orbit coupling by means of the RASSI-SO method.

1.1 Crowded Manifolds of Local Excited States of Luminescent Materials: Limitations of DFT and Periodic-Boundary Conditions Methods

There is a high societal demand of luminescent materials spanning a broad range of properties because they are basic to design a wide variety of solid-state devices. These devices range from white light-emitting diodes or solid-state lighting devices for energy-efficient general illumination, to solid-state lasers with wavelengths from the infrared to the ultraviolet range, liquid crystal displays, functional layers that enhance the efficiency of solar cells, scintillating detectors useful in, e.g., medical imaging or high-energy physics calorimetry, remote pressure and temperature measurement systems, persistent luminescence materials, energy storage phosphors, quantum information processing materials, and more. It is natural that, in these circumstances, luminescent materials are the subject of intense scientific and technological research and it is not strange that computational tools able to aid the discovery and design of new on demand luminescent materials are requested.

Z. Barandiarán et al., *Luminescent Materials*, Springer Series in Materials Science 322,
https://doi.org/10.1007/978-3-030-94984-6_1

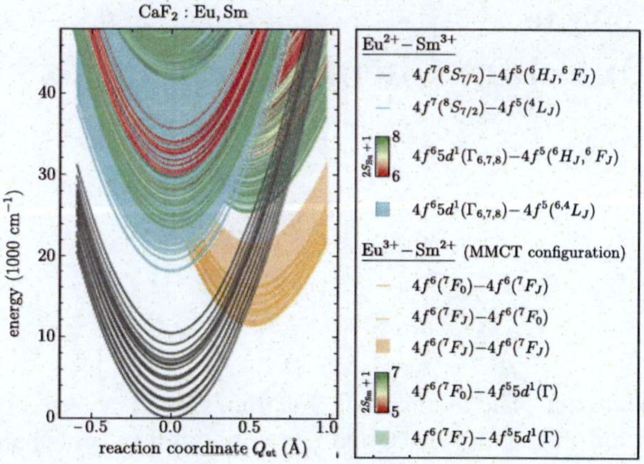

Fig. 1.1 Illustration of rich and crowded manifolds of lanthanide ions. Eu^{2+}/Sm^{3+} pair in CaF_2, showing states of the regular Eu^{2+}-Sm^{3+} pair configuration as well as metal-to-metal charge transfer (MMCT) Eu^{3+}-Sm^{2+} states, colored according to the symmetry type and/or spin of the excited state

The luminescent materials we are talking about must have rich manifolds of excited states that enable fertile and subtle photon engineering. A great deal of them are lanthanide-doped materials. These materials are exceptional because lanthanide ions feature abundant excited sates at reach, they belong to a large family of elements and oxidation states and they can easily be incorporated as impurities in a vast variety of solid hosts. The combination of impurities (single or multiple) and hosts, together with the different natures of the excited manifolds, makes the potential luminescent materials of this kind practically endless.

At this point, it must realized that the computational tools are facing here not only abundant excited states, their numbers ranging from tens to tens of thousands in a single material, but also states with a broad variety of natures depending on their electronic configuration and degree of localization/delocalization (cf. Fig. 1.1). The most common states derive from the familiar $4f^N$ and $4f^{N-1}5d$ electronic configurations of lanthanide ions, but states related with $4f^{N-1}6s$ configurations are also found, together with impurity-trapped-exciton configurations (ITE) $4f^{N-1}\phi_{ITE}$, where a hole is localized in the $4f$ shell and an electron is in molecular orbitals ϕ_{ITE}, partly delocalized outside the first coordination shell of the lanthanide ion. Besides, charge transfer (CT) states in which one electron is totally or partially transferred from one component of the material to another, are central in numerous processes involved in luminescence or its quenching. These states include ligand-to-metal charge transfer states (LMCT), inter-valence charge transfer states (IVCT, one electron is transferred between two different oxidation states of the same element), or metal-to-metal charge transfer states (MMCT, one electron is transferred between two different lanthanide elements, between the lanthanide ions and other dopants, or between the lanthanide ions and the host cations and vice versa).

Empirical theoretical models like Crystal/Ligand Field Theory [1–3] or Judd-Ofelt theory of f-f transitions intensities [4, 5] have been and are fundamental at leading the intense experimental work towards an understanding of these materials' luminescence, but they are specific for selected types of states and no comprehensive empirical model has ever been developed.

Non-empirical methods based on the first-principles of quantum mechanics are general by construction and, contrary to empirical models, they have the potential to set common, practical theoretical and computational grounds to study all materials and states.

First-principles methods based on density functional theory (DFT), [6, 7] both in the molecular-boundary conditions and periodic-boundary conditions (PBC) versions, have been very successful as basic tools to understand the ground state of molecules, solids, and surfaces. And time-dependent DFT [8] can in principle handle excited states. However, DFT-based methods have found severe difficulties to properly calculate complex manifolds of excited states of molecules and materials, mostly as a consequence of their monoconfigurational grounds. In particular, they failed to handle the excited manifolds of lanthanide-containing luminescent materials. This is so because the used single-configuration-based correlation functionals can calculate dynamic correlation with sufficient accuracy in many cases (and its related properties, like ground state structures), but fail to represent the non-dynamic correlations inherent to excited states with a multiconfigurational nature, like those of lanthanide ions. Such a failure is not only quantitative but also qualitative, up to the point that the theory has been unable to reproduce the complex multiplet structure of the unfilled shells, which is, however, natural in wave function theory (WFT) as originally developed from the principles of quantum mechanics [9]. This is important, because this multiplet structure is essential to understand the luminescent behaviour of most lanthanide-based luminescent materials.

The inherent limitations of DFT for complex manifolds of excited states did not overrule its successes at handling monoconfigurational issues, also present in luminescent materials, and, in an attempt to correct the former while taking advantage of the latter, several procedures have been proposed, basically implementing WFT concepts into DFT methods. For instance, after a DFT calculation on an embedded-cluster, the DFT molecular orbitals are used to calculate and diagonalize WFT ligand-field matrices of $4f^N$ and $4f^{N-1}5d$ configurations to give the target excited states of these configurations (something called sometimes Ligand Field DFT, LFDFT), like in $4f \rightarrow 5d$ transitions of Pr^{3+}-doped LaF_3, $LiYF_4$, CaF_2 [10], and Cs_2KYF_6 [11]. Or, in a periodic-boundary conditions context, the one-electron energy difference between the solid host valence band maximum (VBM) and the lanthanide $4f$ orbital levels is empirically tuned by means of an U parameter in DFT+U calculations; the energies of the lanthanide $5d$ orbital levels relative to the host conduction band minimum (CBM) are obtained by substracting, from the calculated band gap of the material, experimental $4f^N \rightarrow 4f^{N-1}5d$ many-electron absorption energies and calculated $4f$-VBM one-electron energy differences; and all of it together has been used to predict the scintillation or non-scintillation of Ce^{3+}-doped and Eu^{2+}-doped inorganic solids [12, 13]. Also, the one-electron transitions between occupied and unoccupied Bloch orbitals obtained as the imaginary part of the dielectric function in PBC DFT calculations have been compared with experimental $4f \rightarrow 4f$ and $4f \rightarrow 5d$ absorption spectra of Ce-doped and co-doped $Y_3Al_5O_{12}$ [14]. In order to improve the correspondence of computed excitation energies, obtained from one-particle excitations between Bloch orbitals, with experimental energies, hybrid functionals that contain a portion of Hartree-Fock exchange are currently gaining popularity [15].

Another approach followed to profit from PBC DFT ground state calculations consists on combining structural calculations of this kind with WFT calculations of excited states. For instance, PBC DFT atomistic structures of $Y_3Al_5O_{12}$:Ce, $Y_3Al_5O_{12}$:Ce,Ga, and $Y_3Al_5O_{12}$:Ce,La combined with embedded-cluster WFT excited states of the $4f^1$ and $5d^1$ manifolds, explained why the blue shift of the Ce^{3+} lowest $4f \rightarrow 5d$ transition upon Ga co-doping becomes a red shift upon La co-doping, in spite of the fact that both co-dopings produce lattice expansions. This procedure was proven to be very fruitful [16–18].

DFT also finds difficulties at calculating emission energies and Stokes shifts, due to its limitations for structural optimization of many-electron excited states. This is partially overcome by forcing the occupations of one electron levels [12, 13]; although this does not produce the optimal structures of many-electron excited states, it can get those of excited electronic configurations and it is not unreasonable to assume that all electronic states of a given configuration have equal atomistic structures. Another way to circumvent this excited state problem when studying the effect of local relaxations on $4f^{N-1}5d \rightarrow 4f^N$ emissions (e.g. in Ce^{3+}-doped $Y_3Al_5O_{12}$ and $Lu_3Al_5O_{12}$) has been to use PBC DFT ground state atomistic structures within the unit cell and reoptimize the positions of a limited number of atoms in every excited state in embedded-cluster WFT calculations [19]. As we see, DFT is very useful but it needs external ad hoc contributions in order to find its way in this environment.

We conclude that the original development of quantum mechanics in the form of wave function theory is the right theoretical framework to handle the complex excited manifolds of luminescent materials. Consequently, the WFT computational tools are the most adequate for comprehensive theoretical studies of luminescence. They are the only ones able to provide a multiconfigurational framework, which is inherent to a vast majority of relevant excited states luminescent materials. This book is about WFT computational tools in luminescence.

1.2 Embedded-Cluster Approximation for Local Properties of Materials

The excited states that play the relevant roles in luminescent materials set a very complex theoretical scenario: they are multiconfigurational because they result from multiple unfilled shells and their interactions, they gather together in complex, crowded manifolds of quasi-degenerate states, they may have different electronic natures, ranging from quasi-atomic $4f^N$ states to relatively delocalized impurity-trapped excitons or metal-to-metal charge transfer states, they are mostly local, non-periodic states of a defective solid, and very often they involve heavy elements, like lanthanides, transition metals and others, with important spin-orbit couplings and other relativistic effects.

The ab initio multiconfigurational methods developed in quantum chemistry for the study of all kinds of ground and excited states of atoms and molecules are able to handle this complex scenario. For that, one needs to choose an active cluster or pseudo-molecule that contains all atoms directly involved in the luminescent processes of interest and consider all bonding interactions between them, but also all the embedding interactions brought about on the active cluster by the rest of the solid host. Quantum chemical methods able to do this are called embedded-cluster methods.

After the 1963 (non-ab initio) embedded-cluster calculation of Shulman and Sugano on $KNiF_3$ [20], several embedding methods have been proposed to handle local electronic states of solids based on the use of wave functions, Green's functions, and density functionals [21]. Especially suitable for the above scenario are embedded-cluster WFT methods derived from McWeeny's *group function theory* for two or more groups of electrons without mutual correlation interactions [22]. This theory, combined with the restricted variational method of Huzinaga, necessary to avoid wave function collapses [23, 24], and the ab initio model potential (AIMP) recipes for the approximation of cumbersome operators in practical calculations [25–27], results in the AIMP embedded-cluster approximation for local states of ionic solids [26, 27]. We summarize this method here. A more general building-block and embedding method for insulators in general, based on the same grounds but avoiding the approximate representation of operators, is described in [28].

Following McWeeny, one can consider wave functions of a defective ionic solid (or not defective) of the type

$$\Phi = M\hat{A}\left[\Phi_{EC}\Phi_{\chi}\Phi_{\eta}\ldots\right], \qquad (1.1)$$

where Φ_{EC} is a multiconfigurational many-electron wave function of an embedded-cluster within a solid, which describes electrons localized in a particular region of the solid close to a set of atoms; Φ_{χ}, Φ_{η} are Hartree-Fock many-electron wave functions of host embedding ions χ, η that are distributed over the rest of the solid; $\hat{A}$ is an intergroup anti-symmetrizing operator and M is a normalization constant.

In order to illustrate this, one can think of electronic states localized around an Eu^{2+} impurity in CaF_2, where Eu^{2+} substitutes for Ca^{2+} and creates a substitutional point-defect. A reasonable computational choice is to define the *active cluster* by one Eu^{2+} ion, its first coordination shell of eight F^- ions, and a second coordination shell of twelve Ca^{2+} ions; this will give a $(EuF_8Ca_{12})^{18+}$ cluster, including 21 nuclei and 357 electrons represented by Φ_{EC}. If all the remaining individual ions of the host are taken as the embedding ions, the set of Φ_{χ}, Φ_{η}, ... wave functions are the Hartree-Fock wave functions of all the Ca^{2+} ions of CaF_2 not included in the cluster, each of them representing 18 electrons, plus the Hartree-Fock wave functions of all the F^- ios of CaF_2 not included in the cluster, each of them representing 10 electrons. In practice, this infinite list of Hartree-Fock wave functions can be truncated when the

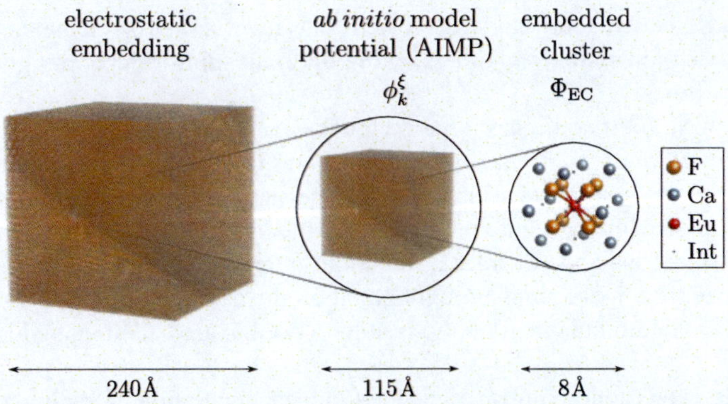

Fig. 1.2 Graphical representation of the simulated system in case of Eu-doped CaF_2. The system is divided in a $(EuF_8Ca_{12})^{2+}$ embedded-cluster (EC) which is treated at the highest all-electron quantum-chemical level, a shell which is represented by the full-ion ab initio model potential (AIMP) and a thick layer of point charges (electrostatic embedding) to terminate the system

Ca^{2+} and F^- ions are sufficiently far from Eu^{2+} so as to make the quantum interactions between their electrons and those of the active cluster negligible. These far distant host ions only introduce electrostatic interactions on the cluster, which are much more simple to consider (cf. Fig. 1.2). Typically, the number of host embedding ions represented by Φ_χ wave functions (and their corresponding quantum-mechanical embedding potentials) is of the order of $\sim 10^4$.

Wave functions like (1.1) take all electron correlation interactions within the cluster plus all mean-field bonding interactions within the cluster and between the host ions and the cluster (also mean-field bonding interactions between host ions, but these will not be relevant here) into account. Let us also consider that the many-electron wave functions of the electronic groups fulfill strong-orthogonality conditions:

$$\int \Phi^*_{EC}(1, i, j, \ldots)\Phi_\chi(1, k, l, \ldots)d\tau_1 = 0\,, \tag{1.2}$$

which is true if the group wave functions are constructed with different sets of orbitals and all orbitals are orthogonal. In that case, the total energy of the defective crystal can be partitioned in group and intergroup contributions, and the wave functions Φ_{EC} and energies E_{EC} of the ground and excited states of the embedded-cluster can be calculated variationally, using an embedded-cluster Hamiltonian $\hat{H}_{EC}$ that is the sum of the isolated cluster and embedding Hamiltonians,

$$\hat{H}_{EC} = \hat{H}_{IC} + \hat{H}_{Emb}\,, \tag{1.3}$$

and the eigenvalue equation

$$(\hat{H}_{EC} + V_{ECnuc-Host})\Phi_{EC} = (\hat{H}_{IC} + \hat{H}_{Emb} + V_{ECnuc-Host})\Phi_{EC} = E_{EC}\Phi_{EC}\,. \tag{1.4}$$

In (1.4), $V_{\text{ECnuc}-\text{Host}}$ is the interaction between the embedded-cluster nuclei and the host (electrons and nuclei), which is independent of the embedded-cluster electrons and does not have any effect on Φ_{EC}; it is equivalent to the inter-nuclear repulsion in the electronic Schrödinger equation of an isolated molecule within the Born-Oppenheimer approximation. $\langle\Phi_{\text{EC}}|\hat{H}_{\text{IC}}|\Phi_{\text{EC}}\rangle$ considers all intra-cluster interactions (including inter-nuclear repulsions) modulated by the embedding effects on the cluster electrons. And $\langle\Phi_{\text{EC}}|\hat{H}_{\text{Emb}}|\Phi_{\text{EC}}\rangle$ includes all interactions between the embedded-cluster electrons and the host. The embedding Hamiltonian $\hat{H}_{\text{Emb}}$ acts on all embedded-cluster electrons and is the sum of 1-electron embedding potentials $\hat{V}_{\text{Emb}}(i)$ that consider the interactions between the host (nuclei and electrons) and one cluster electron,

$$\hat{H}_{\text{Emb}} = \sum_{i}^{N_e^{\text{EC}}} \hat{V}_{\text{Emb}}(i)\,. \tag{1.5}$$

The 1-electron embedding potential acting on the i-th electron of the cluster reads:

$$\hat{V}_{\text{Emb}}(i) = -\sum_{\xi}^{N_{nuc}^{\text{Host}}} \frac{Z_\xi}{r_{\xi i}} + \sum_{\mu}^{N_{occ}^{\text{Host}}} f_\mu \int \frac{\varphi_\mu^*(j)[2-\hat{P}_{ij}]\varphi_\mu(j)}{r_{ij}} d\tau_j + \sum_{\mu}^{N_{occ}^{\text{Host}}} B_\mu |\varphi_\mu\rangle\langle\varphi_\mu|\,. \tag{1.6}$$

The first term is the electrostatic attraction between the cluster electron and the N_{nuc}^{Host} host nuclei; Z_ξ is the charge of nucleus ξ and $r_{\xi i}$ the distance between this nucleus and the electron. The second term contains the Coulomb and exchange interactions between the i-th cluster electron and the host electrons, which occupy N_{occ}^{Host} host orbitals φ_μ with fractional occupancy f_μ; $\hat{P}_{ij}$ is the permutation operator between electrons i and j and r_{ij} is the interelectronic distance. The third term is the Pauli repulsion between the cluster electron and the host electrons. It shows up in the embedding Hamiltonian as a consequence of the restrictions in the variational method [23, 24], which enforce the fulfilling of Pauli antisymmetry conditions between the embedded-cluster electrons and the host electrons [29]; it is a positive contribution when cluster and host orbitals are not orthogonal and becomes zero when they are orthogonal; the B_μ are positive constants and the operator prevents the cluster orbitals from collapsing onto the host orbitals during the self-consistent orbital optimization.

In (1.6), any unitary transformation of the occupied host orbitals φ_μ is valid, but the most convenient choice is a localized set. In ionic solids, localized host orbitals often show a strong character of the individual ions, so that substituting φ_μ orbitals by pure orbitals of the ξ individual ions, ϕ_k^ξ, can be a reasonable approximation. If we adopt this approximation and use ab initio model potential representations of the

Coulomb and exchange operators, we obtain the host AIMP embedding potential [26, 27]:

$$\hat{V}_{\text{Emb}}(i) \rightarrow \hat{V}_{\text{Emb}}^{\text{AIMP}}(i) = \sum_{\xi}^{N_{nuc}^{\text{Host}}} \hat{V}_{\xi}^{\text{AIMP}}(i)\,, \tag{1.7}$$

where the single-ion AIMP embedding potential $\hat{V}_{\xi}^{\text{AIMP}}$ is

$$\hat{V}_{\xi}^{\text{AIMP}}(i) = \hat{V}_{\xi}^{\text{Coulomb}}(i) + \hat{V}_{\xi}^{\text{Exchange}}(i) + \hat{V}_{\xi}^{\text{Pauli Repulsion}}(i)\,, \tag{1.8}$$

with

$$\hat{V}_{\xi}^{\text{Coulomb}}(i) = -\frac{Q_{\xi}}{r_{\xi i}} + \frac{1}{r_{\xi i}} \sum_{p} C_p^{\xi} \exp(-\alpha_p^{\xi} r_{\xi i}^2)\,, \tag{1.9}$$

$$\hat{V}_{\xi}^{\text{Exchange}}(i) = \sum_{p} \sum_{q} |\chi_p^{\xi}\rangle A_{pq}^{\xi} \langle \chi_q^{\xi}|\,, \tag{1.10}$$

and

$$\hat{V}_{\xi}^{\text{Pauli Repulsion}}(i) = \sum_{k} B_k^{\xi} |\phi_k^{\xi}\rangle\langle\phi_k^{\xi}|\,. \tag{1.11}$$

$\hat{V}_{\xi}^{\text{Coulomb}}(i)$ (1.9), represents the Coulomb interaction energy between the i-th cluster electron and ion ξ (i.e. the negative Coulomb potential of ion ξ). The first term is the point-charge contribution of the ion and the second term is the deviation from the point-charge model of the ion, entirely due to the spatial distribution of its electrons. For that, Q_{ξ} is the nominal charge of ion ξ (nuclear charge Z_{ξ} minus number of electrons N_e^{ξ})

$$Q_{\xi} = Z_{\xi} - N_e^{\xi}\,, \tag{1.12}$$

so that $\sum_{\xi}^{N_{nuc}^{\text{Host}}} (-Q_{\xi}/r_{\xi i})$ in (1.7) is the (negative) host Madelung potential minus the contributions of the cluster ions. And the C_p^{ξ} and α_p^{ξ} parameters are chosen so that

$$\frac{1}{r_{\xi i}} \sum_{p} C_p^{\xi} \exp(-\alpha_p^{\xi} r_{\xi i}^2) \approx -\frac{N_e^{\xi}}{r_{\xi i}} + 2\sum_{k}^{N_{occ}^{\xi}} f_k^{\xi} \int \frac{\phi_k^{\xi *}(j)\phi_k^{\xi}(j)}{r_{ij}} d\tau_j\,, \tag{1.13}$$

where the second term in the right hand side is the Coulomb repulsion of the i-th cluster electron with the N_e^{ξ} electrons of ion ξ, distributed in N_{occ}^{ξ} occupied orbitals

ϕ_k^ξ with fractional occupancies f_k^ξ. Calculation of C_p^ξ and α_p^ξ requires the knowledge of the ion orbitals ϕ_k^ξ. These are calculated in *self-consistent embedded ions* (SCEI) calculations on the solid host, which will be described below.

$\hat{V}_\xi^{\text{Exchange}}(i)$ (1.10) represents the exchange interaction energy between the i-th cluster electron and the N_e^ξ electrons of ion ξ. It is represented in the form of a spectral representation (resolution of the identity) of the ion exchange operator on the auxiliary basis set $\{\chi^\xi\}$. This auxiliary basis is an arbitrary set of N_χ^ξ functions; a common, convenient and efficient choice is the set of primitive Gaussian functions used to span the ion orbitals ϕ_k^ξ. Then, the A_{pq}^ξ exchange matrix coefficients in (1.8) are calculated with:

$$A_{pq}^\xi = -\sum_r^{N_\chi^\xi}\sum_s^{N_\chi^\xi}(S^\xi)_{pr}^{-1}\left(\sum_k^{N_{occ}^\xi} f_k^\xi \int\int \frac{\chi_r^{\xi *}(i)\phi_k^{\xi *}(j)\phi_k^\xi(i)\chi_s^\xi(j)}{r_{ij}} d\tau_j d\tau_i\right)(S^\xi)_{sq}^{-1}, \tag{1.14}$$

with the auxiliary basis metrics S^ξ matrix defined by

$$S_{pq}^\xi = \langle \chi_p^\xi | \chi_q^\xi \rangle . \tag{1.15}$$

Calculation of A_{pq}^ξ also requires the knowledge of the ion orbitals ϕ_k^ξ produced in SCEI calculations.

Finally, $\hat{V}_\xi^{\text{Pauli Repulsion}}(i)$ (1.11) represents the Pauli repulsion energy between the i-th cluster electron and the N_e^ξ electrons of ion ξ. This repulsion emerges from the Pauli antisymmetry conditions between such electron and the ion electrons (any correct many-electron wave function describing all these electrons must be antisymmetric under the exchange of the cluster i-th electron with any electron of the ion). Besides an explicit dependence on the ion orbitals ϕ_k^ξ, this term contains the Pauli repulsion constants of the ion B_k^ξ. All of them are obtained in SCEI calculations.

1.2.1 Self-Consistent Embedded Ions (SCEI) Calculations

The components of the AIMP embedding potential used in AIMP embedded-cluster calculations are obtained in *self-consistent embedded ions* (SCEI) calculations on perfect hosts [30]. In the case of CaF_2:Eu^{2+} previously taken for illustration, the SCEI iterative calculations are made on a perfect crystalline lattice of CaF_2 with ions fixed at experimental sites.

At each cycle of the SCEI iterations, AIMP embedded Hartree-Fock calculations are performed on all non-symmetry equivalent ions of the host (one Ca^{2+} and one F^- in CaF_2). In these calculations, each ion is surrounded by a first quantum-mechanical embedding region, which is made of AIMP embedding potentials located at all lattice sites necessary to account for all quantum-mechanical host effects (it has typically

$\sim 10^3$–10^4 lattice ions within $3 \times 3 \times 3$ to $6 \times 6 \times 6$ unit cells). A second electrostatic embedding region guarantees the reproduction of the host Ewald potential [31] around the ions, up to their first-to-second coordination shells. It typically consists of $\sim 10^4$–10^5 fractional charges located at lattice sites, whose values are generated by the zero-multipole method of Gellé-Lepetit [32].

Two ingredients change from cycle to cycle: the embedding potentials used in the quantum-mechanical embedding region (which are input) and the Hartree-Fock orbitals obtained for the ions (which are output). In the first iteration, a very simple representation of the Ca^{2+} and F^- *embedding* ions as +2 and –1 point-charges can be used for the input embedding potential, and ϕ_k^{Ca} and ϕ_k^F Hartree-Fock orbitals are obtained for the Ca^{2+} and F^- *embedded* ions, together with their orbital energies ε_k^{Ca} and ε_k^F. With these, better representations of the Ca^{2+} and F^- *embedding* ions are used in the next cycle, which result from (1.9) and (1.13) for the Coulomb potential, (1.10) and (1.14) for the exchange potential, and (1.11) for the Pauli repulsion operator, with $B_k^{Ca} = -2\varepsilon_k^{Ca}$, $B_k^F = -2\varepsilon_k^F$, according to Huzinaga [23, 24]. The new cycle produces new orbitals and orbital energies, which can be equally used to produce new embedding potentials for the next cycle. When the orbitals energies and the total energies of the embedded ions become constant within a threshold (e.g. 10^{-8} hartree), the SCEI cycles are stopped. The whole SCEI calculation usually takes around 20 cycles per lattice ion.

A final check and an eventual correction of the AIMP embedding potentials of a host are done before they are definitely accepted. This is due to the fact that the Huzinaga choice $B_k^{\xi} = -2\varepsilon_k^{\xi}$ has been found to work very well in a framework of core-valence separation in molecules and solids (core model potentials or effective core potentials); however, it has also been found to fail sometimes in embedding-cluster separation problems and eventually produce anomalous collapse of cluster orbitals onto the host [33]. It is known [28] that the choice of Pauli repulsion constants in (1.11) is arbitrary and any positive values large enough to avoid the collapse of target orbitals onto frozen orbitals (either core orbitals in molecules or host orbitals in solids) are valid when complete basis sets are used for the target orbitals, but there is no theoretical justification for specific values of these constants when truncated basis sets are used in practice. A good practical choice has been found in modulating the B_k^{ξ} of the outermost ion orbitals in order to make the local structures calculated with small embedded-clusters to agree with those calculated with reference embedded clusters of much larger size or with experiments [33]. E.g. choose B_{3p}^{Ca} and B_{2p}^F values not far from $-2\varepsilon_{3p}^{Ba}$ and $-2\varepsilon_{2p}^F$, so that a correlated calculation of CaF_2:$(CaF_8)^{6-}$ gives a Ca–F bond length d(Ca–F) similar to the experiment.

Once the AIMP embedding potentials of a host are produced in SCEI calculations, they are stored in libraries in order to be used for all kind of materials produced by the creation of local defects in the host. The quantum-mechanical and electrostatic embedding regions used in an embedded-cluster calculation on a defective material are defined using the same criteria as in the SCEI calculations. E.g., in CaF_2:Eu^{2+}, the embedding used for Ca^{2+} in the last cycle of the SCEI calculation is also used for the $(EuF_8Ca_{12})^{18+}$ cluster, except for the F_8 and Ca_{12} coordination shells, which are removed from the embedding as they become part of the active cluster.

1.2.2 Relaxation and Polarization of the Cluster Environment

The approximation just described has been shown to be sufficient for the theoretical study of many local states involved in the luminescence of defective solids, as we will see later in the book. This is true as long as the defined cluster involves the whole extension of the relevant electronic states, or the very most of them, and the most relevant distortions produced by the local defect. When the relevant local electronic states significantly extend to regions beyond a chosen cluster, the enlargement of the cluster size to include these regions is a must. Unfortunately, the 3D nature of solids limits very much the enlargements of clusters: one additional coordination shell may involve so many atoms that calculations on the enlarged cluster may be unfeasible.

A different case appears when the target electronic local states do not overrun the limits of the cluster, but the geometrical distortions around the local defect (e.g. produced by an activator) propagate beyond the cluster, or a charged defect produces important long-range electric dipole polarizations, and either the long-range deformations or the long-range polarizations, or both, impact back the electronic structure of the target local states. When this is the case and the cluster size cannot be enlarged, there are still alternatives. One of them is to use an iterative self-consistent embedded-cluster approach, in which cycles of ab initio calculations on a set of embedded-clusters are made [28]. A much simpler alternative is to use an empirical description of the interactions within the host, together with an embedded-cluster description of the interactions within the cluster and between cluster and host, i.e. a QM/MM kind of method [34] (Quantum Mechanical description of a part of a molecular or solid system together with Molecular Mechanics description of other parts of the same system). The AIMP/SM method [35], which results from combining AIMP embedded-cluster calculations with the shell-model description of ionic crystals (SM) [36], is a practical choice of this kind for local states of defective solids.

In brief, in AIMP/SM calculations on an embedded-cluster with relaxation and polarization of its environment, the energy of a defective solid is written as the sum of the embedded-cluster energy E_{EC} (1.4) plus the energy of the cluster environment:

$$E_{\text{Defective Solid}} = E_{\text{EC}} + E_{\text{Environment}} \,. \tag{1.16}$$

The embedded-cluster energy is calculated with quantum mechanics (1.4):

$$E_{\text{EC}} = \langle \Phi_{\text{EC}} | (\hat{H}_{\text{EC}} + V_{\text{ECnuc-Host}}) | \Phi_{\text{EC}} \rangle \tag{1.17}$$

and the environment energy is calculated with molecular mechanics (shell-model):

$$E_{\text{Environment}} = \frac{1}{2} \sum_{\xi}^{N_{nuc}^{\text{Host}}} K_\xi r_{c\xi-s\xi}^2 + \frac{1}{2} \sum_{\xi}^{N_{nuc}^{\text{Host}}} \sum_{\eta \neq \xi}^{N_{nuc}^{\text{Host}}} \left(V_{\xi\eta}^{SM-sr} + V_{\xi\eta}^{SM-lr} \right) , \tag{1.18}$$

where $r_{c\xi-s\xi}$ is the distance between a fixed *core* and a floating *shell* that represent ion ξ in this model; K_ξ is an empirical constant indicating the electric dipole polarizability of ion ξ; and $V_{\xi\eta}^{SM-sr}$ and $V_{\xi\eta}^{SM-lr}$ are the short- and long-range interaction energies between ions ξ and η according to the shell-model, which read:

$$V_{\xi\eta}^{SM-sr} = A_{\xi\eta} \exp\left(-\frac{r_{s\xi-s\eta}}{\rho_{\xi\eta}}\right) - \frac{C_{\xi\eta}}{r_{s\xi-s\eta}^6} \tag{1.19}$$

and

$$V_{\xi\eta}^{SM-lr} = \frac{Q_{c\xi}Q_{c\eta}}{r_{c\xi-c\eta}} + \frac{Q_{c\xi}Q_{s\eta}}{r_{c\xi-s\eta}} + \frac{Q_{s\xi}Q_{c\eta}}{r_{s\xi-c\eta}} + \frac{Q_{s\xi}Q_{s\eta}}{r_{s\xi-s\eta}}\,. \tag{1.20}$$

Here, $A_{\xi\eta}$, $\rho_{\xi\eta}$, and $C_{\xi\eta}$ are empirical parameters for the short-length interaction energy between ionic shells; $Q_{c\xi}$, $Q_{c\eta}$, $Q_{s\xi}$, and $Q_{s\eta}$ are empirical point-charges of the ionic cores and shells (just used for the long-range interaction energies); and the r's are distances between ionic cores and shells. The SM parameters are calculated empirically by making the theoretical results obtained with the model to agree with the experimental data for a list of properties of perfect hosts.

In the AIMP/SM method, the total energy of the defective solid (1.16) is minimized with respect to the many-electron wave functions of the cluster, the positions of the cluster nuclei, and the positions of the environmental cores and shells. This minimization is achieved by means of an iterative procedure with cycles consisting on AIMP embedded-cluster calculations of potential energy curves or surfaces, parametrizations of SM interactions in order reproduce the latter, and SM energy minimizations. The process produces embedded-cluster wave functions, energies and equilibrium geometries, but also equilibrium positions and electric dipole polarizabilities of the environmental ions [35].

1.3 Relativistic Effects for Lanthanides and Other Heavy Elements

A vast majority of luminescent phenomena involve elements of the fourth period of the periodic table and below, in all their stable and metastable oxidation states. Indeed, transition metal ions and, specially, lanthanide ions, occupy a central position in the design of new useful luminescent materials. Elements like these, which do not have a small atomic number, are usually called heavy elements in quantum chemistry and the chemical and physical properties of their molecules and materials experience an important impact from relativity. In order to consider this impact, relativistic Hamiltonians and wave functions have to be used. This is the issue of this section.

The large nuclear charges of lanthanides and other heavy elements of the periodic table generate very large potential energies of their inner electrons; these produce very large kinetic energies (virial theorem) and the high electron velocities make relativistic corrections compulsory. Some of these are spin-independent (or *scalar*) corrections, mostly due to the classic mass-velocity effect and the non-classic Darwin potential, which produce contractions of the *s* orbitals and expansions of the *d* orbitals and have an important impact on the physics and chemistry of low periods' elements. Other corrections are spin-dependent, with spin-orbit coupling as the most important of them. Spin-orbit coupling plays a key role in the excited manifolds of electronic states of many luminescent materials with heavy activators, but it can also be very important for lighter activators (or quenchers) due to its active involvement in non-radiative decay processes.

The corrections materialize in relativistic Hamiltonians. In these, the spin-independent non-relativistic Hamiltonian of a molecule, solid, or embedded-cluster, is corrected and substituted by a new scalar (or spin-independent) relativistic Hamiltonian, and additional spin-dependent terms appear, the most important of them by far being the spin-orbit coupling Hamiltonian.

Several relativistic many-electron Hamiltonians have been used for ab initio calculations of electronic systems. A family of them, called 4-component Hamiltonians [37], result from using the 4-component (kinetic energy) Dirac Hamiltonian for the one-electron terms and different choices of transformations[1] and approximations on these and the potential energy terms. Use of these equations in the so called 4-component relativistic methods give electronic states but also positronic states as solutions, and computational tricks have to be used to get rid of the latter.

Alternatively, it is possible to project out the positronic states by means of algebraic Hamiltonian transformations that lead to 2-component Hamiltonians, whose solutions are only electronic [38]. Different choices led to different methods. All of them, together with other methods that combined transformations of parts of the Hamiltonian with physically sound approximations of other parts [39], made the family of 2-component methods, which account for scalar and spin-orbit coupling relativistic effects at reasonable cost. The fast developments in relativistic quantum chemistry during the last decades of the last century established the second- and third-order Douglas-Kroll-Hess 2-component Hamiltonian [40–42] as the most theoretically sound, accurate and practical 2-component relativistic many-electron Hamiltonian. Quantum chemical ab initio calculations on molecules and embedded-clusters with this Hamiltonian are nowadays routine and it has been used to produce most of the theoretical results shown and discussed in this book. We introduce it in the next section. Extensions of this Hamiltonian to infinite order, where 2-component and 4-component Hamiltonians lead to identical solutions, are possible and feasible [43].

[1] The Dirac (kinetic energy) Hamiltonian for a free electron $\underline{\hat{h}}_{\mathrm{D}}$ is a 4×4 matrix of operators and the one-electron wave function $\underline{\phi}$ (or spinor) is a 4×1 vector of 1-electron functions, so that the Dirac equation for one free electron of energy ε reads $\underline{\hat{h}}_{\mathrm{D}}\,\underline{\phi} = \varepsilon\underline{\phi}$. This equation is equally valid if operator and wave function undergo 4×4 unitary transformations (with an arbitrary 4×4 matrix $\underline{u}$ that fulfils $\underline{u}^{\dagger}\,\underline{u} = \underline{u}\,\underline{u}^{\dagger} = \mathbf{1}$). In effect, $\underline{u}\,\underline{\hat{h}}_{\mathrm{D}}\,\underline{u}^{\dagger}\underline{u}\underline{\phi} = \varepsilon\underline{u}\,\underline{\phi}$, which gives the equivalent, transformed Dirac equation $\underline{\hat{h}}'_{\mathrm{D}}\underline{\phi}' = \varepsilon\underline{\phi}'$, with the transformed Hamiltonian $\underline{\hat{h}}'_{\mathrm{D}} = \underline{u}\,\underline{\hat{h}}_{\mathrm{D}}\,\underline{u}^{\dagger}$ and spinor $\underline{\phi}' = \underline{u}\underline{\phi}$.

1.3.1 Douglas-Kroll-Hess Relativistic Hamiltonian

The many-electron Douglas-Kroll-Hess Hamiltonian $\hat{H}^{\mathrm{DKH}}$ for molecules and embedded-clusters is a 2-component Hamiltonian (its one-electron operators are 2×2 matrices of operators) and it is a no-pair Hamiltonian (its solutions are pure electronic because the formation of virtual electron-positron pairs that collapse the wave function is prevented within the formulation). The Hamiltonian is the sum of two terms: the first term is scalar or spin-independent and the second term is spin-dependent, with the spin-orbit coupling Hamiltonian as its most important contribution,

$$\hat{H}^{\mathrm{DKH}} = \hat{H}^{\mathrm{DKH}}_{\mathrm{scalar}} + \hat{H}^{\mathrm{DKH}}_{\mathrm{SOC}} . \tag{1.21}$$

Regarding the scalar Hamiltonian, its two-electron part is identical to the non-relativistic one, i.e. Coulomb interactions between pairs of electrons. The one-electron part is the one that contains the relativistic corrections. As in the non-relativistic many-electron Hamiltonian, it is the sum of one-electron operators for all electrons of the embedded-cluster N_e^{EC}. All relativistic corrections are included in the kinetic and potential energy one-electron operators.

$$\hat{H}^{\mathrm{DKH}}_{\mathrm{scalar}} = \sum_i^{N_e^{\mathrm{EC}}} \hat{T}_i + \sum_i^{N_e^{\mathrm{EC}}} \underline{\hat{V}}_i^{\mathrm{ext,DKH}} + \frac{1}{2} \sum_i^{N_e^{\mathrm{EC}}} \sum_{j \neq i}^{N_e^{\mathrm{EC}}} \frac{1}{r_{ij}} . \tag{1.22}$$

The kinetic energy operator of the i-th electron is

$$\hat{T}_i = \hat{E}_i - m_e c^2 , \tag{1.23}$$

where m_e is the electron rest mass, c the speed of light in vacuum, and

$$\hat{E}_i = \left(c^2 \hat{p}_i^2 + m_e^2 c^4\right)^{1/2} , \tag{1.24}$$

$\hat{p}_i^2$ being the square electron momentum operator.

The no-pair potential energy operator of the i-th electron of the cluster under the effect of an external potential $\hat{V}_i^{\mathrm{ext}}$ (due to all nuclear attractions within the embedded-cluster plus the host embedding potential (1.6)), is the result of regularizing (preventing it from producing a collapse of the wave functions onto positronic states) the external potential with the use of the so called kinematic factors. At second order, it is the following 2×2 matrix of operators

$$\underline{\hat{V}}_i^{\mathrm{ext,DKH2}} = \hat{A}_i \left(\hat{V}_i^{\mathrm{ext}} + \underline{\boldsymbol{\sigma}}\, \hat{\boldsymbol{R}}_i \hat{V}_i^{\mathrm{ext}} \underline{\boldsymbol{\sigma}}\, \hat{\boldsymbol{R}}_i \right) \hat{A}_i - \frac{1}{2} \left[\hat{W}_i, \left[\hat{W}_i, \hat{E}_i \right] \right] . \tag{1.25}$$

In (1.25),

$$\hat{A}_i = \left(\frac{\hat{E}_i + m_e c^2}{2\hat{E}_i} \right)^{1/2} , \tag{1.26}$$

$$\underline{\boldsymbol{\sigma}}\, \hat{\boldsymbol{R}}_i = \underline{\sigma}_x \hat{R}_{xi} + \underline{\sigma}_y \hat{R}_{yi} + \underline{\sigma}_z \hat{R}_{zi} , \tag{1.27}$$

where $\underline{\sigma}_x$, $\underline{\sigma}_y$, $\underline{\sigma}_z$ are the 2×2 Pauli matrices and

$$\hat{R}_{xi} = \frac{c\,\hat{p}_{xi}}{\hat{E}_i + m_e c^2} , \quad \hat{R}_{yi} = \frac{c\,\hat{p}_{yi}}{\hat{E}_i + m_e c^2} , \quad \hat{R}_{zi} = \frac{c\,\hat{p}_{zi}}{\hat{E}_i + m_e c^2} ; \tag{1.28}$$

and $\hat{W}_i$ is a non-local (integral) operator with kernel

$$\hat{W}(\boldsymbol{p}_i, \boldsymbol{p}_i') = \hat{A}_i \left(\hat{\boldsymbol{R}}_i - \hat{\boldsymbol{R}}_i' \right) \hat{A}_i' \frac{\hat{V}^{\mathrm{ext}}(\boldsymbol{p}_i, \boldsymbol{p}_i')}{\hat{E}_i + \hat{E}_i'} . \tag{1.29}$$

The no-pair potential energy operator at third-order is

$$\underline{\hat{V}}_i^{\mathrm{ext,DKH3}} = \underline{\hat{V}}_i^{\mathrm{ext,DKH2}} + \frac{1}{2} \left[\hat{W}_i , \left[\hat{W}_i , \hat{A}_i \left(\hat{V}_i^{\mathrm{ext}} + \underline{\boldsymbol{\sigma}}\, \hat{\boldsymbol{R}}_i \hat{V}_i^{\mathrm{ext}} \underline{\boldsymbol{\sigma}}\, \hat{\boldsymbol{R}}_i \right) \hat{A}_i \right] \right] . \tag{1.30}$$

Integrals involving these operators are computed with their matrix representations; although this is exact only when a complete, infinite set of basis functions is used, representations on the subspace defined by the set of molecular Gaussian primitive functions are of sufficient quality in practice. Also, integrals with these operators are usually computed in momentum space.

1.3.2 *Spin-Orbit Coupling Hamiltonian*

As all components of the many-electron relativistic Hamiltonian, the spin-orbit coupling Hamiltonian in the no-pair formulation of Douglas, Kroll and Hess, $\hat{H}_{\mathrm{SOC}}^{\mathrm{DKH}}$, results from projecting out positronic states and is a variationally stable Hamiltonian. It is, basically, the many-electron Breit-Pauli spin-orbit Hamiltonian [44] with bracketing kinematic factors that correct the interactions at small distances from the nuclei [42], so that both operators share structure and symmetry properties, and formulas developed for one of them can immediately be applied to the other. The no-pair spin-orbit coupling Hamiltonian is the sum of one-electron terms $\hat{H}_i^{\mathrm{1e-SO}}$ and two-electron terms $\hat{H}_{ij}^{\mathrm{2e-SO}}$:

$$\hat{H}_{\mathrm{SOC}}^{\mathrm{DKH}} = \sum_i^{N_e^{\mathrm{EC}}} \hat{H}_i^{\mathrm{1e-SOC}} + \frac{1}{2} \sum_i^{N_e^{\mathrm{EC}}} \sum_{j \neq i}^{N_e^{\mathrm{EC}}} \hat{H}_{ij}^{\mathrm{2e-SOC}} , \tag{1.31}$$

with

$$\hat{H}_i^{\text{1e-SOC}} = 2c^2 \sum_{\xi}^{N_{nuc}^{\text{EC}}} \frac{\hat{A}_i}{\hat{E}_i + m_e c^2} \frac{Z_\xi}{r_{\xi i}^3} (\boldsymbol{r}_{\xi i} \times \hat{\boldsymbol{p}}_i) \hat{\boldsymbol{s}}_i \frac{\hat{A}_i}{\hat{E}_i + m_e c^2} \tag{1.32}$$

and

$$\hat{H}_{ij}^{\text{2e-SOC}} = -4c^2 \frac{\hat{A}_i \hat{A}_j}{\hat{E}_i + m_e c^2} \left[\frac{1}{r_{ij}^3} (\boldsymbol{r}_{ij} \times \hat{\boldsymbol{p}}_i)(\hat{\boldsymbol{s}}_i + 2\hat{\boldsymbol{s}}_j) \right] \frac{\hat{A}_i \hat{A}_j}{\hat{E}_i + m_e c^2} . \tag{1.33}$$

Handling all one- and two-electron integrals that result from the use of (1.31) with multiconfigurational wave functions is computationally very demanding. However, most of the role of the two-electron terms in this (and in Breit-Pauli's) operator is to include direct and exchange interactions between the outer valence electrons and the inner core electrons and to screen the one-electron terms with two-electron contributions. Fortunately, this is known to be achievable with a mean-field approximation that leads to a much simpler, effective one-electron spin-orbit coupling Hamiltonian. Blume and Watson [45, 46] used this mean-field approach in atoms together with the additional neglect of non-diagonal two-electron contributions within an LS-coupling scheme and monoconfigurational mean-field average. Hess et al. [47] used (1.31) and the mean-field approximation within a multiconfigurational framework and produced a many-electron mean-field spin-orbit coupling Hamiltonian $\hat{H}_{\text{MF-SOC}}^{\text{DKH}}$ that is the sum of one-electron mean-field spin-orbit coupling Hamiltonians $\hat{H}_i^{\text{MF-SOC}}$,

$$\hat{H}_{\text{MF-SOC}}^{\text{DKH}} = \sum_{i}^{N_e^{\text{EC}}} \hat{H}_i^{\text{MF-SOC}} . \tag{1.34}$$

In (1.34) the i-th electron spin-orbit coupling term $\hat{H}_i^{\text{MF-SOC}}$ is defined such that the integral between this operator and any two orbitals, or basis set functions, φ_a and φ_b, includes summations over all occupied molecular orbitals ψ_k with occupation numbers n_k, and reads:

$$\begin{aligned} \int \varphi_a^*(i) \hat{H}_i^{\text{MF-SOC}} \varphi_b(i) d\tau_i &= \int \varphi_a^*(i) \hat{H}_i^{\text{1e-SOC}} \varphi_b(i) d\tau_i \\ &+ \sum_k n_k \int\int \varphi_a^*(i) \psi_k^*(2) \hat{H}_{i2}^{\text{2e-SOC}} \varphi_b(i) \psi_k(2) d\tau_i d\tau_2 \\ &- \frac{3}{2} \sum_k n_k \int\int \varphi_k^*(i) \psi_a^*(2) \hat{H}_{i2}^{\text{2e-SOC}} \varphi_b(i) \psi_k(2) d\tau_i d\tau_2 \\ &- \frac{3}{2} \sum_k n_k \int\int \varphi_a^*(i) \psi_k^*(2) \hat{H}_{i2}^{\text{2e-SOC}} \varphi_k(i) \psi_b(2) d\tau_i d\tau_2 . \end{aligned} \tag{1.35}$$

An additional approximation which is very efficient from the computational point of view consists on neglecting integrals involving different atoms (atomic-mean-field integrals approximation, AMFI) [48]. The mean-field approximation ((1.34) instead of (1.31)) results into ∼1% deviations from pure DKH results (e.g. DKH SOC splittings), whereas ∼5% deviations should be expected for the AMFI approximation [48]. The AMFI approximation of the many-electron Douglas-Kroll-Hess mean-field spin-orbit coupling Hamiltonian has been used in all multiconfigurational ab initio calculations of this book.

Summarizing, in the multiconfigurational ab initio embedded-cluster calculations of this book, the choice of relativistic many-electron isolated-cluster Hamiltonian $\hat{H}_{\mathrm{IC}}$ in (1.3) has been the scalar DKH Hamiltonian $\hat{H}^{\mathrm{DKH}}_{\mathrm{scalar}}$ (1.22) plus the mean-field spin-orbit coupling DKH Hamiltonian $\hat{H}^{\mathrm{DKH}}_{\mathrm{MF-SOC}}$ (1.34),

$$\hat{H}_{\mathrm{IC}} = \hat{H}^{\mathrm{DKH}}_{\mathrm{scalar}} + \hat{H}^{\mathrm{DKH}}_{\mathrm{MF-SOC}} \,. \tag{1.36}$$

This way, the embedded-cluster Hamiltonian $\hat{H}_{\mathrm{EC}}$ (1.3) is written as the sum of a scalar (spin-independent) part, made of the scalar relativist terms and the embedding Hamiltoniana, $\hat{H}_{\mathrm{EC,scalar}} = \hat{H}^{\mathrm{DKH}}_{\mathrm{scalar}} + \hat{H}_{\mathrm{Emb}}$, plus a spin-orbit coupling (spin-dependent) part $\hat{H}_{\mathrm{EC,SOC}} = \hat{H}^{\mathrm{DKH}}_{\mathrm{MF-SOC}}$:

$$\hat{H}_{\mathrm{EC}} = \hat{H}_{\mathrm{EC,scalar}} + \hat{H}_{\mathrm{EC,SOC}} \,. \tag{1.37}$$

This separation into spin-independent and spin-dependent parts will be very convenient at handling together spin-orbit coupling and correlations, as we will see later.

1.4 Electron Correlation for Rich and Crowded Manifolds of Excited States

The set of (simple or complex) luminescence related phenomena that rest on the existence of rich and crowded manifolds of excited states of materials is enormous. This is not strange, because the probability of any of these phenomena to happen depends on the individual probabilities of different photon absorptions, non-radiative decays, and photon emissions, and on their products; and these probabilities increase with the types of excited states that can exist in a material (rich manifolds) and with the number of excited states of each type (crowded manifolds). These manifolds are associated with unfilled (partially filled) shells of orbitals of different nature. E.g. lanthanide ions doped in solid hosts make such a fertile field of luminescent materials because they can join excited manifolds of $4f^N$, $4f^{N-1}5d$, $4f^{N-1}6s$ electronic configurations together with $4f^{N-1}\phi_{ITE}$ ITE configurations and LMCT, IVCT, MMCT and other types of charge transfer configurations; and each of these configurations forms a large, sometimes huge, manifold of excited states.

The energy differences between different manifolds and the energy distribution of the electronic states within the individual manifolds, both depend very much on electron correlation. Moreover, the multiplet structure of a configuration, i.e. the fact that a single electronic configuration can produce an ample set of excited states with different energies instead of a single one, or, in other words, the very structure of an electronic manifold, is a consequence of electron correlation. We need, then, to pay special attention to the concept of electron correlation, often called just *correlation* in our context, and we will do it in this section.

We also need to discuss its separation into dynamic and static correlation and the methodological implications of their respective treatments. Dynamic correlation has a strong impact on the energy separations between manifolds of different configurations, because it results from the Coulomb interelectronic repulsion, which depends very much on the types of orbitals the electrons occupy. Static correlation, however, has its strongest impact on the structure and the energetic distribution of states within the individual manifolds, because it results from an interplay between spins and correlations that are generated by antisymmetry requirements of the many-electron wave functions. We will see here what types of wave functions are capable of accounting for dynamic and static correlations.

Electron correlation is the physical phenomenon that prevents the probability of finding one electron in a differential volume around a first point 1 and, simultaneously, a second electron in a differential volume around a second point 2, from being equal to the product of both individual probabilities [49]. In other words, it is what makes the probability of finding an electron somewhere to depend on where the other electrons are. This is mathematically formulated in terms of two functions that are calculated with the many-electron wave function Ψ: One of the functions is the one-electron density function γ, which shapes the probability of finding an electron somewhere in space [$\gamma(1)$ is the probability density of finding an electron (any electron) at point 1 in space, $\gamma(2)$ is the probability density of finding an electron (again any electron) at point 2, …]. The other function is the two-electron density function Γ, which shapes the probability of finding an electron somewhere in space when another one is somewhere else [$\Gamma(1, 2)$ is the probability density of finding an electron (any electron) at point 1 in space and, simultaneously, another electron (any other electron) at point 2 in space].[2] In a hypothetical independent particle

[2] In an N-electron system, the probability of finding an electron (any of them) in a differential volume dV around a point (x, y, z) is $\gamma(x, y, z)dV$, where γ is known as the one-particle density function, or electron probability density, or electron density, which varies from point to point and has units of V^{-1} (as corresponds to being a density). Equivalently, the probability of simultaneously finding an electron in a differential volume dV_1 around a point (x_1, y_1, z_1) and another in a differential volume dV_2 around a point (x_2, y_2, z_2) is $\Gamma(x_1, y_1, z_1, x_2, y_2, z_2)dV_1dV_2$, where Γ is known as the two-electron probability density or two-electron density function, which depends on the pair of points and has units of V^{-2} (as corresponds to being a 2-particle density). The one- and two-electron density functions γ and Γ of an N-electron system are calculated from its many-electron wave function Ψ (see below), which is a function of the position coordinates x, y, z and spin coordinates σ of all the N electrons.

When $\Gamma(x_1, y_1, z_1, x_2, y_2, z_2) = \gamma(x_1, y_1, z_1)\gamma(x_2, y_2, z_2)$, the wave function Ψ is said to be *uncorrelated*, because the probability of finding an electron at point 1 *when another electron is* case, where the movements of all electrons would be mutually independent, $\Gamma(1, 2)$ would be identical to the product of the two individual probabilities $\gamma(1)$ and $\gamma(2)$, $\Gamma(1, 2) = \gamma(1)\gamma(2)$, and no correlation would exist. In reality, this never happens and the electrons movements are correlated, so that

$$\Gamma(1, 2) = \gamma(1)\gamma(2)[1 + f(1, 2)] \tag{1.38}$$

and the function $f(1, 2)$ is a measure of correlation.

There are two reasons for electron correlation that make $f(1, 2) \neq 0$: Coulomb interelectronic repulsion and Pauli antisymmetry principle. On one hand, interelectronic repulsion lowers the probability of finding an electron around a given point when another electron is at that point (this is known as the Coulomb hole) and, as a consequence of it, electrons are correlated (this is known as Coulomb correlation). On the other hand, Pauli antisymmetry principle establishes that many-electron wave functions describing sets of electrons (fermions in general) must be anti-symmetrical under the exchange of any two of them (they must change sign when the positions of any two electrons are exchanged). One of its implications is that two different electrons of equal spin cannot be found simultaneously at the same point in space (otherwise the many-electron wave function would have been symmetrical under their exchange). This means that Pauli principle lowers the probability of finding an electron around a given point when another electron *of equal spin* is at that point (this is known as the Fermi hole); hence, it causes electron correlation (which is known as Fermi correlation). However, Pauli principle does not prevent any two electrons of opposite spin from being found simultaneously at the same point in space, which means that they do not experience this Pauli principle driven correlation. In other words, Fermi correlation exists only between electrons of equal spins.

In summary, there is (Pauli principle driven) Fermi correlation between pairs of electrons of equal spins and there is (electrostatic repulsion driven) Coulomb correlation between any pair of electrons, regardless of their spins. It is interesting to notice that the so-called independent particle many-electron wave functions (Hartree-Fock) do not contain Coulomb correlation, but they contain Fermi correlation between electrons of equal spins as a consequence of their fulfilling the Pauli

---

already at point 2 (which is calculated substituting $\gamma(x_2, y_2, z_2)$ by 1 in the formula for Γ) is independent of where the second electron is, regardless of its distance from point 1.

Correspondingly, when $\Gamma(x_1, y_1, z_1, x_2, y_2, z_2) \neq \gamma(x_1, y_1, z_1)\gamma(x_2, y_2, z_2)$, i.e. when $\Gamma(x_1, y_1, z_1, x_2, y_2, z_2) = \gamma(x_1, y_1, z_1)\gamma(x_2, y_2, z_2)[1 + f(x_1, y_1, z_1, x_2, y_2, z_2)]$, the probability of finding an electron at point 1 *when another electron is at point 2* depends on the position of the second electron (through the function f) and the wave function Ψ is said to be *correlated*. The function $f(x_1, y_1, z_1, x_2, y_2, z_2)$ is a measure of correlation.

γ is calculated by integrating $\Psi^\star\Psi$ over the spatial x, y, z and spin σ coordinates of all N electrons but one, plus the spin coordinate of the remaining electron, and a factor N due to the fact that all electrons have the same probability of being at a given point. I.e., $\gamma(x_1, y_1, z_1) = N \int_{(x_i,y_i,z_i,\sigma_i)_{i=2,\ldots,N}} \int_{\sigma_1} \Psi^\star\Psi d\sigma_1 d(x_i, y_i, z_i, \sigma_i)_{i=2,\ldots,N}$. Γ is calculated the same way, but integrating the coordinates of all electrons but two: $\Gamma(x_1, y_1, z_1, x_2, y_2, z_2) = N(N - 1) \int_{(x_i,y_i,z_i,\sigma_i)_{i=3,\ldots,N}} \int_{\sigma_1} \int_{\sigma_2} \Psi^\star\Psi d\sigma_1 d\sigma_2 d(x_i, y_i, z_i, \sigma_i)_{i=3,\ldots,N}$.

antisymmetry principle. In other words, even if they are aimed at implementing a model of independent electrons, they have $f(1,2) \neq 0$ and the electrons are not fully independent.

It is agreed to call *uncorrelated* wave functions to those that contain no Coulomb but only Fermi correlation, whose only source of non-null $f(1,2)$ is the Pauli principle, and to call *correlated* wave functions to those containing Coulomb correlation. Correlated wave functions also have a different Fermi correlation than uncorrelated ones. The total energy of the so-called uncorrelated wave function contains exchange interactions, which are due to Fermi correlation; nevertheless, it is also agreed to take such energy as a reference and to define *correlation energy* of a correlated wave function as its excess energy with respect to this reference.

After adopting this naming convention, an uncorrelated N-electron wave function is a Slater determinant $D_0(1,\ldots,N)$, which is an antisymmetrized spin-orbital product, i.e. a product of N *spin-orbitals* $\psi_i (i = 1,\ldots,N)$,[3] each of them with the coordinates of one electron, which is antisymmetrized in order to make the electron indistinguishable and fulfill the Pauli principle. The spin-orbital product $\psi_1(1)\psi_2(2)\ldots\psi_N(N)$ is antisymmetrized by adding all $N!$ possible permutations $\hat{P}$ of the electron coordinates and multiplying each of them by λ_P, which is $\lambda_P = 1$ if there is an even number of electrons changed (even parity) and $\lambda_P = -1$ if the number of electrons changed is odd (odd parity). This is represented with the antisymmetrizing operator $\hat{A} = \sum_P \lambda_P \hat{P}$. Besides, although it is not necessary, it is very convenient to use orthonormalized spin-orbitals ($\langle\psi_i|\psi_j\rangle = \delta_{ij}$) and to normalize D_0. The result is

$$\begin{aligned} D_0(1,\ldots,N) &= |\psi_1\psi_2\ldots\psi_N| \\ &= \frac{1}{\sqrt{N!}}\sum_P \lambda_P \hat{P}[\psi_1(1)\psi_2(2)\ldots\psi_N(N)]\,. \end{aligned} \tag{1.39}$$

This is the expression of a normalized $N \times N$ determinant (hence its name) with the spin-orbitals as rows and the electrons as columns (or vice versa),

$$D_0(1,\ldots,N) = \frac{1}{\sqrt{N!}}\begin{vmatrix} \psi_1(1) & \psi_2(1) & \ldots & \psi_N(1) \\ \psi_1(2) & \psi_2(2) & \ldots & \psi_N(2) \\ \ldots & \ldots & \ldots & \ldots \\ \psi_1(N) & \psi_2(N) & \ldots & \psi_N(N) \end{vmatrix}\,. \tag{1.40}$$

It is an antisymmetric wave function because exchanging any two electrons, which is accomplished by exchanging two rows, changes the sign of the determinant. It also makes clear that a spin-orbital cannot be occupied twice, because two identical columns would make the determinant null.

For the ground state of a molecule or cluster with an even number of electrons, the N spin-orbitals are constructed out of $N/2$ orbitals $\varphi_j (j = 1,\ldots N/2)$ with α and

[3] A spin-orbital ψ is a one-electron function that results from the product of an orbital φ and a spin function α or β: $\psi = \varphi\alpha$ or $\psi = \varphi\beta$.

β spin functions. Such an uncorrelated wave function is known as the (closed-shell) Hartree-Fock wave function Φ^{HF} [50],

$$\Phi^{\mathrm{HF}} = \frac{1}{\sqrt{N!}} \begin{vmatrix} \varphi_1\alpha(1) & \varphi_1\beta(1) & \dots & \varphi_{N/2}\alpha(1) & \varphi_{N/2}\beta(1) \\ \varphi_1\alpha(2) & \varphi_1\beta(2) & \dots & \varphi_{N/2}\alpha(2) & \varphi_{N/2}\beta(2) \\ \dots & \dots & \dots & \dots & \dots \\ \varphi_1\alpha(N) & \varphi_1\beta(N) & \dots & \varphi_{N/2}\alpha(N) & \varphi_{N/2}\beta(N) \end{vmatrix} . \tag{1.41}$$

Regarding the construction of correlated wave functions, there are two basic options: explicitly correlated wave functions and multiconfigurational expansions. In the first case, wave functions are used that depend not only on the electrons' positions but also on their mutual distances r_{ij} [51]; although this option has some accuracy advantages, it has serious computational disadvantages derived from their implementations being cumbersome. The multiconfigurational option consists on using linear combinations of Slater determinants as correlated wave functions [22] and it is the most popular in wave function theory. Its practical development followed multiple avenues depending on the extensions of the expansions and the ways they are organized, as well as on the treatment given to the orbitals. All developmental avenues are very systematic and this is the property that made multiconfigurational wave function theory to dominate the field of quantum chemistry methodology. The calculations reported in this book are multiconfigurational ab initio quantum chemistry calculations.

We discuss next how Slater determinants are going to be mixed in linear combinations that properly and efficiently include the necessary correlation in the calculations. In the process of calculating an uncorrelated many-electron wave function of a molecule or cluster, e.g. the ground state Φ^{HF} in (1.41), the molecular orbitals φ are approximated as linear combinations of basis functions χ for convenience [50],

$$\varphi_i = \sum_{j=1}^{N_B} \chi_j c_{ji} . \tag{1.42}$$

Such basis functions constitute a large set of them, N_B, known as the *basis set*, which is adopted a priori. Basis sets are produced and monitored in systematic manners and their continuous success in molecular calculations ultimately validates them. Most of the times, the basis sets are known Gaussian-type functions centered on the atoms and called atomic basis sets. They range from primitive gaussians or contracted gaussians [52], which are calculated by requiring them to produce the best atomic orbitals in atomic calculations, to atomic orbitals themselves [53]. Contrary to the basis set χ, which is chosen at the start of the molecular calculation, the linear coefficients c_{ji} are unknown; they are obtained in the target molecular calculations, following the variational principle, by minimizing the total energy of the molecule, which is achieved by solving eigenvalue equations in self-consistent field calculations (SCF) [50].

The interesting thing here is that the eigenvalue equations produce the *occupied* orbitals (those used in the uncorrelated wave function, i.e. in the Slater determinant), but they also produce *unoccupied* or *virtual* orbitals. E.g. in a closed-shell, the occupied orbitals are $\varphi_1, \ldots \varphi_{N/2}$ and the virtual orbitals are $\varphi_{N/2+1}, \ldots \varphi_{N_B}$. The virtual orbitals are irrelevant for the independent particle model, but they can play an important role in accounting for correlation. In effect, taking the uncorrelated wave function or Slater determinant D_0 (1.40) as a reference, many new Slater determinants D_i can be constructed with substitutions of occupied spin-orbitals by virtual spin-orbitals (called virtual excitations, Fig. 1.3) and correlated wave functions can be constructed as linear combinations of all the Slater determinants. Each Slater determinant represents an electronic configuration (occupation of a given set of N spin-orbitals with the N electrons), and the correlated wave functions so constructed are called *configuration interaction* (CI) wave functions Φ^{CI}. The set of N_{CI} Slater determinants $D_0, D_1, \ldots, D_{N_{\mathrm{CI}}-1}$ define the CI space (and are often called the *many-electron basis set*), and the CI wave functions read

$$\Phi_k^{\mathrm{CI}} = \sum_{i=0}^{N_{\mathrm{CI}}-1} D_i C_{ik} \,. \tag{1.43}$$

Two issues are of utmost importance for many-electron wave functions of the type (1.43). The first issue is that they are multiconfigurational wave functions and, as a consequence, they represent correlation. The largest contributions to the $f(1,2)$ function that measures correlation come from Coulomb correlation. The second issue

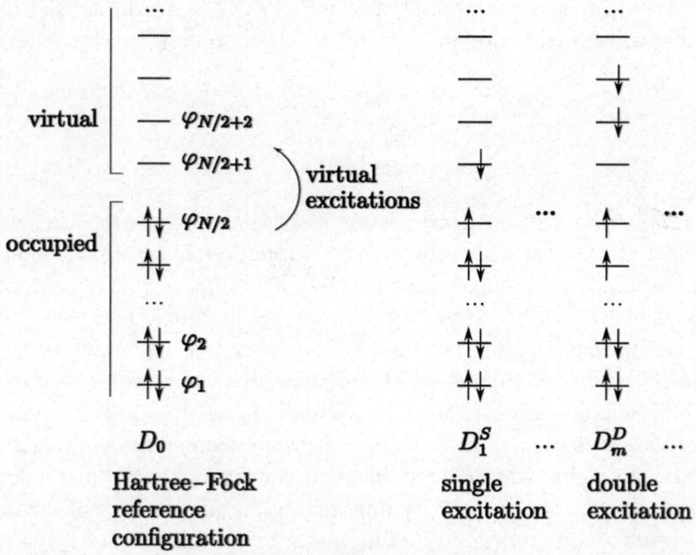

Fig. 1.3 Schematic representation of virtual excitations from a reference configuration

is that they can represent a correlated ground state, but also correlated *excited states*. Neither the ground nor the excited states are of pure configurational character, as it would be the case of monoconfigurational wave functions. Rather, all the states are multiconfigurational. Nevertheless, the largest in the list of C_{ik} coefficients of any state k, Φ_k^{CI}, inform about its dominant configurational character.

It is convenient to organize the multiconfigurational expansion (1.43) according to the degree of virtual excitations, because the number of coefficients can be extremely long, but not all configurations are expected to be important, neither for the correlation of the ground state nor for the main character (nor the correlation) of excited states of interest. So, if taking D_0 as a reference, all single excitations produce the list of N_S determinants D_i^S, all double excitations produce the list of N_D determinants D_i^D, and so on, (1.43) can be rewritten as

$$\Phi_k^{\mathrm{CI}} = D_0 C_{0k} + \sum_{i=1}^{N_S} D_i^S C_{ik} + \sum_{\substack{i=1\\ j=N_S+i}}^{N_D} D_i^D C_{jk} + \sum_{\substack{i=1\\ j=N_S+N_D+i}}^{N_T} D_i^T C_{jk} + \ldots \tag{1.44}$$

A methodological avenue based on (1.43) or (1.44) is the so-called Configuration Interaction methods.[4] In the CI methods, the multiconfigurational expansion is organized in excitations (1.44) and it is truncated. The most common truncation is after the double excitations, which are the most important ones for correlation, and the corresponding wave functions and total energies are called *single and double excitations CI* ones, Φ_k^{SDCI} and E_k^{SDCI}. The occupied and virtual molecular orbitals φ used in the Slater determinants have already been calculated, e.g. in a Hartree-Fock calculation, and are fixed. This way, the CI coefficients C_{ik} are the only unknowns and they are calculated by minimizing the total energies of the molecule or embedded-cluster $E_k^{\mathrm{SDCI}} = \langle \Phi_k^{\mathrm{SDCI}} | \hat{H}_{\mathrm{EC}} | \Phi_k^{\mathrm{SDCI}} \rangle$, following the variational principle. This results in a CI eigenvalue equation

$$\underline{H}^{\mathrm{SDCI}} \underline{C}^{\mathrm{SDCI}} = \underline{C}^{\mathrm{SDCI}} \underline{E}^{\mathrm{SDCI}} , \tag{1.45}$$

where: $\underline{H}^{\mathrm{SDCI}}$ is the matrix of the molecule or embedded-cluster many-electron Hamiltonian $\hat{H}_{\mathrm{EC}}$ on the many-electron basis of Slater determinants,

$$\underline{H}_{ij}^{\mathrm{SDCI}} = \langle D_i | \hat{H}_{\mathrm{EC}} | D_j \rangle , \tag{1.46}$$

D_i and D_j being in the list D_0, D_i^S, D_i^D; $\underline{C}^{\mathrm{SDCI}}$ is the resulting CI matrix coefficients, where the k-th column defines the CI wave function of the k-th state (either ground or excited), and the i-th row multiplies the i-th Slater determinant in the list; and $\underline{E}^{\mathrm{SDCI}}$ is a diagonal matrix of real values whose k-th element is the total energy of the k-th many-electron state in the SDCI approximation.

[4] We must be aware of a potentially confusing notation here: Although all multiconfigurational wave functions of the kind (1.43) are in fact CI functions, the name CI methods (and their corresponding CI wave functions) is commonly reserved to the ones now described.

We may remark here that practical CI calculations take advantage of spin and spatial symmetries in order to drastically reduce the computing time and resources. In order to do so, the expansions (1.43) and (1.44) are reformulated in terms of small linear combinations of them that transform according to those symmetries and can be labeled with their spin quantum numbers and spatial irreducible representations, which are called *configuration state functions* (CSF). Also, when these expansions are not truncated, we have the so-called *full-CI* calculation, which gives the best possible wave functions and total energies for the chosen one-electron basis set χ. If this basis set is complete, which requires an infinite number of one-electron basis functions, the full-CI wave function would be exact. The strong systematics within the CI methods is, then, obvious.

In practice, full-CI expansions are prohibitive, but SDCI expansions are already very long, if not extremely long, and another methodological avenue has been developed: the multiconfigurational self-consistent field methods (MCSCF). The main idea behind these methods is to drastically limit the CI expansion by wise identification of the most important excited configurations for the target problem while the molecular orbitals involved are allowed to vary to reach a balanced description of the energies and wave functions of all resulting multiconfigurational states, rather than the ground configuration alone. Physical and chemical background of the system and properties on focus guide the identification of the configurational space able to incorporate the correlations required for their accurate calculation. Such an educated, albeit unbiased, choice of the configurational space is a key decision underlying the use and success of MCSCF methods and correlation methods based on MCSCF references. In practical MCSCF calculations, the multiconfigurational expansions used are much shorter than in CI methods, but they are still very long, and they are chosen so that their structure facilitates the optimization of the molecular orbitals, which becomes a very important methodological issue. The main virtue of MCSCF methods is that they can capture the essence of static correlation (see below). Their main limitations are in their difficulties to properly consider dynamic correlation.

A third methodological avenue largely explored is that of Many-Body Perturbation Theory methods (MBPT; PT from now on). In these, the target many-electron Hamiltonian $\hat{H}_{\mathrm{EC}}$ is written in terms of a reference Hamiltonian $\hat{H}_0$, of which ground and excited wave functions and energies are known, plus a correction, which is supposed to be small and considered a perturbation,

$$\hat{H}_{\mathrm{EC}} = \hat{H}_0 + \lambda \hat{H}_{\mathrm{pert}} . \tag{1.47}$$

The wave functions and total energies of $\hat{H}_{\mathrm{EC}}$ are expressed in terms of the known wave functions and energies of $\hat{H}_0$, in series expansions of the perturbation, as represented by the parameter λ (which will be taken as unity in the end). When the series include the terms up to λ^2, one refers to the results as second-order PT, which is the most common level of application of Perturbation Theory. When the reference zeroth-order wave function is a non-degenerate single Slater determinant, they include basically the same ingredients used in (variational) SDCI calculations, but in closed formulas much easier and faster to apply. In spite of this advantage of being faster and the fact that size-extensivity is easier to achieve by PT methods, their steady development had to overcome some scepticism towards the accuracy associated with perturbative methods providing no upper bound to the energy. However, the developments have shown that the accuracy of PT methods depends strongly on the capability of the zeroth-order wave function(s) to closely represent the structure of the target electronic system. So, PT methods have been steadily progressing from nondegenerate single reference PT [54] to quasidegenerate PT [55] to multiconfigurational PT methods [56–58] where the zeroth-order wave functions are, respectively, a single nondegenerate Slater determinant, a few quasidegenerate valence configurations, or one or several MCSCF states. These references adapt to the demands of increasingly complex electronic structures of atoms, molecules, and embedded-clusters: from closed-shell systems to the multiplets of heavy element open-shells.

The success of the methodological avenue that combines MCSCF and (low order) PT methods stems from a clean and efficient division of roles: Whereas the MCSCF zeroth-order wave function embodies all static correlation effects (see below), the first order interacting space brings about the dynamic correlation corrections to the first order wave function and second order energy. Ultimately, the quality of the MCSCF PT combination rests upon an adequate choice of the configurational space used to obtain the MCSCF wave functions in the first place, on the lines mentioned above. The multiconfigurational methods chosen for this book use this approach.

Another methodological avenue worth mentioning is the coupled-cluster methods (CC) [59, 60], whose very accurate results become benchmarks for other correlation methods, provided that the CC calculations are feasible. These methods solve the size-extensivity problem of truncated CI expansions like single- and double-excitations CI, SDCI, which fail to represent equally well the N electrons of a molecule when they are together or separated in many possible ways (dissociations, ionizations, …). It was realized that the problem appears because, for instance, some triple excitations are necessary to account for size-extensivity of a SDCI wave function rather than for a better description of correlation (this is so with subsets of higher and lower order excitations in general). The way to include higher-order excitations (e.g. triple and quadruple) to account only for size-extensivity of wave functions that include correlation represented with lower-order excitations (e.g. single and double), so balancing the representation of correlation no matter how the N electrons are distributed, is via an exponential operator on the excitations, which ultimately gives CC wave functions and energies by specific, non-variational CC procedures. Coupled-cluster methods give the best theoretical results but their demands of resources make them useful, nowadays, only to produce benchmark, reference results, in general, but their systematic and continuous use in real problems of luminescence has not been even explored.

1.4.1 Static Correlation and Multiconfigurational Expansions: CASSCF and RASSCF

As we have mentioned, many materials are luminescent because of their rich manifolds of excited states, originating from the multiplet structures of partially filled (open) shells of orbitals, like the $4f^N, 4f^{N-1}5d, 4f^{N-1}6s$ electronic configurations of lanthanide ions in solids, or $4f^{N-1}\phi_{ITE}$ ITE configurations and LMCT, IVCT, MMCT and other types of charge transfer configurations. These open-shell multiplets, like the atomic multiplets, result from an interplay between spins and electron repulsions, driven by the Pauli antisymmetry principle, so that the multiplets are ultimately a consequence of Fermi correlation. Since the electron repulsions depend on the spatial symmetry of the orbitals that constitute the shells, in the end, it is the couplings between spins and spatial symmetries (equivalent to the LS Russell-Saunders coupling in atoms) that determines the structure of the multiplets. The couplings are a means to obtain the simplest many-electron wave functions that fulfill spin and spatial symmetry requirements, which is to say that have well-defined spin quantum numbers and spatial irreducible representations. These spin- and symmetry-adapted wave functions are short linear combinations of Slater determinants. Since they are not single determinants, they are correlated wave functions, i.e. $f(1,2) \neq 0$ in (1.38) and $\Gamma(1,2) \neq \gamma(1)\gamma(2)$, and they describe some Coulomb correlation besides Fermi correlation. But the small size of the expansions does not allow them to properly describe all the Coulomb correlation or a large part of it. The kind of (Fermi plus Coulomb) correlation they represent is called *static* correlation and it is the correlation necessary to account for the multiplet structure of the open-shell. It is the kind of correlation that exists in the simplest molecular orbital theory wave function that can correctly describe H_2 dissociation (i.e. H_2 at bonding, but also at infinite H-H distances, or, in other words, the breaking of the bond made by two electrons of opposite spins), $C_1|\varphi_1\alpha\varphi_1\beta| + C_2|\varphi_2\alpha\varphi_2\beta|$. Although an accurate definition of static correlation has always been elusive, one can safely say that it is the minimum correlation necessary for a correct qualitative representation of a given electronic structure. E.g. H_2 dissociation, which requires two Slater determinants; but also the multiplet structure of atoms, which requires all Slater determinants of a given atomic configuration, like $2p^2$; or, more interesting for the case of luminescent materials, the multiplet structure of molecular open-shells in general, which requires all Slater determinants of a given molecular configuration, like $(\varphi_{\mathrm{Ln}4f})^n$, $\varphi_{\mathrm{Ln}4f}$ being a molecular orbital with a main character of a lanthanide $4f$ orbital.

Complete Active Space Self-Consistent Field wave functions (CASSCF) [61–63] are the most robust and efficient means to consider the static correlation of an electronic problem. They can be described as full-CI wave functions (i.e. including all single-, double-, triple-, etc. excitations) with limitations: the excitations come from a few occupied orbitals only, and they go not to all the virtual orbitals but only to a reduced number of them. The CI space of many-electron basis functions so defined is called Complete Active Space (CAS). The union of the two sets of orbitals together contains all the orbitals whose occupation is variable across the CAS and is called the *active orbital space*; the set of occupied orbitals not experiencing excitations is called *inactive orbital space* and they are doubly occupied in all many-electron functions of the CAS; and the set of virtual orbitals not receiving any excitation and having zero occupation in the CAS is called *virtual orbital space*. Alternatively, CASSCF wave functions can be described as CI wave functions where the CI space is made of all configurations resulting from: double occupancy of a first set of orbitals (the inactive orbitals), distribution of a given number of electrons n_{os} (the open-shell electrons) among a second set of orbitals (the active orbitals) in all possible ways, and zero occupancy of a third set of orbitals (the virtual orbitals). In other words, they are full-CI wave functions within the active orbital space (hence the name Complete Active Space) and Hartree-Fock wave functions within the inactive and virtual orbital spaces. This is represented in Fig. 1.4.

The definition of a multiconfigurational Complete Active Space guarantees that, if all the orbitals that are relevant for the correct description of a given electronic structure are taken as active orbitals, the CASSCF wave functions will necessarily consider all static correlation of the problem under consideration and, consequently, properly describe the multiplets under consideration. Taking the electronic structure of Eu^{3+} in a crystalline host as an example, the above means that, if the seven φ_{Eu4f} molecular orbitals are considered as active orbitals, and six electrons are distributed among them in all possible ways, the constructed CASSCF wave functions will provide a qualitatively correct description of the $4f^6$ multiplets involved in the red luminescence of a given Eu^{3+}-doped phosphor.

Although it is not the case of the Eu^{3+} example just mentioned, it is quite common in practical calculations that the CAS definition contains additional electronic configurations to the ones strictly needed for static correlation, in which case the CASSCF results contain not only static correlation but some amount of dynamic correlation.

In CASSCF wave functions, contrary to standard CI wave functions like SDCI and in common with all MCSCF wave functions, not only the CI coefficients are variationally optimized, but also the molecular orbital coefficients. This is an important strength of the method and it is based on the fact that the *complete* structure of the active space, i.e. the fact that it is a full CI within the active space, enormously facilitates the molecular orbital optimization in a self-consistent field process of the kind followed in Hartree-Fock calculations. Hence the SCF part of the acronym. In general, the orbitals are not obtained to optimize the lowest CASSCF state energy since this would give enormous relaxation corrections to the wave functions and energies of higher CASSCF roots. Rather, they are obtained to optimize a state average (SA) energy. The actual states included and their weight in the average are optional, the goal being that the optimal molecular orbitals give a balanced description of all target CASSCF roots. SA CASSCF calculations will be often referred to as CASSCF calculations throughout this book. (All of this also applies to (SA) RASSCF calculations described below.)

A key question in CASSCF calculations is the correct choice of the CAS. In the above example of the $4f^6$ multiplets of Eu^{3+} in hosts, one has to be sure that the 6 electrons are in fact distributed (in all possible ways) over the seven φ_{Eu4f} molecular

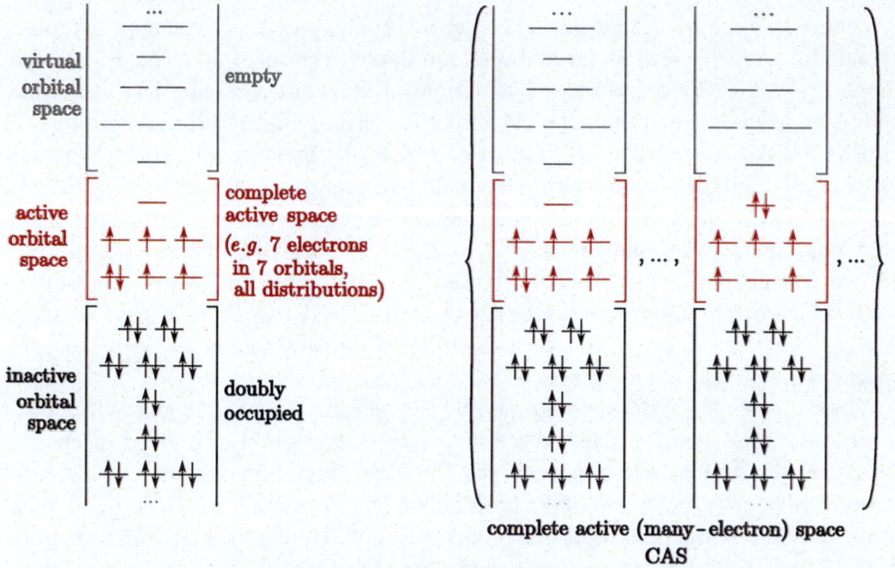

Fig. 1.4 Schematic representation of the orbital spaces used for the definition of the CAS (left) and the set of Slater determinants (or configurational functions) that constitute the many-electron complete-active-space (right)

orbitals. If one or several orbitals are missing, the structure of the whole manifold can largely deviate from the correct one. If they are not missing, but they are substituted by some other orbitals of a different kind, a deformed manifold will also be produced. These undesired substitutions may happen by an uncareful choice at the start of the SCF procedure, but also accidentally during the SCF iterations, which may flip the order of orbitals. A continuous monitoring of the nature of the active orbitals is a necessary condition for a successful CASSCF calculation.

A practical limitation appears in CASSCF calculations when the size of the many-electron Complete Active Space is unaffordable. In effect, being a full CI in a limited orbital space, the size of the CAS increases exponentially with the number of active orbitals. It also increases for many-electron states of lower than maximal total spin quantum number S and when the number of open-shell electrons that are distributed among the active orbitals gets closer to half-occupancy. This makes CASSCF impractical to handle electronic multiplets where the CAS expansions are much larger than necessary for static correlation and prohibitive from the computational point of view. Systems with multiple open-shells usually face this problem. Let us take the so called $4f^65d^1$ manifold of states of Eu^{2+} (either as a free-ion or in a solid) as an example. In this case, a proper consideration of the static correlation that leads to the correct multiplet structure of this electronic problem requires all electronic configurations resulting from distributing 6 electrons among the shell of seven $\varphi_{\mathrm{Eu}4f}$ orbitals in all possible ways and 1 electron among the shell of five $\varphi_{\mathrm{Eu}5d}$ orbitals in all possible ways. However, the CAS $[\varphi_{\mathrm{Eu}4f}, \varphi_{\mathrm{Eu}5d}]^7$, defined by the distribution of 7 electrons

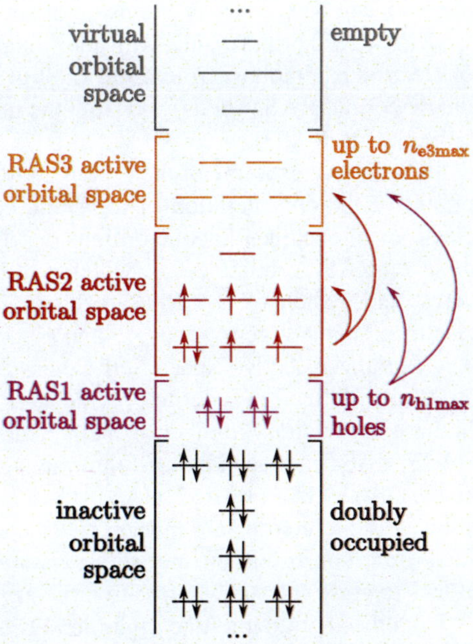

Fig. 1.5 Schematic representation of the orbital spaces used for the definition of the many-electron restricted-active-space (RAS) and restrictions in their occupancies that define the RAS

among the twelve $\varphi_{\text{Eu}4f}$ and $\varphi_{\text{Eu}5d}$ orbitals in all possible ways, contains many more configurations, like e.g. those of the types $\varphi^5_{\text{Eu}4f}\varphi^2_{\text{Eu}5d}$, $\varphi^4_{\text{Eu}4f}\varphi^3_{\text{Eu}5d}$, etc. These configurations do not include static correlation of the Eu^{2+} $4f^65d^1$ manifold, but just a fraction of its dynamic correlation. This fraction is insufficient and it would be included at a prohibitive cost, which makes the so defined CAS inadequate and demands alternatives.

Restricted Active Space Self-Consistent Field wave functions (RASSCF) [64, 65] have been proposed as a means to solve the problem of unaffordable CAS sizes in multiple open-shell systems. The idea behind the RASSCF method is to restrict the configurations that are included in the many-electron active space with respect to the CAS, by not allowing all possible electronic distributions, but limiting the occupations of some orbitals. In the practical implementation of the RASSCF method, the active orbitals are divided in 3 sets called RAS1, RAS2, and RAS3 active orbitals, and the following occupancy rules are defined: the set RAS1 can be either fully occupied or have a maximum of n_{h1max} holes distributed in all possible ways; the set RAS3 can be either empty or have a maximum of n_{e3max} electrons distributed in all possible ways; and the set RAS2 is occupied with all the active electrons that are not in RAS1 nor in RAS3 orbitals, in all possible ways. This is schematically represented in Fig. 1.5.

A RAS can be seen as a restriction imposed on a CAS whose active orbitals were the union of RAS1, RAS2, and RAS3 active orbitals, because the active electrons are not distributed in all possible ways over the RAS1 and RAS3 orbitals; this is why, as a many-electron active space, it is call Restricted instead of Complete. Alternatively, the RAS can be seen as an extension of another, smaller CAS whose active orbitals were the RAS2 orbitals only, with additional configurations resulting from two types of electron excitations: from a full shell of RAS1 orbitals (which can be excited to the RAS2 and RAS3 shells) and from the unfilled shell of RAS2 orbitals (which can be excited to RAS3).

In the above example of the $4f^6 5d^1$ multiplets of Eu^{2+}, a minimum RAS appropriate for the consideration of the static correlation can be defined with 7 active electrons to occupy 12 active orbitals divided as follows: no RAS1 orbitals; a shell of seven RAS2 orbitals $\varphi_{\text{Eu}4f}$, which will be occupied by electrons in all possible ways; and a shell of five RAS3 orbitals $\varphi_{\text{Eu}5d}$ with a maximum occupancy of 1 electron. This will generate $\varphi^7_{\text{Eu}4f}$ and $\varphi^6_{\text{Eu}4f}\varphi^1_{\text{Eu}5d}$ configurations only.

Generalized Active Space Self-Consistent Field wave functions GASSCF [66] are further defined in order to provide additional flexibility to the RAS restrictions. In these: an arbitrary number of shells of active orbitals can be used, GAS1, GAS2, GAS3, GAS4, ...; accumulated minimum and maximum occupancies of shells are given, GAS1, GAS1+GAS2, GAS1+GAS2+GAS3, ...; all intra-shell excitations are allowed (i.e. electrons are always distributed in all possible ways within the GAS shells); and constraints on inter-shell excitations are imposed at will. Such a flexibility allows to grasp the static correlation of very complex open-shell systems.

In all the ab initio multiconfigurational calculations performed on optically active sites of luminescent materials with complex open-shells shown in this book, many-electron multiconfigurational CAS and RAS expansions were sufficient to consider the necessary static correlations and, in consequence, CASSCF and RASSCF calculations were performed for this purpose.

1.4.2 *Static Plus Dynamic Correlation and Multi-references: CASPT2 and RASPT2*

We have just discussed (Sect. 1.4.1) that the correct consideration of static correlation by means of CASSCF, RASSCF, and GASSCF wave functions guarantees a proper qualitative description of the multiplets associated with the open-shells. However, a quantitatively correct description demands the consideration of dynamic correlation as well. And what is dynamic correlation? As with static correlation, its definition is not free from debate, but, having defined static correlation, one can say that it is all the remaining electron correlation, or, in other words, dynamic correlation is all correlation that is not static. Dynamic correlation is not necessary to find the multiplet structure (or the bond breaking) of an electronic system and it cannot account for it. So, it has to be considered together with static correlation. But it is compulsory for an accurate description of the Coulomb (plus Fermi) hole, i.e. the lowering of the probability density of finding an electron around a point in space when another electron is already at that point, both when the two electrons have equal and opposite spins. And this has a strong impact on the energy differences between manifolds and between the states of a given manifold.

In summary, a quantitatively correct description of the various multiplets of a luminescent material active center and of their interplays, i.e. of the complex excited manifolds of the material (like the $4f^7$ and $4f^65d^1$ multiplets of Eu^{2+} in a solid host), demands an accurate consideration of both static and dynamic correlations. Hence any methods capable of retrieving dynamic correlation only are not suitable for luminescent materials. This is the case of methods built upon a single Slater determinant, such as some single reference PT methods and DFT methods.

RASSCF wave functions (and also CASSCF wave functions; we will say RASSCF to refer to both from now on) very often contain a fraction of dynamic correlation when their expansions are larger than necessary for static correlation. In order to consider the remaining dynamic correlation (or all of it, if the RASSCF wave functions contain a null or negligible fraction), one could follow, in principle, the variational CI methodological line, taking the RASSCF wave functions as a multi-reference (MR) for excitations, in the line of MRCI approaches [67–69]. However, feasibility can be seriously compromised when the multi-reference space is large and/or when high order excitations are considered. Multireference Singles and Doubles Configuration Interaction (MR-SDCI) is the simplest alternative limiting the virtual excitations from the RAS multi-reference to one- and two-electron excitations. These wave functions, which are based on the variational principle, would be an accurate alternative to account for the dynamic correlation on top of the static correlation as long as approximations to the size-extensivity problem associated with truncated CI expansions are incorporated [70, 71], but they are still very expensive, if not prohibitive, from the computational point of view in a vast majority of luminescent interesting cases.

Fortunately, practical and sufficiently accurate alternatives emerged by using perturbationally, rather than variationally, the single- and double-excitations from multi-references constituted by CASSCF and RASSCF wave functions. The methods are respectively called Complete Active Space second-order Perturbation Theory (CASPT2) and Restricted Active Space second-order Perturbation Theory (RASPT2) [72–77], and other flavors of the same basic idea have also been developed. The evolution of the RASPT2 methodology along two successive formulations of increasing flexibility and accuracy consolidated its adequacy to the complex open-shell manifolds of excited states that characterize the luminescent materials that matter in this book.

Together with the issues of static and dynamic correlation, comes the problem of near-degeneracy, i.e. the fact that several electronic states can cluster together in a narrow energy region. This is a very common situation in luminescent materials, where the usual thing is finding not several, but many near-degenerate states around a given energy, and many near-degeneracy energy regions crowded of states, either of the same electronic configuration or of a different one. The near-degeneracy problem, which is often but not always related with static correlation, has a tricky interplay with dynamic correlation. Its consideration originated two implementations of the CASPT2 and RASPT2 methods: a first one in which near-degeneracy was not an issue, and a second one just focused on handling it.

In the first proposal, one Single-State (SS) RASSCF wave function Φ_i^{RAS} is used as the zeroth-order single reference for second order perturbation, hence its name SS-RASPT2 (referring to SS-CASPT2 and SS-RASPT2 from now on) [72–74, 77]. In the second proposal, a number of RASSCF states constitute a multidimensional (Multi-State, MS) zeroth-order instead, hence its name MS-RASPT2 [75, 76]. In actual calculations, the use of the MS reference implies three important steps in the calculations with clear physical meanings. (i) In the first calculation step, the use of the SA-RASSCF method allows to obtain as many RASSCF states as needed for a given total spin and spatial symmetry. (ii) In the second step, each of these RASSCF states are used in successive SS-RASPT2 calculations. (iii) In the third step, the ingredients of the RASPT2 calculations are used to solve the MS-RASPT2 problem.

The strength of the first step stems from the capability of the SA-RASSCF wave functions Φ_i^{RAS} to embody the static correlation, as commented above. At this level of theory, the main characteristics of the relevant, crowded excited manifolds show up, very often in a very graphical way (Fig. 1.6): How many states there are for each total spin and symmetry (irreducible representation of the corresponding point symmetry group), in correspondence with the availability of unfilled or open-shells of orbitals; what are their main electronic configurational characters; what are their near-degeneracy patterns; what are their avoided crossings; or what are the patterns of their equilibrium structure properties, like bond lengths and energy curvatures or vibrational frequencies. As mentioned, the weakness of this step is the lack of most or all dynamical correlation effects, which translates into an only qualitative but, in general, non-quantitative reliability of the results.

The strength of the second step stems from a neat separation between static and dynamic correlations. For a particular electronic state i, static correlation is embodied in the single RASSCF state zeroth-order wave function Φ_i^{RAS}. Dynamic correlation is introduced perturbationally to second-order using: (i) a particular partition of the many-electron Hamiltonian into reference and perturbation Hamiltonians which is state specific: $\hat{H}_{\text{EC}} = \hat{H}_{0,i} + \lambda \hat{H}_{\text{pert},i}$; and (ii) a (first-order) interacting space made by the SS reference wave function Φ_i^{RAS}, plus all the states generated by single and double excitations from this wave function (Fig. 1.7).[5]

Neither higher order excitations (triple, quadruple, ...) from the SS reference Φ_i^{RAS}, nor other RASSCF states Φ_j^{RAS} obtained from the previous SA-RASSCF calculations, interact with the SS reference Φ_i^{RAS} via the many-electron Hamiltonian. Then, they do not contribute to the first-order wave function Φ_i^{SSPT2}, which is a linear combination of Φ_i^{RAS} and single- and double-excitations, nor to the second-order energy E_i^{SSPT2}. Eventually, this is precisely the weakness of this second step: the lack of interaction between the RASSCF states. In effect, their interaction becomes crucial

[5] In the single and double virtual electron excitations in RASPT2, the electrons go from inactive and active orbitals (i.e. with double and variable occupancy, respectively, in all RASSCF wave functions) to active and virtual (empty) orbitals. Normally, excitations from the innermost inactive orbitals are excluded; in that case, the orbitals are called frozen and dynamic correlation is not considered for their electrons. Choices that allow to exclude dynamic correlation for core electrons while including it for outer-valence and, optionally, inner-valence electrons are commented in Sect. 2.3 and throughout the applications sections.

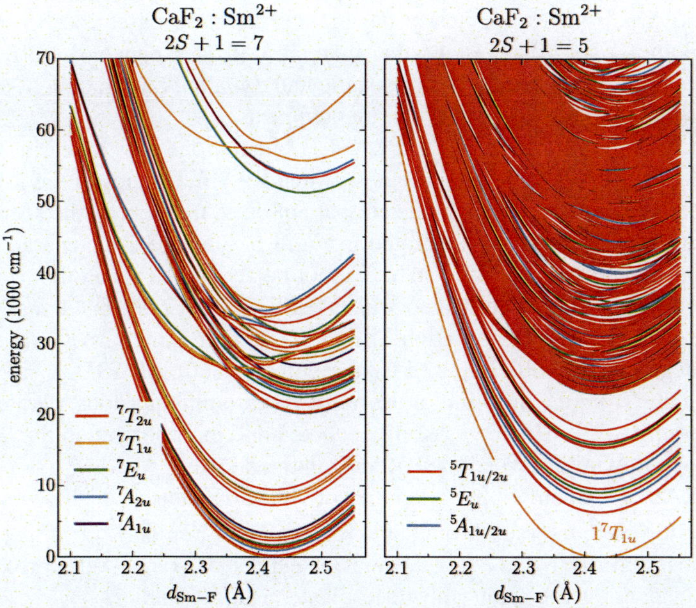

Fig. 1.6 RASSCF excited states of Sm^{2+}-doped CaF_2 in eightfold cubic fluorine coordination. States with different bond lengths (2.42 Å, 2.47 Å, 2.40 Å) and main configurational character ($4f^55de_g^1, 4f^55dt_{2g}^1, 4f^5\phi_{ITE}^1$) avoid crossings above 20×10^3 cm^{-1}

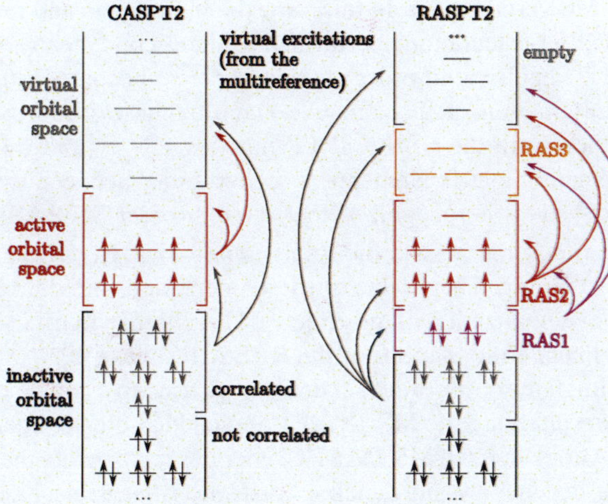

Fig. 1.7 CASPT2 and RASPT2 virtual excitations

when the RASSCF states are nearly degenerate. In these cases the SS-RASPT2 method becomes a subsidiary step rather than the final step capable of offering the best practical alternative to representing static and dynamic correlation. Why this is so becomes clear in Fig. 1.6, as it is commented next.

The results of the first step of SA-RASSCF calculations give an early warning: near-degeneracies and avoided crossings between RASSCF states of the same total spin and spatial symmetry are not exceptional but quite frequent in the spectrum of luminescent materials, as it is illustrated in Fig. 1.6. Such near-degeneracy effects between non-interacting RASSCF states cannot be accounted for in the framework of a single-reference perturbation theory like the SS-RASPT2 one but demands some sort of Quasi Degenerate Perturbation Theory (QDPT) treatment. An alternative to this quasi degeneracy problem posed by the non-interacting RASSCF single state references came from the proposal of the multi-partitioning quasidegenerate perturbation theory [56, 75], which was readily adapted and incorporated in the RASPT2 methodology allowing for MS-RASPT2 calculations [76].

In MS-RASPT2 calculations, a number of RASSCF states $\{\Phi_i^{\mathrm{RAS}}\}$ (the model space) are allowed to interact through an effective Hamiltonian $\hat{H}_{\mathrm{EC}}^{\mathrm{eff}}$. This effective Hamiltonian is defined so that its eigenvalues are expected to be a good approximation to the eigenvalues of the total embedded-cluster Hamiltonian $\hat{H}_{\mathrm{EC}}$ [75, 76]. It is constructed using QDPT to second-order in the multidimensional space of the non-interacting RASSCF functions Φ_i^{RAS}, taking into account a multi-partitioning of the Hamiltonian $\{\hat{H}_{\mathrm{EC}} = \hat{H}_{0,i} + \lambda \hat{H}_{\mathrm{pert},i}\}$, where a different partition into zeroth-order plus perturbation Hamiltonians corresponds to each different RASSCF state i in the MS zeroth-order. In this way, the ingredients and results of all independent SS-RASPT2 calculations performed in the second step (i.e. first-order wave functions Φ_i^{SSPT2} and second-order energies E_i^{SSPT2}) become instrumental to build the matrix representation of the effective Hamiltonian in the MS model space: Its diagonal elements are the SS-RASPT2 energies E_i^{SSPT2} and the off-diagonal elements are interaction matrix elements at second-order between two different RASSCF states. This way, it is recognized that the first-order SS-RASPT2 wave functions Φ_i^{SSPT2} do in fact interact via the total Hamiltonian $\hat{H}_{\mathrm{EC}}$ even though the RASSCF states Φ_i^{RAS} do not. The effects of these interactions are delivered to the total energies by the diagonalization of the effective Hamiltonian matrix. The eigenfunctions obtained are linear combinations of the RASSCF states $\{\Phi_i^{\mathrm{MSRAS}}\}$. They are a new effective multi-state zeroth-order set that can be used to evaluate one-electron properties. The eigenvalues are the MS-RASPT2 embedded-cluster energies $\{E_i^{\mathrm{MSPT2}}\}$.

Altogether, the MS-RASPT2 energies incorporate the effects of static correlation, near-degeneracy, and dynamic correlation that are necessary to describe the dense and complex manifolds of excited states of luminescent materials. This is the strength of the third step of MS-RASPT2 calculations. Its weakness is that it uses a spin-free Hamiltonian, in particular, a spin-orbit-free Hamiltonian and this is a serious drawback when heavy elements are involved. We deal with this methodological demand in Sect. 1.5.

Table 1.1 Electron correlation

With Ψ, one calculates $\Gamma(1, 2) = \gamma(1)\gamma(2)[1 + f(1, 2)]$, (1.38)

If ...	Correlation origin	Correlation type		Ψ is said to be ...	Example
$f(1, 2) = 0$, uncorrelated electrons	None	None		Uncorrelated	Hartree-Fock for $2e^-$ of opposite spins
$f(1, 2) \neq 0$, correlated electrons	Pauli antisymmetry (between e^- of equal spins)	Fermi correlation		Uncorrelated	Hartree-Fock
	+ Coulomb repulsion	+ Coulomb correlation (dominant)	Static correlation[a]	Correlated	RASSCF
			+ Dynamic correlation[b]	Correlated	+ MS-RASPT2

[a]Minimum correlation necessary for a correct qualitative representation of the nature of a given electron problem. E.g. H_2 dissociation (bond breaking); multiplet structure of atoms; multiplet structure of molecular open-shells
[b]All the remaining correlation

We end this chapter on correlation summarizing in Table 1.1 the correlation types and examples of many-electron wave functions and methods that take them into account. Our choice to handle static and dynamic correlation in excited manifolds of active sites of luminescent materials is to perform multiconfigurational multi-reference RASSCF/MS-RASPT2 calculations (or, occasionally, CASSCF/MS-CASPT2). We used throughout this book RASSCF (or CASSCF) wave functions in order to account for static correlation and added on top of it near-degeneracy effects and second-order dynamic correlation corrections in MS-RASPT2 (or MS-CASPT2) calculations. The MS-RASPT2 corrected RASSCF wave functions were used to calculate radiative transition moments and oscillator strengths, as well as any other properties.

1.5 Electron Correlation and Spin-Orbit Coupling Together: RASSI-SO

We have discussed (Sect. 1.3) the importance of relativity, both of spin-independent and spin-orbit coupling relativistic interactions, for the complex excited manifolds of optically active centers of luminescent materials, specially when they contain heavy elements like lanthanide ion activators. And we have seen a theoretically sound many-electron relativistic Hamiltonian that accurately accounts for these effects in practice, the Douglas-Kroll-Hess Hamiltonian. We have also discussed (Sect. 1.4) the key role of electron correlation, both static and dynamic, for the qualitative and quantitative energetic distribution of the excited multiplets in the luminescent materials. And we have seen multiconfigurational CASSCF and RASSCF wave functions and multireference second-order perturbation theory MS-CASPT2 and MS-RASPT2 methods, specially designed to account for such correlation effects in practical calculations. However, a serious methodological problem shows up when relativistic spin-orbit coupling and static and dynamic correlation have to be handled together, which is paradoxically solved in a very simple manner. This is the issue in this section.

The problem of handling together spin-orbit coupling and electron correlation appears because the very efficient methods to consider correlation with multiconfigurational expansions are based on the use of spin-independent Hamiltonians. When spin-orbit coupling is included in the Hamiltonian, new methods have to be developed; these are necessarily much more cumbersome and, in spite of some partial achievements [78], none of these methods has been able to reach the high efficiency of the correlation methods for spin-independent Hamiltonians to date. As a result, the amounts of correlation that can be included in practice in a molecule or embedded-cluster with a spin-orbit coupling Hamiltonian are much lower than with a spin-independent Hamiltonian.

Specifically, in the study of an optically active center containing a lanthanide ion with important spin-orbit coupling interactions, ab initio embedded-cluster RASSCF/MS-RASPT2 calculations of an adequate size to accurately consider static and dynamic correlation effects, cannot be done with the full Hamiltonian that includes spin-orbit coupling. The question, then, is: how can these correlation effects be included together with spin-orbit coupling?

This problem found an accurate solution with the so called two-step spin-orbit coupling methods. These methods take advantage of the fact that the full many-electron Hamiltonian $\hat{H}_{\mathrm{EC}}$ of an embedded-cluster or molecule is the sum of two parts (1.37): A spin-independent $\hat{H}_{\mathrm{EC,scalar}}$ part that contains the two-electron repulsions responsible for correlation, and a spin-orbit coupling part $\hat{H}_{\mathrm{EC,SOC}}$, which is an effective one-electron operator and has a negligible impact on correlation.

Then, in a first step, spin-orbit coupling is neglected and accurate scalar and dynamic correlations are calculated for the so called spin-orbit-free states with the spin-independent Hamiltonian $\hat{H}_{\mathrm{EC,scalar}}$, keeping the total spin as a good quantum number. The wave functions and energies calculated to account for static correlation can be called Φ_i^{S} and E_i^{S}, the superindex S standing for static; those that incorporate dynamic correlation as well, can be called Φ_i^{D} and E_i^{D}, the superindex D standing for dynamic. These spin-orbit-free solutions can be obtained variationally by MRCI methods or perturbationally by RASSCF/SS-RASPT2 or RASSCF/MS-RASPT2 methods. In all cases, the configurational space S needed to retrieve static correlation is much smaller than the space D needed to retrieve dynamic correlation as well.

In a second step, the full Hamiltonian $\hat{H}_{\mathrm{EC}}$ $(= \hat{H}_{\mathrm{EC,scalar}} + \hat{H}_{\mathrm{EC,SOC}})$ is used and spin-orbit coupling interactions between the spin-orbit-free states of the first step are included. Here, one has to consider that two conditions must be fulfilled in this step in order to calculate accurate spin-orbit *splittings*, which are required for the ultimate goal of obtaining good excited manifolds: The first condition is that the energy differences between spin-orbit-free states must be accurate. This is guaranteed with the E_i^{D} energies, which contain all necessary correlation. The second condition is that the spin-orbit *couplings* between spin-orbit-free states must also be accurate. At this point, the fact that $\hat{H}_{\mathrm{EC,SOC}}$ is an effective one-electron operator becomes important, because it means that accurate spin-orbit couplings can be achieved with CI wave functions that are relatively short expansions, like the Φ_i^{S} wave functions that represent static correlation; the much longer expansions required for wave functions representing dynamic correlation are not necessary.

Then, in the second step, we need to use an effective many-electron Hamiltonian such that its eigenvalues and eigenfunctions prior to activating the spin-orbit coupling, i.e. when the spin-orbit coupling component is removed, are, respectively, E_i^{D} and Φ_i^{S}, i.e. the energies with dynamic correlation and the short expansion wave functions that represent static correlation (cf. right panel of Fig. 1.8). Such an effective Hamiltonian is the *spin-free-state shifted* (sfss) Hamiltonian [79]

$$\hat{H}_{\mathrm{EC}}^{\mathrm{sfss}} = \hat{H}_{\mathrm{EC}} + \hat{P}_{\mathrm{sfss}} = \hat{H}_{\mathrm{EC,scalar}} + \hat{H}_{\mathrm{EC,SOC}} + \hat{P}_{\mathrm{sfss}}\,, \tag{1.48}$$

with the spin-free-state shifting operator $\hat{P}_{\mathrm{sfss}}$ defined as

$$\hat{P}_{\mathrm{sfss}} = \sum_i |\Phi_i^{\mathrm{S}}\rangle(E_i^{\mathrm{D}} - E_i^{\mathrm{S}})\langle\Phi_i^{\mathrm{S}}|\,. \tag{1.49}$$

The role of this operator is to project onto the S space and shift the eigenvalues from the static correlation energies E_i^{S} to the static plus dynamic correlation energies E_i^{D}. In other words, it acts onto the static correlation wave functions Φ_i^{S} shifting their energies to E_i^{D}: If the spin-orbit coupling part is removed from the effective Hamiltonian (1.48) and the matrix of the resulting spin-independent Hamiltonian $\hat{H}_{\mathrm{EC,scalar}} + \hat{P}_{\mathrm{sfss}}$ in the S many-electron space is calculated and diagonalized, the eigenfunctions are the Φ_i^{S} wave functions, but the eigenvalues are the E_i^{D} energies (this is schematically represented in Fig. 1.8). I.e., $\hat{H}_{\mathrm{EC,scalar}} + \hat{P}_{\mathrm{sfss}}$ is the spin-orbit-free Hamiltonian that, in the S space, provides: (i) the spin-orbit-free state wave functions adequate for the calculation of spin-orbit couplings and (ii) the spin-orbit-free state energies adequate for the energy differences. These are the properties that make it adequate to give accurate spin-orbit splittings when the spin-orbit coupling part $\hat{H}_{\mathrm{EC,SOC}}$ is added. When this is done, the resulting Hamiltonian that is used in the second step is the sfss Hamiltonian $\hat{H}_{\mathrm{EC}}^{\mathrm{sfss}}$.

The use of the sfss Hamiltonian $\hat{H}_{\mathrm{EC}}^{\mathrm{sfss}}$ in two-step spin-orbit CI calculations, requires to do spin-orbit-free calculations (with $\hat{H}_{\mathrm{EC,scalar}}$) on the large (D) and small (S) CI spaces in the first step, leading to E_i^{D}, Φ_i^{D}, E_i^{S}, and Φ_i^{S}. In the second step, the spin-orbit coupling operator $\hat{H}_{\mathrm{EC,SOC}}$ is activated; the $\hat{P}_{\mathrm{sfss}}$ operator (1.49) is constructed with E_i^{D}, E_i^{S}, and Φ_i^{S}, and added; and the spin-orbit CI matrix of the

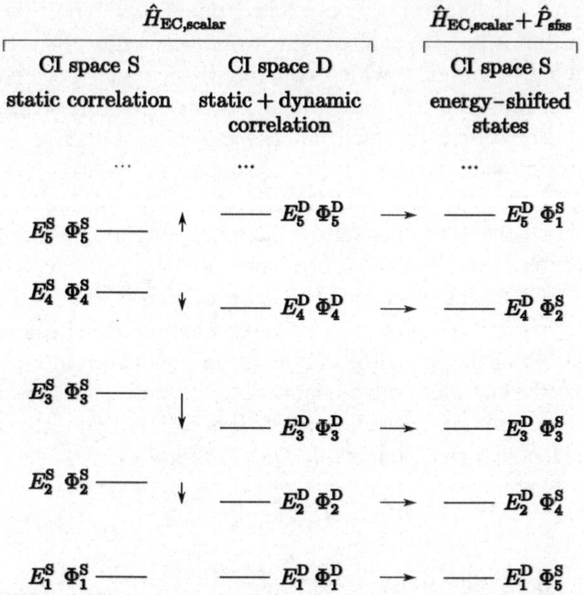

Fig. 1.8 Schematic representation of the CI eigenfunctions and eigenvalues of a spin-independent Hamiltonian $\hat{H}_{\text{EC,scalar}}$ (including static correlation only and both static and dynamic correlation) and a spin-free-state shifted spin-independent Hamiltonian $\hat{H}_{\text{EC,scalar}} + \hat{P}_{\text{sfss}}$

sfss Hamiltonian $\hat{H}_{\text{EC}}^{\text{sfss}}$ on the small CI space S is calculated and diagonalized, which leads to the final results that include the necessary correlation together with spin-orbit coupling.

In a RASSCF/RASPT2 methodological framework to handle static and dynamic correlation, the first step consists of SA-RASSCF calculations followed by RASPT2 calculations (either SS-RASPT2 or MS-RASPT2) with the spin-orbit-free Hamiltonian $\hat{H}_{\text{EC,scalar}}$. Then, the $\hat{P}_{\text{sfss}}$ operator (1.49) is built with static correlation wave functions and energies $\Phi_i^{\text{RAS}}(=\Phi_i^{\text{S}})$ and $E_i^{\text{RAS}}(=E_i^{\text{S}})$, and static plus dynamic energies, either $E_i^{\text{SSPT2}}(=E_i^{\text{D}})$ or $E_i^{\text{MSPT2}}(=E_i^{\text{D}})$. In the second step, CI calculations in a RAS subspace are made with the correct spin-orbit coupling Hamiltonian $\hat{H}_{\text{EC}}^{\text{sfss}}$ (1.48). The RAS subspace to be used in this step is defined by a set of RASSCF wave functions $\{\Phi_i^{\text{RAS}}\}$, which are usually chosen so as to include an energy range above the ground state larger than the one of interest, depending on the luminescent material and properties under study. All the Φ_i^{RAS} wave functions used in this spin-orbit CI must be included in the $\hat{P}_{\text{sfss}}$ operator, which is allowed to include additional ones. Such second step calculations are called Restricted Active Space State Interaction spin-orbit coupling calculations (RASSI-SO) [80, 81]. The whole two-step calculation is called RASSCF/RASPT2/RASSI-SO, with several admissible variants and options. These are the calculations that have been used throughout this book. All of them have been performed with the MOLCAS suite of programs [82].

Before we end this section, it is worth mentioning an apparently minor methodological detail of RASSI-SO calculations that can have, however, detrimental consequences when configuration coordinate energy diagrams are calculated. In effect, as discussed in Sect. 1.4.2, when near-degeneracies and avoided crossings can be excluded, SS-RASPT2 calculations might be sufficient in the first step; however, in the landscape of the excited states of luminescent materials, near-degeneracies and avoided crossings are not exceptional, but common, and MS-RASPT2 calculations become necessary. In this landscape, the spin-free-state-shifting operator is defined as

$$\hat{P}_{\rm sfss} = \sum_i |\Phi_i^{\rm RAS}\rangle (E_i^{\rm MSPT2} - E_i^{\rm RAS}) \langle \Phi_i^{\rm RAS}| \tag{1.50}$$

and the correspondence between states in the RASSCF calculation and in the MS-RASPT2 calculation becomes an issue: For states of given spin and symmetry, an energy driven correspondence (where i is an ordinal in energy increasing lists of $E_i^{\rm RAS}$ and $E_i^{\rm MSPT2}$) [80] may lead to anomalous results when avoided crossings exist at different nuclear configurations in the RASSCF and the MS-CASPT2 calculations, but the problems disappear when the level shifts are assigned according to the characters of the wave functions (e.g. an impurity state of $4f^{N-1}5d$ configuration and impurity-trapped exciton of $4f^{N-1}\phi_{ITE}$ configuration) rather than according to their energy order [83]. Since finding correspondences of this type is far from obvious in mass production calculations, it is important to use methodological procedures with intrinsically correct correspondences [84].

In this respect, we may first remark that two CI bases can be used with identical results in RASSI-SO calculations, because they span the same many-electron space: the set of RASSCF wave functions $\{\Phi_i^{\rm RAS}\}$ and the set of multi-state zeroth-order wave functions $\{\Phi_i^{\rm MSRAS}\}$ that diagonalize the effective MS-RASPT2 Hamiltonian $\hat{H}_{\rm EC}^{\rm eff}$. Both sets transform into each other by unitary transformations. With this in mind, an optional procedure with intrinsically correct correspondences is to use, in the second step, the $\{\Phi_i^{\rm MSRAS}\}$ basis with an *effective* spin-orbit coupling Hamiltonian $\hat{H}_{\rm EC}^{\rm eff} + \hat{H}_{\rm EC,SOC}$ [81]. The spin-orbit-free contribution to the resulting matrix is diagonal, with non-null values $\{E_i^{\rm MSPT2}\}$, hence equivalent to using $\hat{H}_{\rm EC}^{\rm sfss}$ (1.48) and (1.50). An equivalent alternative is to use $\hat{H}_{\rm EC}^{\rm eff} + \hat{H}_{\rm EC,SOC}$ with the $\{\Phi_i^{\rm RAS}\}$ basis, in which case the spin-orbit-free contribution is non-diagonal [84].[6] These two options give identical spin-orbit coupling results. In them, the shifting of the static correlation spin-orbit-free states from static correlation energies to static+dynamic energies is implicit rather than explicit, implied in the substitution of the real spin-orbit-free Hamiltonian $\hat{H}_{\rm EC}$ by the effective MS-CASPT2 Hamiltonian $\hat{H}_{\rm EC}^{\rm eff}$, and, in consequence, they are spin-free-state-shifting calculations.

Finally, let us comment that the option of using the effective spin-orbit coupling Hamiltonian $\hat{H}_{\rm EC}^{\rm eff} + \hat{H}_{\rm EC,SOC}$ and the CI basis $\{\Phi_i^{\rm MSRAS}\}$ in the second step, has an additional advantage besides avoiding the correspondence problem: the possibility of a partial correction of the consequences of slight, yet undesirable, wave function symmetry breakings. These are almost unavoidable in high symmetry luminescent

[6] Interestingly, this option is formally a one-step spin-orbit coupling calculation, because the diagonalization of the matrix of $\hat{H}_{\rm EC}^{\rm eff}$ on the $\{\Phi_i^{\rm RAS}\}$ basis, which is required to obtain the spin-orbit-free MS-CASPT2 energies, is not necessary.

centers (e.g. of the O_h point symmetry group), which have to be calculated using a subgroup (e.g. D_{2h}) when the program is bound to use abelian symmetry groups [82]. The small symmetry breakings produce broken degeneracies in the E_i^{MSPT2} spectrum than can be magnified in the spin-orbit coupling results. These breakings can be corrected before the second step by convenient averaging of the E_i^{MSPT2} values with broken degeneracy and, although this procedure cannot avoid totally the contamination of the spin-orbit results, because it does not symmetrize the spin-orbit-free wave functions, it does minimize it, largely facilitating the identification of spin-orbit states and energies. This way, it becomes an important step in the symmetry control protocol.

References

1. H. Bethe, Ann. Phys. **395**, 133 (1929)
2. R. Finkelstein, J.H. Van Vleck, J. Chem. Phys. **8**, 790 (1940)
3. M.F. Reid, F.S. Richardson, J. Chem. Phys. **83**, 3831 (1985)
4. B.R. Judd, Phys. Rev. **127**, 750 (1962)
5. G.S. Ofelt, J. Chem. Phys. **37**, 511 (1962)
6. P. Hohenberg, W. Kohn, Phys. Rev. B **136**, B864 (1964)
7. W. Kohn, L.J. Sham, Phys. Rev. **140**, A1133 (1965)
8. E. Runge, E.K.U. Gross, Phys. Rev. Lett. **52**, 997 (1984)
9. J.C. Slater, *Quantum Theory of Matter* (McGraw-Hill, New York, 1968)
10. T. Ishii, M.G. Brik, K. Ogasawara, J. Alloys Compd. **380**, 136 (2004)
11. H. Ramanantoanina, W. Urland, F. Cimpoesu, C. Daul, Phys. Chem. Chem. Phys. **15**, 13902 (2013)
12. A. Canning, A. Chaudhry, R. Boutchko, N. Grønbech-Jensen, Phys. Rev. B **83**, 125115 (2011)
13. A. Chaudhry, R. Boutchko, S. Chourou, G. Zhang, N. Grønbech-Jensen, A. Canning, Phys. Rev. B **89**, 155105 (2014)
14. K. Choudhary, A. Chernatynskiy, K. Mathew, E.W. Bucholz, S.R. Phillpot, S.B. Sinnott, R.G. Hennig, Appl. Phys. Lett. **107**, 112109 (2015)
15. A.D. Becke, J. Chem. Phys. **98**(2), 1372 (1993)
16. J. Wen, L. Ning, C.K. Duan, Y. Chen, Y. Zhang, M. Yin, J. Phys. Chem. C **116**, 20513 (2012)
17. L. Ning, F. Yang, C. Duan, Y. Zhang, J. Liang, Z. Cui, J. Phys.: Condens. Matter **24**, 05502 (2012)
18. L. Ning, X. Huang, Y. Huang, P.A. Tanner, J. Mater. Chem. C **6**, 6637 (2018)
19. Q.M. Phung, Z. Barandiarán, L. Seijo, Theor. Chim. Acc. **134**, 37 (2015)
20. S. Sugano, R.G. Shulman, Phys. Rev. **130**, 517 (1963)
21. P. Huang, E.A. Carter, Ann. Rev. Phys. Chem. **59**, 261 (2008)
22. R. McWeeny, Rev. Mod. Phys. **32**, 335 (1960)
23. S. Huzinaga, A.A. Cantu, J. Chem. Phys. **55**, 5543 (1971)
24. S. Huzinaga, D. McWilliams, A.A. Cantu, Adv. Quantum Chem. **7**, 187 (1973)
25. S. Huzinaga, L. Seijo, Z. Barandiarán, M. Klobukowski, J. Chem. Phys. **86**, 2132 (1987)
26. Z. Barandiarán, L. Seijo, J. Chem. Phys. **89**, 5739 (1988)
27. L. Seijo, Z. Barandiarán, *Computational Chemistry: Reviews of Current Trends*, vol. 4 (World Scientific, Singapore, 1999), pp. 55–152
28. L. Seijo, Z. Barandiarán, J. Chem. Phys. **121**, 6698 (2004)
29. L. Seijo, S.P. Feofilov, Z. Barandiarán, J. Phys. Chem. Lett. **10**, 3176 (2019)
30. L. Seijo, Z. Barandiarán, J. Chem. Phys. **94**, 8158 (1991)
31. P.P. Ewald, Ann. Phys. **369**, 253 (1921)

32. A. Gellé, M.B. Lepetit, J. Chem. Phys. **128**, 244716 (2008)
33. J.L. Pascual, N. Barros, Z. Barandiarán, L. Seijo, J. Phys. Chem. A **113**, 12454 (2009)
34. A. Warshel, M. Levitt, J. Mol. Biol. **103**, 227 (1976)
35. J.L. Pascual, L. Seijo, J. Chem. Phys. **102**, 5368 (1995)
36. B.G. Dick, A.W. Overhauser, Phys. Rev. **112**, 90 (1958)
37. T. Saue, L. Visscher, *Theoretical Chemistry and Physics of Heavy and Superheavy Elements* (Springer, Dordrecht, 2003), pp. 211–267
38. L.L. Foldy, S.A. Wouthuysen, Phys. Rev. **78**, 29 (1950)
39. L. Seijo, J. Chem. Phys. **102**, 8078 (1995)
40. M. Douglas, N.M. Kroll, Ann. Phys. (N.Y.) **82**, 89 (1974)
41. B.A. Hess, Phys. Rev. A **32**, 756 (1985)
42. B.A. Hess, Phys. Rev. A **33**, 3742 (1986)
43. T. Nakajima, K. Hirao, J. Chem. Phys. **113**, 7786 (2000)
44. H. Bethe, E.E. Salpeter, *Quantum Mechanics of One- and Two-Electron Atoms* (Springer, Berlin, 1957)
45. M. Blume, R.E. Watson, Proc. R. Soc. London, Ser. A **270**, 127 (1962)
46. M. Blume, R.E. Watson, Proc. R. Soc. London, Ser. A **271**, 565 (1963)
47. B.A. Hess, C.M. Marian, U. Wahlgren, O. Gropen, Chem. Phys. Lett. **251**, 365 (1996)
48. C.M. Marian, Wiley Interdiscip. Rev. Comput. Mol. Sci. **2**, 187 (2012)
49. R. McWeeny, *Methods of Molecular Quantum Mechanics*, 2nd edn. (Academic Press, London, 1989). (Ch. 14)
50. C.C.J. Roothaan, Rev. Mod. Phys. **23**, 69 (1951)
51. J. Rychlewski (ed.), *Explicitly Correlated Wave Functions in Chemistry and Physics: Theory and Applications, Progress in Theoretical Chemistry and Physics*, vol. 13 (Springer Science & Business Media, New York, 2013)
52. S. Huzinaga, J. Andzelm, E. Radzio-Andzelm, Y. Sakai, H. Tatewaki, M. Klobukowski, *Gaussian Basis Sets for Molecular Calculations* (Elsevier, Amsterdam, 2012)
53. P.O. Widmark, P.A. Malmqvist, B.O. Roos, Theor. Chim. Acta **77**, 291 (1990)
54. C. Møller, M.S. Plesset, pr **46**, 618 (1934)
55. B.H. Brandow, Rev. Mod. Phys. **39**, 771 (1967)
56. J.P. Malrieu, J.L. Heullly, A. Zaitzevskii, Theor. Chim. Acta **90**, 167 (1995)
57. B.O. Roos, Acc. Chem. Res. **32**, 137 (1999)
58. K. Hirao, *Recent Advances in Multireference Methods* (World Scientific, Singapore, 1999)
59. R.J. Bartlett, Annu. Rev. Phys. Chem. **32**, 359 (1981)
60. I. Shavitt, R.J. Bartlett, *Many-Body Methods in Chemistry and Physics: MBPT and Coupled-Cluster Theory* (Cambridge University Press, Cambridge, 2009)
61. B.O. Roos, P.R. Taylor, P.E.M. Siegbahn, Chem. Phys. **48**, 157 (1980)
62. P.E.M. Siegbahn, A. Heiberg, B.O. Roos, B. Levy, Phys. Scr. **21**, 323 (1980)
63. P.E.M. Siegbahn, A. Heiberg, J. Almlöf, B.O. Roos, J. Chem. Phys. **74**, 2384 (1981)
64. J. Olsen, B.O. Roos, P. Jørgensen, J.A. Jensen, J. Chem. Phys. **89**, 2185 (1988)
65. P.A. Malmqvist, A. Rendell, B.O. Roos, J. Phys. Chem. **94**, 5477 (1990)
66. D. Ma, G.L. Manni, L. Gagliardi, J. Chem. Phys. **135**, 044128 (2011)
67. R.J. Buenker, S.D. Peyerimhoff, Theor. Chim. Acta **35**, 33 (1974)
68. H. Werner, P.J. Knowles, J. Chem. Phys. **89**, 5803 (1988)
69. P.J. Knowles, H. Werner, Theor. Chim. Acta **84**, 95 (1992)
70. R. Ahlrichs, P. Scharf, C. Ehrhardt, J. Chem. Phys. **82**, 890 (1985)
71. R.J. Gdanitz, R. Ahlrichs, Chem. Phys. Lett. **143**, 413 (1988)
72. B.O. Roos, P. Linse, P.E.M. Siegbahn, M.R.A. Blomberg, Chem. Phys. **66**, 197 (1982)
73. K. Andersson, P.A. Malmqvist, B.O. Roos, A.J. Sadlej, K. Wolinski, J. Phys. Chem. **94**, 5483 (1990)
74. K. Andersson, P.A. Malmqvist, B.O. Roos, J. Chem. Phys. **96**, 1218 (1992)
75. A. Zaitsevskii, J.P. Malrieu, Chem. Phys. Lett. **233**, 597 (1995)
76. J. Finley, P.A. Malmqvist, B.O. Roos, L. Serrano-Andrés, Chem. Phys. Lett. **288**, 299 (1998)

77. P.A. Malmqvist, K. Pierloot, A.R.M. Shahi, C.J. Cramer, L. Gagliardi, J. Chem. Phys. **128**, 204109 (2008)
78. S. Yabushita, Z. Zhang, R.M. Pitzer, J. Phys. Chem. A **103**, 5791 (1999)
79. R. Llusar, M. Casarrubios, Z. Barandiarán, L. Seijo, J. Chem. Phys. **105**, 5321 (1996)
80. P.A. Malmqvist, B.O. Roos, B. Schimmelpfennig, Chem. Phys. Lett. **357**, 230 (2002)
81. J. Paulovic, T. Nakajima, K. Hirao, R. Lindh, P.A. Malmqvist, J. Chem. Phys. **119**, 798 (2003)
82. G. Karlström, R. Lindh, P.A. Malmqvist, B.O. Roos, U. Ryde, V. Veryazov, P.O. Widmark, M. Cossi, B. Schimmelpfennig, P. Neogrady, L. Seijo, Comput. Mater. Sci. **28**, 222 (2003)
83. G. Sánchez-Sanz, Z. Barandiarán, L. Seijo, Chem. Phys. Lett. **498**, 226 (2010)
84. Z. Barandiarán, L. Seijo, The Multi-State CASPT2 Spin-Orbit Method (2010). arXiv:1012.5572v1

Chapter 2
Feasibility and Accuracy: Criteria and Choices

Abstract The theoretical methods combined in this work to study the excited states of luminescent materials are well rooted in the ab initio quantum chemical methodology, as described in Chap. 1. As such, they are expressed in terms of complete or infinite expansions that become feasible by wise, educated, and systematic truncations that ultimately ensure systematic accuracies. The so-called details of the calculations specify the actual truncations assumed to perform the theoretical study on a given electronic system; these truncations can become recommended standard choices after repeated validations of theoretical results versus experiments. It is convenient to remark that, following an ab initio route, agreement with experiments should never guide truncations even if agreement with experiments is a necessary goal. Also, it is possible that the results of calculations performed with less demanding expansions achieve similar or acceptable accuracy but this will not be the general case. In this chapter we use common and theoretically demanding luminescence activators (Sm, Eu, Yb) as sample cases to describe criteria or standard choices for truncations applicable, e.g. to define the boundaries of a defect cluster and limit the various types of interactions with its infinite solid environment, to expand the molecular orbitals of the embedded-clusters, to limit the infinite multiconfigurational expansions of the many-electron wave functions, or to limit spin-orbit interaction to only some of all possible spin multiplicities. The accuracy achieved using the recommended criteria and standard choices is reviewed and evaluated throughout this book, especially in the chapters and sections referred below.

2.1 Gaussian Basis Sets

In quantum chemical calculations, it is common to span the molecular orbitals as linear combinations of given functions, known as basis set functions. The experience gathered over decades of molecular calculations led to the establishment of basis set standards that guarantee acceptable balances of accuracy and economy. Gaussian basis sets [1] are by far the most popular and all the calculations in this book used them. A Gaussian basis set for a quantum chemical calculation on a molecule or on an embedded-cluster is made of a set of functions centered on the constituent

Z. Barandiarán et al., *Luminescent Materials*, Springer Series in Materials Science 322,
https://doi.org/10.1007/978-3-030-94984-6_2

atoms (called atomic functions) plus, perhaps, a small number of some additional functions. Usually, the atomic functions are called *contracted* functions because they are linear combinations of *primitive* gaussians (or primitives), so that an atomic basis set is composed of a set of primitives and a set of contraction coefficients. The primitives are products of radial functions and angular functions of the types s, p, d, f, g, ..., so that the contacted functions are also of the types s, p, d, f, g, ..., like the atomic orbitals. However, depending on the contraction method of choice, the contracted functions or basis set functions, may or not be atomic orbitals. For instance, in *segmented contraction* basis sets, each s-type basis set function is a linear combination of a few s primitives, whereas in *general contraction* basis sets each s basis set function is a linear combination of all s primitives. The same applies to basis set functions of any angular momentum, p, d, ... This means that, as a rule, general contraction basis sets are made of atomic orbitals, whereas segmented contraction functions are not atomic orbitals, but pieces that can be combined to get atomic orbitals. In the former case of basis sets, the molecular orbitals are, then, linear combinations of atomic orbitals, but in the latter they are linear combinations of segments.

Besides the general quantum chemical knowledge on the performance of atomic Gaussian basis sets in molecular calculations, accumulated in decades of experience, specific basis set features for excited manifolds of $4f^N$, $4f^{N-1}5d$, $4f^{N-1}6s$, $4f^{N-1}\phi_{ITE}$, and charge transfer types in lanthanide compounds (and their equivalent in actinide compounds) demanded particular systematic basis set studies. The lowest number of basis set functions of each angular momentum s, p, ..., that are needed to guarantee a minimum quality of the mentioned excited manifolds in f-element compounds, was systematically studied initially in relativistic effective-core potential calculations [2], later in relativistic all-electron calculations with double f-shell effects [3], and confirmed in a number of calculations on lanthanide- and actinide-doped materials [4].

Typically, for all-electron calculations on lanthanides, relativistic atomic natural orbital basis sets with general contraction are recommended of the type ANO-RCC $(25s22p15d11f4g2h)[9s8p5d4f3g2h]$ [5]. In these, the set of primitives is made of 25 of s-type, 22 of p-type, and so on, and the set of contracted functions or basis set functions is made of 9 of s-type (which are natural atomic orbitals from $1s$ to $9s$), 8 of p-type (which are natural atomic orbitals from $2p$ to $9p$), and so on. These basis sets allow for a very good representation of the changes experienced in the embedded-cluster by the molecular orbitals of main $4f$ and $5d$ atomic characters, which are the most important ones in the manifolds of interest, but they also give a good description of other orbitals, specially those of main $6s$ and $4d$ characters, which are found to be necessary for good accuracies. These basis sets also contain g and h functions. They are not meant to represent $5g$ and $6h$ orbitals, which are very high in energy and will not be occupied in the calculations, but to provide important, additional flexibility to the basis set, allowing for angular polarizations of the molecular orbitals, especially when static and dynamic correlation are included in the calculations, and to represent nominally empty or virtual molecular orbitals that are, however, occupied in doubly excited electronic configurations important for dynamic correlation. These basis sets are said to be of quadruple-zeta with polarization quality, meaning that, in each angular momentum channel involved in bonding or in the excited manifold, four functions are used to provide radial flexibility to the atomic orbitals when they mix in the molecule in order to form molecular orbitals, and that polarization functions are used in order to provide additional angular flexibility. For the ligands, a similar quality is recommended. For instance, a $(14s9p4d)[5s4p3d]$ set for F or O, or a $(17s12p5d)[6s5p3d]$ set for Cl or S. Here, besides the d polarization functions, f and g functions can be used, but they are not compulsory and they largely increase the computational demands.

In embedded-cluster calculations of local properties of materials, contrary to molecular calculations, the basis set functions of all constituent atoms still have to be supplemented with additional basis set functions. E.g., orthogonalization functions may be required in order to improve the approximate fulfillment of Pauli antisymmetry requirements between cluster and host wave functions, or interstitial functions may be required for a good description of states where electronic density localized in sites not occupied by atoms is important, like in impurity-trapped-exciton states.

2.1.1 Orthogonalization Functions

In embedded-cluster calculations of local properties of solids, an eventual lack of fulfillment of Pauli antisymmetry conditions between the many-electron wave functions of a cluster and the host that embeds it, immediately results in a strong cluster-size dependence of structural properties like bond lengths, vibrational frequencies and others. Then, either these Pauli antisymmetry conditions are enforced or very large clusters, often of a prohibitive size, are required for good structural results.

In order to fulfill such conditions, two things are necessary: the embedded-cluster Hamiltonian must enforce them and the basis set in use must be able to provide cluster molecular orbitals that are orthogonal to the host orbitals. The AIMP embedded-cluster Hamiltonian (Sect. 1.2) used in the AIMP embedded-cluster multiconfigurational ab initio calculations discussed in this book have the cluster-host Pauli antisymmetry conditions built-in; however, for them to be efficiently enforced, the cluster basis set cannot just be as a regular basis set used in an isolated molecule, but it needs additional flexibility.

In an ab initio study of the Jahn-Teller coupling in Cu^{2+}- and Ag^{2+}-doped MgO, CaO, and SrO, [6] it was shown that Jahn-Teller couplings, structures and stabilization energies calculated with small and large embedded-clusters were very similar as long as *orthogonalization* functions were used in the cluster first-neighbors. These are basis set functions that enable the cluster molecular orbitals to be orthogonal to the outermost valence orbitals of the neighboring host ions and, this way, the cluster-host Pauli antisymmetry conditions to be fulfilled. Although the final orthogonality achieved in a real embedded-cluster calculation is not complete, it is sufficient to remove the structural anomalies found when the Pauli principle is simply omitted. An example of such orthogonality functions in, e.g., calculations on MgO:Cu^{2+}, where

Cu substitutes for Mg in a MgO host with rock-salt structure, are $[1s2p]$ contracted Gaussian basis functions located at the (100) next-neighbor Mg sites of a $(CuO_6)^{10-}$ cluster, where the single s contracted function is a $2s$ atomic orbital of Mg^{2+} and the two p contracted functions have the ability to produce the $2p$ atomic orbital of Mg^{2+} by linear combination (e.g. one of them is the $2p$ atomic orbital of Mg^{2+} itself and the other is an outermost primitive Gaussian, or, alternatively, the two of them are two segments of the $2p$ atomic orbital). The use of these additional orthogonalization basis set functions enables the $(CuO_6)^{10-}$ embedded-cluster to produce very similar results to the $(CuO_6Mg_6)^{2+}$ embedded-cluster. Similar agreements between smaller clusters with orthogonalization basis set functions and larger clusters, were found in calculations of adsorbate-surface distance and binding energy in AIMP embedded-cluster ab initio studies of adsorption of CO on MgO [7].

2.1.2 *Interstitial Functions*

It is not uncommon in solids with interstices that they can accommodate impurities or host constituent ions. But they can also eventually accommodate partial electronic charge. This is the case, for instance, of impurity-trapped-excitons. In such cases, the cluster basis set extended with orthogonalization functions can still be insufficient and *interstitial* basis set functions able to adequately represent the electronic charge at the interstices are necessary.

The production of interstitial functions is something to do prior to the multiconfigurational ab initio calculations of a material and, in principle, specific to the material. The process to follow in order to obtain the functions may be general, but the functions themselves are expected to be material specific. In order to illustrate this, the calculation of interstitial functions able to represent impurity-trapped-excitons in Yb^{2+}-doped $SrCl_2$ followed in [8] is discussed next.

In $SrCl_2$:Yb^{2+}, Yb substitutes for Sr in a cubic site of 8-fold coordination (Fig. 2.1(left)). The $YbCl_8$ moiety is surrounded by six empty interstices at $(\frac{1}{2},0,0)$ sites, forming an octahedron around the Yb impurity, and a number n of s-type primitive Gaussian functions were located at these sites. These have been chosen as the n outermost primitives of a $5s$ orbital of Sr^+ in $SrCl_2$ (the five outermost s-type primitive exponents are 0.0100, 0.0221, 0.0537, 0.3156, and 0.7000). Delocalization of electronic charge towards the interstices appears to be mandatory only for the $2\,{}^3A_{2u}$ state of $(YbCl_8)^{6-}$ at short Yb-Cl distance regime. This state has $4f^{13}\phi_{ITE,a_{1g}}$ ITE configurational character at the short Yb-Cl distance regime and $4f^{13}6s$ configurational character at the long Yb-Cl distance regime (Fig. 2.2). As shown in the right panel of Fig. 2.1, the ${}^1A_{1g}\rightarrow 2\,{}^3A_{2u}$ absorption energy, i.e. the ITE energy, is stabilized by improving the ability of the basis set to represent electronic charge of the ITE state inside the interstices, and the stabilization appears to be converged for $n = 4$. The electronic density of the $4f^{13}\phi_{ITE,a_{1g}}$ ITE state found at the interstices is shown in Fig. 2.2 (bottom left), where it can be compared with its absence in the

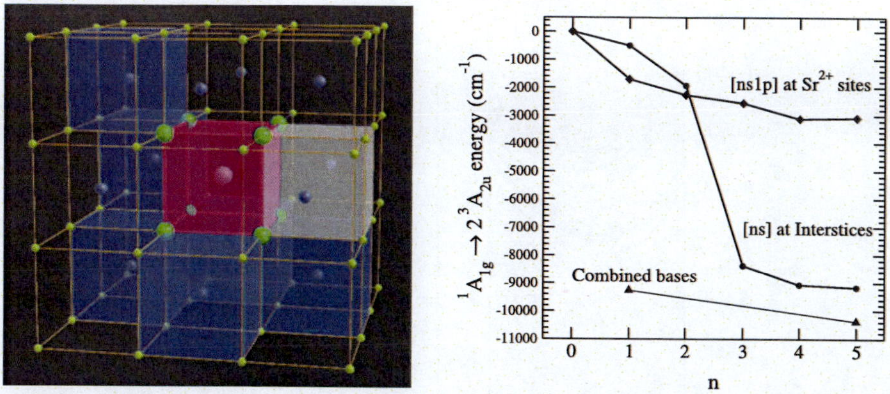

Fig. 2.1 Left: Surroundings of the $YbCl_8$ moiety in $SrCl_2{:}Yb^{2+}$. The interstitial site at $(\frac{1}{2}, 0, 0)$ is shaded in grey. Right: ITE energy $(4f^{14}\text{-}^1A_{1g} \rightarrow 4f^{13}\phi_{ITE,a_{1g}}\text{-}2\ ^3A_{2u}$ excitation energy) as a function of extensions on the $(YbCl_8)^{6-}$ embedded-cluster basis set. CASSCF results at $d_{\text{Yb-Cl}} = 2.75$ Å

$4f^{13}6s$ state (bottom right). In consequence with this, a regular basis set for a multiconfigurational ab initio calculation on the excited manifolds of $SrCl_2{:}Yb^{2+}$ must include at least the four outermost s primitives shown above as interstitial functions.

2.2 Host Embedding Potentials

The host embedding potential of CaF_2 used, for example, in multiconfigurational ab initio calculations of $CaF_2{:}Yb^{2+}$ [9] and $CaF_2{:}Eu^{2+}$ [10] is a good example of a typical definition of an AIMP host embedding potential (cf. Fig. 1.2).

The AIMP embedding potentials of individual Ca^{2+} and F^- ions in CaF_2 have been obtained in SCEI Hartree-Fock calculations on CaF_2 (Sect. 1.2.1) [9]. When they are used in the many-electron Hamiltonians of the embedded-clusters, like $(YbF_8)^{6-}$, $(YbF_8Ca_{12})^{2+}$, and $(EuF_8Ca_{12})^{2+}$ in the examples, they account for several interactions between cluster and host ions: classical Coulomb interactions between host nuclei and electrons and cluster nuclei and electrons; quantum-mechanical exchange interactions between host electrons and cluster electrons; and quantum-mechanical Pauli repulsion interactions between host electrons and cluster electrons, which are derived from the fulfillment of Pauli antisymmetry conditions between host and cluster electrons. Then, the facts that exchange and Pauli repulsion interactions are short-range and that the Coulomb potentials are divided into long-range point-charge contributions and short-range density contributions, are taken into account. Then, as it is illustrated in Fig. 1.2, the clusters are surrounded by: (1) a close region of total-ion AIMP embedding potentials of Ca^{2+} cations and F^- anions located at experimental sites of the CaF_2 host lattice [Group 225, $Fm\bar{3}m$, $a_0 = 5.46294$ Å] [11] within a

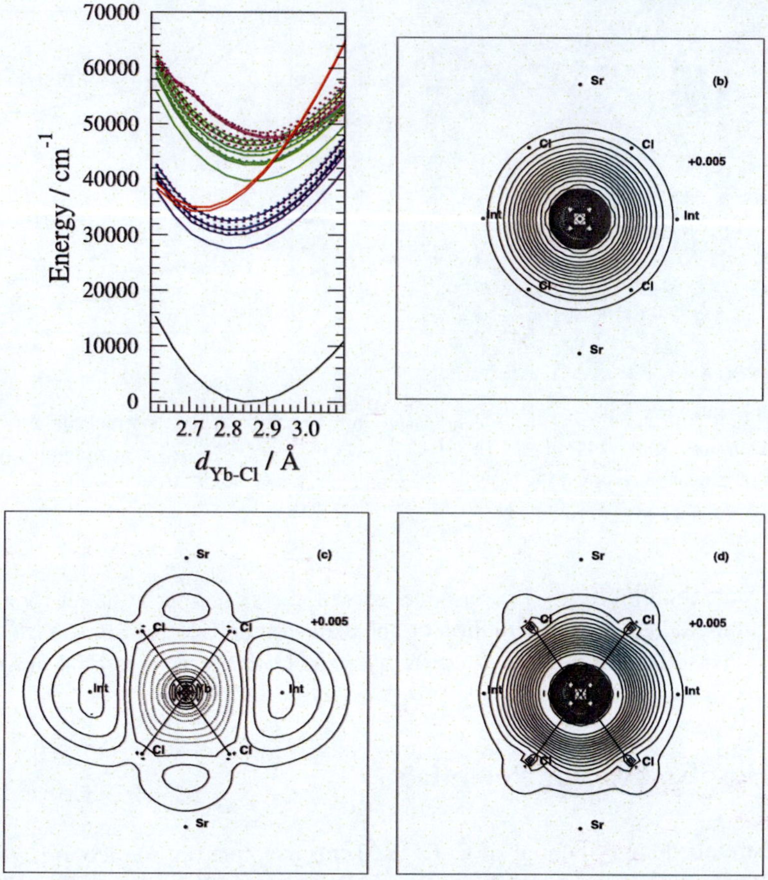

Fig. 2.2 $4f^{13}\phi_{ITE,a_{1g}}$ impurity-trapped-exciton and $4f^{13}6s$ states of $SrCl_2{:}Yb^{2+}$. Top left: spin-orbit coupling free potential energy curves (black: $4f^{14}$ ground state; blue: $4f^{13}5de_g$ states; green: $4f^{13}5dt_{2g}$ states; magenta: $4f^{13}a_{1g}$ dual states, $4f^{13}\phi_{ITE,a_{1g}}$ at short $d_{\text{Yb-Cl}}$ and $4f^{13}6s$ at long $d_{\text{Yb-Cl}}$. Top right: $6s$ atomic natural orbital in a $4f^{13}6s\ ^3F^o$ state of the free-ion Yb^{2+}. Cl, Sr, Int (for interstice) labels are indicated where they appear in the lower panels, for reference. Bottom panels: a_{1g} molecular natural orbitals with $\phi_{ITE,a_{1g}}$ (left) and $6s$ (right) characters of the dual $4f^{13}a_{1g}$ $2\ ^3A_{2u}$ state, plotted across the diagonal plane of the Cl_8 cube

cube made of $7 \times 7 \times 7$ unit cells centered at the impurity site; and (2) a more distant region of 2781 point-charges, also located at experimental sites, generated by the zero-multipole method [12] in order to ensure the Ewald potential [13] is well reproduced within the cluster volume.

2.3 Multiconfigurational Expansions of Many-Electron Wave Functions

Chapter 1 described methods of quantum chemistry suitable to calculate the numerous excited states associated with the multivalent open-shell and heavy element nature of lanthanide-based luminescent materials. They deal with static and dynamic correlation and with relativistic effects including spin-orbit coupling, within an embedded-cluster containing the luminescent center. The methods recognize that electron correlation and spin-orbit coupling can be calculated separately in two steps, a first one focused on the difficulty of retrieving static and dynamic electron correlation using spin-free Hamiltonians and extensive multiconfigurational spaces, and a second one allowing spin-orbit coupling interactions over a reduced multi-state space including all possible spin multiplicities associated with the activator's open-shells. The calculations involve the SA-RASSCF/MS-RASPT2 methods in the first step and the RASSI-SO method in the second, as discussed.

But, in spite of the evident benefits of such two-step decoupling procedure in terms of feasibility, lanthanide activators pose a tremendous challenge because the number of excited states is very large and sometimes too large; in the case of Eu, for instance, their number reaches the order of 10^4. In addition, the spectrum of levels is extremely dense showing in many cases basically no energy gaps, which suggests that a dangerous energy-based criterion for truncation of the multiconfigurational expansions might eventually be the only possible way towards feasibility. Yet, the two-step strategy, together with analyses of exploratory calculations on several oxidation states of the activators, serve the purpose of providing clues and strategies towards unbiased truncations pertaining what total spin multiplicities, active orbital spaces, number of states, and correlated shells must be considered in order to make the final SA-RASSCF/MS-RASPT2/RASSI-SO results robust and reliable, and their accuracy versus experiment systematic, acceptable, and useful to provide both good data and insight.

Some of the common clues and strategies that end up in recommendations of standard methodological choices or truncations are discussed next. The cases of impurity states of Sm and Eu activators are used as examples, given that their computational demands are very challenging even when they occupy sites with high local symmetry, probably among the highest in the lanthanide series. The choices are valid for impurity states of other lanthanide activators (cf. Chap. 6). However, when charge transfer states are on focus, additional orbital shells with main ligand character must be included in the active space (cf. Chap. 7). In the calculation of inter-valence and metal-to-metal charge transfer states, impurity states like the ones addressed here are building-blocks for diabatic theoretical approaches.

2.3.1 Restricting Spin Multiplicities of the Wave Functions

Very often the number of possible total spin values (S) that can contribute, in principle, to the many-electron multiconfigurational wave functions of luminescent materials is relatively large. Samarium and Europium, whose most stable and relevant oxidation states are Sm^{2+}, Sm^{3+}, and Eu^{2+}, Eu^{3+}, are good examples of that, with their respective $4f^6$ ($2S$+1 = 7, 5, 3, 1), $4f^5$ ($2S$+1 = 6, 4, 2), and $4f^7$ ($2S$+1 = 8, 6, 4, 2), $4f^6$ ($2S$+1 = 7, 5, 3, 1) ground configurations. Moreover, the number of states that can be obtained from symmetry and spin couplings grows significantly as total spin decreases and this is so for all significant open-shell configurations ($4f^N$, $4f^{N-1}5d$, $4f^{N-1}6s$, $4f^{N-1}\phi_{ITE}$). Hence, imposing restrictions to the total spin values contributing to the target spin-orbit coupling many-electron wave functions becomes absolutely necessary. The point is whether such restrictions can be justified, systematic, and compatible with the necessary accuracy the final spin-orbit results must achieve.

The total spin S remains a good quantum number of the wave functions in the SA-RASSCF/MS-RASPT2 calculations of the first step, where spin-orbit-free Hamiltonians are used. This enables to adopt reasonable spin-based truncations in cases with more than two spin multiplicities: Spin-orbit-free calculations readily reveal that the dense manifolds of excited states are formed by overlapping groups of states with a given spin-multiplicity, whose energy onset is higher the lower their value of S is. As a consequence of this, the spin-orbit coupling effects of these groups of states on the lower excitation bands and lower lying excited states responsible for luminescence are increasingly weakened as their spin-multiplicity goes down.[1] Hence, an acceptable spin restriction is to calculate states with only the two higher spin values at the first spin-orbit-free step and allow their spin-orbit coupling in the second step. Then, only spin-orbit-free states with $2S$+1 = 7, 5 for Sm^{2+}, $2S$+1 = 6, 4 for Sm^{3+}, and $2S$+1 = 8, 6 for Eu^{2+}, $2S$+1 = 7, 5 for Eu^{3+}, are calculated at the spin-orbit-free SA-RASSCF/MS-RASPT2 step and are allowed to interact via spin-orbit coupling at the final RASSI-SO calculations. This obviously means a tremendous saving in the number of states to be calculated.

Ultimately, the quality of the limitation imposed that restricts spin-orbit coupling to states from the two higher spin-multiplicities should be revealed by the comparisons between theoretical and experimental spectra (cf. Chap. 9).

[1] To illustrate this trend, the energies of the Eu^{2+} $4f^7$ and Eu^{3+} $4f^6$ states were computed in the BaF_2 host at a fixed Eu–F distance (2.60 Å for Eu^{2+}; 2.38 Å for Eu^{3+}), using the simplest SA-RASSCF calculations possible, i.e. those where the $4f$ shell creates a CAS in D_{2h} symmetry with 7 or 6 active open-shell electrons, respectively. The energies (in cm^{-1}) of the lowest $4f^7$ states for each spin multiplicity of Eu^{2+} in BaF_2 are: 0 (octet); 37000 (sextet); 60000 (quartet); 70000 (doublet). In the case of BaF_2:Eu^{3+} states they vary as follows: 0 (septet), 25000 (quintet); 45000 (triplet); 55000 (singlet).

2.3.2 Criteria to Define the Restricted Active Space and the Target States in SA-RASSCF Calculations

In addition to the just discussed spin-multiplicity based restrictions, the complexity of the luminescent defects demands additional criteria to limit the multiconfigurational expansions of the SA-RASSCF calculations. In what follows the focus is put on what orbital shells of the defect cluster must be included in the restricted active space and what electron excitations among them could give near CASSCF quality, as well as what strategies or criteria are followed to efficiently limit the number of many-electron wave functions and energies to be actually calculated (i.e. the target states, usually referred to as *roots*) and transferred to subsequent MS-RASPT2 and RASSI-SO calculations.

Restricted Active Space of Near-CAS Quality

The first decisions to be made relative to the multiconfigurational space of any SA-RASSCF calculation are: (i) to choose what orbital shells are to be considered active and (ii) whether a complete or restricted CI space is going to be produced from the active shells for the particular lanthanide defect with N open-shell electrons. These decisions are instrumental in SA-RASSCF calculations and the accuracy of the SA-RASSCF/MS-RASPT2 calculations of the first spin-orbit-free step depends substantially on them.

Well before ab initio calculations of any kind could give information about the excited states of lanthanide-based luminescent materials, an abundant literature on experimental spectroscopy was gathered, including crystal or ligand field theory analyses, labeling the ground and excited electronic states as $4f^N$, $4f^{N-1}5d$, and $4f^{N-1}6s$ when charge transfer from or to the lanthanide activator could be excluded. The reference made to the *atomic* orbital shells of the lanthanide ($4f$, $5d$, $6s$), suggesting the expected dominant character of the molecular orbitals of the defects on the one hand, and the reference made to a *configurational* character, suggesting what type of electronic coupling captures the essence of groups of many-electron states ($4f^N$, $4f^{N-1} \times 5d$, $4f^{N-1} \times 6s$) on the other hand settled an extremely useful and significant nomenclature that served to claim that the lanthanide ions in condensed phase create a wealth of *local* electronic states. This nomenclature highlights a parentage with the multiplets originating from the open-shells of the free lanthanide ions in the gas phase. These results have a strong appeal and serve, at the same time, as an essential guide to quantum chemistry researchers' minds.

Analyses of multiconfigurational ab initio embedded-cluster wave functions, at any level of accuracy, corroborate over and over the existence of the $4f^N$, $4f^{N-1}5d$, or $4f^{N-1}6s$ manifolds. Therefore, in a defect cluster containing an optically active lanthanide ion, the molecular orbitals with main contributions from the lanthanide $4f$, $5d$, and $6s$ atomic orbitals, should be considered active.

In addition, an unoccupied $5f$ shell (i.e. a shell of molecular orbitals with main character of the lanthanide $5f$ orbitals) must also be considered as an active shell for the sake of accurately representing the strong $4f$ radial correlation.[2] The multiconfigurational space suggested by the $4f^N, 4f^{N-1}5d, 4f^{N-1}6s$ nomenclature, i.e. the space of many-electron wave functions obtained by allowing any occupations in the $4f$ shell and zero or one electrons in the $5d$ and $6s$ shells out of the N open-shell electrons, would be sufficient to produce the correct number of states associated with the $4f^N$, $4f^{N-1} \times 5d$, $4f^{N-1} \times 6s$ couplings, but it would be a too small space to deal with the whole static correlation as required to obtain accurate spectra. This latter methodological requirement would be fulfilled by the multiconfigurational complete-active-space (CAS) generated by allowing all possible occupations of the $4f, 5d, 6s, 5f$ shells with the N open-shell electrons, which are compatible with the total spin and site symmetry irreducible representations of the many-electron wave functions. However, such a complete-active-space is usually too large and needs to be truncated. So, a feasible restricted-active-space can be one where all occupations of the $4f$ shell are allowed, whereas only single, double, triple, and quadruple (SDTQ) excitations from the $4f$ to the $5d$, $6s$, and $5f$ shells are permitted, compatible with the total spin and site symmetry of the lanthanide-containing embedded-cluster wave functions. This truncation should be very close to the CAS limit, hence near-CAS quality should be attained with the SA-RASSCF calculations [3]. This RAS can be called RAS($4f$/$5d6s5f$-$4e$).[3]

Once the decision on the RAS is made, it is necessary to determine the number of SA-RASSCF roots (many-electron wave functions and energies) to be calculated per symmetry+spin-multiplicity block and the particular sate average (SA) where the occupied embedded-cluster molecular orbitals are going to be optimized. This is discussed next using two meaningful and demanding cases involving the $4f^{N-1}5d$ manifolds of Sm and Eu luminescent materials.

Target SA-RASSCF States for Sm Luminescent Materials

The $4f^55d$ states of Sm^{2+} in cubic sites of CaF_2 offer an opportunity to illustrate the procedure to calculate $4f^{N-1}5d$ excited manifolds in general. Like the vast majority of them, except perhaps for the $4f5d$ and $4f^{13}5d$ cases, the $4f^55d$ manifold of Sm^{2+} in CaF_2 is extremely complex: It is made of interacting states of different $4f^55de_g$, $4f^55dt_{2g}$, and $4f^5\phi_{ITE}$ configurations (see below), and several spin-multiplicities; the number of states composing the manifold is very large; they make a very crowded set of states with only very few windows or energy gaps between them; and the gaps are geometry dependent, i.e. they appear and disappear, or appear in different

[2] A double f-shell ($4f, 5f$) should be used to study atomic-like and electron transfer states using the SA-RASSCF/MS-RASPT2 method in order to properly account for strong radial correlation changes in the $4f$ shell when its occupation varies upon excitation (cf. Sect. 2.4).

[3] The RAS notation specifies the RAS1, RAS2, and RAS3 orbital subspaces as follows: RAS(shells in RAS1-nh/shells in RAS2/shells in RAS3-ne), where the maximum total number of holes nh allowed to be created in the fully occupied RAS1 shells and the maximum total number of electrons ne allowed to be excited to the empty RAS3 shells are made explicit. Each of the RAS2 orbitals can allocate any occupation (0, 1, or 2) such that the total occupation of RAS1, RAS2, and RAS3 must agree with twice the number of RAS1 orbitals plus the formal number of open-shell electrons (e.g. 7 for Eu^{2+}; 6 for Eu^{3+}). Note that total spin and site symmetry restrictions apply. In RAS($4f$/$5d6s5f$-$4e$) there is no RAS1.

positions of the state list, depending on the Sm-F distance. This complexity makes the computational task specially difficult. In effect, on one hand, the very high computing times and general computational demands make the goal of computing all states of the manifold impractical. This does not pose an important restriction in principle, because only a limited number of states is involved in optical absorptions and emissions that are of interest. However, the complexity of the manifold causes serious problems to the task of computing a number of states only. This is so because, although MS-RASPT2 calculations exhibit normal convergence when there is an energy gap between two states within a block of given spin-multiplicity and spatial symmetry irreducible representation, they run into severe, if not unsolvable, convergence problems when a crowded set of states is arbitrarily cut into a lower set (which is the target of the calculation) and an upper set (which is discarded in the calculation). In addition, in the improbable case that such convergence problems can be solved, either at the cost of an important increase of the computing time or by any other means, it is most likely that the calculated upper states are not the same at all Sm-F distances, which prevents from a correct calculation of potential energy surfaces and curves. In these circumstances, before embarking into a costly calculation of the $4f^5 5d$ manifold at a high theoretical level, it is important to make exploratory calculations at a lower theoretical level; they are expected to identify energy gaps and to drive the selection of target states that will be calculated at higher theoretical levels.

SA-RASSCF exploratory calculations have been done using a minimal configurational space. In these, 6 electrons are distributed among the shells of molecular orbitals with main character Sm $4f$, $5d$, and $6s$, allowing all possible occupations of the Sm $4f$ shell and only one electron in the Sm $5d$ and $6s$ shells. It has been found for fluorite-type hosts that, when the basis set is sufficiently flexible to represent a Sm $6s$ orbital that can be deformed by the interactions on the host, a natural orbital whose electron density is spread over the closest interstitial sites occurs rather than a Sm $6s$ orbital localized within the boundaries of the SmF_8 moiety. This delocalized active molecular orbital is referred to as ϕ_{ITE}, given the similarity of $4f^5\phi_{ITE}$ states with the so-called impurity-trapped excitons, or ITEs. Since the lowest states in the $4f^5 5d$ manifold that are involved in absorption and emission processes have high spin-multiplicities and their spin-orbit couplings with states of low spin-multiplicities and much higher energies are expected to be very small, we only calculate spin-septet ($2S + 1 = 7$) and spin-quintet ($2S + 1 = 5$) multiplicities, as recommended in Sect. 2.3.1. These calculations are referred to as SA-RASSCF($4f/5d6s$-$1e$). The results are summarized in Fig. 2.3, where all energies are referred to the lowest $4f^5 5de_g$ spin-orbit-free state at this level, $1\,^7T_{1u}$.

The analysis of spin-septets (left box in Fig. 2.3) is straightforward: The states correspond to the expected couplings between the $4f^5$ subshell and the active $5de_g$, $5dt_{2g}$, and ϕ_{ITE} orbital shells. Only the states of the $4f^5$ subshell with maximum spin-multiplicity can contribute: $4f^5(^6H{+}^6F)$; the $4f^5$ spin-quartets will contribute to the $4f^5 5d$ quintets, but not to the septets. All states of the $4f^5 5de_g$ configuration show similar curvature (hence vibrational frequency) and Sm-F equilibrium distance (which is shorter than in the $4f^6$ states at the same level of theory); and the same is

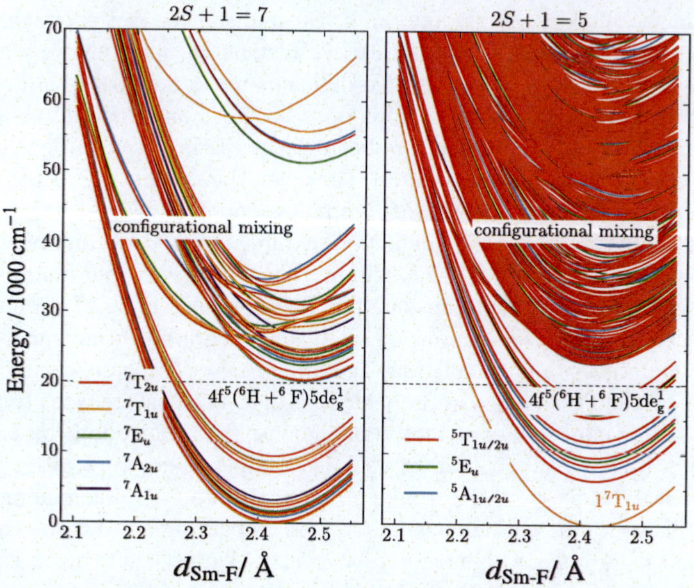

Fig. 2.3 Potential energy curves of CaF_2:Sm^{2+}excited states of main $4f^55de_g$, $4f^55dt_{2g}$, and $4f^5\phi_{ITE}$ configurations, calculated at the exploratory SA-RASSCF($4f/5d6s$-$1e$) level of theory. Energies are referred to the lowest $4f^55de_g$ spin-orbit-free state at this level, $1\ ^7T_{1u}$

true for $4f^55dt_{2g}$ states (with longer Sm-F distance than the $4f^6$ states) and $4f^5\phi_{ITE}$ states (with significantly shorter Sm-F distance, only slightly longer than the Sm^{3+}-F^- distance). The $4f^55de_g$ states span the lowest energy range, below 20000 cm^{-1} except for a few around 30000 cm^{-1}. The $4f^55dt_{2g}$ states start at 20000 cm^{-1} and the $4f^5\phi_{ITE}$ states at around 27000 cm^{-1} above the first $4f^55de_g$ state. It is around this energy that important configurational mixing between the three configurations are developed, which is obvious from the potential energy curves; it is interesting to note, though, that the configurational mixing is important only near the crossing points. The found energy gaps are strongly dependent on the Sm-F distance.

The analysis of spin-quintets (right box in Fig. 2.3) gives a far more complicated landscape. Only the lowest $4f^5(^6H+^6F)5de_g$ states are well separated from the rest. Starting at 23000 cm^{-1}, a basically continuous set of states with very important configurational mixing and high density of states is found, with no relevant energy gaps.

Overall, septets and quintets of $4f^5(^6H+^6F)5de_g$ character span an energy range of around 20000 cm^{-1} and, above an energy gap, the rest of septets and quintets gather together in a dense band of states. Spin-triplets and spin-singlets are not considered in the calculation, but they would occur at higher energies and increase the density of states even more.

From the exploratory calculations, it is reasonable to assume that SA-RASSCF calculations using more flexible and reliable RAS spaces, and their corresponding

MS-RASPT2 calculations, will be doable only on the $4f^5(^6H+^6F)5de_g$ septet and quintet states, which become the target states for RAS($4f/5d6s5f$-$4e$) calculations. It is based on these considerations that the state-averages, shown in Sect. 6.2.3, were chosen. Fortunately, the lowest $4f$–$5d$ absorption bands of CaF_2:Sm^{2+} (Sect. 9.3) correspond well with these lower states. Should higher absorption bands of CaF_2:Sm^{2+} be of interest, the same type of problems as those encountered in the calculation of the $4f^{N-1}5d$ states of CaF_2:Eu^{2+} crystals described next would probably be faced.

Target SA-RASSCF States for Eu Luminescent Materials

Luminescence of Eu^{2+} and Eu^{3+} activators has since long been associated with $4f^65d \rightarrow 4f^7$ and $4f^6 \rightarrow 4f^6$ emissions, respectively, putting the focus on these manifolds. It is also known that the $4f^7 \rightarrow 4f^65d$ absorption spectrum of Eu^{2+} in fluorites and other high-symmetry hosts contains two broad bands separated by a large energy gap. The lowest absorption band appears to be more or less structured, often referred to as a staircase structure, depending on the host chosen (i.e. CaF_2, SrF_2, BaF_2) whereas the second band, lying higher in energy, shows no such structure. A detailed understanding of the microscopic nature of the emitting states and of both absorption bands and the gap that seems to separate them is, obviously, an important goal for theory given that it yields an opportunity to provide insights that are difficult to extract from experiment and empirical models alone. However, calculating states up to such high energies is a challenge. In particular, similar criteria as those applied to limit the number of target states in the CaF_2:Sm^{2+} case as discussed above would not allow to reach the second absorption band. Therefore, in this case, it is necessary to venture into the continuous and dense part of the energy spectrum of Eu^{2+} to study and offer feasible and sufficiently accurate strategies to calculate high-energy states [10].

The use of site symmetry is very valuable in this quest. It allows to calculate the $4f^N$ and $4f^{N-1}5d/4f^{N-1}6s$ manifolds separately even if they occur in the same energy regions, provided that the Eu^{2+} ion occupies a site with inversion symmetry, which is the case for cubic fluorite-type and rocksalt-type host crystals. In that case, it is feasible to calculate, independently, all O_h-adapted electronic states of the $4f^N$ ($N = 7, 6$) configurations up to the final, spin-orbit coupling second-step without further restrictions than that of limiting the spin-orbit coupling to the two higher spin-multiplicities mentioned above.

However, the number of Eu^{2+} *gerade* states resulting from the $4f^6 \times 5d$ and $4f^6 \times 6s$ couplings is too large, even within the spin restrictions imposed and the available high site symmetry blocks. So, the results of the $4f^6$ manifold of Eu^{3+} can be used to establish a reasonable $4f^6$-subshell-based truncation criterion that balance feasibility and accuracy. In particular, the energy gap above the lowest Eu^{3+} $4f^6$ term ^7F, suggests restricting the number of states to be calculated (*roots* in the diagonalization language) to that number resulting from the following couplings: $4f^6(^7F) \times 5d$, $4f^6(^7F) \times 6s$.[4] This criterion only decides the number of states to

[4] In CaF_2:Eu^{3+}, the lowest $4f^6$ crystal states are the O_h spin-orbit-free 1 $^7A_{2g}$ (ground state), 1 $^7T_{1g}$, and 1 $^7T_{2g}$ (in energy order). All these three states are emparented with the lowest $4f^6$

be calculated, not their actual nature. Hence it does not preclude intercalation of states of other configurations such as those corresponding to $4f^6(^5D) \times 5d$ coupling.

The chosen $4f^6(^7F) \times 5d, 4f^6(^7F) \times 6s$ couplings determine that the following number of O_h terms are the target; they are used to define the state-average where the cluster molecular orbitals are optimized: one $^8A_{1g}$, two $^8A_{2g}$, and three ^8E_g (9 roots from the ^8A_g D_{2h} symmetry block); six $^8T_{1g}$ and five $^8T_{2g}$ (11 roots from each of the $^8B_{1g}$, $^8B_{2g}$, $^8B_{3g}$ blocks)[5]; thirty nine $^6A_{1g}$, thirty six $^6A_{2g}$, and seventy five ^6E_g (225 roots from the ^6A_g block); and one hundred eight $^6T_{1g}$ and one hundred eleven $^6T_{2g}$ (219 roots from the $^6B_{1g}$, $^6B_{2g}$, and $^6B_{3g}$ blocks). As discussed above, only spin octets and sextets are considered and all of these roots are obtained using the multiconfigurational space generated by the 7 open-shell electron RAS($4f/5d6s5f$-$4e$) in SA-RASSCF calculations on the different spin and D_{2h} symmetry blocks. The energy curves resulting from these calculations for the CaF_2:Eu^{2+} material up to 70000 cm^{-1} are plotted in Fig. 2.4 under the label RASSCF.

Analyses of the SA-RASSCF wave functions and energy curves show the occurrence of impurity-states ascribable to Eu^{2+} $4f^65de_g$, $4f^65dt_{2g}$, and $4f^66s$ configurations, but they also show the presence of more delocalized states that can be called impurity-trapped-excitons (ITE) because the excited electron density spreads close but beyond first neighbors: $4f^6\phi_{ITE,a_{1g}}$. These ITE states (of a_{1g} symmetry) were found in the studied energy range in the difluoride hosts (cf. CaF_2:Eu^{2+} results in Fig. 2.4), whereas $4f^66s$ states were obtained in the sulfides.

States of different configurations show significant differences in bond lengths, as can be seen in Fig. 2.4 for CaF_2:Eu^{2+} where the configurational labels are used. The bond lengths follow the usual trend found for impurity states when states have pure configurational character around their energy minima: $d_{e,\text{Eu-F}}(4f^6\phi_{ITE,a_{1g}}) < d_{e,\text{Eu-F}}(4f^65de_g) < d_{e,\text{Eu-F}}(4f^65dt_{2g})$, but the considerable difference in bond length between the ITE states and close lying $4f^65dt_{2g}$ states favours avoided crossings clearly observed above the lower lying $4f^65de_g$. The avoided crossings are very sharp, this meaning that the corresponding wave functions experience a sharp change of configurational character at both sides of the avoided crossing structure and the corresponding energy curves are strongly anharmonic (cf. Fig. 2.4). Here again, the very high density of sextet states above 50000 cm^{-1} poses further difficulties beyond

Eu^{3+} free ion term ^7F. Above the 1 $^7T_{2g}(^7F)$ level lies the $^7E_g(^5D)$, which is the first of the two crystal components of the ^5D term. An energy gap of about 20000 cm^{-1} is found between them at SA-RASSCF($4f/5d6s5f$-$4e$) and MS-RASPT2 levels of theory. Spin-orbit coupling reduces this gap but it is still a large gap of about 14000 cm^{-1}. It is this large gap what makes the lowest crystal component of ^5D a metastable emitting state of the Eu^{3+} activator. Analogous results are obtained in other fluorides (SrF_2, BaF_2) and sulfides (CaS, SrS, BaS) due to the very local nature of the $4f^6$ manifold.

[5] Note that the roots of these three blocks should be degenerate within numerical accuracy as they correspond to subspecies of the irrep of a particular O_h term, $^8T_{1g}$ or $^8T_{2g}$ in this case, and the Hamiltonian used in SA-RASSCF calculations is spin-orbit-free and is totally symmetric within the O_h point group. However they must be calculated if the term is to be included in the RASSI-SO input given that the spin-orbit coupling Hamiltonian is totally symmetric within the octahedral double group $\bar{O}_h$ for which the O_h terms are reducible, leading to interacting components.

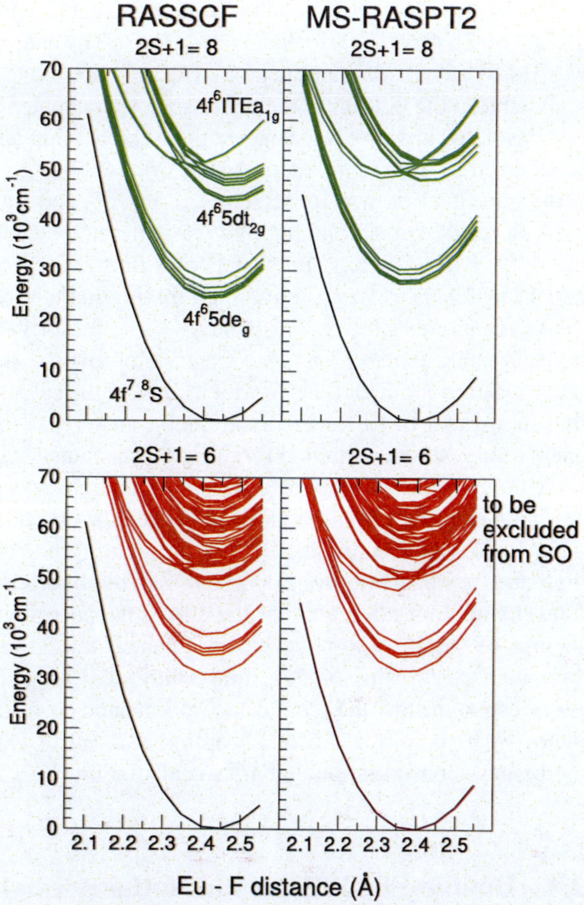

Fig. 2.4 Potential energy curves of $CaF_2:Eu^{2+}$ ground and excited states of main $4f^7, 4f^65de_g, 4f^65dt_{2g}$, and $4f^6\phi_{ITE}$ configurations, calculated at the SA-RASSCF($4f/5d6s5f$-$4e$) and MS-RASPT2 levels of theory. Spin octets in green (except ground state in black), sextets in red

the SA-RASSCF level of calculations probably due to the high density of states resulting in deep and widespread configurational mixing through multiple avoided crossings.

The number of sextets resulting from the couplings chosen (39 $^6A_{1g}$, 36 $^6A_{2g}$, 75 ^6E_g, 108 $^6T_{1g}$ and 111 $^6T_{2g}$) is still too high to be calculated at the MS-RASPT2 level of theory. Moreover, no energy gaps are found above 50000 cm^{-1} to guide further limitation in the number of roots to be computed. This means that an arbitrary truncation in the number of roots seemed to be the only way out. Even though the MS-RASPT2 method is prepared to deal with near-degeneracy and avoided crossings, and it generally does, some energy curves showed irregular shapes in the form of small spikes and/or bumps close to the sharp avoided crossings of their parent SA-RASSCF states when the number of roots per block were limited following some sort of energy cut off. The irregularities smoothed slightly when the number of states allowed to interact was increased, but they did not disappear when the MS-RASPT2 roots calculated got the following demanding, yet feasible numbers: 28 $^6A_{1g}$, 29 $^6A_{2g}$, 22 ^6E_g (×2), 44 $^6T_{1g}$ (×3) and 44 $^6T_{2g}$ (×3). The energy curves resulting from these MS-RASPT2 calculations are plotted up to 70000 cm^{-1} in Fig. 2.4 where some spikes are visible at the bottom of the dense sextet set above 50000 cm^{-1}. These results can be considered unstable and suggest that sextets lying above 50000 cm^{-1} must not be transferred to the second step of spin-orbit coupling with RASSI-SO calculations. The same situation was found for Eu^{2+} in SrF_2 and BaF_2.

As a result, spin-orbit coupling was calculated using as input all components of the $4f^6(^7F)5de_g$ octets and sextets plus those of the octets showing $4f^6(^7F)5dt_{2g}$ and $4f^6(^7F)\phi_{ITE,a_{1g}}$ configurational mixing in the three fluorite crystals.

As mentioned above, $4f^66s$ states were found in the MS:Eu^{2+} (M = Ca, Sr) crystals rather than $4f^6\phi_{ITE,a_{1g}}$ states. Moreover, a small gap of some 3000 cm^{-1} was found around 57000 cm^{-1} which allowed the calculation of all states lying below that energy at the MS-RASPT2 and their transference to spin-orbit coupling. Smooth energy curves were obtained for all of those states.

Although validation of the different truncations accepted (over spin-multiplicities and spin-orbit-free roots) can only come from comparisons with experiment (cf. Sect. 9.4), it is meaningful to consider that spin-orbit coupling exclusion of the high energy sextets (above 50000 cm^{-1} in the fluorides and 57000 cm^{-1} in the sulfides) should not affect neither the emission spectra nor the first and the inter-band regions of the absorption spectra dominated by the low energy high and low spin $4f^7 \rightarrow 4f^65d$ transitions ($5de_g$ in the fluorides; $5dt_{2g}$ in the sulfides). Moreover, the energies spanned by the second absorption band correspond to spin-orbit states dominated by high-spin (i.e. octet) coupling. All of this explains the accuracy achieved by the multiconfigurational ab initio calculations [10].

2.4 Double-Shell Effect in Lanthanides: 4*f* Radial Correlation

The *double-shell* quantum chemical concept was developed in the context of multiconfigurational ab initio calculations on transition metal compounds, as a means to handle specific static correlation effects, called radial correlations, which become very important when energy differences between states of different configurations are involved [14, 15]. In this Section we discuss its application in the context of lanthanide-based luminescent materials.

Let us start by considering the case of Yb^{2+}-doped phosphors, where $Yb^{2+}4f^{14} \rightarrow Yb^{3+}4f^{13}$ ionizations and $Yb^{2+}4f^{14} \rightarrow 4f^{13}5d$ excitations are relevant. The electronic ground state of an Yb^{2+} dopant is predominantly a full 4*f* shell configuration, $4f^{14}$. Therefore, its spectrum is characterized by interconfigurational transitions where, roughly speaking, one electron is excited from the 4*f* shell to higher shells of which the closest in energy is the 5*d* one, this leading to the $4f^{13}5d$ excited manifold. As a result, Yb^{2+} is expected to luminesce from the lowest of $4f^{13}5d$ states. As a matter of fact, the origin of the luminescent level of Yb^{2+} in $SrCl_2$:Yb^{2+} has been found to be at 25000 cm^{-1} (3.1 eV) from single crystal low temperature absorption and emission spectra [16–18]. This evidences a large energy gap above the $4f^{14}$ ground state.[6]

There are reasons to believe that this large energy gap is dominated by electron correlation effects: (i) The electron excitations from the $4f$ shell to the more diffuse $5d$ shell of Yb^{2+}, characterizing the $4f^{14} \rightarrow 4f^{13}5d$ interconfigurational transitions, change the $4f$ shell occupation and break many $4f$–$4f$ electron pairs, hence electron correlation stabilization should be much stronger in the $4f^{14}$ ground state where the number of tight $4f$–$4f$ electron pairs is larger than in the luminescent level with its $4f^{13}$ sub-shell. (ii) Uncorrelated open-shell Hartree-Fock-like calculations give an extremely small energy gap (5400 cm^{-1}); i.e. the energy gap is not found at HF level, hence it is due to electron correlation (cf. Sect. 1.4). A way to illustrate this is calculating the energy difference between the ground and excited states using a minimal restricted active space that can be referred to as RAS($4f$/$5d6s$-$1e$) [3][7]: The ground state is described by the single Slater determinant corresponding to the closed-shell $4f^{14}$ configuration; the excited states are calculated using a RAS where only one electron out of the 14 open-shell electrons is allowed to occupy the molecular orbitals with dominant $5d$ and $6s$ character so that only the $4f^{13}5d$ and $4f^{13}6s$ open-shell excited configurations are used. These minimal RASSCF($4f$/$5d6s$-$1e$) (multiconfigurational open-shell HF like) calculations give a 5400 cm^{-1} energy gap. Therefore, corrections beyond Hartree-Fock are substantial. Furthermore, subsequent RASPT2 calculations lead to an energy gap of 41700 cm^{-1} which means correlation corrections are very large (36300 cm^{-1}); too large to be reliable at second order perturbation level; too large to give a quantitative account of the experimental observation (25000 cm^{-1}; about 29500 cm^{-1} without spin-orbit coupling). The results of these calculations for the lowest $4f^{13}5d$ excited states of $SrCl_2$:Yb^{2+} are shown in Fig. 2.5(top).

The failure of the RASSCF($4f$/$5d6s$-$1e$)/MS-RASPT2 calculations suggest an important conclusion and some questions. The conclusion: The restricted active space used, made of the open-shells involved ($4f$, $5d$, and $6s$), does not capture the essence of the electron correlation changes that occur upon *interconfigurational* excitations (e.g. $4f^N \rightarrow 4f^{N-1}5d, 4f^{N-1}6s$) and ionizations ($4f^N \rightarrow 4f^{n-1}$), which change the occupation of the $4f$ shell. This, in the end, results in too large and hence unreliable

---

[6] In the $SrCl_2$:Yb^{2+} crystals, Yb^{2+} substitutes for Sr^{2+} ions occupying cubic sites in eightfold chlorine coordination. Its first excited state is the luminescent level $4f^{13}5de_g$–$1E_u$ (25000 cm^{-1} above the ground $4f^{14}$–$1A_{1g}$). According to SA-RASSCF/MS-RASPT2/RASSI-SO calculations the spin-orbit coupling state $1E_u$ is constituted dominantly (97%) by the lowest spin-orbit-free excited state of Yb^{2+} $4f^{13}5de_g$–1^3T_{1u}, whose energy is around 4500 cm^{-1} higher [8, 19]. If this spin-orbit correction is applied to the experimental value (25000 cm^{-1}), the spin-orbit-free energy gap above the ground state would be even higher: around 29500 cm^{-1}.

[7] The RAS notation described in Sect. 2.3.2 is used here, RAS(shells in RAS1-*nh*/shells in RAS2/shells in RAS3-*ne*), to the 14 and 13 open-shell electron cases pf Yb^{2+} and Yb^{3+}, subject to spin and O_h site symmetry restrictions. As an example, in the space RAS($4f$/$5d6s$-$1e$), where no RAS1 orbital subspace is used, if the spin multiplicity and O_h irrep of the many-electron wave function are $^1A_{1g}$, only the $4f^{14}$ *gerade* configuration is produced. If, however, the wave function is $^3T_{1u}$, the $4f^{14}$ configuration does not appear and only a list of $4f^{13}5d$ and $4f^{13}6s$ *ungerade* configurations are produced.

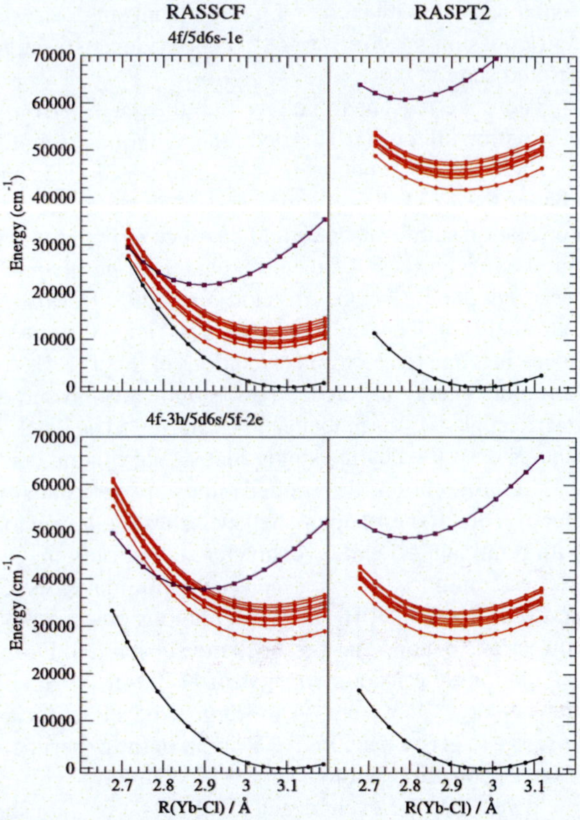

Fig. 2.5 Double f-shell effect on the lowest electronic transitions $4f^{14} \rightarrow 4f^{13}5de_g$ and the first ionization potential $4f^{14} \rightarrow 4f^{13}$ of the $SrCl_2{:}Yb^{2+}$ luminescent material. Top: RASSCF and MS-RASPT2 results without double-f shell. Bottom: Corresponding results with double-f shell. Black: $4f^{14}$ ground state. Red: $4f^{13}5de_g$ excited states. Magenta: $4f^{13}$ ionized state

second-order perturbation corrections. Therefore, some important contribution to the total static correlation is missing in the $4f5d6s$ RAS. The questions: (i) What type of electron correlation is missing in the $4f5d6s$ RAS? (ii) Is there a sound recipe to incorporate the missing correlation? (iii) Does it affect intraconfigurational energy differences? (iv) Is it an atomic contribution? Does it affect the local structure of the embedded-clusters? (v) How does it vary across the lanthanide series?

The answers to these questions can be found in the work of Froese-Fischer [14] and Dunning et al. [15] on transition metal compounds in 1977 and 1980. Their conclusions were adapted to the RASSCF/RASPT2 methodology and were further explored in the context of transition metal compounds [20–24]. In this Section we show the adaptations found to be needed to apply their ideas and recipes to the computation of interconfigurational transitions and ionization potentials of lanthanide activated luminescent materials using RASSCF/RASPT2 methods [3].

Regarding radial correlations in open-shells, Froese-Fischer [14] and Dunning et al. [15] pointed out the inadequacy of the "equivalence" restriction of the Hartree-Fock method to properly describe the electronic structure of states of the $4s^13d^{n+1}$ and $3d^{n+2}$ configurations and their energy relative to the $4s^23d^n$ states for Ti and

Ni [15], and Cu [14]. They showed that, as one or two electrons are added to the $3d$ shell, one of the orbitals in the shell (called $3d'$ orbital) or two orbitals (called $3d'$ and $3d''$) are "functionally inequivalent" and more diffuse than the rest of orbitals in the $3d$ shell. According to this, proper orbital configurations of these states are rather $4s^1 3d^n 3d'$ and $3d^n 3d' 3d''$. The authors showed the significance of this correlation contributions performing multiconfigurational Hartree-Fock calculations using an extra $4s3d^n4d$ configuration involving not only the regular $3d$ shell of five orbitals, but also a second d-shell of five orbitals, called $4d$ shell.

In order to realize the role that the double-shell can play in radial polarization, let us think, for instance, of an uncorrelated two-electron wave function like the Slater determinant $|\phi_a^\alpha \phi_b^\alpha|$, in which ϕ_a and ϕ_b are two orbitals belonging to a shell. If we take an atom, for simplicity, the shell could be the $3d$ shell, or any other degenerate set of orbitals, and the determinant $|3d_2^\alpha 3d_1^\alpha|$, for instance. The atomic symmetry imposes the radial components of $3d_2$ and $3d_1$ to be identical, as well as the other members of the $3d$ shell. This restriction limits the radial polarizations of both orbitals, which must be the same if an uncorrelated wave function is used. The only way the two orbitals could have a different radial polarization is to use different radial components: two-electron uncorrelated wave function $|3d_2^\alpha 3d_1^{\prime\alpha}|$. There is, however, a way of maintaining the right symmetry of the many-electron wave function while allowing different orbitals of a shell to have different radial polarizations: to use a double-shell. In effect, since the $3d_1'$ orbital can be written in terms of the $3d_1$ orbital and the corresponding orbital of a second d shell ($4d$): $3d_1' = c_1 3d_1 + c_2 4d_1$, the Slater determinant is also written as a linear combination of Slater determinants: $|3d_2^\alpha 3d_1^{\prime\alpha}| = c_1 |3d_2^\alpha 3d_1^\alpha| + c_2 |3d_2^\alpha 4d_1^\alpha|$. In other words, CI wave functions in the small CI space resulting from using the complete $3d$ and $4d$ shells (the double-shell) can achieve different radial polarizations of the electrons when they occupy different orbitals of the shell. These CI wave functions fulfill the atomic symmetry, but they necessarily include some sort of correlation (the two-electron probability density cannot be written as the product of two one-electron probability densities). This is the smallest amount of correlation able to give correct, orbital specific radial polarizations. It does not account for the Coulomb hole. Hence, it is static correlation and is called radial correlation.

Andersson and Roos [20] faced the same basic type of radial correlation interaction in the context of a CASPT2 study of excitation energies of the Ni atom, where they showed the need to include a second d-shell in the CAS space. Further work by Pierloot and collaborators [21–24] allowed to conclude that the double d-shell effect is more important: (a) for the heavier transition metal ions (d^n; $n > 5$) than for the lighter, (b) for transition energies involving a change of the d occupation, and (c) for first-row than for second- and third-row transition metals.

Following the track of these authors, the effects of introducing a second f-shell in the RAS in order to account for $4f$ radial correlation on the interconfigurational transitions of $SrCl_2$:Yb^{2+} and on the ionization potential of Yb^{2+} in gas phase and doped in $SrCl_2$ were analyzed in [3] using RASSCF/MS-RASPT2 calculations.

The effects of introducing a second f-shell in the RAS resulted to be substantial. Large corrections were observed at the static correlation RASSCF (variational) step, together with corresponding small corrections at the dynamic correlation MS-RASPT2 (perturbational) step. This is a desirable situation, because second order perturbation treatments show better accuracies for smaller corrections. Altogether, the spin-orbit free results get closer to the experimental or CCSD(T) benchmarks when a second f-shell is used [3].

The results on $SrCl_2$:Yb^{2+} using three meaningful spaces are summarized in Table 2.1: Space (1) is the minimal RAS($4f/5d6s$-$1e$) referred above and does not include the double f-shell. Spaces (2) and (3) do include it: space (2) is like (1) plus $4f$-to-$5f$ SD excitations only; space (3) is like (1) plus $4f$-to-$5d6s5f$ SD excitations. Energy curves obtained with spaces (1) and (3) are plotted in Fig. 2.5 (top and bottom, respectively). Unpublished results for CaF_2:Eu^{2+} are also included in Table 2.1 for two near-CAS spaces where SDTQ excitations from the $4f$ shells are allowed to the $5d6s$ shells and optionally to the second f-shell $5f$: (I) RAS($4f/5d6s$-$4e$) and (II) RAS($4f/5d6s5f$-$4e$). We comment first the results on $SrCl_2$:Yb^{2+}.

The RASSCF $4f^{14}$–1^1A_{1g}→$4f^{13}5de_g$–1^3T_{1u} interconfigurational transition energy from minimum to minimum (T_e) grows from 5427 cm^{-1} to 27242 cm^{-1} when the minimal RAS($4f/5d6s$-$1e$) space (1) is increased with $4f$-to-$5d6s5f$ SD excitations as can be deduced from Table 2.1. The corresponding T_e MS-RASPT2 results change from 41682 cm^{-1} to 27581 cm^{-1}, respectively. This means that the dynamic correlation corrections operated by the MS-RASPT2 step changed dramatically: from 36255 cm^{-1} (without double f-shell) to 339 cm^{-1} (with double f-shell in the RAS). And comparison of the final MS-RASPT2 values, (3)–(1), shows that the effects associated with the $4f$-to-$5d6s5f$ SD excitations amounts to –14101 cm^{-1}, of which –12773 cm^{-1} are due to $4f$-to-$5f$ SD excitations only. The double f-shell brings the discrepancy between the best MS-RASPT2 result (27581 cm^{-1}) and the estimated experimental value (~29500 cm^{-1} without spin-orbit coupling) down to an acceptable 6.5% error.

The results of the calculations on all the $SrCl_2$:Yb^{2+} $4f^{14}$→$4f^{13}5de_g$ excited states show they are uniformly shifted –14223±80 cm^{-1} as a result of the double f-shell effects (cf. Fig. 2.5 and $4f^{13}5de_g$ average correction values in Table 2.1). In other words, the effects on intraconfigurational energy differences are insignificant, which explains the accuracy of the MS-RASPT2 spectrum obtained for $SrCl_2$:Yb^{2+} without double f-shell versus experimental spectra, when their origins are set to the first excited state [19]. It is also apparent that the double f-shell corrections due to $4f$-shell radial correlation do not affect the local structure of the Yb^{2+} center.

Altogether, $4f$ radial correlation is basically an atomic interaction that is expected to increase with the number of $4f$-$4f$ electron pairs, hence with the occupancy of the $4f$ shell. In agreement with this, double f-shell effects on the lowest $4f$→$5d$ interconfigurational transitions of Eu^{2+} and on its ionization potential, calculated in CaF_2:Eu^{2+} as the difference between RASSCF($4f/5d6s$-$4e$)/MS-RASPT2 and RASSCF($4f/5d6s5f$-$4e$)/MS-RASPT2 results (entries (I) and (II) in Table 2.1, respectively), are smaller than in Yb^{2+}. In any case, the double f-shell corrections in Eu^{2+} are still significant.

The important double f-shell effects on the ionization potentials of Yb^{2+} and Eu^{2+} in the two phosphor materials $SrCl_2$:Yb^{2+} and CaF_2:Eu^{2+} shown in Table 2.1, indicate that these static, radial correlation effects are also important in metal-to-metal

Table 2.1 Double f-shell effects on $4f^n \rightarrow 4f^{n-1}5de_g$ interconfigurational transitions and $4f^n \rightarrow 4f^{n-1}$ ionization potentials of Yb^{2+}-doped $SrCl_2$ ($n = 14$) and Eu^{2+}-doped CaF_2 ($n = 7$), calculated with the MS-RASPT2 method using different restricted active spaces in the RASSCF calculations

	MS-RASPT2[a]							Double f-shell
	State		$d_{Yb-Cl,e}$[b]		$\omega_{a_{1g}}$[c]		T_e[d]		effect on T_e
$(YbCl_8Sr_{12})^{18+}$ and $(YbCl_8Sr_{12})^{19+}$ clusters in $SrCl_2$									
(1) RAS *without* double f-shell: RAS($4f/5d6s$-$1e$)									
	Yb^{2+}	$4f^{14}$–1^1A_{1g}	2.969	(–0.132)	179	(+10)	0	(0)	
		$4f^{13}5de_g$–1^3T_{1u}	2.915	(–0.147)	180	(+7)	41682	(+36255)	
		$\langle 4f^{13}5de_g \rangle$	2.904	(–0.153)	181	(+7)		(+35539±439)	
	Yb^{3+}	$4f^{13}$–1^2A_{2u}	2.787	(–0.105)	232	(+9)	60766	(+39305)	
(2) RAS *with* double f-shell: RAS (1) plus $4f$-to-$5f$ SD excitations									(2)–(1)
	Yb^{2+}	$4f^{13}5de_g$–1^3T_{1u}	2.917		176		28949		–12733
(3) RAS *with* double f-shell: RAS (1) plus $4f$-to-$5d6s5f$ SD excitations									(3)–(1)
	Yb^{2+}	$4f^{14}$–1^1A_{1g}	2.977	(–0.122)	180	(+10)	0	(0)	0
		$4f^{13}5de_g$–1^3T_{1u}	2.915	(–0.147)	180	(+7)	27581	(+339)	–14101
		$\langle 4f^{13}5de_g \rangle$	2.909	(–0.148)	180	(+7)		(–721±571)	–14223±80
	Yb^{3+}	$4f^{13}$–1^2A_{2u}	2.793	(–0.097)	233	(+10)	48771	(+10921)	–11995
$(EuF_8Ca_{12})^{18+}$ and $(EuF_8Ca_{12})^{19+}$ clusters in CaF_2									
(I) RAS *without* double f-shell: RAS($4f/5d6s$-$4e$)									
	Eu^{2+}	$4f^7$-1^8A_{1u}					0	(0)	
	Eu^{2+}	$4f^65de_g$-1^8T_{2g}					31027	(+12815)	
	Eu^{3+}	$4f^6$-1^7A_{2g}					21553	(+17427)	
RAS *with* double f-shell: RAS($4f/5d6s5f$-$4e$)									(II)–(I)
	Eu^{2+}	$4f^65de_g$-1^8T_{2g}					27329	(+2985)	–3698
	Eu^{3+}	$4f^6$-1^7A_{2g}					17494	(+6841)	–4059

[a] Changes from RASSCF to MS-RASPT2 are given in parentheses
[b] Yb–Cl bond length, in Å
[c] Totally symmetric (breathing) harmonic vibrational frequency, in cm^{-1}
[d] Minimum-to-minimum (adiabatic) energy difference relative to the ground state ($4f^{14}$-$^1A_{1g}$ for Yb^{2+}; $4f^7$-$^8A_{1u}$ for Eu^{2+}), in cm^{-1}

charge transfer (MMCT) states of donor/acceptor pairs in luminescent materials activated with lanthanide ions. These MMCT states are very important to understand the luminescence of materials because their energy levels intercalate between the impurity levels of the donor and the acceptor centers in quantitative configurational energy level diagrams, which makes them to actively participate in energy and charge transfer processes, either enhancing, quenching, or modifying expected luminescences, and producing new luminescences. And the MMCT energies are controlled by the ionization potential of the donor and the electron affinity of the acceptor (i.e. the negative ionization potential of the reduced acceptor).

In summary, a double f-shell should be used to study interconfigurational and electron transfer states using the RASSCF/MS-RASPT2 method. Given that SDTQ excitations from the $4f$ shell to the relevant excited open shells $5d6s$ and to the second $5f$ shell define a restriction that should yield near-CAS quality [3], the use of RAS($4f/5d6s5f$-$4e$) in embedded-cluster SA-RASSCF/MS-RASPT2 calculations is recommended for any lanthanide based material.

References

1. S. Huzinaga, J. Andzelm, E. Radzio-Andzelm, Y. Sakai, H. Tatewaki, M. Klobukowski, *Gaussian Basis Sets for Molecular Calculations* (Elsevier, Amsterdam, 2012)
2. L. Seijo, Z. Barandiarán, J. Chem. Phys. **118**, 5335 (2003)
3. Z. Barandiarán, L. Seijo, J. Chem. Phys. **138**, 074102 (2013)
4. Z. Barandiarán, L. Seijo, in *Computational Methods in Lanthanide and Actinide Chemistry*. ed. by M. Dolg (Wiley, New York, 2015), pp. 217–240
5. B.O. Roos, R. Lindh, P.A. Malmqvist, V. Veryazov, P.O. Widmark, A.C. Borin, J. Phys. Chem. A **112**, 11431 (2008)
6. J.L. Pascual, L. Seijo, Z. Barandiarán, J. Chem. Phys. **98**, 9715 (1993)
7. M.A. Nygren, L.G.M. Pettersson, Z. Barandiarán, L. Seijo, J. Chem. Phys. **100**, 2010 (1994)
8. G. Sánchez-Sanz, L. Seijo, Z. Barandiarán, J. Chem. Phys. **133**, 114509 (2010)
9. Z. Barandiarán, L. Seijo, J. Chem. Phys. **141**, 234704 (2014)
10. J.J. Joos, P.F. Smet, L. Seijo, Z. Barandiarán, Inorg. Chem. Front. **7**, 871 (2020)
11. R.W.G. Wyckoff, *Crystal Structures*, vol. 1, 2nd edn. (Interscience Publishers, 1982)
12. A. Gellé, M.B. Lepetit, J. Chem. Phys. **128**, 244716 (2008)
13. P.P. Ewald, Ann. Phys. **369**, 253 (1921)
14. C. Froese-Fischer, J. Phys. B **10**, 1241 (1977)
15. T.H. Dunning Jr., B.H. Botch, J.F. Harrison, J. Chem. Phys. **72**, 3419 (1980)
16. T.S. Piper, J.P. Brown, D.S. McClure, J. Chem. Phys. **46**, 1353 (1967)
17. H. Witzke, D.S. McClure, B. Mitchell, in *Luminescence of Crystals, Molecules, and Solutions*. ed. by F.E. Williams (Plenum, New York, 1973), p. 598
18. Z. Pan, C. Duan, P.A. Tanner, Phys. Rev. B **77**, 085114 (2008)
19. G. Sánchez-Sanz, L. Seijo, Z. Barandiarán, J. Chem. Phys. **133**, 114506 (2010)
20. K. Andersson, P.A. Malmqvist, B.O. Roos, J. Chem. Phys. **96**, 1218 (1992)
21. P.A. Malmqvist, K. Pierloot, A.R.M. Shahi, C.J. Cramer, L. Gagliardi, J. Chem. Phys. **128**, 204109 (2008)
22. K. Pierloot, in *Advances in Chemical Physics: New Methods in Computational Quantum Mechanics*. ed. by T.R. Cundari (Deker, New York, 2001)
23. K. Pierloot, S. Vancoillie, J. Chem. Phys. **125**, 124303 (2006)
24. V. Sauri, L. Serrano-Andrés, A.R. Mougal Shahi, L. Gagliardi, S. Vancoillie, K. Pierloot, J. Chem. Theory Comput. **7**, 153 (2011)

Chapter 3
Calculations of Local Properties of Luminescent Materials

Abstract The quantum chemical methods used in this book give electronic energies and wave functions as eigenvalues and eigenfunctions of the electronic Schrödinger equation at fixed nuclei positions, within the Born-Oppenheimer approximation. This process is typically repeated for multiple nuclear arrangements, leading to so-called potential energy surfaces. Their analyses, much helped with the use of symmetry, which may reduce the number of relevant normal vibrational modes to one in high-symmetry sites, leads to important properties of the luminescent centers such as equilibrium geometries, vibrational frequencies and spectral shapes for optical absorption and emission. This chapter summarizes how these properties are derived from quantum chemical calculations.

3.1 Potential Energy Surfaces and Curves

The multiconfigurational ab initio calculation of optically relevant properties of a luminescent material usually rely on the application of the Born-Oppenheimer approximation to the local embedded-cluster adopted as a model. In this approximation, the solutions of the total time-independent Schrödinger equation of the embedded-cluster,

$$\hat{H}_{\mathrm{TOT}}(\boldsymbol{r}, \boldsymbol{Q})\Psi(\boldsymbol{r}, \boldsymbol{Q}) = E\Psi(\boldsymbol{r}, \boldsymbol{Q}), \tag{3.1}$$

are obtained in two steps. In (3.1), E is the energy of an electronic-vibrational stationary state, Ψ its wave function, which depends on all electronic coordinates, $\boldsymbol{r}$, and all nuclear coordinates, $\boldsymbol{Q}$, and $\hat{H}_{\mathrm{TOT}}$ is the total Hamiltonian of the embedded-cluster, which includes the kinetic energy operators of all electrons, $\hat{T}_{\mathrm{e}}$, and all nuclei, $\hat{T}_{\mathrm{nuc}}$, plus the potential energy operators of all interactions (electron-electron, electron-nucleus, nucleus-nucleus, electron-host, nucleus-host) and the spin-orbit coupling operator. Collecting all operators but $\hat{T}_{\mathrm{nuc}}$ in an electronic Hamiltonian, $\hat{H}_{\mathrm{elec}}$, results in $\hat{H}_{\mathrm{TOT}}(\boldsymbol{r}, \boldsymbol{Q}) = \hat{T}_{\mathrm{nuc}}(\boldsymbol{Q}) + \hat{H}_{\mathrm{elec}}(\boldsymbol{r}, \boldsymbol{Q})$.[1] Adopting such structure of the total

[1] Note that the electronic Hamiltonian $\hat{H}_{\mathrm{elec}}$ is $\hat{H}_{\mathrm{EC}} + V_{\mathrm{ECnuc-Host}}$ in (1.4).

Z. Barandiarán et al., *Luminescent Materials*, Springer Series in Materials Science 322,
https://doi.org/10.1007/978-3-030-94984-6_3

Hamiltonian is the first step towards the Born-Oppenheimer approximation. The second step is assuming a factorization of the total wave function Ψ into an electronic wavefunction Φ times a vibrational wave function ξ, $\Psi = \Phi\xi$, the former depending explicitly on the electron coordinates $\boldsymbol{r}$ and parametrically on the nuclear coordinates $\boldsymbol{Q}$, $\Phi(\boldsymbol{r};\boldsymbol{Q})$, and the latter depending only on the nuclear coordinates $\boldsymbol{Q}$, $\xi(\boldsymbol{Q})$; i.e. $\Psi(\boldsymbol{r},\boldsymbol{Q}) = \Phi(\boldsymbol{r};\boldsymbol{Q})\xi(\boldsymbol{Q})$. This simplified ansatz is not an exact solution of (3.1), but the larger masses of nuclei compared to electrons make it often a good approximation, except when Jahn-Teller dynamic coupling and other vibronic couplings are important.

Under the Born-Oppenheimer approximation, a separation of variables can be performed and the electronic and vibrational wave functions can be found in two steps. In the first step, the so-called electronic Schrödinger equation is solved

$$\hat{H}_{\text{elec}}(\boldsymbol{r};\boldsymbol{Q})\Phi_i(\boldsymbol{r};\boldsymbol{Q}) = E_i(\boldsymbol{Q})\Phi_i(\boldsymbol{r};\boldsymbol{Q}) \,. \tag{3.2}$$

This is the very demanding equation that multiconfigurational ab initio calculations handle; note that this equation is (1.4). It is solved for a set of distributions of all the nuclei in space, i.e. for a set of fixed $\boldsymbol{Q}$ values, hence the parametric dependences on $\boldsymbol{Q}$. Its solution gives a number of electronic states with wave functions Φ_i and energies E_i. These electronic energies depend on the nuclear positions, as explicitly recognized in (3.2), and the $E_i(\boldsymbol{Q})$ functions are called Potential Energy Surfaces (PES).

They are called surfaces as an abbreviation for hyper-surfaces, because they are functions of $3N_{\text{nuc}}$ nuclear coordinates $Q_1, Q_2, \ldots, Q_{3N_{\text{nuc}}}$, represented with $\boldsymbol{Q}$.[2] Each of these coordinates represents an internal vibration of the embedded-cluster. Formally, they are constructed as symmetry-adapted linear combinations of the nuclear displacement vectors from a reference structure (cf. Fig. 8.5 and Table 8.2 in Sect. 8.2.1.3, for instance).

They are called potential energies because they are the potential energies acting on the nuclei and determining their movement, i.e. their vibration, as it is explicitly recognized in the second step of the Born-Oppenheimer approximation,

$$\left[\hat{T}_{\text{nuc}}(\boldsymbol{Q}) + E_i(\boldsymbol{Q}) - E_{ie}\right]\xi_{im}(\boldsymbol{Q}) = E_{im}^{\text{vib}}\xi_{im}(\boldsymbol{Q}) \,. \tag{3.3}$$

In (3.3), $E_{ie} = \min E_i(\boldsymbol{Q})$, i.e. the electronic energy of state Φ_i at its equilibrium nuclear configuration, so that $E_i(\boldsymbol{Q}) - E_{ie}$ is the vibrational potential energy in the electronic state Φ_i; ξ_{im} is the m-th vibrational wave function of the electronic state Φ_i; and E_{im}^{vib} is its vibrational energy. $E_{ie} + E_{im}^{\text{vib}}$ is one of the eigenvalues E in (3.1), provided that the non-adiabatic terms are neglected, as prescribed by the Born-Oppenheimer approximation.

---

[2] In non-linear free molecules, the number of internal nuclear coordinates is $3N_{\text{nuc}}$–6, because three coordinates describe free translations and three coordinates describe free rotations (two in linear molecules); however, free translations and rotations do not exist in a cluster embedded in a solid host, hence the number of $3N_{\text{nuc}}$ internal nuclear coordinates.

Important properties that emerge from the topological analysis of the potential energy surfaces are the equilibrium structures of the active site in each of its electronic states, its harmonic vibrational frequencies in the different electronic states, and the potential energy differences between states.

Some specific properties may require to release the Born-Oppenheimer approximation and go beyond it. This is the case of electronic degeneracies and near-degeneracies, when electronic and vibrational energy differences are comparable. E.g. in the dynamic Jahn-Teller effect, when vibronic interactions between degenerate electronic states are important, the simplest reasonable wave functions have to be written as expansions including all degenerate electronic states Φ_{ia} $(a = 1, \ldots, d_i)$: $\Psi = \sum_{a=1}^{d_j} \sum_m c_{am} \Phi_{ia} \xi_{ia,m}$. In cases like this, calculation of the Born-Oppenheimer wave functions and potential energy surfaces are often a first step towards more elaborate solutions.

3.1.1 Ground and Excited State Local Structures and Vibrations

A topological search for the minima of the potential energy surfaces $E_i(\boldsymbol{Q})$, computed in the first step of the Born-Oppenheimer approximation (3.2), yields interatomic equilibrium distances and angles in all calculated electronic states, i.e. the local structures of the active site in its ground and excited states.

Calculation of energy derivatives with respect to the nuclear displacements at the equilibrium structures allows to perform Taylor expansions of the vibrational potential energies around their respective minima, leading to

$$E_i(\boldsymbol{Q}) = E_{ie} + \frac{1}{2} \sum_{k=1}^{3N_{\text{nuc}}} \left(\frac{\partial^2 E_i}{\partial Q_k^2} \right)_e Q_k^2 + \ldots \tag{3.4}$$

for every electronic state. In (3.4), a particular choice among all valid sets of vibrational coordinates was made for each electronic state: One (1) in which they vanish at the potential energy minimum, $Q_{ke} = 0$, where the first derivatives also vanish, $(\partial E_i / \partial Q_k)_e = 0$; and (2) which makes the second energy derivative matrix diagonal, $\left(\partial^2 E_i / \partial Q_k \partial Q_{l \neq k}\right)_e = 0$. Vibrational coordinates fulfilling these conditions are called normal vibrational coordinates or normal vibrational modes. Their second derivatives, or curvatures, are the *force constants* of the respective vibrations, which, together with the effective masses of the vibrations M_k, give the *normal mode harmonic vibrational frequencies*,

$$\omega_k = \frac{1}{2\pi} \sqrt{\frac{\left(\partial^2 E_i / \partial Q_k^2\right)_e}{M_k}} \, . \tag{3.5}$$

3.1.2 State Energy Differences and Potential Energy Crossings

Representing the potential energies of all electronic states of an optically active center in a common energy diagram, where the energies vary with the positions of the nuclei, is necessary to understand excitations, non-radiative decays, and emissions. Such energy diagram requires the use of a single set of nuclear or vibration coordinates.

The usual choice is the set of ground state's normal modes. With this choice, the ground state potential energy surface is expressed as in (3.4) but the excited states may still have non-null linear components and residual off-diagonal quadratic components,

$$E_i(Q) = E_{ie} + \sum_{k=1}^{3N_{\text{nuc}}} \left(\frac{\partial E_i}{\partial Q_k}\right)_e Q_k + \frac{1}{2}\sum_{k,\ell=1}^{3N_{\text{nuc}}} \left(\frac{\partial^2 E_i}{\partial Q_k \partial Q_l}\right)_e Q_k Q_\ell + \ldots. \tag{3.6}$$

Let us focus now on the linear components $(\partial E_i/\partial Q_k)_e\, Q_k$ in (3.6). A non-null value of the $(\partial E_i/\partial Q_k)_e$ coefficient creates a horizontal shift of the minimum of $E_i(Q)$ with respect to the ground state's minimum along the Q_k axis. On the contrary, a null $(\partial E_i/\partial Q_k)_e$ makes $E_i(Q)$ basically parallel to the ground state energy $E_{\text{GS}}(Q)$ along Q_k, except for minor contributions of quadratic and higher orders. Having in mind that only vibrational modes with horizontal shifts contribute to the absorption or emission band shapes between the two states and to direct non-radiative decay processes via potential energy crossings,[3] it is important to concentrate the computational efforts on vibrational coordinates with horizontal offsets. Then, the important question is: what vibrations of the active center have horizontal offsets between electronic states? Or, in other words, what Q_k coordinates fulfill $\partial E_i/\partial Q_k \neq 0$?

Symmetry plays a key role to answer this question and to largely simplify the calculations that are necessary to calculate the relevant properties of an optically active center in a luminescent material. In effect, first, according to Hellmann-Feynman theorem [1], the derivative fulfills

$$\frac{\partial E_i}{\partial Q_k} = \langle \Phi_i | \frac{\partial \hat{H}_{\text{elec}}}{\partial Q_k} | \Phi_i \rangle, \tag{3.7}$$

and second, the right-hand side integral vanishes unless the integrand is totally symmetric, which requires the product of the irreducible representations (irreps) of the three components to contain the totally symmetric irrep of the active center point group, Γ_1 (e.g. A_{1g} in the O_h group, see Sect. 4.1.5). Since the irrep of the Hamilto-

---

[3] Strictly speaking, contributions to the band shapes exist from all vibrational coordinates that make the energy differences non-parallel, but the differences in curvatures (i.e. vibrational frequencies) between states is usually so small that the only relevant contribution to the non-parallelism is the horizontal offset.

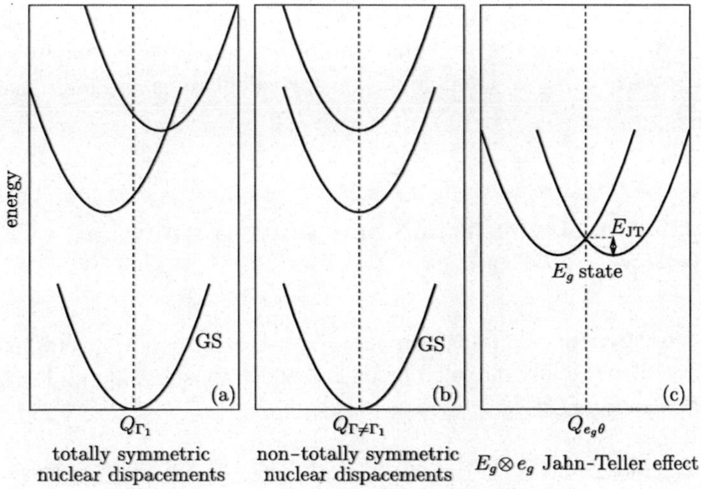

Fig. 3.1 Schematic representation of horizontal shifts of the electronic states in energy diagrams

nian derivative is that of Q_k, the previous condition reads $\Gamma(\Phi_i) \otimes \Gamma(Q_k) \otimes \Gamma(\Phi_i) \ni \Gamma_1$. For non-degenerate states, $\Gamma(\Phi_i) \otimes \Gamma(\Phi_i) = \Gamma_1$ and the condition becomes $\Gamma(Q_k) = \Gamma_1$; in other words, only totally symmetric vibrational coordinates are important (cf. Fig. 3.1).[4]

In high-symmetry active sites it is common to find only one totally symmetric vibrational mode. This is the case of octahedral LnL_6 and cubic LnL_8 centers of O_h symmetry (Ln and L represent the central lanthanide ion and the ligands, respectively), in which only the breathing mode has A_{1g} symmetry. Since the breathing is unambiguously described by the distance between the lanthanide and the ligands, $d_{\text{Ln-L}}$, the potential energy *curves* as a function of $d_{\text{Ln-L}}$ are sufficient to get the most relevant information for luminescence processes and the much more complex potential energy surfaces, with many variables, are not necessary. These representations of the potential energy curves of the electronic states as a function of one relevant coordinate, representing a simultaneous displacement of all nuclei, is usually called the configuration energy diagram or configurational coordinate diagram. A large number of optically active centers in this book belong to these cases.

[4] In the case of degenerate electronic states, there are always degenerate vibrations that give non-null linear coefficients. E.g. in the O_h group, $E_g \otimes E_g \otimes E_g \ni A_{1g}$ and E_g electronic states $\Phi_{E_g,\theta}$ and $\Phi_{E_g,\epsilon}$ show horizontal offsets along the two E_g vibrations $Q_{E_g,\theta}$ and $Q_{E_g,\epsilon}$. Besides, the two degenerate electronic states are coupled with first and second derivative operators, all of it resulting in the so called $E_g \otimes e_g$ Jahn-Teller or vibronic coupling [2]. The stabilizations achieved as a consequence of the horizontal shift along these degenerate vibration coordinates, known as Jahn-Teller energies E_{JT}, are usually small, as the horizontal shifts are themselves. This makes the Jahn-Teller active or vibronic modes important to understand some detailed spectroscopic features. Multiconfigurational ab initio calculations of Jahn-Teller problems demand the calculation of the potential energy surfaces along them [3, 4]. The result of the $E_g \otimes e_g$ Jahn-Teller effect on a degenerate electronic state is displayed in Fig. 3.1c.

In lower symmetry sites, the number of totally symmetric vibrations grows. For instance, in the D_2 LnL_8 site of lanthanide-doped garnets, six A_1 totally symmetric vibrational modes exist, which complicates the calculations and their analyses (cf. Sect. 8.2.1).

3.2 Oscillator Strengths and Spontaneous Emission Lifetimes

The oscillator strength of a transition between two electronic states of an optically active center, theoretically modeled by an embedded-cluster, an initial state Φ_i and a final state Φ_j, $\Phi_i \rightarrow \Phi_j$, is defined as [5, 6]

$$f_{i\rightarrow j} = \frac{2}{3} \Delta E_{ij} \mu_{ij}^2 \,. \tag{3.8}$$

In (3.8), ΔE_{ij} is the vertical transition energy, i.e. the energy difference between the states evaluated at the geometry of the minimum of the initial state, in hartree or atomic units (1 hartree = 219474.624 cm^{-1}). μ_{ij}^2 is the square electric dipole transition moment, in atomic units (au). If the initial and final states are non-degenerate, it is $\mu_{ij}^2 = |\langle \Phi_i | \hat{\mu}_e | \Phi_j \rangle|^2$, where $\hat{\mu}_e$ is the electronic dipole moment operator of all the N_e^{EC} electrons of the embedded-cluster, $\hat{\mu}_e = -\sum_{i=1}^{N_e^{EC}} e\,\mathbf{r}_i$. If Φ_i or Φ_j are degenerate, with respective degeneracies d_i and d_j and their respective components being Φ_{ia} $(a = 1, \ldots, d_i)$ and Φ_{jb} $(b = 1, \ldots, d_j)$, it is

$$\mu_{ij}^2 = \sum_{b=1}^{d_j} |\langle \Phi_{ia} | \hat{\mu}_e | \Phi_{jb} \rangle|^2 \,, \tag{3.9}$$

with all choices of a giving identical result.

With the same ingredients, the Einstein coefficient for a spontaneous emission $\Phi_i \rightarrow \Phi_j$ is calculated as [5, 6]

$$A_{ij} = \frac{2\omega_{ij}^3 \mu_{ij}^2}{3\epsilon_o h c^3} = \frac{16\pi^3 \mu_{ij}^2}{3\epsilon_o h \lambda_{ij}^3} \,. \tag{3.10}$$

Here, ω_{ij} is the emission angular frequency $\omega_{ij} = 2\pi c/\lambda_{ij}$, λ_{ij} is the emission wavelength, h the Planck constant, c the velocity of light, and ϵ_o the permittivity of vacuum. Local field corrections can be included here using appropriate values of the materials' permittivities [7]. If ΔE_{ij} is given in cm^{-1} and μ_{ij} in au, the expression of A_{ij} in ns^{-1} is $A_{ij} = 2.02608 \times 10^{-15} (\Delta E_{ij})^3 \mu_{ij}^2$.

And the lifetime for spontaneous emission from Φ_i is

$$\tau_i = \frac{1}{\sum_{j<i} A_{ij}}, \tag{3.11}$$

where the sum $\sum_{j<i}$ runs over all electronic states Φ_j with lower energy than Φ_i.

3.3 Absorption and Emission Spectra Profiles

The band shape of an electronic absorption or emission transition $\Phi_i \rightarrow \Phi_j$ is mainly due to its vibrational structure, i.e. to the different relative intensities of the individual transitions between the vibrational levels of Φ_i and the vibrational levels of Φ_j, although temperature, disorder, and other factors also contribute. If, for the sake of simplicity of the description, we assume non-degenerate electronic states and vibrational states, the individual lines of the band correspond to $\Phi_i\xi_{im} \rightarrow \Phi_j\xi_{jn}$ transitions, where ξ_{im} and ξ_{jn} stand for the vibrational wave functions of the m-th vibrational level of the initial electronic state Φ_i and the n-th vibrational level of the final electronic state Φ_j, respectively. The intensities of the individual lines that shape the band are hence proportional to the squares of the electric dipole transition moment integrals $|\langle\Phi_i\xi_{im}|\hat{\mu}_e|\Phi_j\xi_{jn}\rangle|^2$, where $\hat{\mu}_e$ is the electronic dipole moment operator $\hat{\mu}_e = -\sum_{i=1}^{N_e^{EC}} e\,\mathbf{r}_i$. The fact that this operator only contains electronic coordinates, simplifies the transition moment integrals into products of a purely electronic integral and purely vibrational integrals:

$$|\langle\Phi_i\xi_{im}|\hat{\mu}_e|\Phi_j\xi_{jn}\rangle|^2 = |\langle\Phi_i|\hat{\mu}_e|\Phi_j\rangle|^2 |\langle\xi_{im}|\xi_{jn}\rangle|^2 . \tag{3.12}$$

Here, besides the Born-Oppenheimer approximation leading to the assumption of factorized electronic and nuclear wave functions of the type $\Phi_i\xi_{im}$, it is also assumed that the parametric dependence on the nuclear coordinates of Φ_i and Φ_j, or more exactly of the electronic transition moment integral $\langle\Phi_i|\hat{\mu}_e|\Phi_j\rangle$, is very small and can be neglected.

In (3.12), the square of the electronic transition moment integral is proportional to the oscillator strength of the electronic transition and determines its overall intensity and the band size. On the other hand, the band shape is determined by the squares of the overlaps between initial and final vibrational wave functions $\langle\xi_{im}|\xi_{jn}\rangle$, or Frank-Condon factors. At sufficiently low temperatures, only the population of the ground vibrational level of the initial electronic state ξ_{i0} is relevant, and the band shape is determined by the overlaps between this vibrational wave function and all the final state vibrational wave functions ξ_{jn}. In a harmonic oscillator approximation for the potential energy curves of Φ_i and Φ_j with equal vibrational frequencies, simple closed formulas exist for these overlaps which are very useful from a phenomenological point of view [7, 8].

In an ab initio approach, it is also possible to use the same procedure and calculate band shapes out of overlaps between vibrational levels of the initial and final electronic states. An equivalent alternative is the semiclassical time-dependent or wave-packet propagation approach of Heller [9, 10], which gives identical results, but is very convenient when the potential energy curves and surfaces have a general rather than quadratic (harmonic) form, and when several vibrational modes are important at determining the band shapes. According to this method, the intensity profile of an individual electronic absorption or emission band, i.e. one with only one electronic origin, is given by:

$$I(\omega) = C\omega^n \int_{-\infty}^{\infty} e^{i\omega t} \langle \xi | \xi(t) \rangle \, dt \,, \tag{3.13}$$

where ω is the frequency of the absorbed or emitted radiation, $n = 1$ in absorption and $n = 3$ in emission, ξ is the initial vibrational wave function (wave-packet) of the initial electronic state, and $\xi(t)$ is its propagation on the potential energy surface of the final electronic state. Such vibrational propagation results from the vibrational time-dependent Schrödinger equation

$$i\hbar \frac{\partial \xi(t)}{\partial t} = \hat{H}_{\text{vib},j} \xi(t) \quad (\xi(0) \equiv \xi) \,, \tag{3.14}$$

where $\hat{H}_{\text{vib},j}$ stands for the vibrational Hamiltonian of the active center on its final electronic state Φ_j. Ingredients of this procedure are potential energy surfaces of Φ_i and Φ_j and the energies of the adiabatic electronic transitions (minimum-to-minimum). In an absorption or emission band with multiple electronic origins, when several electronic transitions are superimposed, the intensity profiles of the individual transitions (3.13) are added up, weighted with their respective oscillator strengths, to generate the full spectrum.

The calculation of individual absorption and emission profiles requires, in general, the potential energy surfaces of the electronic states involved. In practice, the key topological features of set of potential energy surfaces are: horizontal offsets between energy minima, which are obtained from bond length differences in high-symmetry active centers, like octahedral LnL_6 and cubic LnL_8; curvatures or second-derivatives at the minima, which determine vibrational frequencies; and vertical offsets between energy minima, or adiabatic transition energies. The calculation of real absorption and emission profiles made of several individual transitions requires, in addition, the absorption and emission electric dipole transition moments.

The line broadening effects of temperature, disorder, and other factors are taken into account with a phenomenological approach in this method, by the introduction of a factor $e^{-\Gamma^2 t^2}$ in the overlap between the initial and the propagated vibrational functions $\langle \xi | \xi(t) \rangle$, which determines the width of the vibrational lines; typical values range between 5 cm^{-1} and 300 cm^{-1}. If only one totally symmetric or breathing mode Q_k is responsible for the band shape, a common value of its vibrational frequency ω_k is used for the initial and final electronic states, and a harmonic oscillator approximation is assumed, the overlap $\langle \xi | \xi(t) \rangle$ is strongly simplified [11],

$$\langle \xi | \xi(t) \rangle = \exp \left\{ -\frac{\mu_k \omega_k (\Delta Q_k)^2}{2\hbar} \left(1 - e^{-i\omega_k t}\right) - i \left(E_0 + \frac{\omega_k}{2}\right) t - \Gamma^2 t^2 \right\} \quad (3.15)$$

In (3.15), μ_k is the effective mass of the vibrational mode Q_k, ω_k its vibrational frequency, and ΔQ_k its horizontal offset or displacement between initial and final electronic states[5]; and E_0 is the vertical offset or energy difference between the minima of the upper and lower potential energy surfaces.

References

1. R.P. Feynman, Phys. Rev. **56**, 340 (1939)
2. I.B. Bersuker, *The Jahn-Teller Effect and Vibronic Interactions in Modern Chemistry* (Plenum Press, New York and London, 1984)
3. L. Seijo, Z. Barandiarán, J. Chem. Phys. **94**, 8158 (1991)
4. J.L. Pascual, L. Seijo, Z. Barandiarán, J. Chem. Phys. **98**, 9715 (1993)
5. J. Oddershede, Phys. Scr. **20**, 587 (1979)
6. R.C. Hilborn, Am. J. Phys. **50**, 982 (1982)
7. B. Henderson, G.F. Imbusch, *Optical Spectroscopy of Inorganic Solids* (Clarendon, Oxford, 1989)
8. M. de Jong, L. Seijo, A. Meijerink, F. Rabouw, Phys. Chem. Chem. Phys. **17**, 16959 (2015)
9. E.J. Heller, J. Chem. Phys. **62**, 1544 (1975)
10. E.J. Heller, Acc. Chem. Res. **14**, 368 (1981)
11. J.I. Zink, K.S. Shin, *Advances in Photochemistry*, vol. 16 (Wiley, New York, 1991), pp. 119–214

[5] E.g., in a LnL_8 cubic moiety, the effective mass is $\mu_k = \mu_{a_{1g}} = m_L$ and the breathing mode distortion of the LnL_6 moiety is given by $\Delta Q_k = \Delta Q_{a_{1g}} = \sqrt{8}\, \Delta d_{\text{Ln-L}}$.

Part II
Tutorial: Performing *Ab Initio* Calculations on Complex Manifolds of Excited States of Lanthanides in Solids

This Part is a tutorial that can be followed in order to perform multiconfigurational ab initio calculations on complex manifolds of excited states of lanthanide ions in solid hosts using the suite of programs MOLCAS. The calculations of potential energy curves of several manifolds of Ce and Pr ions doped in the fluorite-like BaF_2 host are described in full detail. It is shown how to calculate, step by step, the $4f^0$ state of $BaF_2:Ce^{4+}$; the $4f^1$, $5d^1$, and $\phi^1_{ITE,a_{1g}}$ manifolds of $BaF_2:Ce^{3+}$ and $BaF_2:Pr^{4+}$; and the $4f^2$, $4f^15d^1$, and $4f^1\phi^1_{ITE,a_{1g}}$ manifolds of $BaF_2:Pr^{3+}$.

Chapter 4
Symmetry Handling

Abstract Symmetry is often exploited in physics and chemistry to derive fundamental principles, reduce computational workloads or obtain physical insights and information without the need for elaborate calculations. Also in the quantum chemical study of luminescent systems and their spectroscopy, group theory, which is the mathematical framework that enables one to exploit symmetries, is ubiquitous and indispensable to keep calculations feasible. While no expert knowledge is required, a selection of basic ingredients from group and representation theory and symmetry aspects from crystal or ligand field theory are reviewed here that are useful in drafting program input and to correctly interpret and process intermediate and final program output. These concepts are applied to generate the input and analyze the output needed to obtain the potential energy curves of cubic Pr^{3+} defects in BaF_2 from multiconfigurational ab initio calculations. The latter section serves as a support for the symmetry handling in the tutorial of Chap. 5.

4.1 Group and Representation Theoretical Ingredients

In this section, a narrow selection of concepts from group and representation theory is reviewed. This is a minimal selection that should enable one to use symmetry handling as a tool during quantum chemical calculations. For a more complete and in-depth survey, we redirect readers to established or more recent standard works, e.g. those of Tinkham or Dresselhaus [1, 2]. Readers who have a working knowledge on group and representation theory of rotations and point groups can perfectly skip this chapter or go directly to Sect. 4.3.1.

Z. Barandiarán et al., *Luminescent Materials*, Springer Series in Materials Science 322,
https://doi.org/10.1007/978-3-030-94984-6_4

4.1.1 Definitions

A group, G, is constructed of several elements, $g_1, \ldots, g_n$ which can be combined with the *multiplication* $\circ$, satisfying following conditions:

- Closure: $g_i \circ g_j \in G$.
- Associativity: $(g_i \circ g_j) \circ g_k = g_i \circ (g_j \circ g_k)$.
- Unity element: $g_i \circ E = E \circ g_i = g_i$.
- Inverse element: $g_i \circ g_i^{-1} = g_i^{-1} \circ g_i = E$.

A group is not necessarily commutative. The latter is examined through the group commutator, $g_1 g_2 g_1^{-1} g_2^{-1}$, which yields E for a commutative group. Commutative groups are also known as Abelian groups, after the tragically short-lived Norwegian mathematician Niels Henrik Abel.

The number of elements, n, can be finite, countable finite or uncountable finite, yielding finite, infinite discrete and continuous groups respectively. In the last case, which is often referred to as Lie groups,[1] a set of algebraically defined and continuously varying parameters $\boldsymbol{a} = a_1, \ldots, a_r$ can be used to represent the group elements.[2] Applied to quantum chemistry, mostly discrete groups, describing the symmetry of molecules and crystals, and some elements of Lie groups, describing atomic states, are required. In the case of finite groups, the number of elements n is called the order of the group, while for Lie groups, the order refers to the number of parameters, r.

A group G can be mapped onto/into another group G'. This is formally written by ϕ for which $g' = \phi(g)$. Some special mappings are important and have hence received a distinct name. If a mapping is bijective, i.e. one-to-one, it is said to be faithful and the inverse mapping can be defined. When ϕ conserves the group multiplication, $\phi(g_i \circ g_j) = \phi(g_i) \circ \phi(g_j)$, it is called a homomorphism. An isomorphic mapping is a homomorphism which is also faithful.

A group G can contain a subgroup H, denoted as $H \subset G$. Here, H is composed of a selection of elements of G which fulfill the group axioms for the same multiplication and inversion operation as G. A subgroup is called normal when each element of G commutes with all the elements of H.

4.1.2 Representations

Assume that a physical system is invariant for a group G of symmetry operations g_i. These can be geometrical symmetries, such as rotations, translations or reflections, but are not restricted to these. These symmetry operations are quantum mechanically encoded as unitary transformations in Hilbert space, mapping a state vector onto

---

[1] Named after the Norwegian, 19th century mathematician Sophus Lie.

[2] All possible rotations along a fixed axis form a Lie group, parametrized by the rotation angle. For all rotations in 3D space, three parameters are required, e.g. the Euler angles.

another state vector. Unitarity of U is required as the norm of state vectors is conserved during the symmetry operations:

$$|\Psi'\rangle = U(g_i)\,|\Psi\rangle. \tag{4.1}$$

If a basis for the Hilbert space is introduced, $|\psi_i\rangle$, each symmetry operation can be represented by a matrix:

$$U(g_i)\,|\psi_n\rangle = \sum_m [\Gamma(g_i)]_{mn}\;|\psi_m\rangle. \tag{4.2}$$

These matrices necessarily fulfill the group axioms:

$$\Gamma(g_i \circ g_j) = \Gamma(g_i)\Gamma(g_j), \tag{4.3}$$

where the group multiplication is simply the matrix multiplication. The dimension of these matrices is equal to the dimension of Hilbert space, while the order of the matrix group equals the order of G. In general, a group of matrices that satisfies (4.2) is called a matrix representation, or in short a representation of the group G. A representation is at least homomorphic with G. When the mapping is isomorphic in addition, the representation is called faithful.

Representations which are connected through a similarity transformation[3] are considered as equivalent. If a representation can be brought into block diagonal form upon a similarity transformation, i.e. the same transformation matrix S is used for all group operations and the same block structure is found for all group operations, the representation is said to be reducible:

$$\Gamma' = S^{-1}\Gamma S = \begin{pmatrix} \Gamma_1 & \mathbf{0} & \cdots \\ \mathbf{0} & \Gamma_2 & \cdots \\ \vdots & \vdots & \ddots \end{pmatrix} = \Gamma_1 \oplus \Gamma_2 \oplus \ldots, \tag{4.4}$$

with $\mathbf{0}$ matrices containing only zeros. The direct sum, $\oplus$, was introduced to simplify notation and to distinct it from a regular matrix addition.

By reducing representations, Hilbert space is divided in invariant subspaces for the group operations $U(g_i)$. If a group operation works on a state vector in an invariant subspace, the result will necessarily be a linear combination of the states in the same invariant subspace. If a representation cannot be brought into block diagonal form, it is called an irreducible representation, abbreviated as irrep.

The direct or Kronecker product of representations is an often encountered construction. When $\psi_n^{(1)}$ and $\psi_k^{(2)}$ are the basis functions of two representations, Γ_1 and Γ_2 of a group G and are considered to be linearly independent, e.g. depending on different variables, the direct product representation is found from:

[3] $\Gamma' = S^{-1}\Gamma S$, with S an invertible matrix.

$$
\begin{aligned}
U(g_i)(|\psi_n^{(1)}\rangle\,|\psi_k^{(2)}\rangle) &= \left(\sum_m [\Gamma_1(g_i)]_{mn}\,|\psi_m^{(1)}\rangle\right)\left(\sum_l [\Gamma_2(g_i)]_{lk}\,|\psi_l^{(2)}\rangle\right) \\
&= \sum_{ml} [\Gamma(g_i)]_{ml,\,nk}\,|\psi_m^{(1)}\rangle\,|\psi_l^{(2)}\rangle
\end{aligned}
\tag{4.5}
$$

where Γ is the direct product of Γ_1 and Γ_2 and is itself a representation of G:

$$
\Gamma_{ml,\,nk} = (\Gamma_1 \otimes \Gamma_2)_{ml,\,nk} = \Gamma_{1,\,mn}\,\Gamma_{2,\,lk}. \tag{4.6}
$$

The product representation is in general reducible, also when it is composed of irreps.

4.1.3 Symmetry Group

Consider a physical observable, given by the operator $\mathcal{A}$, with eigenvalues a_n,

$$
\mathcal{A}\,|\psi_n\rangle = a_n\,|\psi_n\rangle. \tag{4.7}
$$

The eigenstates of $\mathcal{A}$ can be used as basis for the Hilbert space, which is divided in invariant subspaces, the different eigenspaces of $\mathcal{A}$, corresponding to the different eigenvalues. If it is further assumed that the commutators of the observable $\mathcal{A}$ and the operations of G vanish:

$$
[\mathcal{A}, U(g_i)] = \mathcal{A}U(g_i) - U(g_i)\mathcal{A} = 0, \qquad \forall i, \tag{4.8}
$$

it is found that:

$$
\mathcal{A}U(g_i)\,|\psi_n\rangle = a_n U(g_i)\,|\psi_n\rangle. \tag{4.9}
$$

So, $U(g_i)\,|\psi_n\rangle$ and $|\psi_n\rangle$ have the same eigenvalue and they therefore are part of the same invariant subspace. In most physical problems, the operator for which the spectrum is inspected, is the Hamiltonian, $\mathcal{A} = \mathcal{H}$. With a similar argument, it is easy to show that if the commutator of two operators vanishes, they can be diagonalized simultaneously.

Summarized, Hilbert space is divided in several invariant subspaces which form irreducible representations of the symmetry group G. The states within the invariant subspace correspond to the same eigenvalue of the considered operator, i.e. the energy in typical quantum mechanical problems. The irreps of the symmetry groups can in other words be used to label the energy eigenvalues of the Hamiltonian and the eigenfunctions of the Hamiltonian within each eigenspace can be used as basis for the representation spaces. Every basis function or partner is characterized by a so-called row index, denoted γ. The degeneracy of the eigenvalue is then given by the dimension of the associated representation space.

These results are very useful when the effect of symmetry lowering on the energy eigenvalues is studied. A representation which is irreducible for a group G does not necessarily remain irreducible when the symmetry is lowered to $H \subset G$. When the representation has to be reduced, the invariant subspaces of G are divided in new invariant subspaces of H. This is associated with a lifting of degeneracy in the eigenvalues. Group theoretical arguments can consequently be used to predict the degeneracies and number of physical eigenstates, i.e. energy levels in a quantum chemical context. This is hence an important asset to compose input files, ultimately facilitating computations.

4.1.4 Spherical Symmetry

Angular Momentum Algebra

Consider a system which shows spherical symmetry. This means that rotations, R, form the symmetry operations of the group:

$$\boldsymbol{r}' = R(\alpha, \beta, \gamma)\boldsymbol{r}. \tag{4.10}$$

When applied on a function of the spatial coordinates,

$$U(R)(\alpha, \beta, \gamma) f(x, y, z) = f'(x, y, z) = f(x', y', z'). \tag{4.11}$$

Rotations in three-dimensional space can be specified by three angles, e.g. the Euler angles. Consider here the simplified case of an infinitesimal rotation about the z axis,

$$R_z(\delta\phi) = \lim_{\phi \to \delta\phi} \begin{pmatrix} \cos\phi & \sin\phi & 0 \\ -\sin\phi & \cos\phi & 0 \\ 0 & 0 & 1 \end{pmatrix} = \begin{pmatrix} 1 & \delta\phi & 0 \\ -\delta\phi & 1 & 0 \\ 0 & 0 & 1 \end{pmatrix} \tag{4.12}$$

yielding

$$\begin{aligned} U(R_z)(\delta\phi) f(x, y, z) &= f(x + y\delta\phi, y - x\delta\phi, z) \\ &= f(x, y, z) + \delta\phi \left[y\frac{\partial}{\partial x} - x\frac{\partial}{\partial y} \right] f(x, y, z) \\ &= (1 - \mathrm{i}\delta\phi L_z)\, f(x, y, z) \end{aligned} \tag{4.13}$$

where a Taylor expansion was used in the first step and the definition of orbital angular momentum, $\boldsymbol{L} = \boldsymbol{r} \times \boldsymbol{p}$, was used in the second step. Composing a finite and an infinitesimal rotation yields:

$$
\begin{aligned}
U(R_z)(\phi + \delta\phi) f(x, y, z) &= U(R_z)(\delta\phi) U(R_z)(\phi) f(x, y, z) \\
&= (1 - \mathrm{i}\delta\phi L_z)\, U(R_z)(\phi) f(x, y, z)
\end{aligned} \tag{4.14}
$$

upon rearranging and taking the limit $\delta\phi \to 0$, a derivative of the operator is found:

$$
\frac{\mathrm{d}U(R_z)(\phi)}{\mathrm{d}\phi} = \lim_{\delta\phi \to 0} \frac{U(R_z)(\phi + \delta\phi) - U(R_z)(\phi)}{\delta\phi} = -\mathrm{i}L_z U(R_z)(\phi), \tag{4.15}
$$

which can be integrated to yield

$$
U(R_z)(\phi) = \exp(-\mathrm{i}\phi L_z). \tag{4.16}
$$

This shows that the components of the angular momentum vector are the generators for rotations. Their commutation relation can be found from the definition of $\boldsymbol{L}$:

$$
\left[L_i, L_j\right] = \mathrm{i}\epsilon_{ijk} L_k \tag{4.17}
$$

yielding immediately the structure of the associated Lie algebra. For this reason, SO(3) is often referred to as the rotation group.[4] This is not surprising because rotation matrices are indeed orthogonal. From Noether's theorem, it follows that angular momentum is conserved for spherically symmetric systems.

The dimension of the Lie algebra is three, given the three generators L_x, L_y and L_z. None of these generators commute. Conventionally, L_z is chosen as the so-called Cartan generator, whose eigenvalues label the partners (rows) of the irreps. Furthermore, it is easy to show that L^2 is the so-called Casimir operator as it commutes with all L_i and its eigenvalues label the irreps. The Cartan and Casimir operator can be diagonalized simultaneously. Their eigenvalues and eigenvectors are denoted as:

$$
L^2 \left|\ell m_\ell\right\rangle = \ell(\ell + 1)\hbar^2 \left|\ell m_\ell\right\rangle \tag{4.18}
$$

$$
L_z \left|\ell m_\ell\right\rangle = m_\ell \hbar \left|\ell m_\ell\right\rangle. \tag{4.19}
$$

The raising and lowering (also called Weyl operators) are found as $L_\pm = L_x \pm \mathrm{i}L_y$:

$$
L_\pm \left|\ell m_\ell\right\rangle = \sqrt{\ell(\ell + 1) - m_\ell(m_\ell \pm 1)}\hbar \left|\ell m_\ell \pm 1\right\rangle. \tag{4.20}
$$

This angular momentum algebra was derived for orbital angular momentum. For spin angular momentum the same commutation relations are found, as well as for composite angular momenta. Now, the irreps for the rotation group can be written down. These are labeled by the eigenvalue of the Casimir operator, ℓ:

$$
U(R) \left|\ell m\right\rangle = \sum_{m'} \mathcal{D}^{(\ell)}_{m'm}(R) \left|\ell m'\right\rangle \tag{4.21}
$$

[4] SO(3) is the matrix group of all orthogonal (i.e. $O^{-1} = O^T$) matrices for which $\det(O) = 1$.

and are called the Wigner-$\mathcal{D}$ matrices. When the direct product of two irreps is reduced, the expansion is known as the Clebsch-Gordan series:

$$(\mathcal{D}^{(\ell_1)} \otimes \mathcal{D}^{(\ell_2)})_{m_1m_2,\,m_1'm_2'} = \mathcal{D}^{(\ell_1)}_{m_{\ell 1}m_{\ell 1'}}(R)\,\mathcal{D}^{(\ell_2)}_{m_{\ell 2}m_{\ell 2'}}(R) \tag{4.22}$$

$$= \sum_{\ell=|\ell_1-\ell_2|}^{\ell_1+\ell_2} \sum_{m,m'=-\ell}^{\ell} \langle \ell_1\ell_2 m_1 m_2\,|\ell_1\ell_2\ell m\rangle$$

$$\times \langle \ell_1\ell_2 m_1' m_2'\,|\ell_1\ell_2\ell m'\rangle \mathcal{D}^{(\ell)}_{mm'}(R)$$

where the coefficients $\langle \ell_1\ell_2 m_1 m_2\,|\ell_1\ell_2\ell m\rangle$ are known as the Clebsch-Gordan coefficients.

Homomorphism Between SO(3) and SU(2)

In previous section, rotations were represented by real orthogonal 3×3 matrices, i.e. elements of the matrix group SO(3). Alternatively, one can also use complex-valued unimodular unitary matrices, i.e. elements of the matrix group SU(2).[5]

A general SU(2) matrix can be written as

$$u = \begin{pmatrix} a & b \\ -b^* & a^* \end{pmatrix} \quad \text{with} \quad |a|^2 + |b|^2 = 1 \tag{4.23}$$

and can be expanded in terms of the unit matrix and the three Pauli matrices,

$$\sigma_1 = \begin{pmatrix} 0 & 1 \\ 1 & 0 \end{pmatrix}, \quad \sigma_2 = \begin{pmatrix} 0 & -\mathrm{i} \\ \mathrm{i} & 0 \end{pmatrix}, \quad \sigma_3 = \begin{pmatrix} 1 & 0 \\ 0 & -1 \end{pmatrix}, \tag{4.24}$$

which are the generators of SU(2). The Pauli matrices satisfy the same algebraic structure as the angular momentum operators:

$$\left[\frac{\sigma_i}{2}, \frac{\sigma_j}{2}\right] = \mathrm{i}\epsilon_{ijk}\frac{\sigma_k}{2} \tag{4.25}$$

Furthemrore, it can be shown that a general rotation in three dimensional space about an angle ϕ around an arbitrary axis $\boldsymbol{n} = (n_1, n_2, n_3)$ (with $|n| = 1$) is equivalent to a rotation in two-dimensional complex space, given by the SU(2) matrix:

$$u(R(\boldsymbol{n}, \phi)) = I_2 \cos\frac{\phi}{2} + \mathrm{i}\,(n_1\sigma_1 + n_2\sigma_2 + n_3\sigma_3)\sin\frac{\phi}{2} \tag{4.26}$$

The correspondence is however not one-to-one. If the angle is altered according to $\phi \to \phi + 2\pi$, the same result is obtained for the 3×3 orthogonal matrix, while $u \to -u$ is found for the 2×2 unitary matrix. Only after a rotation over 4π, the same

---

[5] SU(2) is the matrix group of all unitary (i.e. $U^{-1} = U^{\dagger}$) matrices for which $\det(U) = 1$.

result is found. For this reason, the map between SO(3) and SU(2) is not isomorphic, it is a 2:1 homomorphism. The double group of SO(3) can be used to obtain an isomorphism nevertheless (see Sect. 4.1.6).

4.1.5 *Discrete Symmetry*

A point group is a collection of symmetry operations that keep one point fixed. They describe the symmetry of molecules or point defects in crystals. Here, discrete point groups are discussed where the number of symmetry elements, h, called the order, is a finite number. The character table of a point group is an indispensable tool to study the effect of symmetry lowering on atomic states when they are in a molecule or form a defect in a crystal. Below, the necessary ingredients to use and interpret character tables is discussed.

Character Tables

Two group elements, g_1 and g_2 are said to be conjugate if they are connected by a third group element, g_3 such that:

$$g_2 = g_3^{-1} g_1 g_3. \tag{4.27}$$

Conjugate elements are hence equivalent. Conjugacy classes are formed by elements which are all mutually conjugate.

Consider now a matrix representation for the point group. As all elements of a class are equivalent up to a similarity transformation, the representation matrices of elements of the same class will automatically have the same trace which is an invariant for similarity transformations:

$$\chi^{(\Gamma)}(R) = \mathrm{Tr}\,[\Gamma(R)] = \sum_{i=1}^{l_\Gamma} \Gamma(R)_{ii}, \tag{4.28}$$

with l_Γ the dimension of the representation Γ. Because of its distinguished role in point group representation theory, the trace gets a special name in this context, i.e. the character.

A discrete point group has a finite number of irreps. From representation theory it can be shown that the number of irreps is exactly equal to the number of conjugacy classes, r. The classes, irreducible representations and their characters are the work horses when point groups are applied. For this reason, so-called character tables for the different point groups can be found anywhere.

As an example, Table 4.1 shows the character table of the O_h point group, i.e. the point group that leaves a cube or an octahedron invariant. This group has 48 symmetry elements ($h = 48$), grouped in 10 conjugacy classes ($r = 10$). These represent the

Table 4.1 Character table of the point group O_h

O_h	E	$8C_3$	$6C_2'$	$6C_4$	$3C_2$	i	$8S_6$	$6\sigma_d$	$6S_4$	$3\sigma_h$
A_{1g}	1	1	1	1	1	1	1	1	1	1
A_{2g}	1	1	−1	−1	1	1	1	−1	−1	1
E_g	2	−1	0	0	2	2	−1	0	0	2
T_{1g}	3	0	−1	1	−1	3	0	−1	1	−1
T_{2g}	3	0	1	−1	−1	3	0	1	−1	−1
A_{1u}	1	1	1	1	1	−1	−1	−1	−1	−1
A_{2u}	1	1	−1	−1	1	−1	−1	1	1	−1
E_u	2	−1	0	0	2	−2	1	0	0	−2
T_{1u}	3	0	−1	1	−1	−3	0	1	−1	1
T_{2u}	3	0	1	−1	−1	−3	0	−1	1	1

unit operation (E), threefold rotations along the body diagonal of the cube/octahedron ($8C_3$), twofold rotations along the middle of every pair of opposite sides ($6C_2'$), fourfold/twofold rotations along the middle of the sides ($6C_4/3C_2$), an inversion center (i), sixfold improper rotations along the body diagonal of the cube/octahedron ($8S_6$), dihedral mirror planes ($6\sigma_d$), fourfold improper rotations along the middle of every pair of opposite sides ($6S_4$) and horizontal mirror planes ($3\sigma_h$). Ten different irreps can be identified of which four are one-dimensional[6] (A_{1g}, A_{2g}, A_{1u} and A_{2u}), two are two-dimensional (E_g and E_u) and four are three-dimensional ($T_{1g}, T_{2g}, T_{1u}, T_{2u}$).[7] As the O_h group contains an inversion center (i), all irreps have a well-defined parity.

Energy Level Splitting

A straightforward, but important application of point groups is to inspect the splitting of energy levels upon symmetry lowering. Group theoretically, the energy level is initially labeled by an irrep of the supergroup, Γ (see Sect. 4.1.2). Upon lowering of the symmetry, Γ will in general be a reducible representation of the subgroup. Upon reducing, multiple irreps of the subgroup are found,

$$\Gamma = \bigoplus_{j=1}^{r} a_j \Gamma^{(j)}, \tag{4.29}$$

corresponding to the splitting of the energy level. It can be shown that the integer coefficients can be found from the characters:

[6] The dimension of an irrep is given by the character belonging to the unit operation, $l_\Gamma = \chi^{(\Gamma)}(E)$.

[7] According to the Mulliken notation, one-dimensional irreps are denoted by A or B, two-dimensional irreps by E and three-dimensional irreps by T.

Table 4.2 Characters for $\mathcal{D}^{(\ell)}$ in O_h

	E	$8C_3$	$6C_2'$	$6C_4$	$3C_2$	i	$8S_6$	$6\sigma_d$	$6S_4$	$3\sigma_h$
s: $\mathcal{D}^{(\ell=0)}$	1	1	1	1	1	1	1	1	1	1
p: $\mathcal{D}^{(\ell=1)}$	3	0	−1	1	−1	−3	0	1	−1	1
d: $\mathcal{D}^{(\ell=2)}$	5	−1	1	−1	1	5	−1	1	−1	1
f: $\mathcal{D}^{(\ell=3)}$	7	1	−1	−1	−1	−7	−1	1	1	1

$$a_j = \frac{1}{h}\sum_{i=1}^{h} \chi(g_i)\left[\chi^{(j)}(g_i)\right]^* . \tag{4.30}$$

Often, reductions can be found *on sight*, without the need to use (4.30), using the fact that when a reduction holds, the characters of the reducible representation equal the sum of the corresponding characters of the irreps in the reduction:

$$\chi(g_i) = \sum_{j=1}^{r} a_j \chi^{(j)}(g_i). \tag{4.31}$$

When the supergroup corresponds to the rotation or rotation-inversion group, which is the case for atomic states that are labeled by angular momentum quantum numbers, the representations of SU(2) can be used for which the characters are:

$$\chi^{(j)}(\alpha) = \frac{\sin(2j+1)\frac{\alpha}{2}}{\sin\frac{\alpha}{2}} . \tag{4.32}$$

for a general angular momentum j.

As an example, we apply (4.32) to a single electron in a hydrogen-like orbital, which is labeled by the orbital angular momentum ℓ (s ($\ell = 0$), p ($\ell = 1$), d ($\ell = 2$) or f ($\ell = 3$)), corresponding to the irreps of the rotation group. If the symmetry is lowered from spherical to O_h, these representations are in general reducible in O_h. First, the characters of representation ℓ are calculated for the operations (classes) of O_h using (4.32). These are listed in Table 4.2.

Now, these representations are reduced by means of the character table of O_h and (4.30), leading to following reductions:

$$\mathcal{D}^{(\ell=0)} = A_{1g} \tag{4.33}$$
$$\mathcal{D}^{(\ell=1)} = T_{1u} \tag{4.34}$$
$$\mathcal{D}^{(\ell=2)} = E_g \oplus T_{2g} \tag{4.35}$$
$$\mathcal{D}^{(\ell=3)} = A_{2u} \oplus T_{1u} \oplus T_{2u}. \tag{4.36}$$

This shows that the s and p energy levels do not split for the given symmetry lowering, however the d and f levels show a splitting in two and three levels, respectively. In Sect. 4.2, this is further elaborated for more complex electronic states.

Chronecker Product and Integrals
A new representation is formed by the chronecker (or direct) product of several (irreducible) representations (see (4.6)). The characters of the new representation are easily found by the corresponding multiplication:

$$\chi^{\Gamma_1 \otimes \Gamma_2}(g_i) = \chi^{\Gamma_1}(g_i) \cdot \chi^{\Gamma_1}(g_i). \tag{4.37}$$

This property is of relevance for the coupling of angular momenta, but also to evaluate whether integrals are zero by symmetry. For an integral of the form

$$\langle \Psi_1 | \hat{O} | \Psi_2 \rangle, \tag{4.38}$$

the result will be zero unless the totally symmetric irrep (usually denoted as A_{1g}, A_1 or A, depending on the point group) occurs in the reduction of the direct product $\Gamma(\Psi_1) \otimes \Gamma(\hat{O}) \otimes \Gamma(\Psi_2)$.

This is for instance useful to obtain selection rules for electronic transitions (3.9). In case of electric dipole transitions, the operator is given by the total electronic electric dipole moment, $\hat{\mu}_e$, which, in O_h symmetry, transforms according to T_{1u}.

4.1.6 Double Point Groups

If the angular momentum in (4.32) is an integer number, no problems are encountered. However, when j is half-integer, e.g. in cases where spin angular momenta are used, a complication occurs. Consider a rotation about $\alpha + 2\pi$. The associated character is:

$$\chi^{(j)}(\alpha + 2\pi) = (-1)^{2j} \chi^{(j)}(\alpha). \tag{4.39}$$

For half-integer j, the sign of χ changes after a rotation of 2π. This is a consequence of the 2:1 isomorphism of SU(2) and SO(3) (see Sect. 4.1.4).

To circumvent this problem, a new group element is introduced, R, which signifies a rotation about 2π and which is explicitly different to the unity operation E:

Table 4.3 Character table of the double point group $O_h{}'$

O_h'	E	R	$8C_3$	$8RC_3$	$6C_2'+$ $6RC_2'$	$6C_4$	$6RC_4$	$3C_2+$ $3RC_2$	i	Ri	$8S_6$	$8RS_6$	$6\sigma_d+$ $6R\sigma_d$	$6S_4$	$6RS_4$	$3\sigma_h+$ $3R\sigma_h$
Γ_{6g}	2	−2	1	−1	0	$\sqrt{2}$	$-\sqrt{2}$	0	2	−2	1	−1	0	$\sqrt{2}$	$-\sqrt{2}$	0
Γ_{7g}	2	−2	1	−1	0	$-\sqrt{2}$	$\sqrt{2}$	0	2	−2	1	−1	0	$-\sqrt{2}$	$\sqrt{2}$	0
Γ_{8g}	4	−4	−1	1	0	0	0	0	4	−4	−1	1	0	0	0	0
Γ_{6u}	2	−2	1	−1	0	$\sqrt{2}$	$-\sqrt{2}$	0	−2	2	−1	1	0	$-\sqrt{2}$	$\sqrt{2}$	0
Γ_{7u}	2	−2	1	−1	0	$-\sqrt{2}$	$\sqrt{2}$	0	−2	2	−1	1	0	$\sqrt{2}$	$-\sqrt{2}$	0
Γ_{8u}	4	−4	−1	1	0	0	0	0	−4	4	1	−1	0	0	0	0

$$R \neq E \tag{4.40}$$
$$R^2 = E. \tag{4.41}$$

The introduction of this operation, and the combination of this operation with the other original operation of the symmetry group G gives rise to the double group G'. The order of G' is doubled with respect to the order of G. This does however not mean that the number of classes/irreps doubles as some of the new symmetry operations can be added to original classes. This is governed by the rules of Opechowski. As in the case of single groups, character tables can be constructed for the double groups. Their application is completely analogous. The extension of the character table from O_h to $O_h{}'$ is shown in Table 4.3 as illustration.

4.2 Crystal Field Theory from a Representation Theoretical Perspective

The fundamental ideas of crystal field theory (CFT) were devised in the 1930s by Hans Bethe and John Hasbrouck Van Vleck [3, 4]. This theory allowed them to qualitatively understand the colors and magnetic properties of transition metal salts.

The original theory was characterized by two major lines of reasoning. First, group theoretical arguments were used to predict the splitting of degenerate states due to the symmetry lowering in crystals. These arguments are explained below for different CF coupling schemes. Second, splitting energies could be calculated by the parametrization of a symmetry-constrained crystal field interaction by a point charge model where the atoms of the host crystal, or more specifically the anions in the first coordination shell, are modeled as point charges. Later on, more advanced parameterization schemes have been established, including ligand field theory (LFT) [5]. In what follows, we will not be bothered by the parameterization of the effective CFT Hamiltonian, but restrict ourselves to the symmetry properties of its eigenstates.

From our modern, quantum chemical perspective, CFT can be regarded as a minimal multiconfigurational description of the relevant d or f valence electrons of the luminescent center in the effective field originating from the host crystal and the central field of the core electrons. As discussed in the previous chapters, the minimal active space that corresponds to CFT is by far insufficient to even grasp the static electron correlation. However, the symmetry analysis that underlies CFT and yields information on splittings, degeneracies, and number of levels is useful to construct and analyze the targeted quantum chemical calculations themselves. It is evident that, while we use the same symmetry concepts, the level of theory of multiconfigurational ab initio calculations, and hence their applicability, transcend that of CFT by far.

In CFT, the interaction of the valence electrons of the metal ion with the host crystal is described by an effective Hamiltonian, $\hat{H}_{\mathrm{cf}}$, which can be added to the standard atomic Hamiltonian:

$$\hat{H}_0 = \hat{H}_{\text{centr. field}} + \hat{H}_{\text{Coulomb}} + \hat{H}_{\text{so}}. \tag{4.42}$$

The terms in (4.42) represent the following:

- $\hat{H}_{\text{centr. field}}$, Central field Hamiltonian: Collects all single-electron and spherically symmetric interactions. The solutions are the atomic electron configurations (e.g. d^N, f^N or $f^{N-1}d^1$). By including all inner-shell electrons in the central field, the other Hamiltonians pertain exclusively to the valence electrons.
- $\hat{H}_{\text{Coulomb}}$, Coulomb repulsion between the valence electrons. This interaction is responsible for term splitting of the given configuration according to the quantum numbers L and S.
- $\hat{H}_{\text{so}}$, spin-orbit interaction: A single-electron interaction that couples the orbital angular momentum of an electron to its spin momentum, causing a splitting of every LS term into multiplets LSJ.

Symmetry operations for the metal atom in a crystal are very much restricted with respect to the atom in a free state. The remaining symmetry operations are those of the point group of the lattice site, i.e. $\mathcal{H}_{\text{cf}}$ commutes with these symmetry operations. $\hat{H}_0$ obviously commutes as well with these operations as the point group is a subgroup of the rotation group. On the other hand, $\hat{H}_{\text{cf}}$ does not commute with all operations of SO(3). Therefore, angular momentum labels cease to be good quantum numbers and the irreducible representations (irreps) of the point group have to be used to label the eigenstates of $\hat{H}_0 + \hat{H}_{\text{cf}}$. If the different Hamiltonians are treated consecutively in first order perturbation theory, solutions are found where the angular momentum labels can still have a meaning.

If Russell-Saunders (RS) coupling[8] is assumed, three possible CF coupling schemes are thinkable:

- weak crystal field: $\langle\hat{H}_{\text{cf}}\rangle \ll \langle\hat{H}_{\text{so}}\rangle \ll \langle\hat{H}_{\text{RS}}\rangle$,
- intermediate crystal field: $\langle\hat{H}_{\text{so}}\rangle \ll \langle\hat{H}_{\text{cf}}\rangle \ll \langle\hat{H}_{\text{RS}}\rangle$,
- strong crystal field: $\langle\hat{H}_{\text{so}}\rangle \ll \langle\hat{H}_{\text{RS}}\rangle \ll \langle\hat{H}_{\text{cf}}\rangle$.

These are discussed in the following. It should be stressed that in the targeted ab initio calculations no perturbation sequence needs to be chosen. In fact, the distinction between atomic and crystal field interactions in CFT is an artificial one and in ab initiocalculations, all electronic interactions in the embedded cluster are treated on the same footing. The value of discussing the various CFT perturbation schemes is that they give rise to different, but equivalent basis states for the configurational space spanned by the given electron configuration, i.e. all are relatable by unitary transformations. These basis states and their associated symmetry reductions facilitate the input/output handling and analysis of ab initio calculations.

---

[8] Atomic perturbation sequence where $\hat{H}_{\text{Coulomb}}$ is treated first, followed by $\hat{H}_{\text{so}}$. This corresponds to physical reality for most elements where the effect of spin-orbit coupling (multiplet splitting) is small compared to the effect of electron repulsion (term splitting). The opposite case is referred to as jj coupling. In general, both Hamiltonians must be diagonalized simultaneously according to the atomic intermediate coupling scheme.

4.2.1 Weak Crystal Field

In the weak field perturbation scheme, the crystal field is regarded as a perturbation on the free ion multiplets originating from the considered electron configuration. This coupling scheme is well suited for $4f^N$ configurations.

For a given atomic $^{2S+1}L_J$ multiplet, the crystal field levels are found by reducing the total angular momentum J,

$$\mathcal{D}^{(J)} = \bigoplus_{i=1}^{r} a_i \Gamma^{(i)}, \tag{4.43}$$

with r the number of irreps $\Gamma^{(i)}$ of the relevant point group. As J is composed of spin and orbital contributions, the reduction is performed for the double group (see Sect. 4.1.6). The labels of the crystal field levels are $|\ell^N \ \alpha LSJ \ a\Gamma\rangle$ where a is a branching multiplicity. If some degeneracy remains, the different partners of the irreps can be labeled by an additional row index γ. These symmetry-adapted linear combinations (SALCs) can be obtained from,

$$|\ell^N \ \alpha LS \ Ja\Gamma\gamma\rangle = \sum_{M=-J}^{J} \langle JM| \ Ja\Gamma\gamma\rangle \, |\ell^N \ \alpha LSJM\rangle, \tag{4.44}$$

upon calculation of the reduction coefficients.

4.2.2 Intermediate Crystal Field

In the intermediate CF perturbation scheme, term splitting is dominant. Therefore, atomic ^{2S+1}L terms are used as starting point. The crystal field is spin-independent and acts only on the orbital angular momentum:

$$\mathcal{D}^{(L)} = \bigoplus_{i=1}^{r} a_i \Gamma_L^{(i)}. \tag{4.45}$$

While L ceases to be a good quantum number at this point, S is unaffected and the resulting levels are $(2S+1)d_\Gamma$ degenerate, with d_Γ the dimension of irrep Γ, and labeled by $^{2S+1}\Gamma$.

Subsequently, spin-orbit coupling can be considered as an additional perturbation on the $^{2S+1}\Gamma$ levels. The final spin-orbit states are obtained in two steps. First, in the double group, $\mathcal{D}^{(S)}$ shows following reduction:

$$\mathcal{D}^{(S)} = \bigoplus_{j=1}^{s} a_j \Gamma_S^{(j)}. \tag{4.46}$$

Second, the splitting of the ${}^{2S+1}\Gamma$ levels is obtained according to:

$$\begin{aligned} {}^{2S+1}\Gamma_L^{(i)} \quad \rightarrow \quad \Gamma_L^{(i)} \otimes \mathcal{D}^{(S)} &= \bigoplus_{j=1}^{s} a_j \left(\Gamma_L^{(i)} \otimes \Gamma_S^{(j)} \right) \\ &= \bigoplus_{j=1}^{s} a_j \bigoplus_{k=1}^{t} a_k \Gamma^{(k)} \end{aligned} \tag{4.47}$$

The associated basis functions can be obtained from the Clebsch-Gordan coefficients of the (double) point group and the reduction coefficients:

$$\begin{aligned} |\ell^N \, \alpha(La_L\Gamma_L, \; Sa_S\Gamma_S) \, a\Gamma\gamma\rangle = \sum_{M_L M_S \gamma_L \gamma_S} & \langle LM_L | \, La_L\Gamma_L\gamma_L\rangle\langle SM_S | \, Sa_S\Gamma_S\gamma_S\rangle \\ & \times \langle \Gamma_L\gamma_L\Gamma_S a_S | \, a\Gamma\gamma\rangle \, |\ell^N \, \alpha LM_L SM_S\rangle. \end{aligned} \tag{4.48}$$

The intermediate crystal field perturbation scheme is common in the literature on the electronic structure of transition metal d^N shells.

4.2.3 *Strong Crystal Field*

In this CF perturbation scheme, the crystal field is the dominant interaction. If we assume that the crystal field is described by a single-electron Hamiltonian, it will accordingly act on the individual orbital angular momenta:

$$\mathcal{D}^{(\ell)} = \bigoplus_{i=1}^{r} a_i \gamma^{(i)}. \tag{4.49}$$

Here, the notation that is commonly used in the atomic case, where single-electron labels are small characters and multi-electron labels are capital characters, is adopted. The label for the irrep, $\gamma^{(i)}$ should not be confused with the symbol of partners used before.

If there are N electrons in the considered active space, these can be used to fill the crystal field orbitals according to an *aufbau* principle, $(a_1\gamma_1)^{n_1}(a_2\gamma_2)^{n_2} \ldots (a_r\gamma_r)^{n_r}$ with $n_1 + n_2 + \cdots + n_r = N$. Inter-electronic Coulomb repulsion subsequently couples the orbitals to terms:

$$\bigotimes_{i=1}^{r}(\gamma^{(i)})^{n_i} = \bigoplus_{j=1}^{s} a_j \Gamma_L^{(j)}. \tag{4.50}$$

Spin labels are added while taking care of the required antisymmetry of the multi-electron wave function according to Pauli's exclusion principle. Finally, spin-orbit coupling can be accounted in a similar way as in (4.47). The strong field states are found using recoupling and Clebsch-Gordan coefficients:

$$|\ell^N \alpha \left[\left((a_1\gamma_1)^{n_1}(a_2\gamma_2)^{n_2}\ldots(a_r\gamma_r)^{n_r}\right) a_L\Gamma_L,\ a_S\Gamma_S\right] a\Gamma\gamma\rangle. \tag{4.51}$$

Crystal-field-split orbitals are often used for qualitative interpretation of transition metal compounds, even though term splitting can seldom be ignored. From the quantum chemical perspective, crystal-field-split orbitals show a qualitative resemblance with the molecular orbitals with a considerable central-metal contribution.

4.3 Example: Pr^{3+} in BaF_2

In Chap. 5, a tutorial on how to perform quantum chemical calculations for Ce^{4+}, Ce^{3+} and Pr^{3+} impurities on a cubic Ba^{2+} site in BaF_2 is given. In this tutorial, symmetry properties of orbital and multi-electron states are often exploited to simplify the calculations. Furthermore, the O_h states, which emerge naturally from the physical point symmetry need to be translated back and forth to D_{2h} symmetry, the highest achievable symmetry in the used MOLCAS software. These considerations are elaborated here, building on the representation theoretical basics that were discussed above.

Results are elaborated here only for Pr^{3+}, which is a representative example for all group-theoretical analyses underlying the calculations in this book. These considerations can be transferred to any other point symmetry. The cases of Ce^{3+} and Ce^{4+} are considerably simpler and not explicitly elaborated in this chapter, given that they contain only one or zero electrons in the active space.

4.3.1 *Connecting O_h with D_{2h} Irreps*

The character table of the O_h point group is given in Table 4.1, it lists ten different irreps, five with even parity (g, gerade) and five with odd parity (u, ungerade). All orbital and multi-electron states will hence be labeled by one of these irreps for spin-free calculations.

MOLCAS is restricted to D_{2h} as highest symmetry. When the symmetry is lowered from O_h to D_{2h}, the O_h irreps will no longer be irreducible:

Table 4.4 Character table of the point group D_{2h}

D_{2h}	E	C_2 (z)	C_2 (y)	C_2 (x)	i	σ (xy)	σ (xz)	σ (yz)
A_g	1	1	1	1	1	1	1	1
B_{1g}	1	1	−1	−1	1	1	−1	−1
B_{2g}	1	−1	1	−1	1	−1	1	−1
B_{3g}	1	−1	−1	1	1	−1	−1	1
A_u	1	1	1	1	−1	−1	−1	−1
B_{1u}	1	1	−1	−1	−1	−1	1	1
B_{2u}	1	−1	1	−1	−1	1	−1	1
B_{3u}	1	−1	−1	1	−1	1	1	−1

$$
\begin{aligned}
a_1 &= a \\
a_2 &= a \\
e &= a \oplus a \\
t_1 &= b_1 \oplus b_2 \oplus b_3 \\
t_2 &= b_1 \oplus b_2 \oplus b_3
\end{aligned}
\tag{4.52}
$$

where the parity label (g or u) was not omitted for brevity. This was obtained using the character table of D_{2h} (see Tab 4.4). This result indicates that states with different O_h irreps can end up in the same symmetry block (e.g. a_{1g} and e_g end up in a_g) whereas different basis functions of the same O_h irrep can end up in different symmetry blocks (e.g. the three basis functions of t_{1g} end up in b_{1g}, b_{2g} and b_{3g}). Degeneracies between these blocks are hence expected due to the real O_h symmetry.

In summary, following symmetry blocks are specified by MOLCAS:

block	D_{2h} irrep	auxiliary name	O_h basis function
1	a_g	ag	a_{1g}, a_{2g}, $e_{g,1}$, $e_{g,2}$
2	b_{3u}	$tu2$	$t_{1u,1}$, $t_{2u,1}$
3	b_{2u}	$tu3$	$t_{1u,2}$, $t_{2u,2}$
4	b_{1g}	$tg4$	$t_{1g,1}$, $t_{2g,1}$
5	b_{1u}	$tu5$	$t_{1u,3}$, $t_{2u,3}$
6	b_{2g}	$tg6$	$t_{1g,2}$, $t_{2g,2}$
7	b_{3g}	$tg7$	$t_{1g,3}$, $t_{2g,3}$
8	a_u	au	a_{1u}, a_{2u}

where the corresponding partners of the three-dimensional irreps of O_h are explicitly given as they occur in different D_{2h} blocks. For convenience during the calculations, and to recall the implicit O_h symmetry, auxiliary labels for the different D_{2h} symmetry blocks have been introduced (see Chap. 5).

4.3.2 Symmetry of Molecular Orbitals

Molecular orbitals (MOs, one-electron states) are generated by linear combination of atomic orbitals (LCAO), which, in turn, are constructed from linear combinations of Gaussian basis functions (1.42). Finding appropriate MOs is performed by MOLCAS under the specified point group restrictions, i.e. with D_{2h} as highest achievable symmetry.

It is essential to identify the correct symmetry label for every MO, e.g. to specify supersymmetry, active spaces, or to analyze the roots of multiconfigurational calculations. This can be achieved by inspecting the atomic contributions, specifically those originating from the central metal (lanthanide) ion. As an example, the t_{1u} or t_{2u} character of block-tu2/tu3/tu5 MOs can be distinct by identifying the presence of atomic Ln-p-character in the MO expansion, which is expected for t_{1u} but not for t_{2u} (4.34). Analogously, a_{1g} character in block a1 can be identified by the presence of Ln-s-character in the MO (4.33). MOs of symmetry-type e_g can be easily identified thanks to their degeneracy in the orbital energy or due to the presence of Ln-d-character (4.35). MOs of symmetry-type t_{2g} can be identified in the tg4/tg6/tg7 blocks by their Ln-d-character (4.35). Using this reasoning, O_h symmetry labels can be attached to every MO, irrespective of its D_{2h} block.

4.3.3 Active Space

For these calculations, the active space is composed of the Ln $4f$, $5d$ and $6s$ shells, corresponding to the reductions to O_h in (4.33)–(4.36). Reducing this further to D_{2h} i.e. using (4.52) leads to:

$$4f : \mathcal{D}^{(3)} \overset{O_h}{=} a_{2u} \oplus t_{1u} \oplus t_{2u} \overset{D_{2h}}{=} a_u \oplus 2b_{1u} \oplus 2b_{2u} \oplus 2b_{3u} \tag{4.53}$$

$$5d : \mathcal{D}^{(2)} \overset{O_h}{=} t_{2g} \oplus e_g \overset{D_{2h}}{=} 2a_g \oplus b_{1g} \oplus b_{2g} \oplus b_{3g} \tag{4.54}$$

$$6s : \mathcal{D}^{(0)} \overset{O_h}{=} a_{1g} \overset{D_{2h}}{=} a_g, \tag{4.55}$$

which can be further summed and arranged to respect the order of the irreps used by MOLCAS. The active space is hence fully specified by the numbers:

$$3 \;\; 2 \;\; 2 \;\; 1 \;\; 2 \;\; 1 \;\; 1 \;\; 1, \tag{4.56}$$

which is directly used for the CASSCF (or RASSCF) calculations.

4.3.4 Spin-Free Roots

Input

Spin-free calculations can be parallelized by calculating the various spin/symmetry ($^{2S+1}\Gamma$) blocks separately. For every block, a meaningful number of desired roots (i.e. multi-electron states) needs to be given. Depending on the situation, this number can correspond to a set of states belonging to a given electron configuration or manifold, or be user-defined by certain truncation criteria (see Sect. 2.3).

In the former case, two methods can be discerned to obtain the number of roots for every $^{2S+1}\Gamma$ block. First, following e.g. Griffith for transition metals [6], it is possible to construct the multi-electron symmetry-adapter terms $^{2S+1}\Gamma$ from a given electron configuration (e.g. f^2 for Pr^{3+}) by arranging these electrons into the symmetry-adapted orbitals (e.g. a_{2u}, t_{1u}, t_{2u} for an f shell, cf. Sect. 4.2.3 and (4.36)), taking the restrictions posed by the Pauli exclusion principle into account and coupling the spin and orbital angular moments individually to obtain the allowed S-Γ combinations. This is however rather cumbersome. It is more convenient to start from the atomic LS (^{2S+1}L) Russell-Saunders terms which are in reality always available for the f^N configurations as they have been derived by Racah [7–10] and can e.g. be found in the tables of Nielson and Koster [11]. For every LS term, the orbital momentum L is then reduced to the point group (cf. Sect. 4.2.2). The $^{2S+1}\Gamma$ labels for the other relevant configurations (i.e. $4f^{N-1}5d$ and $4f^{N-1}6s$) can be obtained by direct coupling of the two subshells, utilizing the $4f^{N-1}$ LS terms.

For the Pr^{3+} $4f^2$ configuration, following reductions are easily obtained using (4.30)–(4.32):

$$\begin{aligned}
{}^3H &= {}^3E_g \oplus 2{}^3T_{1g} \oplus {}^3T_{2g} \\
{}^3F &= {}^3A_{2g} \oplus {}^3T_{1g} \oplus {}^3T_{2g} \\
{}^3P &= {}^3T_{1g} \\
{}^1I &= {}^1A_{1g} \oplus {}^1A_{2g} \oplus {}^1E_g \oplus {}^1T_{1g} \oplus 2{}^1T_{2g} \\
{}^1G &= {}^1A_{1g} \oplus {}^1E_g \oplus {}^1T_{1g} \oplus {}^1T_{2g} \\
{}^1D &= {}^1E_g \oplus {}^1T_{2g} \\
{}^1S &= {}^1A_{1g},
\end{aligned} \tag{4.57}$$

where it was used that the parity of an atomic term depends only on the electron configuration from which it stems, i.e. $\hat{\mathcal{P}}\,|4f^2,\ LS\rangle = |4f^2,\ LS\rangle$, irrespective of L and S, with $\hat{\mathcal{P}}$ the parity operator.[9]

For the Pr^{3+} $4f5d$ configuration, the Russell-Saunders terms from both subshells can be straightforwardly coupled without risking a violation of the Pauli exclusion principle, i.e.

[9] $\hat{\mathcal{P}}\boldsymbol{r} = -\boldsymbol{r}$.

$$\begin{aligned} 4f \otimes 5d &= {}^2F \otimes {}^2D \\ &= {}^3H \oplus {}^3G \oplus {}^3F \oplus {}^3D \oplus {}^3P \oplus {}^1H \oplus {}^1G \oplus {}^1F \oplus {}^1D \oplus {}^1P. \end{aligned} \tag{4.58}$$

These are reduced to O_h using (4.30)–(4.32):

$$\begin{aligned} {}^3H &= {}^3E_u \oplus 2{}^3T_{1u} \oplus {}^3T_{2u} \\ {}^3G &= {}^3A_{1u} \oplus {}^3E_u \oplus {}^3T_{1u} \oplus {}^3T_{2u} \\ {}^3F &= {}^3A_{2u} \oplus {}^3T_{1u} \oplus {}^3T_{2u} \\ {}^3D &= {}^3E_u \oplus {}^3T_{2u} \\ {}^3P &= {}^3T_{1u} \\ {}^1H &= {}^1E_u \oplus 2{}^1T_{1u} \oplus {}^1T_{2u} \\ {}^1G &= {}^1A_{1u} \oplus {}^1E_u \oplus {}^1T_{1u} \oplus {}^1T_{2u} \\ {}^1F &= {}^1A_{2u} \oplus {}^1T_{1u} \oplus {}^1T_{2u} \\ {}^1D &= {}^1E_u \oplus {}^1T_{2u} \\ {}^1P &= {}^1T_{1u}, \end{aligned} \tag{4.59}$$

where it was used that the $4f5d$ configuration has odd parity, i.e. $\hat{\mathcal{P}}\,|4f5d,\ LS\rangle = -\,|4f5d,\ LS\rangle$, irrespective of L and S.

For the Pr^{3+} $4f6s$ configuration, it is found similarly,

$$\begin{aligned} 4f \otimes 6s &= {}^2F \otimes {}^2S \\ &= {}^3F \oplus {}^1F. \end{aligned} \tag{4.60}$$

If all roots are collected and further reduced to the D_{2h} irreps, one arrives at following dimensions for every ${}^{2S+1}\Gamma$ block,

$$3\ \ 11\ \ \ 11\ \ 6\ \ \ 11\ \ 6\ \ 6\ \ 9 \quad \text{and} \quad 10\ \ 11\ \ \ 11\ \ 6\ \ \ 11\ \ 6\ \ 6\ \ 9, \tag{4.61}$$

for spin-triplets and spin-singlets, respectively, respecting the MOLCAS order of symmetry blocks. It is convenient that, thanks to the presence of inversion symmetry, the states belonging to the $4f^2$ configuration are calculated separately from states of the $4f5d$ and $4f6s$ configurations. This is obviously valid for any pair of $4f^N$ and $4f^{N-1}5d/4f^{N-1}6s$ configurations, provided that the system is centrosymmetric.

It must be noted that the labels of the spin-free states, ${}^{2S+1}\Gamma$, can be confusing, especially in case of $4f^N$ manifolds for which weak crystal field labels (Sect. 4.2.1) are well established. In CFT, this represents a perturbation sequence where first spin-orbit coupling is evaluated and then the crystal field Hamiltonian. However, the computational strategy of our quantum chemical framework as described in Chap. 1 does not rely on a perturbation sequence. The intermediate-CFT-like labels are meant to allow efficient parallelization and analysis of intermediate results.

Output

The output of CASSCF (or RASSCF) calculations consists of the specified number of roots (energies) and the corresponding multi-electron wavefunctions. The latter are given by multiconfigurational expansions, i.e. linear combinations of Slater determinants (see Chap. 1). When an effective higher O_h symmetry is computed, it is essential to be able to distinct between roots belonging to different O_h irreps in the same D_{2h} symmetry block. This can be done by inspecting the symmetry of the Slater determinants that occur in the different multiconfigurational expansions.

The Slater determinants are built from the molecular orbitals (MOs) that make up the active space. Before this analysis can be performed, it is hence necessary to correlate the MOs with a O_h symmetry label as described in Sect. 4.3.2.

As an example, the 3Tg4 block which occurs when computing the Pr^{3+} $4f^2$ configuration contains roots of ${}^3T_{1g}$ or ${}^3T_{2g}$ symmetry that have to be distinct. The two-electron Slater determinant,

$$|\psi^{\alpha}_{4ft_{1u}}\psi^{\alpha}_{4fa_{2u}}|, \tag{4.62}$$

where α denotes spin-up, belongs to the symmetry class ${}^3T_{2g}$ because $t_{1u} \otimes a_{2u} = t_{2g}$. If the CI coefficient for the determinant, (4.62), is nonzero, the ${}^3T_{2g}$ label can be attached to the corresponding root. In this way, possibly by inspecting multiple Slater determinants, all calculated roots can be analyzed in terms of the O_h-D_{2h} chain of point groups.

4.3.5 Spin-Orbit Coupling

In a final step of the calculation, all roots, irrespective of their ${}^{2S+1}\Gamma$ block are merged into one RASSI-SO calculation. The result is a (typically large) number of spin-orbit-split roots for which the total spin, S, has ceased to be a good quantum number. Double group labels are needed as discussed in Sect. 4.1.6.

Double Group Labels

From the spin-free ${}^{2S+1}\Gamma$ labels, the orbital and spin angular momenta are coupled to obtain the double group irrep labels and their expected number following the procedure of Sect. 4.2.2. As an example for the Pr^{3+} blocks, we start from the spin wave functions,

$$\mathcal{D}^{(S=1)} = T_{1g} \quad \text{and} \quad \mathcal{D}^{(S=0)} = A_{1g}, \tag{4.63}$$

which yields for the roots of the $4f^2$ manifold:

$$
\begin{aligned}
{}^3A_{1g} &= T_{1g} \otimes A_{1g} &&= T_{1g} \\
{}^3A_{2g} &= T_{1g} \otimes A_{2g} &&= T_{2g} \\
{}^3E_{g} &= T_{1g} \otimes E_{g} &&= T_{1g} \oplus T_{2g} \\
{}^3T_{1g} &= T_{1g} \otimes T_{1g} &&= A_{1g} \oplus E_{g} \oplus T_{1g} \oplus T_{2g} \\
{}^3T_{2g} &= T_{1g} \otimes T_{2g} &&= A_{2g} \oplus E_{g} \oplus T_{1g} \oplus T_{2g} \\
{}^1A_{1g} &= A_{1g} \otimes A_{1g} &&= A_{1g} \\
{}^1A_{2g} &= A_{1g} \otimes A_{2g} &&= A_{2g} \\
{}^1E_{g} &= A_{1g} \otimes E_{g} &&= E_{g} \\
{}^1T_{1g} &= A_{1g} \otimes T_{1g} &&= T_{1g} \\
{}^1T_{2g} &= A_{1g} \otimes T_{2g} &&= T_{2g}
\end{aligned}
\tag{4.64}
$$

and for the roots of the $4f5d$ and $4f6s$ manifolds:

$$
\begin{aligned}
{}^3A_{1u} &= T_{1g} \otimes A_{1u} &&= T_{1u} \\
{}^3A_{2u} &= T_{1g} \otimes A_{2u} &&= T_{2u} \\
{}^3E_{u} &= T_{1g} \otimes E_{u} &&= T_{1u} \oplus T_{2u} \\
{}^3T_{1u} &= T_{1g} \otimes T_{1u} &&= A_{1u} \oplus E_{u} \oplus T_{1u} \oplus T_{2u} \\
{}^3T_{2g} &= T_{1g} \otimes T_{2u} &&= A_{2u} \oplus E_{u} \oplus T_{1u} \oplus T_{2u} \\
{}^1A_{1u} &= A_{1g} \otimes A_{1u} &&= A_{1u} \\
{}^1A_{2u} &= A_{1g} \otimes A_{2u} &&= A_{2u} \\
{}^1E_{u} &= A_{1g} \otimes E_{u} &&= E_{u} \\
{}^1T_{1u} &= A_{1g} \otimes T_{1u} &&= T_{1u} \\
{}^1T_{2u} &= A_{1g} \otimes T_{2u} &&= T_{2u} .
\end{aligned}
\tag{4.65}
$$

In combination with the number of calculated spin-free roots (4.61), one obtains

$$7\ A_{1g},\ 3\ A_{2g},\ 9\ E_g,\ 9\ T_{1g},\ 12\ T_{2g},\ 7\ A_{1u},\ 7\ A_{2u},\ 14\ E_u,\ 21\ T_{1u},\ 21\ T_{2u},$$

i.e. a total of 40 + 70 = 130 spin-orbit levels. Alternatively, the correct number of spin-orbit levels and their double group symmetry label can be obtained from the atomic Russell-Saunders ${}^{2S+1}L_J$ multiplets, where the reduction of the quantum number J is performed according to Sect. 4.2.1.

Output

In the output of a RASSI-SO calculation, every spin-orbit state is expanded in the spin-free multi-electron basis. These expansions, next to the degeneracies of the resulting states, are a useful aid to assign double-group symmetry labels to every state.

As an example, the distinction between T_{1g} and T_{2g} spin-orbit states can be made by identifying contributions of ${}^3A_{1g}$ or ${}^1T_{1g}$ (for T_{1g}) or ${}^3A_{2g}$ or ${}^1T_{2g}$ (for T_{2g}) according to the direct products in (4.64). In reality, is often needed to combine information on degeneracies, spin-free expansions and even oscillator strengths to achieve a satisfactory labeling of all spin-orbit states.

References

1. M. Tinkham, *Group Theory and Quantum Mechanics* (McGraw-Hill, New York, 1964)
2. A.J.M.S. Dresselhaus, G. Dresselhaus, *Group Theory- Application to the Physics of Condensed Matter* (Springer, Berlin Heidelberg, 2010)
3. H. Bethe, Ann. Phys. **395**, 133 (1929)
4. J.H. Van Vleck, Phys. Rev. **41**, 208 (1932)
5. J.S. Griffith, L.E. Orgel, Q. Rev, Chem. Soc. **11**(4), 381 (1957)
6. J.S. Griffith, *The Theory of Transition Metal Ions* (University Press, Cambridge, 1961)
7. G. Racah, Phys. Rev. **61**, 186 (1942)
8. G. Racah, Phys. Rev. **62**, 9 (1942)
9. G. Racah, Phys. Rev. **63**, 367 (1943)
10. G. Racah, Phys. Rev. **76**, 1352 (1949)
11. C.W. Nielson, G.F. Koster, *Spectoscopic Coefficients for the p^n, d^n, and f^n Configurations* (MIT Press, Cambridge, 1963)

Chapter 5
Configuration Coordinate Energy Diagrams of Optically Active Sites in BaF_2

Abstract This is a tutorial on how to perform the calculations of the book. A ship's log (*cuaderno de bitácora*) approach is followed: we write all the steps we do like navigation notes, so that somebody else can follow the track. Then, the basic document is the *cuaderno de bitácora*, in brief *bitácora* (though the *bitácora* is only the binnacle, where the compass and the ship's log book are stored.) Ce and Pr activators in a fluorite host (BaF_2) have been chosen because they have 0, 1, and 2 active electrons only. This way, explanations relative to input/output/analyses keep short but meaningful.

5.1 The Project Directory

MOLCAS manual [1] describes standard procedures for regular runs. They are simple and efficient. However, we prefer to follow our own procedures in order to perform the calculations on the excited manifolds of lanthanides in solids. We have developed them in the course of the years of calculations and we feel comfortable with them. More importantly, they allow to minimize the number of errors associated with preparing the calculations. A large number of excited states must be calculated, and they need to be analyzed and labelled according to their irreducible representations (irreps) in symmetry groups of higher symmetry than the one actually used in the calculations. E.g., it is normal to carry out the calculations of an octahedral system using D_{2h} symmetry because of the limitations of MOLCAS, which means that states of different O_h irreps will be together in the same D_{2h} irrep and they need to be identified. Often, it is very hard, if not impossible, to go on to the next step in the calculations without having completed the symmetry analysis in the previous step. Also, we usually need potential energy curves of many states along a geometry variable. And there is a vast amount of information generated during the calculations that has to be very well organized in order to make all the analyses feasible.

In summary, in this tutorial we will use a particular structure of subdirectories in the project directory and we will use script shells for analysis of the results of a given step and for input preparation of the next step. This is by no means the only way of doing these calculations, but it is an efficient way that we have been using,

Z. Barandiarán et al., *Luminescent Materials*, Springer Series in Materials Science 322,
https://doi.org/10.1007/978-3-030-94984-6_5

developing, and debugging for years. We will stick to it in this document. The reader may find it more cumbersome than necessary in the early stages of the calculations, but may also find it very helpful in the final steps and in more complex calculations than the ones chosen as sample cases here. In this tutorial, references are made to directories, script shells, and input/output files that are available on-line from [2].

We assume we start from a project directory. Normally, in this directory we have defined an active cluster, its embedding potential, its symmetry, its basis set, and the restricted or complete active space that will be used. The following subdirectories will be used: `./shells`, `./inputs`, `./libs`, `./printouts`, and `./vectors`. It is convenient to use also `./results`. We will refer to these as the `shells`, `inputs`, `libs`, `printouts`, `vectors`, and `results` directories regardless of the directory we are located at the time of reference. Depending on the user style of work, other subdirectories can be created, although they will not be used by the script shells.

We will start with a template project directory and we will follow the procedures described in the next bitacora sections.

5.2 $4f^1$ and $5d^1$ States of Ce^{3+} and $4f^0$ State of Ce^{4+} in BaF_2

These are the *navigation* notes made during the calculation of the $4f^1$ and $5d^1$ states of Ce^{3+} in BaF_2, assuming it substitutes for Ca^{2+} and there is no short-range charge compensation, so that it has an 8-fold F^- coordination with O_h local symmetry. We assume we start with a template project directory that contains the necessary information to perform the calculations, both script shells and input files.

5.2.1 Calculating One- and Two-Electron Integrals with SEWARD

Prepare SEWARD **input**

- Check files `seward.input.head` and `seward.input.embedding` in the `inputs` directory.
- Symmetry used: D_{2h}; actual symmetry: O_h.
- Relativistic Hamiltonian: second order Douglas-Kroll-Hess.
- The cluster: basis sets.
  - Ce `Ce.ano-rcc.Roos.25s22p15d11f4g2h.9s8p5d4f3g2h.`
    QZP for the lanthanide activator; all-electron basis set
  - F8 `F.ano-rcc.Roos.14s9p4d3f2g.5s4p3d.`
    QZP-3f2g for the ligands; all-electron basis set

- 12Ba2+ `Ba.ECP.Seijo.27s23p.1s1p.0e-AIMP-BaF2-Ba.`
  basis functions for orthogonalization between cluster MOs and 12Ba neighbors
- Int6 `5s` optimized for BaF_2:Yb^{2+} 1^3A_{2u} state
  basis functions at interstices to allow delocalitation towards them

- The cluster: geometry.
  - Ce is at 0, 0, 0.
  - F is at :xB:, :yB:, :zB:; the geometry script shell will give the value $d_{Ce-F}/\sqrt{3}$ to all these three parameters.
  - Ba is at lattice sites.
  - The three symmetry planes of D_{2h} produce the whole CeF_8Ba_{12} cluster.

- The embedding:
  - Ab initio model potentials (AIMP): one-electron operators in the Hamiltonian. The AIMPs incorporate the quantum mechanical effects of (Ba^{2+}; F^-) frozen closed-shell Hartree Fock wave functions obtained in preparatory self-consistent-embedded-ions calculations. Note that orthogonality functions are used in the nearest 12 Ba^{2+} to fulfill strong orthogonality with occupied Ba^{2+} external wave functions. They are taken from *libs/AIMPLIB* library

```
*********************************************************************
* Embedding potential
*********************************************************************
* BaF2 (a=6.2001 A)
* Structure:
*       R.W.G. Wyckoff, Crystal Structures, 2nd. ed.
*       Interscience Publisher, 1967, Vol. 1.
*       (Fluorite; Oh5-Fm3m; a=6.2001 A)
* Embedding potentials:
*       Z.Barandiaran and L.Seijo, J.Chem.Phys.,141,234704(2014)
*       Out of SCEI DKH-HF calculations
*       /Ba.ECP.Seijo.0s.0s.0e-AIMP-BaF2-Ba.
*       /Ba.ECP.Seijo.27s23p.1s1p.0e-AIMP-BaF2-Ba.
*       /F.ECP.Seijo.0s.0s.0e-AIMP-BaF2-F.
*********************************************************************
```

  - External lattice ions occupy experimental crystal structure sites.
  - All ions within a cube of 7 × 7 × 7 unit cells, centered at the impurity site, are represented by total ion embedding AIMPs [3, 4]. They have been obtained in Self-Consistent-Embedded-Ion (SCEI) Hartree-Fock calculations [5].
  - An additional set of 2697 symmetry-independent point charges is used to ensure that the Ewald potential is reproduced within the cluster volume. These charges were obtained following the zero multipole method of Gellé and Le Petit [6].

Run `SEWARD` **at different Ce-F distances**

- Go to the *shells* directory.

- Use the shells to submit jobs and to calculate integrals setting the *$GeomLab* parameter to actual $d_{\mathrm{Ce-F}}$ values in Angstrom. Use Shannon-Prewitt ionic radii mismatch to set a reasonable initial grid for the energy-distance curves.
  - Update file *CLUSTER* with the name given to the embedded-cluster, e.g. CeF8Ba12.
  - Update file *GEOMSHELL* with the name of the script shell used to handle the actual geometry of the cluster. We use *AB2C2D2.D2h.geom.sh* in this case. This script shell is in directory *./geom* and will be invoked by the job file, prior to the calculation of SEWARD in order to prepare its final input file.
  - Edit *j.calculation* and uncomment only the "ksh run.integ.sh" line among all the lines that run programs.
  - Edit *prepare.jobs*, update the variable *$Distance_list* with the list of $d_{\mathrm{Ce-F}}$ values in Angstrom, and invoke it. It prepares a number of job files *$Cluster.$Geomlab.job* that are ready to be submitted to the batch queue system for calculation. It also updates file *do.submit*; it can be updated by commenting out lines corresponding to previous runs. Invoke this file to submit all the jobs. Note that this file is useful to keep a record of the jobs submitted.

Printout of SEWARD

- When the calculations of integrals are done, go to the *printouts* directory.
- Grep the rc= value of all printouts. It should be 0. It is advisable to move all the **.integ.output* files to a convenient subdirectory, like *printouts/INTEG*.
- Check basis set data (cf. MOLCAS ANO-RCC and local AIMPLIB libraries)
- Note symmetry adaptation of the atomic basis sets at the end of the printout.
- Occupied AOs transform as follows:

```
Symmetry adapted orbitals:
     ---- D2h blocks and irreps ---     ---------- Oh irreps ----------
     1   2   3   4   5   6   7   8
     ag  b3u b2u b1g b1u b2g b3g au
Ce
s    1   0   0   0   0   0   0   0     a1g
p    0   1   1   0   1   0   0   0              t1u
d    2   0   0   1   0   1   1   0     eg                     t2g
f    0   2   2   0   2   0   0   1              t1u,t2u                   a2u
8F
s    1   1   1   1   1   1   1   1     a1g      t1u           t2g         a2u
p    3   3   3   3   3   3   3   3     a1g,eg   2t1u,1t2u     2t2g,t1g    a2u,eu

With the following Oh-D2h correspondence:
Oh irreps     D2h  blocks  irreps
a1g,a2g,eg          1       ag
t1u,t2u             2       b3u
                    3       b2u
                    5       b1u
t1g,t2g             4       b1g
                    6       b2g
                    7       b3g
a1u,a2u,eu          8       au
```

Integral files

- The integral files and other important files that are produced by MOLCAS up to this level must be available for any of the following calculations of wave functions and energies. Depending on the case, they can be either kept in disk or recalculated again. This is a lot of storage.

5.2.2 *Obtaining Good Initial Orbitals for Multireference Wave Function Calculations*

- A closed-shell is needed to make sure all initial orbitals show full O_h symmetry in D_{2h} calculations. This is important to avoid symmetry breakings from the beginning.
- Calculating all the electrons at once tends to lead to symmetry breakings. An alternative is suggested next where orbitals for the target $4f^0$-$1A_{1g}$ closed-shell state of BaF_2:Ce^{4+} are obtained to be used as initial orbitals for all RASSCF calculations in the BaF_2:Ce and BaF_2:Pr projects.[1]
- The procedure described next can be skipped given that the closed-shell vectors are available in the `vectors` directory in `*.input.RasOrb` files.

Procedure of increasing occupancy by defining successive closed-shells

- Programs to be used: `SCF` and `RASSCF`. Integrals should have been calculated.
- A bit tedious, but extremely clean. The orbitals obtained will be used extensively: it pays.
- Use the following initial vectors and closed-shell occupations and look at the printouts to note how the unoccupied orbitals rearrange to appear in the expected order as more electrons (interactions) are included from calculation 1 to 5 below. (Number of electrons = 2 × closed-shell orbitals).

```
* Ce at 0,0,0; F at xF,xF,xF.     xF = 2.700 bohr in this case.
*
* calc.  init. vectors    closed-shell used      prog.  iter conv symmetry
*                                                             (1)    (2)
* 0.     core-Ham        5  2  2  1  2  1  1  0  scf        5  no     ok
*                         Ce: 1-3s,2-3p,3d
* 1.     from 0          5  2  2  1  2  1  1  0  rasscf  20   yes    ok
* 2.     from 1          9  3  3  3  3  3  3  1  rasscf  20   yes    ok
*                         Ce: 1-4s,2-3p,3-4d
*                         F:  1s
* 3.     from 2         14  4  4  3  4  3  3  1  rasscf  20   no     ok
```

[1] For heavier lanthanides, using the Yb^{2+} $4f^{14}$ closed-shell state to generate good initial orbitals is a good alternative, even if Yb is not the target. This option implies calculating the integrals and the closed-shell wave function at a given significant embedded-cluster geometry for the Yb^{2+}-doped material. The closed-shell molecular orbitals (vectors) obtained this way are then useful for any other lanthanide as long as the embedded-cluster basis set is formally the same.

```
*                        Ce: 1-5s,2-4p,3-4d
*                        F:  1s,2s2p(some)
* 4.     from 3       14  6  6  4  6  4  4  2  rasscf  20   no   ok
*                        Ce: 1-5s,2-5p,3-4d
*                        F:  1s,2s2p(some)
* 5.     from 4       14  9  9  7  9  7  7  5  rasscf  200  yes  ok
*                        Ce: 1-5s,2-5p,3-4d
*                        F:  1s,2s2p
* this is it: 4f(0)-1A1g ground state of (CeF8Ba12)20+ embedded in BaF2.
```

1. Convergence does not matter along the procedure. It matters only at the end.
2. Check symmetry of MOs in the printouts:
   - O_h degeneracies: e_g (block 1), t_{1u}+t_{2u} (blocks 2,3,5), t_{1g}+t_{2g} (blocks 4,6,7), e_u (block 8).
   - Use the lanthanide basis set "to see" the symmetry of MOs:
     Ce *s* functions $\neq 0$ a_{1g}; =0 other or negligible contribution;
     Ce *p* functions $\neq 0$ t_{1u}; =0 other or negligible contribution;
     Ce *d* functions in block 1 to detect e_g; in blocks 4,6,7 t_{2g} components
     etc.

Produce "supersymmetry" input data

- MOs of different O_h irreps appear in each D_{2h} block and we should prevent they mix because mixing could result in symmetry breakings. This is the purpose of using "supersymmetry" (SUPSYM keyword in program RASSCF).
- MOs (with correct symmetry) and their corresponding supersymmetry data should in general be used as input for wave function calculations in other geometries and/or other states and serve to make sure that the symmetry is kept so that the new output MOs symmetry is ok.
- Supersymmetry classifies the MOs so that only those in the same class (irrep) are allowed to mix along the MO optimizations.
- Use the printout of the last calculation in Sect. 5.2.2 to define supersymmetry data as explained in file *vectors/Supersymmetry_data/Supsym.data*.
- Note that the content of *Supsym.data* should be included in *rasscf.input* file to be used.
- These data will be useful for all wave function calculations in Ce^{3+}/Ce^{4+} and Pr^{3+}/Pr^{4+} (or other Ln with the same type and size of basis set). So, even if it seems to be tedious, it is done only once.
- Note that supersymmetry classification is usually maintained by output MOs, which is good when restart is needed due to non-convergence.

5.2.3 *Performing CASSCF Calculations*

Prepare and submit all the CASSCF calculations using the program RASSCF.

- Goal: to calculate wave functions and energies of relevant states to produce the energy curves in the chosen grid of $d_{\text{Ce-F}}$ distances (totally symmetric mode of the

CeF_8 cube). External lattice is frozen beyond CeF_8. The CASSCF wave functions are multiconfigurational SCF wave functions. So:

- The MOs calculated are variational for an average state energy. They are expanded in a variational linear combination of symmetry adapted orbitals (the symmetry adapted orbitals are constructed from the atomic basis sets using symmetry couplings only by SEWARD).
- The CI expansion associated with the complete active space (see below) is also variational.

Note that in the simple Ce^{3+}/Ce^{4+} cases, the wave functions are single reference (one-configuration wave functions) like Hartree-Fock wave functions, but, since they correspond to open-shells, they might demand the mixing of one or several determinants to obtain the particular single-configuration wave function.

- In the `inputs` directory, check the `$Block.rasscf.input` files.
In Ce^{4+} there is only ${}^1A_g$, which contains the ${}^1A_{1g}$ irrep. In Ce^{3+} there are the following D_{2h} blocks and corresponding O_h irreps [2]:

D_{2h} block	O_h irreps	
$4f^1$ manifold		
${}^2A_u$	${}^2A_{2u}$	
${}^2T_u2$	${}^2T_{1u}$	${}^2T_{2u}$
${}^2T_u3$	${}^2T_{1u}$	${}^2T_{2u}$
${}^2T_u5$	${}^2T_{1u}$	${}^2T_{2u}$
$5d^1$ and $6s^1$ manifolds		
${}^2A_g$	${}^2A_{1g}$	${}^2E_g$
${}^2T_g4$	${}^2T_{2g}$	
${}^2T_g6$	${}^2T_{2g}$	
${}^2T_g7$	${}^2T_{2g}$	

- Defining the multielectronic wave functions in RASSCF:
  - Inactive orbital space (= doubly occupied): Both Ce^{4+} and Ce^{3+} have the same set of inactive orbitals:

```
Inactive 14 9 9 7 9 7 7 5
```

    derived from: Ce 1-5s^2, 2-5p^6, 3-4d^{10}, plus 8F 1-2s^2, 2p^6: total 134 electrons.

  - Complete active orbital space and spin and symmetry of the target multielectronic wave functions:

```
Ln    config. Terms   2S+1  D2h    roots   -------- active ---------
      atomic  Oh            block            e-      orbitals  (RAS2)
```

[2] Note that, for convenience, the D_{2h} block notation has been arbitrarily modified: e.g. B_{3u} is labelled T_u2; the latter recalls the triple degeneracy the D_{2h} blocks 2, 3, 5 (B_{3u}, B_{2u}, B_{1u}) must fulfill for a given spin multiplicity (doublet in this case) when actual site symmetry is O_h. The same applies to blocks 4, 6, 7 (T_g4, T_g6, T_g7).

```
Ce4+  4f0-1S   1A1g          1   1       1     2          1 0 0 0 0 0 0 0
Ce3+  4f1-2F   2A2u          2   8       1     1          3 2 2 1 2 1 1 1
               2T1u,2T2u     2   2,3,5   2     1                "
      5d-2D    2Eg           2   1       3     1                "
               2T2g          2   4,6,7   1     1                "
```

Note that program CASPT2 fails if there are 0 electrons and 0 active orbitals. For this reason the a_{1g} active orbital is moved from the Inactive to the RAS2 active space of Ce^{4+} so that

`Inactive 13 9 9 7 9 7 7 5` and `RAS2 1 0 0 0 0 0 0 0`

in this case. Of course, the total energies obtained this way are the same as those obtained in the last calculation in Sect. 5.2.2 described above.

In the case of Ce^{3+}, the same active space must be used for all states that will be mixed by spin-orbit coupling. This is a requisite of RASSI program used later on for RASSI-SO calculations. Actually, many of the active orbital will not be occupied given the wave function's multiplicity and symmetry (note the warnings issued in the printouts of RASSCF). Active orbitals for $4f^1$-$^2A_{2u}$, for example, could be: `0 0 0 0 0 0 1`; so, `3 2 2 1 2 1 1 1` is a waste, but does not harm!

- Submit the RASSCF calculations.
  - Go to the *shells* directory.
  - Edit *j.calculation* and uncomment only the "ksh run.rasscf.sh $Block" lines of the target blocks.
  - Invoke *prepare.jobs* to prepare all the job files, update *do.submit*, and invoke it to submit all the jobs.

Analyze and plot the results of RASSCF.
It is very important to check that everything is correct before moving to the next step (MS-CASPT2 calculations).

- Go to the *printouts* directory.
- Grep the rc= value in all printouts. It should be 0. Check some printouts and move all of them to some meaningful subdirectory to process them, like *./RASSCF*. The printout files are *$Cluster.$GeomLab.$Block.rasscf.output*.
- Use the RASSCF analyzers to assign the D_{2h} states to O_h irreps and to partially prepare input for EFIT program (which is not a part of MOLCAS, but one of the tools for the present procedures).
  - Go to *./RASSCF*.
  - Check *$Block.key_configurations* files.
  Key configurations are used by the RASSCF analyzer to find out the irrep of states. The analyzer produces assignment files to be used later on by the analyzer of CASPT2 and, eventually, to prepare input for MS-CASPT2 calculations. You can use grep command to simply extract the total energies from printouts when only one root has been calculated (^1A_g, ^2A_u, $^2T_g4$). Copy and adapt assignments file to these cases.

- Update and invoke *do.analyze.rasscf*.
  This script shell calls *shells/pre-post/analyze.rasscf.ksh*, the analyzer.
- Check the assignments made in the *$Block.rasscf.assignments* files. Note that **.summary* and **.efit.dat* files have also been created; the former are useful to detect errors if something goes wrong and the latter will be used in the next step.

- Use EFIT to fit the CASSCF energy curves and to prepare input to plot them. Fitting the energy curves to a polynomial (using only points close to the minimum) allows to obtain bond lengths, totally symmetric vibrational energies, and minimum-to-minimum energy differences. XMGRACE is a good choice for plotting the energy curves.
  - Go to the *results* directory.
  - Update un run *prep.rasscf.efit*. Check file *rasscf.efit.out*.
  - To plot the CASSCF energy curves, open *Ce.RAS.PT2.SO.agr* with XMGRACE and import *rasscf.curves.txt* in the window of the CASSCF results.
  - Edit *rasscf.efit.inp* and run "efit.ksh rasscf" until the fitting of the energy curves has a good quality. The data in *rasscf.efit.out* inmmediatly after the keyword SUMMARY will be later read to make a LaTeX table for a report.

5.2.4 *Performing Multi-State CASPT2 Calculations*

Prepare and submit all MS-CASPT2 calculations using the program CASPT2**.**

- Goal: including dynamic electron correlation by a second order perturbation method where the CASSCF states are the zero order. Only valence electron correlation. Note that Ln $5s5p$ shells must be considered valence. "Frozen" in this context means: no excitations will be allowed from the frozen shells. Core shells should not be correlated. Hence Ce 1-4s,2-4p,3-4d; F 1s should be frozen. Translate that to symmetry MOs and check: `Frozen 9 4 4 3 4 3 3 1`.
- In Ce^{3+} we will do a MS-CASPT2 calculation over the whole block because the mixing of irreps due to broken symmetry is negligible. This will be different in the Pr case, where we will do MS-CASPT2 calculations for sub-blocks containing O_h irreps only.
- In directory *inputs/CASPT2*, check file *ms-caspt2.input*.
- Submit the irrep-dependent MS-CASPT2 calculations for all points in the $d_{\text{Ce-F}}$ grid.
  For that, in the *shells* directory: Edit *j.calculation* and uncomment only the "ksh run.caspt2.sh $Block $ms-roots $Caspt2Dir" lines of the target blocks. Invoke *prepare.jobs* to prepare all the job files, update *do.submit*, and invoke it to submit all the jobs.

Analyze and plot the results of `CASPT2`.

- Go to directory *`printouts/CASPT2`*.
- Grep the rc= value in all printouts. It should be 0. Check some printouts.
- Use the `CASPT2` analyzers to assign the D_{2h} states to O_h irreps and to partially prepare input for `EFIT` program. Again, when only one root has been calculated, you can simply grep to extract the energies.
  - Update and invoke *`do.analyze.ms-caspt2`*.
    This script shell links the *`$Block.rasscf.assignments`* files in the `RASSCF` printouts directory *`../RASSCF`* and calls the analyzer *`shells/pre-post/analyze.ms-caspt2.ksh`*, which uses the files:
    · *`*.$Block.*.caspt2.output`*
    · *`$Block.rasscf.assignments`*

    and produces the files:

    · *`*.summary`*, which are useful to detect errors if something goes wrong
    · *`*.MS.efit.dat`* files for `EFIT`
    · *`$Block.ms-caspt2.assignments`*, which must be checked
    · *`*.$Block.*.msin.energies`* which will be used to prepare the input of subsequent RASSI-SO calculations
    · some other files that will not be used in the present procedure
- Use `EFIT` to fit the MS-CASPT2 energy curves and to prepare input to plot them.
  - Go to the *`results`* directory.
  - Update un run *`prep.MS.efit`*. Check file *`MS.efit.out`*.
  - To plot the MS-CASPT2 energy curves, open *`Ce.RAS.PT2.SO.agr`* with `XMGRACE` and import *`MS.curves.txt`* in the window of the MS-CASPT2 results.
  - Edit *`MS.efit.inp`* and run "efit.ksh MS" until the fitting of the energy curves has a good quality. The data in *`MS.efit.out`* inmmediatly after the keyword SUMMARY will be later read to make a LaTeX table in a report.

Make a basic report with the CASSCF and MS-CASPT2 results

- Make the body of a LaTeX table that combines CASSCF and MS-CASPT2 spectroscopic constants calculated with `EFIT`.
  - In the *`results`* directory, run *`do.SF.table`*. It will combine the spectroscopic constants in *`rasscf.efit.out`* and *`MS.efit.out`* and produce the core of a useful LaTeX table in *`SF.table.tex.dat`*.
- Make the report.
  - In the *`results/latex`* directory, issue make. It will make the report in *`report.pdf`*, which can be visualized.
- Conclude on the dynamic electron correlation effects by comparing MS-CASPT2 results (with) and CASSCF results (without).

5.2.5 *Performing RASSI-Spin-Orbit Calculations*

- In the previous MS-CASPT2 step, the matrix of an effective Hamiltonian in the CAS basis $\{\Phi_i^{\mathrm{RAS}}\}$ was diagonalized per symmetry block (cf. Sect. 1.4.2). The eigenvectors form the $\{\Phi_i^{\mathrm{MSRAS}}\}$ basis; the eigenvalues are the MS-CASPT2 energies $\{E_i^{\mathrm{MSPT2}}\}$. Both are transferred to `RASSI` through the *JobMix* file. `RASSI` will compute and add the spin-orbit couplings $\langle\Phi_i^{\mathrm{MSRAS}}|\hat{H}_{\mathrm{EC,SOC}}|\Phi_j^{\mathrm{MSRAS}}\rangle$ to the input (diagonal) MS-CASPT2 effective Hamiltonian matrix in the $\{\Phi_i^{\mathrm{MSRAS}}\}$ basis and will diagonalize it to get the wave functions and energies which contain SOC effects in addition to electron correlation.
- MSIN option in RASSI-SO calculations with `RASSI`.
Due to residual symmetry breakings, some MS-CASPT2 eigenvalues (MS-CASPT2 energies) that should be degenerate are not, or are only almost-degenerate. The MSINPUT option provides the energies to do a "manual shift" of the input eigenvalues so that the unphysically symmetry-broken eigenvalues are substituted by their average value; this avoids a lot of problems with identifying the SO results later on. This is why the tail...file (see below) to be used with the "msin" option specifies which blocks should be degenerate. Note that in case of non-degeneracy beyond a threshold the input preparator will issue a warning. The MSIN option is not available in standard releases of MOLCAS; slight modifications of the RASSI code are needed which can be found in [2].

Prepare and submit all RASSI-SO calculations using the program `RASSI`**.**

- Spin-orbit coupling will not mix *gerade* and *ungerade* states and two independent SO calculations could be done. However, if one is interested in obtaining the electric dipole transition moments between *gerade* and *ungerade* states, both sets should be included in the RASSI-SO calculations at once. We will do this. This means that all the wave functions we have calculated so far (i.e. for all spin and symmetry blocks) should be invoked.
- Go to directory *inputs/RASSI*.
- To prepare a RASSI-SO calculation for *gerade* and *ungerade* states using the MSIN option (*$RassiStates=g.and.u*, *$RassiType=msin*), prepare the head and tail files

    head.g.and.u.msin.rassi.input
    tail.g.and.u.msin.rassi.input.

  - The head file specifies the number of *JobIph* (and *JobMix*) files stored in the *vectors* directroy that will be used by `RASSI` and the number of roots to be used from each. Note that spin-orbit coupling will mix different spins and different irrep subspecies. You will find in the template file:

```
NROF JOBIPHS= 8    2 2 2  1  3  1 1 1
```

- The tail file specifies the labels that identify the *JobIph* (and *JobMix*) files, as well as degeneracy information. You will find in the template file:

```
RASSI LIST
DEGENERATE 3
2Tu2.ms-2
2Tu3.ms-2
2Tu5.ms-2
2Au.ms-1
2Ag.ms-3
DEGENERATE 3
2Tg4.ms-1
2Tg6.ms-1
2Tg7.ms-1
END OF RASSI LIST
```

  Any order is possible in the RASSI LIST, but once a given input order is chosen, the info in the head and tail files must match!
- Between the geometry-independent head and tail info, the RASSI input preparator used in the next step will insert the MSINPUT information to finish the preparation of an input file for each geometry.

• Edit and run *do.prepare.rassi*.
This will invoke *shells/pre-post/prepare.rassi.ksh*, the RASSI input preparator, which will prepare the geometry-dependendt RASSI input files *$GeomLab.$RassiStates.msin.rassi.input*, by using:

  - the above mentioned geometry-independent head and tail files and
  - the MSINPUT information available in the files **.msin.energies* in the *printouts/CASPT2/* directory.

  The MSIN information is the set of MS-CASPT2 energies $\{E_i^{\mathrm{MSPT2}}\}$ corresponding to the $\{\Phi_i^{\mathrm{MSRAS}}\}$ wave functions in the RASSI list. These energies are averaged as demanded by the DEGENARATE n keyword used in the tail file. Should the non-degeneracy be larger than a threshold, the shell will issue a warning. For this reason, the directory where the MS-CASPT2 energies are should be given as a parameter to the input preparator shell. Check some input files. Note the average values obtained from actual quasi-degenerate energies by comparing the values in the **.msin.energies* or **.caspt2.output* files.
• Submit the RASSI-SO calculations for all points in the $d_{\mathrm{Ce-F}}$ grid.
For that, in the *shells* directory: Edit *j.calculation* and uncomment only the "ksh run.rassi.msin.sh g.and.u $RassiDir" line. Invoke *prepare.jobs* to prepare all the job files, update *do.submit*, and invoke it to submit all the jobs.

Analyze and plot the results of `RASSI`**.**

- Go to directory *`printouts/RASSI`*.
- Grep the rc= value in all printouts. It should be 0. Check some printouts.
- Use the `RASSI` analyzer to analyze the RASSI-SO calculations (i.e. to assign O_h' double group irreps) and to partially prepare input for `EFIT` program.
  - Update and invoke *`do.analyze.rassi`*.
    This script shell links the *`$Block.ms-caspt2.assignments`* files in the `CASPT2` printouts directory *`../CASPT2`* and calls the analyzer *`shells/pre-post/analyze.rassi.ksh`*.
  - The following files will be used by the analyzer:
    - *`*.msin.rassi.output`* files
    - *`../../inputs/RASSI/*.msin.rassi.input`* files
    - *`*.ms-caspt2.assignments`*
    - *`SO.REDUCTION.TABLE`* file, which is self-explanatory
    - *`*.msin.rassi.fixed`* files (see below)
  - The following files are produced by the analyzer:
    - *`*.msin.rassi.summary`* files with the results of the analyses of the SO wave functions read from `RASSI` printouts
    - *`*.msin.efit.dat`* file for `EFIT`
    - A file *`rassi.table.tex.dat`* with SO wave function analyses made at all geometries, preformatted for later use to produce a LaTeX table.
  - The *`*.summary`* files must be checked very carefully. They contain the spin-orbit vectors (states) specifying their spin-orbit-free components. On the basis of this table and using *`SO.REDUCTION.TABLE`*, the analyzer tries to assign each SO state to a double-group O_h' irrep in this case. Once it has done that it looks for broken degeneracies larger than a threshold (300 cm^{-1}) and issues warnings. This must be checked! After that, it counts and prints out the total number of SO levels obtained (called terms in the analyzer). Again, this must be checked because if it is different from what is expected this could reveal wrong assignments. If this is the case, the assignments of these particular states must be reset in the *`*.fixed`* file and the analysis rerun. This is not the case here, though, but it should be expected in other, more complex, manifolds of states; Pr, for instance.
    - The total number of SO states adapted to O_h' double-group irreps for the Ce^{3+} case we have calculated ($4f^1$,$5d^1$,$6s^1$) are:

```
                    Oh double-group irreps
                  G6u G7u G8u    G6g G7g G8g  (G stands for Gamma)
                   2   2   4      2   2   4  <-- degeneracies
SF Oh irreps
states  irrep
1       2T1u       1   0   1      0   0   0
1       2T2u       0   1   1      0   0   0
1       2A2u       0   1   0      0   0   0
total:             1   2   2                 <-- to be checked in
```

```
                                                        *.summary files

1        2Eg      0   0   0     0   0   1
1        2T2g     0   0   0     0   1   1
1        2A1g     0   0   0     1   0   0
total:                          1   1   2  <-- to be checked in
                                                 *.summary files
```

- Use EFIT to fit the RASSI-SO energy curves and to prepare input to plot them.
  - Go to the *results* directory.
  - Update and run *prep.msin.efit*. Check file *msin.efit.out*.
  - To plot the RASSI-SO energy curves, open *Ce.RAS.PT2.SO.agr* with XMGRACE and import *msin.curves.txt* in the window of the RASSI-SO results.
  - Edit *msin.efit.inp* and run "efit.ksh msin" until the fitting of the energy curves has a good quality. The data in *msin.efit.out* inmmediatly after the keyword SUMMARY will be later read to make a LaTeX table in a report.

Calculate absorption and emission electric dipole oscillator strengths and radiative lifetimes.
In the RASSI calculation for *gerade* and *ungerade* states together, the program calculates and prints out the components of the electric dipole transition moment between all states above a threshold, as well as the Einstein emission coefficients. The correct combination of these ingredients yields the absorption and emission electric dipole oscillator strengths and the radiative lifetimes.

- Go to the *printouts/RASSI/OS* directory.
- Update un run *do.oscillator.strength.rassi*.
  Given a geometry to calculate the oscillator strengths (*$GeomLab*), this script shell links the neccessary
  ../.$GeomLab.*.msin.rassi.output* and
  ../.$GeomLab.*.msin.rassi.summary* files
  and calls *shells/pre-post/oscstr.from.rassi.ksh*.
  The script shell *oscstr.from.rassi.ksh* calculates the total absorption or emission oscillator strengths associated with the electronic transition from **one** state of the initial SO level to **all** states of the final SO level. The parameter *$initL* specifies the initial SO level; its sign sets a flag for:

initL	> 0	absorptions from *$initL* to all possible higher-lying levels
	< 0	emissions from *$initL* to all possible lower-lying levels; the emission lifetime of this state will also be calculated using the Einstein coefficients;
	$= 0$	total oscillator strengths are not calculated.

  It produces the files:

  - **.msin.rassi.OscStr.$initL*, with absolute and relative oscillator strengths (and radiative lifetime for emission), preceded by the table of oscillator strength values between individual SO states read from the RASSI printout.

  – *ABSORPTION.OS* or *EMISSION.OS* files, which will be later read to make a LaTeX table in a report.

Make a basic report with the RASSI-SO results.

- Make the body of a LaTeX table that combines RASSI-SO spectroscopic constants calculated with `EFIT` with the analyses of the SO wave functions (made at a given geometry) and, optionally, with the absorption and/or oscillator strengths (made at their own given geometries).
  – In the *results* directory, run *do.SO.table*. It will use the files:
    · *msin.efit.out*, with RASSI-SO spectroscopic constants
    · *printouts/RASSI/rassi.table.tex.dat*, with the SO wave function analyses
    · *printouts/RASSI/OS/ABSORPTION.OS* and *printouts/RASSI/OS/EMISSION.OS*, with oscillator strengths

  It will produce:
    · *SO.table.tex.dat*, with the core of a LaTeX table
    · *SO.banda.dat*, with the basic data to plot absorption and emission spectra profiles with the program `BANDA-LS`.
- Make the report.
  – In the *results/latex* directory, issue make. It will make the report in *report.pdf*, which can be visualized.

Plot absorption and emission spectra profiles.

- Go to *results/spectrum-profile*.
- Update *abs*.inp* and/or *emi*.inp* files by inserting *results/SO.banda.dat* and changing the parameters of the spectrum simulations.
- Update and run *do.spectrum-profile*.
- Run `XMGRACE`, read *spectrum-profile.agr*, and import the appropriate **.dat* files.

5.3 $4f^2$ and $4f^1 5d^1$ States of Pr^{3+} and $4f^1$ and $5d^1$ States of Pr^{4+} in BaF_2

These are the *navigation* notes made during the calculation of the $4f^2$ and $4f^1 5d^1$ states of Pr^{3+} in BaF_2, and of $4f^1$ states of Pr^{4+} in BaF_2, assuming it substitutes for Ca^{2+} and there is no short-range charge compensation, so that it has an 8-fold F^- coordination with O_h local structure. We assume we start with a template project directory that contains the necessary information to perform the calculations, both script shells and input files.

5.3.1 *Calculating One- and Two-Electron Integrals with* SEWARD

Prepare `SEWARD` **input**

- Check files *`seward.input.head`* and *`seward.input.embedding`* in the *`inputs`* directroy.
- Symmetry used: D_{2h}; actual symmetry: O_h.
- Relativistic Hamiltonian: second order Douglas-Kroll-Hess.
- The cluster: basis sets. [3]
  - Pr `Pr.ano-rcc.Roos.25s22p15d11f4g2h.9s8p5d4f3g2h.` QZP
    the lanthanide activator; all-electron basis set
  - F8 `F.ano-rcc.Roos.14s9p4d3f2g.5s4p3d.` QZP-3f2g
    the ligands; all-electron basis set
  - 12Ba2+ `Ba.ECP.Seijo.27s23p.1s1p.0e-AIMP-BaF2-Ba.`
    basis functions for orthogonalization between cluster MOs and 12Ba neighbors
  - Int6 `5s` optimized for BaF_2:Yb^{2+} 1^3A_{2u} state
    basis functions at interstices to allow delocalization towards them
- The cluster: geometry.
  - Pr is at 0, 0, 0.
  - F is at :xB:, :yB:, :zB:; the geometry script shell will give the value $\sqrt{3}\times d_{\text{Pr-F}}$ to all these three parameters.
  - Ba is at lattice sites.
  - The three symmetry planes of D_{2h} producc thc whole PrF_8Ba_{12} cluster.
- The embedding:
  - The AIMP embedding is identical to that used for BaF_2:Ce as described in Sect. 5.2.1.

Run `SEWARD` **at different Pr-F distances**

- Go to the *`shells`* directory.
- Use the shells to submit jobs and to calculate integrals setting the *`$GeomLab`* parameter to actual $d_{\text{Pr-F}}$ values in Angstrom. Use Shannon-Prewitt ionic radii mismatch to set a reasonable initial grid for the energy-distance curves.

---

[3] Note that the only difference between the CeF_8Ba_{12} (above, cf. Sect. 5.2.1) and PrF_8Ba_{12} cluster basis sets is the lanthanides basis. However, both Ce and Pr basis sets have the same structure: `25s22p15d11f4g2h.9s8p5d4f3g2h`, i.e. the same type and number of gaussian-type functions (GTF, `25s22p15d11f4g2h`) and of contracted GTF, which are actually specific atomic natural orbitals (ANO, `9s8p5d4f3g2h`). The fact that both clusters share the same basis set structure allows to use the same initial input molecular orbitals for cluster wave function calculations, see below.

– Update file *CLUSTER* with the name given to the embedded-cluster, e.g. PrF8Ba12.
– Update file *GEOMSHELL* with the name of the script shell used to handle the actual geometry of the cluster. We will use *AB2C2D2.D2h.geom.sh* in this case. This script shell is in directory *./geom* and will be invoked by the job file, prior to the calculation of SEWARD in order to prepare its final input file.
– Edit *j.calculation* and uncomment only the "ksh run.integ.sh" line among all the lines that run programs.
– Edit *prepare.jobs*, update the variable *$Distance_list* with the list of d_{Pr-F} values in Angstrom, and invoke it. It prepares a number of job files *$Cluster.$Geomlab.job* that are ready to be submitted to the batch queue system for calculation. It also updates the file *do.submit*; lines corresponding to previous runs should be commenting out manually. . Invoke this file to submit all the jobs. Note that this file is useful to keep a record of the jobs submitted.

Printout of SEWARD

- When the calculations of integrals are done, go to the *printouts* directory.
- Grep the rc = value of all printouts. It should be 0. It is advisable to move all the **.integ.output* files to a convenient subdirectory, like *printouts/INTEG*.
- Check basis set data (cf. MOLCAS ANO-RCC and local AIMPLIB libraries)
- Note symmetry adaptation of the atomic basis sets at the end of the printout.
- Occupied AOs transform as shown above for the CeF_8Ba_{12} embedded-cluster, with obvious substitution of Ce by Pr and the same O_h to D_{2h} correspondence (cf. Sect. 5.2.1).

Integral files

- The integral files and other important files that are produced by MOLCAS up to this level must be available for any of the following calculations of wave functions and energies. Depending on the case, they can be either kept in disk or recalculated again. This is a lot of storage.

5.3.2 *Initial Orbitals*

The basis set size and structure is the same in this calculation on PrF_8Ba_{12} and in the previous calculation on CeF_8Ba_{12} embedded-clusters. Besides, the change experienced by the orbitals from one system to another is not dramatic. So, we can use here the same initial orbitals we used before, which have been obtained for the BaF_2:Ce^{4+} closed-shell cluster.

- Copy the BaF_2:Ce^{4+} closed-shell vectors *$RasBlock.input.RasOrb* to the *vectors* directory.
- Use the "supersymmetry" input data corresponding to those vectors (cf. Sect. 5.2.2).

5.3.3 *Performing CASSCF Calculations*

Prepare and submit all the CASSCF calculations using `RASSCF`

- Goal: to calculate wave functions and energies of relevant states to produce the energy curves in the chosen grid of Pr-F distances d_{Pr-F} (totally symmetric mode of the PrF_8 cube). External lattice is frozen beyond PrF_8. The CASSCF wave functions are multiconfigurational SCF wave functions. So:
  - The MOs calculated are variational for an average state energy. They are expanded in a variational linear combination of symmetry adapted orbitals (the symmetry adapted orbitals are constructed from the atomic basis sets using symmetry couplings only by `SEWARD`).
  - The CI expansion associated with the complete active space (see below) is also variational.
- In the *`inputs`* directory, check *`$Block.rasscf.input`* files.
  Pr^{4+} has the same blocks and irreps as Ce^{3+}. Pr^{3+} blocks and irreps follow:

D_{2h} block	O_h irreps	D_{2h} block	O_h irreps
	$4f^2$ manifold		
^3A_g	$^3A_{1g}$ $^3A_{2g}$ ^3E_g	^1A_g	$^1A_{1g}$ $^1A_{2g}$ ^1E_g
$^3T_g4$	$^3T_{1g}$ $^3T_{2g}$	$^1T_g4$	$^1T_{1g}$ $^1T_{2g}$
$^3T_g6$	$^3T_{1g}$ $^3T_{2g}$	$^1T_g6$	$^1T_{1g}$ $^1T_{2g}$
$^3T_g7$	$^3T_{1g}$ $^3T_{2g}$	$^1T_g7$	$^1T_{1g}$ $^1T_{2g}$
	$4f^15d^1$ and $4f^16s^1$ manifolds		
^3A_u	$^3A_{1u}$ $^3A_{2u}$ ^3E_u	^1A_u	$^1A_{1u}$ $^1A_{2u}$ ^1E_u
$^3T_u2$	$^3T_{1u}$ $^3T_{2u}$	$^1T_u2$	$^1T_{1u}$ $^1T_{2u}$
$^3T_u3$	$^3T_{1u}$ $^3T_{2u}$	$^1T_u3$	$^1T_{1u}$ $^1T_{2u}$
$^3T_u5$	$^3T_{1u}$ $^3T_{2u}$	$^1T_u5$	$^1T_{1u}$ $^1T_{2u}$

- Defining the multielectronic wave functions in `RASSCF`:
  - Inactive orbital space (= doubly occupied): Both Pr^{4+} and Pr^{3+} have the same set of inactive orbitals

```
Inactive 14 9 9 7 9 7 7 5
```

  derived from: Pr 1-5s^2, 2-5p^6, 3-4d^{10}, plus 8F 1-2s^2, 2p^6: total 134 electrons
  - Complete active orbital space and spin and symmetry of the target multielectronic wave functions:

```
              Oh Terms            2S+1  D2h     roots  ----- active -----
                                        blocks         e-   orbitals(RAS2)
Pr4+
         A1u A2u Eu  T1u T2u
        -------------------
        |    2S+1=2        |
4f1-2Fu|     1         1   1 |-> 2    8         1      1  0 2 2 0 2 0 0 1
                                      2,3,5     2      "        "
```

```
5d1-2D is very high in energy (lowest atomic level: 115052 cm-1);
it will not be calculated nor the terms derived from 6s-2S.

Pr3+
         A1g A2g Eg  T1g T2g
        --------------------
4f2     |    2S+1=3         |
    3Hg|         1   2   1 |
    3Fg|    1        1   1 |
    3Pg|             1     |
  Total|0   1    1   4   2 |-> 3     1      3       2  3 2 2 1 2 1 1 1
                                     4,6,7  6       "        "
        --------------------
        |   2S+1=1          |
    1Gg|1        1   1   1 |
    1Dg|         1       1 |
    1Ig|1   1    1   1   2 |
    1S |1                  |
  Total|3   1    3   2   4 |-> 1     1      10      "        "
                                     4,6,7  6       "        "


4f5d,4f6s
         A1u A2u Eu T1u T2u
        --------------------
        |   2S+1= 1,3       |
4f1     |  x 5deg-2Eg       |
   2A2u|         1         |
   2T1u|            1   1  |
   2T2u|            1   1  |
        |  x 5dt2g-2T2g     |
   2A2u|            1      |
   2T1u|     1   1  1   1  |
   2T2u| 1       1  1   1  |
        |  x 6s-2A1g        |
   2A2u|     1             |
   2T1u|            1      |
   2T2u|                1  |
  Total| 1   2   3  6   5  |-> 1,3   8      9       "        "
                                     4,6,7  11      "        "
```

Note that the same active space must be used for all states that will be mixed by spin-orbit coupling. This is a requisite of `RASSI` program used later on for RASSI-SO calculations.

- Submit the `RASSCF` calculations.
  - Go to the *`shells`* directory.
  - Edit *`j.calculation`* and uncomment only the "ksh run.rasscf.sh $Block" lines of the target blocks.
  - Invoke *`prepare.jobs`* to prepare all the job files, update *`do.submit`*, and invoke it to submit all the jobs.

Analyze and plot the results of `RASSCF`

It is very important to check that everything is correct before moving to the next step (MS-CASPT2 calculations).

- Go to the *printouts* directory.
- Grep the rc= value in all printouts. It should be 0. Check some printouts and move all of them to some meaningful subdirectory to process them, like *./RASSCF*. The printout files are *$Cluster.$GeomLab.$Block.rasscf.output*.
- Use the RASSCF analyzers to assign the D_{2h} states to O_h irreps, to partially prepare input for EFIT program (which is not a part of MOLCAS, but one of the tools for the present procedures), and to prepare input files for subsequent MS-CASPT2 calculations on irreps rather than on blocks.
  - Go to *./RASSCF*.
  - Check *$Block.key_configurations* files.
    Key configurations are used by the RASSCF analyzer to find out the irrep of states. The analyzer produces assignment files to be used by the analyzer of CASPT2 and, eventually, to prepare input for MS-CASPT2 calculations.
  - Edit the *inputs/CASPT2/head.ms-caspt2.input* file so that the RASSCF analyzer can prepare irrep and geometry dependent inputs for MS-CASPT2 calculations with the program CASPT2 (see below: **).
  - Update and invoke *do.analyze.rasscf*.
    This script shell calls the analyzer
    shells/pre-post/analyze.rasscf.subblocks.ksh, which uses the files:
    · **.rasscf.output*
    · *$Block.key_configurations*
    · *inputs/CASPT2/head.ms-caspt2.input*

    and produces the files:

    · **.rasscf.assignments*, which must be checked.
    · **.rasscf.summary*, which are useful to detect errors if something goes wrong.
    · **.rasscf.efit.dat*, which will be used in the next step.
    · *inputs/CASPT2/$Irrep.$GeomLab.ms-caspt2.input*, which are input files for subsequent MS-CASPT2 calculations on the selected roots of a block that belong to the same irrep.

- Use EFIT to fit the CASSCF energy curves and to prepare input to plot them. Fitting the energy curves to a polynomial (using only points close to the minimum) allows to obtain bond lengths, totally symmetric vibrational energies, and minimum-to-minimum energy differences. XMGRACE is a good alternative to plotting the energy curves.

- Go to the `results` directory.
- Update un run *prep.rasscf.efit*. Check file *rasscf.efit.out*.
- To plot the CASSCF energy curves, open *Pr.RAS.PT2.SO.agr* with XMGRACE and import *rasscf.curves.txt* in the window of the CASSCF results.
  - Use colors to identify states and label them trying to find out: (1) in the case of states of the f^2 manifold, their parent atomic terms: ^3H, ^3F, ^3P (increasing energy) ^1G, ^1D, ^1I, ^1S (increasing energy); (2) in the other states, their $4f5de_g$, $4f5dt_{2g}$, $4f\phi_{ITE,a_{1g}}$, ... configurational character. For this, use the tables of atomic-to-O_h correspondence above or from Chap. 4. You can also check the printouts to observe the occupations of natural orbitals to make sure, for instance, whether the $5d$ electron occupies the $5de_g$ or $5dt_{2g}$ subshell. Note also that the bond length of states follows these trends:

    $4f5de_g$ < $4f^2$ < $4f5dt_{2g}$
    Pr^{4+} < Pr^{3+} $4f\phi_{\text{ITE}}$ < other states of Pr^{3+}.

    And, of course, you need to check the shape of the curves, which can be either regular (and common to all states of the same dominant configuration) or correspond to avoided crossings, which indicates that states of different dominant configurations and same symmetry avoid crossings and change configurational character from one side to the other of the crossing.
- Edit *rasscf.efit.inp* and run "efit.ksh rasscf" until the fitting of the energy curves has a good quality. The data in *rasscf.efit.out* immediately after the keyword SUMMARY will be later read to make a LaTeX table in a report.

5.3.4 Performing Multi-state CASPT2 Calculations

Prepare and submit all MS-CASPT2 calculations using the program CASPT2

- Goal: including dynamic electron correlation by a second order perturbation method. Once single state (SS) CASPT2 calculations are done using each individual CASSCF root Φ_i^{RAS} of a given symmetry block as a reference for second order perturbation, the multi-state section begins. The matrix of a second order effective hamiltonian is calculated on the CAS reference basis. The diagonal elements of this matrix are the single-state (SS) CASPT2 energies already available $\{E_i^{\text{SSPT2}}\}$; the off diagonal elements correspond to the interactions of two CAS states through second-order perturbation theory. The effective Hamiltonian matrix is then diagonalized to obtain the MS-CASPT2 results: the eigenvalues are the MS-CASPT2 energies $\{E_i^{\text{MSPT2}}\}$ and the eigenvectors, the $\{\Phi_i^{\text{MSRAS}}\}$ wave functions, which are linear combinations of the CAS wave functions $\{\Phi_i^{\text{RAS}}\}$, are the new effective zeroth order wave functions (cf. Sect. 1.4.2).
- Only valence electron correlation. Note that Ln $5s5p$ shells must be considered valence. "Frozen" in this context means: no excitations will be allowed from the

frozen shells. Core shells should not be correlated. Hence Pr 1-4s,2-4p,3-4d; F 1s. Translate that to symmetry MOs and check: `Frozen 9 4 4 3 4 3 3 1`.

- In Pr^{4+}, as in Ce^{3+}, we will do the MS-CASPT2 calculation over D_{2h} blocks rather than over O_h irreps, because the mixing of irreps is negligible.
- (**) In the case of Pr^{3+} we will do the MS-CASPT2 calculation over the states of each O_h irrep separately. So, for example, at RASSCF level we have calculated 9 roots in D_{2h} block 8, which we call ^3A_u block. This block contains one $^3A_{1u}$ root, two $^3A_{2u}$ roots, and $3 \times 2 = 6$ ^3E_u roots. Now, to avoid further symmetry breakings at MS-CASPT2 level, we will calculate separately

3Au.3A1u	1 root
3Au.3A2u	2 roots
3Au.3Eu	6 roots

  This is why the RASSCF analyzer must create a geometry-dependent and Irrep-dependent input for MS-CASPT2 calculations. For this reason, the following files must be created in *inputs/CASPT2* directory:

  caspt2.input
  ms-caspt2.input, and
  head.ms-caspt2.input (**)

  The latter is used by the RASSCF analyzer *shells/pre-post/analyze.rasscf.subblocks.ksh* to append the info necessary to create *$Irrep.$GeomLab.ms-caspt2.input*
  It is necessary to supervise the action is done correctly by the analyzer.
- Submit the irrep-dependent MS-CASPT2 calculations for all points in the $d_{\mathrm{Pr-F}}$ grid.
  For that, in the *shells* directory: Edit *j.calculation* and uncomment only the "ksh run.caspt2-sub.sh $Block $Irrep $ms-roots $Caspt2Dir" lines of the target blocks. Run *prepare.jobs* to prepare all the job files, update *do.submit*, and rut it to submit all the jobs.

Analyze and plot the results of CASPT2

- Go to directory *printouts/CASPT2*.
- Grep the rc= value in all printouts. It should be 0. Check some printouts.
- Use the CASPT2 analyzers to assign the D_{2h} states to O_h irreps and to partially prepare input for EFIT program.
  - Update and invoke *do.analyze.ms-caspt2*.
    This script shell links the *$Block.$Irrep.rasscf.assignments* files in the RASSCF printouts directory *../RASSCF* and calls the analyzer *shells/pre-post/analyze.ms-caspt2.ksh*, which uses the files:
    · **.$Block.$Irrep.*.caspt2.output*
    · *$Block.$Irrep.rasscf.assignments*

    and produces the files:

- `*.summary`, which are useful to detect errors if something goes wrong
- `*.MS.efit.dat` files for EFIT
- `$Block.$Irrep.ms-caspt2.assignments`, which must be checked and will be used to prepare the input of subsequent RASSI-SO calculations
- `*.$Block.$Irrep.*.msin.energies` which will be used to prepare the input of subsequent RASSI-SO calculations
- some other files that will not be used in the present procedure

- Use EFIT to fit the MS-CASPT2 energy curves and to prepare input to plot them.
  - Go to the `results` directory.
  - Update un run `prep.MS.efit`. Check file `MS.efit.out`.
  - To plot the MS-CASPT2 energy curves, open `Pr.RAS.PT2.SO.agr` with XMGRACE and import `MS.curves.txt` in the window of the MS-CASPT2 results.
    - Use colors to identify states as in the CASSCF step. Check `README.agr` file.
  - Edit `MS.efit.inp` and run "efit.ksh MS" until the fitting of the energy curves has a good quality. The data in `MS.efit.out` inmmediatly after the keyword SUMMARY will be later read to make a LaTeX table in a report.
    It is important to check the shape of numerical energy curves: interaction between states of the same symmetry but different configurational character will produce avoided crossings which may result in very anharmonic (even double-minimum) curves. This should guide the fitting and should serve to point out its accuracy when very irregular curves are obtained.

Make a basic report with the CASSCF and MS-CASPT2 results

- Proceed as in Sect. 5.2.4

5.3.5 Performing RASSI-Spin-Orbit Calculations

- In the previous MS-CASPT2 step, the matrix of an effective Hamiltonian in the CAS basis $\{\Phi_i^{\text{RAS}}\}$ was diagonalized per symmetry block (cf. Sect. 1.4.2). The eigenvectors form the $\{\Phi_i^{\text{MSRAS}}\}$ basis; the eigenvalues are the MS-CASPT2 energies $\{E_i^{\text{MSPT2}}\}$. Both are transferred to RASSI through the *JobMix* file. RASSI will compute and add the spin-orbit couplings $\langle\Phi_i^{\text{MSRAS}}|\hat{H}_{\text{EC,SOC}}|\Phi_j^{\text{MSRAS}}\rangle$ to the input (diagonal) MS-CASPT2 effective Hamiltonian matrix in the $\{\Phi_i^{\text{MSRAS}}\}$ basis and will diagonalize it to get the wave functions and energies which contain SOC effects in addition to electron correlation.
- MSIN option in RASSI-SO calculations with RASSI.
  Due to residual symmetry breakings, some MS-CASPT2 eigenvalues (MS-CASPT2 energies) that should be degenerate are not, or are only almost-degenerate. The MSINPUT option provides the energies to do a "manual shift" of the input

eigenvalues so that the close-but-non degenerate eigenvalues are substituted by their mean value; this avoids a lot of problems of identification of SO results later on. This is why the tail...file (see below) to be used with the "msin" option specifies which blocks should be degenerate. Note that in case of non-degeneracy beyond a threshold the input preparator will issue a warning.

Prepare and submit all RASSI-SO calculations using `RASSI`

- Spin-orbit coupling will not mix *gerade* and *ungerade* states and two independent SO calculations could be done. However, if one is interested in obtaining the electric dipole transition moments between *gerade* and *ungerade* states, both sets should be included in the RASSI-SO calculations at once. We will do this. This means that all the wave functions we have calculated so far should now be invoked. Note that spin-orbit coupling will mix different spins and different irrep subspecies. You will find this info in the files below.
- Go to directory *inputs/RASSI*.
- To prepare a RASSI-SO calculation for *gerade* and *ungerade* states using the MSIN option (*$RassiStates=g.and.u*, *$RassiType=msin*), prepare the head and tail files

  head.g.and.u.msin.rassi.input and
  tail.g.and.u.msin.rassi.input

  – The head file specifies the number of *JobIph* (and *JobMix*) files stored in the *vectors* directroy that will be used by `RASSI` and the number of roots to be used from each. Note that spin-orbit coupling will mix different spins and different irrep subspecies. You will find in the template file:

```
NROF JOBIPHS= 35 4 4 4 2 2 2 2 1 2 2 2 4 4 4 6 1 3 6 6 6 5 5 5 6 2 1 6 6 6 5 5 5 6 2 1
```

  – The tail file specifies the labels that identify the *JobIph* (and *JobMix*) files, as well as degeneracy information. All 35 $Block.$Irrep cases we have calculated should appear. You will find in the template file:

```
RASSI LIST
DEGENERATE 3
3Tg4.3T1g.ms-4
3Tg6.3T1g.ms-4
3Tg7.3T1g.ms-4
DEGENERATE 3
3Tg4.3T2g.ms-2
3Tg6.3T2g.ms-2
3Tg7.3T2g.ms-2
3Ag.3Eg.ms-2
```

```
3Ag.3A2g.ms-1
DEGENERATE 3
1Tg4.1T1g.ms-2
1Tg6.1T1g.ms-2
1Tg7.1T1g.ms-2
DEGENERATE 3
1Tg4.1T2g.ms-4
1Tg6.1T2g.ms-4
1Tg7.1T2g.ms-4
1Ag.1Eg.ms-6
1Ag.1A2g.ms-1
1Ag.1A1g.ms-3
DEGENERATE 3
1Tu2.1T1u.ms-6
1Tu3.1T1u.ms-6
1Tu5.1T1u.ms-6
DEGENERATE 3
1Tu2.1T2u.ms-5
1Tu3.1T2u.ms-5
1Tu5.1T2u.ms-5
1Au.1Eu.ms-6
1Au.1A2u.ms-2
1Au.1A1u.ms-1
DEGENERATE 3
3Tu2.3T1u.ms-6
3Tu3.3T1u.ms-6
3Tu5.3T1u.ms-6
DEGENERATE 3
3Tu2.3T2u.ms-5
3Tu3.3T2u.ms-5
3Tu5.3T2u.ms-5
3Au.3Eu.ms-6
3Au.3A2u.ms-2
3Au.3A1u.ms-1
END OF RASSI LIST
```

Any order is possible in the `RASSI LIST`, but once a given input order is chosen, the info in the head and tail files must match!

- Between the geometry-independent head and tail info, the `RASSI` input preparator used next will insert the MSINPUT information to finish the preparation of an input file for each geometry.

- Edit and run *do.prepare.rassi*.
  This will invoke the RASSI input preparator
  shells/pre-post/prepare.rassi.subblocks.ksh, which will use the files:
  - *head.g.and.u.msin.rassi.input*
  - *printouts/CASPT2/*.msin.energies*
  - *tail.g.and.u.msin.rassi.input*

  and prepare the files:
  - *$GeomLab.$RassiStates.msin.rassi.input*, which are geometry dependent input files for RASSI.

  The MSIN information is the set of MS-CASPT2 energies corresponding to the Φ_i^{MSRAS} wave functions in the RASSI list. These energies are averaged as demanded by the DEGENARATE n keyword used in the tail file. Should the non-degeneracy be larger than a threshold, the shell will issue a warning. For this reason, the directory where the MS-CASPT2 energies are should be given as a parameter to the input preparator shell. Check some input files. Note the average values obtained from actual quasi-degenerate energies by confronting the values in the **.msin.energies* or **.caspt2.output* files.
- Submit the RASSI-SO calculations for all points in the $d_{\text{Pr}-\text{F}}$ grid.
  For that, in the *shells* directory: Edit *j.calculation* and uncomment only the "ksh run.rassi.msin.sh g.and.u $RassiDir" line. Invoke *prepare.jobs* to prepare all the job files, update *do.submit*, and invoke it to submit all the jobs.

Analyze the results of RASSI

- Go to directory *printouts/RASSI*.
- Grep the rc= value in all printouts. It should be 0. Check some printouts.
- Use the RASSI analyzer to analyze the RASSI-SO calculations (i.e. to assign O_h' double group irreps) and to partially prepare input for EFIT program.
  - Update and invoke *do.analyze.rassi*.
    This script shell links the *$Block.$Irrep.ms-caspt2.assignments* files in the CASPT2 printouts directory *../CASPT2* and calls the analyzer
    shells/pre-post/analyze.rassi.ksh.
  - The following files will be used by the analyzer:
    - **.msin.rassi.output* files
    - *../../inputs/RASSI/*.msin.rassi.input* files
    - **.ms-caspt2.assignments*
    - *SO.REDUCTION.TABLE* file, which is self explanatory
    - **.msin.rassi.fixed* files (see below)
  - The following files are produced by the analyzer:
    - **.msin.rassi.summary* files with the results of the analyses of the SO wave functions read from RASSI printouts
    - **.msin.efit.dat* file for EFIT

- A file `rassi.table.tex.dat` with SO wave function analyses made at all geometries, preformatted for a later use to produce a LaTeX table.

– The `*.summary` files must be checked very carefully. They contain the spin-orbit vectors (states) specifying their spin-orbit-free components. On the basis of this table and using *SO.REDUCTION.TABLE*, the analyzer tries to assign each SO state to a double-group O_h' irrep in this case. Once it has done that it looks for broken degeneracies larger than a threshold (300 cm^{-1}) and issues warnings. This must be checked! After that, it counts and prints out the total number of SO levels (called terms in the analyzer) obtained. Again, this must be checked because if it is different from what is expected this could reveal wrong assignments. In this case, the assignment of these particular states must be reset in the `*.fixed` file and the analysis rerun. This happens for several geometries in this case:

- The total number of SO terms and states one should get for Pr^{3+} can easily be calculated with `printouts/RASSI/SO_states.xlsx` or similar (see also Chap. 4). Here is a copy of the g.and.u case:

```
4f2 manifold:

Spin-orbit-free Terms          Spin-orbit Terms
Terms   No.of Terms       A1g  A2g  Eg   T1g  T2g
1A1g        3              1    0    0    0    0
1A2g        1              0    1    0    0    0
1Eg         3              0    0    1    0    0
1T1g        2              0    0    0    1    0
1T2g        4              0    0    0    0    1
3A1g        0              0    0    0    1    0
3A2g        1              0    0    0    0    1
3Eg         1              0    0    0    1    1
3T1g        4              1    0    1    1    1
3T2g        2              0    1    1    1    1
                           7    3    9    9   12  (a)
Total      49                                  91 SO states

4f5d and 4f6s manifolds:

Spin-orbit-free Terms          Spin-orbit Terms
Terms   No.of Terms       A1u  A2u  Eu   T1u  T2u
1A1u        1              1    0    0    0    0
1A2u        2              0    1    0    0    0
1Eu         3              0    0    1    0    0
1T1u        6              0    0    0    1    0
1T2u        5              0    0    0    0    1
3A1u        1              0    0    0    1    0
3A2u        2              0    0    0    0    1
3Eu         3              0    0    0    1    1
3T1u        6              1    0    1    1    1
3T2u        5              0    1    1    1    1
                           7    7   14   21   21  (a)
Total      84                                 168 SO states

g.and.u   133              110 SO terms       259 SO states (a)

(a) to be checked in *.summary files
```

· After the first run of `do.analyze.rassi`, we see that the number of expected SO terms for each irrep is correct at the d_{Pr-F} distances 2.100, 2.150, and 2.200 Å, but wrong at longer distances. Checking the `*.summary` file at 2.250 Å, for instance, we see that SO states 73, 74, and 75 have almost pure $^3T_{1g}$ character (`*****` appears in the place of a 1.0000 because of a format issue, but this is not a problem). According to `SO.REDUCTION.TABLE`, four O_h' double-group irreps could have contributions from $^3T_{1g}$:

```
3T1g (9 states)  -->  A1g, Eg, T1g, T2g  (9 states)
```

so that the analyzer cannot solve the assignment. Then, the analyzer writes down "unknown" as a first option and A_{1g} as a second one. It adopts the second option and the number of A_{1g} states is larger than expected by 3, while the number of T_{1g} states is smaller by 1. At shorter distances there is a tiny contribution from $^1T_{1g}$ which allows the analyzer to assign the three states as the components of the T_{1g} term because of its reduction table:

```
1T1g (3 states)  -->  T1g  (3 states)
```

· Having in mind that the three states appear to be degenerate (25127 cm^{-1}), and that they seem to correspond clearly to the same states at shorter Pr-F distances, the correct assignment should be T_{1g}. Then, we set the assignments accordingly in file `*.2.250.*.fixed` and similarly in the `*.fixed` files of other geometries, and run `do.analyze.rassi` again.
· Something similar happens with states 39, 40, and 41 at distances 2.550 and 2.600 Å. At 2.450 Å, some broken symmetry at the crossing point between two T_{1u} and T_{2u} terms prevents the automatic identification of states 181, 182, 183 as T_{2u} and 184, 185, 186 as T_{1u}. At 2.600 Å, some broken symmetry prevents the automatic identification of state 42 as A_{2g}.
· After all these "fixed" assignments, the numbers of SO terms provided by the analyzer are all correct. Further checking of the `*.summary` files does not show signs of further troubles.

Plot and fit the RASSI-SO results

- Use `EFIT` to fit the RASSI-SO energy curves and to prepare input to plot them.
  - Go to the `results` directory.
  - Update un run `prep.msin.efit`. Check file `msin.efit.out`.
  - To plot the RASSI-SO energy curves, open `Pr.RAS.PT2.SO.agr` file with `XMGRACE`, and import `msin.curves.txt` in the window of the RASSI-SO results.

· The plot with all CASSCF/MS-CASPT2/RASSI-SO results shows the effects of dynamic correlation and SO coupling progressively.
· Another interesting plot is that of SO results with all the terms together in one graph and separated according to O_h' double-group irreps in other graphs.
· Use colors in the plots to show the parentage of states with either atomic SO terms or crystal field splitting of the $4f5d$ configuration using the analyses described next.

– Edit `msin.efit.inp` and run "efit.ksh msin" until the fitting of the energy curves has a good quality. The data in `msin.efit.out` inmmediatly after the keyword SUMMARY will be later read to make a LaTeX table in a report.

It is important to check the shape of numerical energy curves: interaction between states of the same symmetry but different configurational character will produce avoided crossings which may result in very anharmonic (even double-minimum) curves. This should guide the fitting and serve to point out when the derived spectroscopic constants are uncertain (as indicated usually by footnotes in the tables).

Calculate absorption and emission electric dipole oscillator strengths and radiative lifetimes

• Proceed as in Sect. 5.2.5

Make a basic report with the RASSI-SO results

• Proceed as in Sect. 5.2.5

Analysis of SO terms

File `printouts/RASSI/rassi.table.tex.dat` contains the analyses of the SO wave functions for each geometry analyzed and the report of the RASSI-SO results contains the analysis at one given geometry, usually the closest to the ground state minimum.

As we show next it is possible further to analyze the final results by combining the spin-orbit analyses we have just made with the previous spin-orbit free (SF) analyses and with auxiliary symmetry couplings to deduce parentage from free ion and/or crystal field terms (whenever possible). This allows to organize and group the numerous calculated states according to relevant characteristics of their electronic structure. All of this facilitates extracting physical insights from the calculated wave functions and energies while at the same time it allows to bridge the obtained theoretical information with available experimental data.

• Combine the following to trace back the free ion parentage of the $4f^2$ manifold:

– The spin-free components of the spin-orbit terms:
they refer to the MS-CASPT2 table of results where the parentage with atomic SF terms has been worked out.
– The splitting of atomic J quantum number in O_h field (see below).

Here is an example:
The analysis of the lowest four SO terms at $d_{Pr-F} = 2.400$ Å:

```
              ----------- 3H -------------
SO term   1.3T1g  2.3T1g  1.3T2g  1.3Eg
1.T2g     67                      26        (in %)
1.A1g     28      69
1.Eg      27      27      42
1.T1g     17      36      31      13
```

allows to associate them with SF atomic ^3H. In addition, atomic ^3H can lead to $J=$ 4, 5, and 6: $^3H_4$, $^3H_5$, $^3H_6$ spin-orbit terms which in an octahedral field transform as follows: (this auxiliary table is in `SO_states.xlsx` file and can be obtained by applying (4.43))

```
J   A1  A2  E   T1  T2
0   1
1               1
2           1       1
3       1       1   1
4   1       1   1   1    <<< which allows to associate the four SO terms
5           1   2   1        mentiond with atomic J=4 of 3H: 3H4
6   1   1   1   1   2
```

Similarly, the next

$2E_g, 2T_{1g}, 2T_{2g}, 3T_{1g}$ are associated with ^3H, J=5: $^3H_5$,
and $2A_{1g}, 3T_{2g}, 3E_g, 4T_{1g}, 4T_{2g}, 1A_{2g}$ are associated with ^3H, J=6: $^3H_6$,
and so on. In some cases, the parentage with free ion terms is not possible due to spin-orbit mixing.

- For the higher *ungerade* states, combine the following:
  - the SF components whose $4f5de_g$, $4f5dt_{2g}$, or $4f6s$ character has been worked out above.
  - the trends of bond lengths:

$$\begin{array}{lclcl} 4f5de_g & < & 4f^2 & < & 4f5dt_{2g} \\ \text{Pr}^{4+} & < & \text{Pr}^{3+}\ f\phi_{ITE,a_{1g}} & < & \text{other states of Pr}^{3+} \end{array}$$

  - the shape of the curves, which can be either regular (and common to all states of the same dominant configuration) or correspond to avoided crossings, which indicate that states of different dominant configurations and same symmetry avoid crossings and change configurational character from one side to the other of the crossing,
  - the matching with the number of SO terms one can expect from each configuration:

```
Spin-orbit states for 4f5deg states only
                  A1u      A2u      Eu       T1u      T2u
Terms    No.
1A1u     0        1        0        0        0        0
1A2u     0        0        1        0        0        0
1Eu      1        0        0        1        0        0
1T1u     2        0        0        0        1        0
1T2u     2        0        0        0        0        1
3A1u     0        0        0        0        1        0
3A2u     0        0        0        0        0        1
3Eu      1        0        0        0        1        1
3T1u     2        1        0        1        1        1
3T2u     2        0        1        1        1        1
                  2        2        5        7        7 <<<---

Spin-orbit states for 4f5dt2g states only
                  A1u      A2u      Eu       T1u      T2u
Terms    No.
1A1u     1        1        0        0        0        0
1A2u     1        0        1        0        0        0
1Eu      2        0        0        1        0        0
1T1u     3        0        0        0        1        0
1T2u     2        0        0        0        0        1
3A1u     1        0        0        0        1        0
3A2u     1        0        0        0        0        1
3Eu      2        0        0        0        1        1
3T1u     3        1        0        1        1        1
3T2u     2        0        1        1        1        1
                  4        3        7        11       10 <<<---

Spin-orbit states for 4f6s states only
                  A1u      A2u      Eu       T1u      T2u
Terms    No.
1A1u     0        1        0        0        0        0
1A2u     1        0        1        0        0        0
1Eu      0        0        0        1        0        0
1T1u     1        0        0        0        1        0
1T2u     1        0        0        0        0        1
3A1u     0        0        0        0        1        0
3A2u     1        0        0        0        0        1
3Eu      0        0        0        0        1        1
3T1u     1        1        0        1        1        1
3T2u     1        0        1        1        1        1
                  1        2        2        3        4 <<<---
```

Plot absorption and emission spectra profiles

- Go to `results/spectrum-profile`.
- Update files `abs*.inp` and/or `emi*.inp` by inserting `results/SO.banda.dat` and changing the parameters of the spectrum simulations.
- Update and run `do.spectrum-profile`.
- Run XMGRACE, read `spectrum-profile.agr`, and import the appropriate `*.dat` files.

References

1. MOLCAS manual. http://www.molcas.org/documentation/manual/. Accessed: 2021-11-22
2. Performing ab initio calculations on complex manifolds of excited states of lanthanides in solids. https://sites.google.com/site/howtodoabinitioln/home. Accessed: 2021-11-22
3. Z. Barandiarán, L. Seijo, J. Chem. Phys. **89**, 5739 (1988)
4. L. Seijo, Z. Barandiarán, in *Computational Chemistry: Reviews of Current Trends*, vol. 4, ed. by J. Leszczyński (World Scientific, Singapore, 1999), pp. 55–152
5. Z. Barandiarán, L. Seijo, J. Chem. Phys. **141**, 234704 (2014)
6. A. Gellé, M.B. Lepetit, J. Chem. Phys. **128**, 244716 (2008)

Part III
Excited State Manifolds of Luminescent Materials

In this Part, results of multiconfigurational *ab initio* calculations on several types of excited state manifolds of inorganic luminescent materials are discussed. It is organized in two Chapters. A first one covers impurity states of atomic-like $4f^N$, $4f^{N-1}5d$, and $4f^{N-1}6s$characters, and of impurity-trapped exciton $4f^{N-1}\phi_{ITE}$character. A second chapter is dedicated to charge transfer states of ligand-to-metal, inter-valence, and metal-to-metal characters; this chapter also introduces the inter-valence and metal-to-metal charge transfer diabatic configuration coordinate energy diagrams.

Chapter 6
Impurity States

Abstract Upon doping an inorganic crystal with lanthanide ions, excited states that alter the local electronic structure of the lanthanide moiety are introduced. The de-excitation of these local excited states underlies the fascinating luminescent properties of these materials. In this Chapter, the numerous states belonging to the (atomic-like) $4f^N$ and $4f^{N-1}5d$ configurations of the dopants are discussed for various lanthanide-activated materials. In addition, states of so-called impurity trapped exciton (ITE) character, where the dopant is partially oxidized, are described. The complete oxidation or reduction of the dopants in charge-transfer states requires the involvement of two optical centers and is reserved for Chap. 7. Special attention is paid to the quantum chemical and computational recipes and choices that enable one to obtain accurate information on these states from first principles.

6.1 $4f^N$ Manifolds

The unique optical properties of crystals doped with lanthanide (Ln) ions are a direct consequence of the $4f$ valence electrons of the latter. The ground state of most but a few exceptions of the Ln^{2+}, $Ln^{3+}$and Ln^{4+} ions corresponds to an incompletely filled $4f^N$ shell ($N = 0 - 14$). If $1 \leq N \leq 13$, inter-$4f^N$ transitions give rise to remarkably sharp, atomic-like spectral peaks that have intrigued scientists from the first half of the 20th century. The energies of the transitions are reasonably unaffected by host (crystal field) effects, explaining the popularity of the so-called Dieke diagram which summarizes the atomic energy levels of the Ln^{3+} ions [1]. Other properties are host-dependent, e.g. (small) Stark splittings of atomic levels, variations in oscillator strengths, bond lengths and local vibrational frequencies. Furthermore, accurately assessing the excited states of the $4f^N$ configuration is essential to understand more complex phenomena that affect the performance of functional materials such as luminescence quenching, charge transfer, etc. This section summarizes the most important findings and systematics of the various $4f^N$ configurations resulting from quantum chemical ab initio calculations.

Z. Barandiarán et al., *Luminescent Materials*, Springer Series in Materials Science 322,
https://doi.org/10.1007/978-3-030-94984-6_6

6.1.1 $4f^1$ Manifold of Ce^{3+}

The $4f^1$ manifold of Ce^{3+} in hosts is made by a very well known set of seven Kramers doublets with additional degeneracies, or not, depending on the symmetry of the site occupied by Ce^{3+}. In octahedral sites, like the ones it occupies in elpasolites (e.g. Cs_2NaYCl_6) [2], the lower ${}^2F_{5/2}$ and higher ${}^2F_{7/2}$ atomic levels split respectively under the effect of a weak external field into Γ_{7u}, Γ_{8u}, and Γ_{7u}, Γ_{8u}, Γ_{6u} (Fig. 6.1), with respective additional degeneracies (over the double degeneracy of the Kramers doublets) 1, 2, 1, 2, and 1. In cubic sites the splitting of the atomic levels is identical but in reverse order. In lower fields, they split further into seven Kramers doublets with no additional degeneracies (see Fig. 8.3 in Sect. 8.2.1).

Multiconfigurational ab initio calculations on the $4f^1$ manifold of Ce^{3+} in octahedral sites of chloroelpasolites have been performed in [2, 3]. In CaF_2, perfect cubic sites with long-range charge compensation have been calculated in [4] and locally compensated distorted sites in [5]. They have also been studied in a number of other hosts, like $LuAlO_3$ [6], YF_3, LaF_3, $KMgF_3$, $LiYF_4$, K_2YF_5, and KY_3F_{10} [4], or $Y_3Al_5O_{12}$ [7] and other garnets [8]. A general agreement with experiments has been found in all cases. A large splitting of the $4f^1$ manifold found in garnets that is partly hidden from experiments because of the low intensity of the emission to the upper $4f$ level is discussed in Sect. 8.1.

Structurally speaking, all the states of the $4f^1$ manifold have basically the same structure, with very minor deviations. E.g. in $Cs_2LiLuCl_6$:Ce^{3+} the Ce-Cl distance

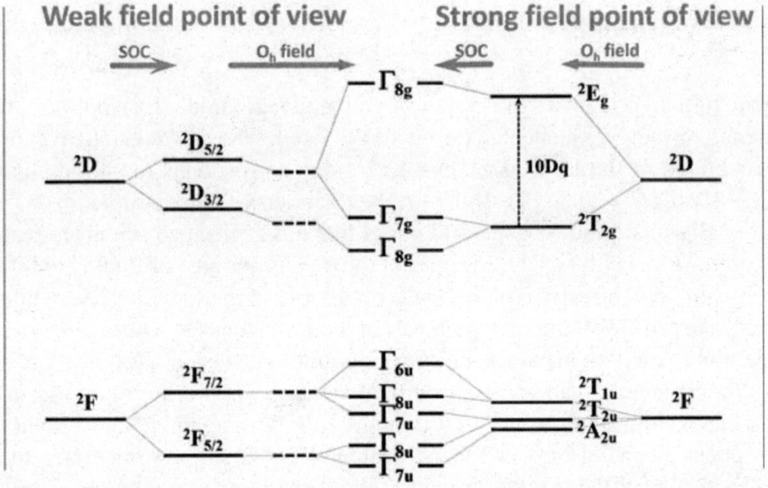

Fig. 6.1 Schematic representation of the $4f^1$ and $5d^1$ energy levels in an octahedral site of a material. Their parentages with the free-ion energy levels are shown from the *weak field* and *strong field* points of view. The dominant characters of intra-atomic spin-orbit coupling interactions in the $4f^1$ manifold and interactions with the host in the $5d^1$ manifold justify the use of the respective notations, e.g. $4f\,{}^2F_{5/2}\,\Gamma_{8u}$ and $5dt_{2g}\,{}^2T_{2g}\,\Gamma_{7g}$

of the octahedral $CeCl_6$ moiety is calculated to range between 2.660 Å and 2.663 Å in the five $4f^1$ levels, which also share a breathing mode vibrational frequency value ranging between 313 cm^{-1} and 314 cm^{-1} [3]. In Cs_2NaYCl_6:Ce^{3+} [2], the theoretical Ce-Cl distance ranges between 2.723 Å and 2.727 Å and the vibrational frequency between 305 cm^{-1} and 307 cm^{-1}. This value agrees well with the 300 ± 2 cm^{-1} experimentally determined value from the analysis of the vibrational progressions in the emission spectra of this material [2].

6.1.2 $4f^2$ Manifold of Pr^{3+}

The calculations done in [9] on the ground and excited states of Pr^{3+} as a free-ion and as a dopant in CaF_2 have been performed with an earlier variant of the relativistic method later established as a standard and described in Sect. 1.3.1. In it, the spin-orbit-free relativistic Hamiltonian used in the first step was the effective-core-potential AIMP Cowan-Griffin-Wood-Boring Hamiltonian (WB-AIMP) [10] rather than the all-electron DKH Hamiltonian. In the second step, the spin-orbit-coupling DKH Hamiltonian was used.

A [Kr]-core relativistic AIMP was combined with a $(14s10p10d8f3g)[6s5p6d4f3g]$ valence basis set for Pr [11]. For F, a [He]-core relativistic AIMP was combined with a $(5s5p1d)[3s4p1d]$ valence basis set [12]. For second neighbor Ca^{2+} ions, additional $(13s7p)[1s1p]$ basis sets [13] were used in order to improve cluster-host orthogonality.

Pr^{3+} Free-Ion

The energy levels of an ion in a solid host are derived from its energy levels as a free-ion, which are fewer and more degenerate. Hence, the calculation of the free-ion state manifolds give us a first hint on the accuracies that can be achieved in the calculations on optically active centers of ion-activated phosphor materials, because the errors are expected to propagate from the free-ion to the ion in the host. We must have in mind that the propagation is not straightforward, though. For instance, the molecular orbitals formed by the ion in the host are not as compact as its atomic orbitals as a free-ion, which can reduce in part electron correlation contributions to intraconfigurational and interconfigurational energy differences. And splittings induced by interactions between ion and ligands in the host can significantly alter interconfigurational energy differences. But, having these preventions in mind, preliminary studies on free-ions are convenient and enlightening.

The results of SA-CASSCF/MS-CASPT2/RASSI-SO calculations on the $4f^2$ manifold of Pr^{3+} free-ion are shown in Table 6.1, where they are compared with experiments [14].[1] (The $4f5d$ and $4f6s$ manifolds are discussed in Sects. 6.2 and

[1] These calculations are performed in the D_{2h} point symmetry group due to the limitations of the program, which is bound to use abelian groups [15]. The use of a lower symmetry than the real atomic symmetry originates 100–200 cm^{-1} maximum broken degeneracies in the spin-orbit-free first step. These broken degeneracies are corrected in the spin-orbit-coupling second step by proper

Table 6.1 Energy levels of the $4f^2$ configuration of the Pr^{3+} free-ion: Experiments [14] and SA-CASSCF/MS-CASPT2/RASSI-SO calculations (deviations from experiment are given in parentheses). All values given in cm^{-1}

Level	Experiment	Calculation
$^3H_4$	0	0
$^3H_5$	2152	2282 (+130)
$^3H_6$	4389	4642 (+253)
$^3F_2$	4997	5112 (+115)
$^3F_3$	6415	6561 (+146)
$^3F_4$	6855	6858 (+3)
$^1G_4$	9921	9910 (−11)
$^1D_2$	17334	18981 (+1647)
$^3P_0$	21390	24188 (+2798)
$^3P_1$	22007	24882 (+2875)
$^1I_6$	22212	25365 (+3153)
$^3P_2$	23161	26108 (+2947)
$^1S_0$		50667

6.3.) They have been performed with a [Kr]-core relativistic WB-AIMP combined with a $(14s10p10d8f3g)[6s5p6d4f3g]$ valence basis set; a CAS space made of 2 electrons distributed in 13 atomic orbitals $4f$, $5d$, and $6s$; including correlation for all valence electrons except the $4d$ shell; and allowing all spin singlet and triplet states to mix via spin-orbit coupling.

The results show rather uniform deviations for levels with similar energy, which are very small up to 10000 cm^{-1} (up to $^1G_4$) and range between 1600 cm^{-1} and 3200 cm^{-1} for the upper levels from $^1D_2$ and above. These overestimations are very similar and in line with multiconfigurational Dirac-Coulomb calculations [16]. Only atomic oriented Coupled-Cluster Dirac-Coulomb calculations [17] achieve a better agreement with experiment; nevertheless, the improvement is not large [9] and the technique is hardly applicable to optically active centers of materials. The experimental energy of the $^1S_0$ level has not been reported; the location of this level in materials, either below or above the onset of the $4f5d$ manifold, is important, because it determines if the emission is of $4f \rightarrow 4f$ narrow type or $5d \rightarrow 4f$ broad type. The present result of 50667 cm^{-1} is quite similar to the referenced benchmark Coupled-Cluster Dirac-Coulomb calculation, of 50090 cm^{-1}.

Pr^{3+} in CaF_2, Cubic Site

The energy levels of the cubic defect of Pr^{3+} in CaF_2, where praseodymium substitutes for a Ca^{2+} ion with a long-range charge compensation, have also been computed in [9]. SA-CASSCF/MS-CASPT2/RASSI-SO calculations were performed on the

averaging. E.g. the energy of a ^3P term used in the second step is the average of the three broken degeneracy energies of the $^3T_{1g}$ components obtained in the first step.

$(PrF_8)^{5-}$ cluster under the effects of an AIMP embedding potential of CaF_2 host [13] with structure: group 225, $Fm\overline{3}m$, $a_0 = 5.46294$ Å [18]. The above indicated effective core potential and basis set of Pr was used, together with a [He]-core relativistic WB-AIMP combined with a $(5s5p1d)[3s4p1d]$ valence basis set [12] for F. Additional $(13s7p)[1s1p]$ Gaussian basis sets [13] were used in Ca^{2+} second neighbor ions in order to improve cluster-host orthogonality. A CAS space was used resulting from the distribution of 2 electrons in 13 molecular orbitals with main characters of Pr $4f$, $5d$, and $6s$, which belong to the O_h irreducible representations: a_{2u} (1), t_{1u} (3), t_{2u} (3), e_g (2), t_{2g} (3), and a_{1g} (1). In order to optimize the molecular orbitals, the following average energies were minimized in SA-CASSCF calculations: For the $4f^2$ configuration, one $^3A_{2g}$ and one ^3E_g levels; four $^3T_{1g}$ and two $^3T_{2g}$; three $^1A_{1g}$, one $^1A_{2g}$, and three ^1E_g; and two $^1T_{1g}$ and four $^1T_{2g}$. For the $4f5d$ and $4f6s$ configurations, one $^3A_{1u}$, two $^3A_{2u}$, and three ^3E_u levels; six $^3T_{1u}$ and five $^3T_{2u}$; one $^1A_{1u}$, two $^1A_{2u}$, and three ^1E_u levels; and six $^1T_{1u}$ and five $^1T_{2u}$. All valence electrons were correlated except those with main character Pr $4d$. All spin singlet and triplet states were allowed to couple in the RASSI-SO calculations.

The potential energy curves versus the Pr-F distance d_{Pr-F} (i.e. the configuration coordinate diagram along the totally symmetric breathing vibrational mode a_{1g}) are shown in Fig. 6.4 (red lines). The states group themselves in several manifolds according to their main configurational character: $4f^2$, $4f5de_g$, $4f5dt_{2g}$, and $4f\phi_{ITE,a_{1g}}$. Here, $4f$ and $5d$ stand for molecular orbitals with main character of Pr $4f$ and $5d$ respectively, and $\phi_{ITE,a_{1g}}$ is the a_{1g} active molecular orbital, whose inspection gives a Pr impurity-trapped-exciton (ITE) character, with significant electron density on the interstitial sites along the <100> directions. In this section we focus on the $4f^2$ manifold, whose spectroscopic constants and electronic structure analyses are summarized in Table 6.2.

The $4f^2$ levels are grouped in three sets. They can be identified attending to the expected O_h splittings of the atomic levels, their energies and the spin-orbit-free characters. The lowest set extends up to 11000 cm^{-1} and is made of the Stark levels derived from the free-ion $^3H_4$ to $^1G_4$ levels. The second set derives from $^1D_2$ to $^3P_2$. The third set contains only the A_{1g} level related to $^1S_0$. This is well above the two previous sets and is embedded in the $4f5de_g$ manifold.

It is interesting to see that all $4f^2$ levels have the same Pr-F bond length within 0.001 Å (2.302±0.001 Å) and breathing mode harmonic frequency within 1 cm^{-1} (505±1 cm^{-1}). The bond length has a very small contraction of –0.06 Å with respect to the Ca-F distance in CaF_2, 2.366 Å. This is in line with Pr^{3+} and Ca^{2+} having very similar ionic radii (1.126 Å and 1.12 Å respectively [19]) and a larger positive charge of the impurity. Regarding the breathing mode vibrational frequency, a 450 cm^{-1} vibrational progression was attributed to the cubic Pr site of CaF_2:Pr^{3+} [20]; if this is the breathing mode harmonic vibrational frequency of the cubic defect, the calculated value of 505 cm^{-1} would mean an overestimation and indicate that the calculation level used here leads to slightly too hard energy curves.

Table 6.2 Calculated $4f^2$ energy levels of Pr^{3+} in CaF_2: Spectroscopic constants $d_{e,\text{Pr-F}}$, $\bar{\nu}_{a_{1g}}$, and T_e; percentage spin-triplet characters; and spin-orbit-free state characters of the wave functions

State	$d_{e,\text{Pr-F}}$[a]	$\bar{\nu}_{a_{1g}}$[b]	T_e[c]	%($S=1$)	Spin-orbit-free character (%)[d]							
$4f^2$ configuration: $\langle d_{e,\text{Pr-F}}\rangle = 2.302 \pm 0.001$ Å; $\langle \bar{\nu}_{a_{1g}}\rangle = 505 \pm 1$ cm^{-1}												
Derived from $^3H_4$												
1 T_{2g}	2.301	505	0	97.8	62.3	1 $^3T_{1g}$	29.1	1 ^3E_g				
1 A_{1g}	2.301	504	193	97.8	63.0	2 $^3T_{1g}$	34.3	1 $^3T_{1g}$				
1 E_g	2.302	505	386	95.6	39.3	1 $^3T_{2g}$	29.9	2 $^3T_{1g}$	26.3	1 $^3T_{1g}$		
1 T_{1g}	2.301	504	448	97.3	31.5	2 $^3T_{1g}$	26.7	1 $^3T_{2g}$	22.4	1 $^3T_{1g}$	16.4	1 ^3E_g
Derived from $^3H_5$												
2 E_g	2.301	505	2121	99.5	68.8	1 $^3T_{1g}$	28.7	1 $^3T_{2g}$				
2 T_{1g}	2.301	505	2411	99.9	50.1	1 $^3T_{1g}$	38.0	1 $^3T_{2g}$				
3 T_{1g}	2.301	504	2695	100.0	46.3	2 $^3T_{1g}$	39.0	1 ^3E_g				
2 T_{2g}	2.302	504	2750	99.9	50.1	2 $^3T_{1g}$	34.1	1 $^3T_{2g}$	15.3	1 ^3E_g		
Derived from $^3H_6$												
2 A_{1g}	2.301	505	4391	99.8	64.6	1 $^3T_{1g}$	35.2	2 $^3T_{1g}$				
3 T_{2g}	2.300	504	4664	99.6	46.4	1 ^3E_g	35.9	1 $^3T_{1g}$	15.9	2 $^3T_{1g}$		
4 T_{1g}	2.301	505	4773	99.7	36.3	1 ^3E_g	25.7	1 $^3T_{1g}$	21.3	1 $^3T_{2g}$	16.2	2 $^3T_{1g}$
3 E_g	2.301	504	4930	98.3	66.8	2 $^3T_{1g}$	25.7	1 $^3T_{2g}$				
4 T_{2g}	2.301	504	5065	99.5	57.4	1 $^3T_{2g}$	26.1	2 $^3T_{1g}$				
1 A_{2g}	2.301	504	5212	99.8	96.5	1 $^3T_{2g}$						
Derived from $^3F_2$												
4 E_g	2.303	505	5634	96.5	68.7	3 $^3T_{1g}$	25.6	2 $^3T_{2g}$				
5 T_{2g}	2.303	504	5844	98.5	32.3	2 $^3T_{2g}$	30.6	3 $^3T_{1g}$	24.8	1 $^3A_{2g}$		
Derived from $^3F_4$, $^3F_3$												
5 E_g	2.301	506	6769	36.2	63.8	1 ^1E_g	27.5	2 $^3T_{2g}$				
6 T_{2g}	2.301	506	7037	99.4	54.7	3 $^3T_{1g}$	42.8	1 $^3A_{2g}$				
5 T_{1g}	2.303	505	7180	99.5	53.5	3 $^3T_{1g}$	46.0	2 $^3T_{2g}$				
2 A_{2g}	2.305	508	7442	100.0	96.7	2 $^3T_{2g}$						
3 A_{1g}	2.303	505	7494	66.7	64.6	3 $^3T_{1g}$	33.4	1 $^1A_{1g}$				
6 T_{1g}	2.303	507	7499	61.5	39.2	1 $^1T_{1g}$	35.8	2 $^3T_{2g}$	22.6	3 $^3T_{1g}$		
7 T_{2g}	2.302	508	7755	69.6	42.9	2 $^3T_{2g}$	30.4	1 $^1T_{2g}$	15.7	1 $^3A_{2g}$		
Derived from $^1G_4$												
6 E_g	2.302	507	9742	71.1	45.9	2 $^3T_{2g}$	29.0	1 ^1E_g	23.6	3 $^3T_{1g}$		
7 T_{1g}	2.302	506	10526	42.0	57.2	1 $^1T_{1g}$	22.0	3 $^3T_{1g}$	19.7	2 $^3T_{2g}$		
4 A_{1g}	2.303	505	10733	35.5	64.5	1 $^1A_{1g}$	34.9	3 $^3T_{1g}$				
8 T_{2g}	2.301	506	10887	33.7	66.3	1 $^1T_{2g}$	16.9	2 $^3T_{2g}$	12.7	1 $^3A_{2g}$		
Derived from $^1D_2$												
7 E_g	2.299	505	17820	6.5	93.5	2 ^1E_g						
9 T_{2g}	2.302	504	20271	10.4	89.6	2 $^1T_{2g}$						

(continued)

Table 6.2 (continued)

State	$d_{e,\mathrm{Pr\text{-}F}}$[a]	$\bar{\nu}_{a_{1g}}$[b]	T_e[c]	%($S=1$)	Spin-orbit-free character (%)[d]							
Derived from $^1I_6, ^3P_0, ^3P_1, ^3P_2$												
3 A_{2g}	2.299	504	23476	0.3	99.7	$1\,^1A_{2g}$						
8 E_g	2.299	504	23849	0.4	99.6	$3\,^1E_g$						
5 A_{1g}	2.301	505	24080	98.5	98.5	$4\,^3T_{1g}$						
8 T_{1g}	2.301	505	24774	100.0	100.0	$4\,^3T_{1g}$						
6 A_{1g}	2.304	506	25462	0.6	99.5	$2\,^1A_{1g}$						
9 E_g	2.301	505	25896	95.9	95.8	$4\,^3T_{1g}$						
10 T_{2g}	2.301	505	25983	84.5	84.5	$4\,^3T_{1g}$						
11 T_{2g}	2.299	504	26668	0.2	99.7	$3\,^1T_{2g}$						
9 T_{1g}	2.303	505	27142	0.2	99.8	$2\,^1T_{1g}$						
12 T_{2g}	2.304	508	27327	6.9	92.7	$4\,^1T_{2g}$						
Derived from $^1S_0$												
7 A_{1g}	2.299	507	49455	1.2	98.8	$3\,^1A_{1g}$						

[a] Pr-F equilibrium distance, in Å
[b] Breathing mode harmonic vibrational frequency, in cm^{-1}
[c] Adiabatic (minimum-to-minimum) transition energy, in cm^{-1}
[d] Weights calculated at $d_{\mathrm{Pr-F}} = 2.291$ Å; only those larger than 15% are shown

6.1.3 $4f^5$ Manifold of Sm^{3+}

Sm^{3+} in CaF_2, Cubic Site

The $4f^5$ energy levels of the cubic defect of Sm^{3+} in CaF_2, where it substitutes for a Ca^{2+} ion with a long-range charge compensation, have been computed in [21]. They do not only have an intrinsic value, but they are also important to analyze and understand the numerous levels of Sm^{2+} that result from the $4f^5 \times 5d$ coupling.

SA-RASSCF/MS-RASPT2/RASSI-SO calculations with the DKH relativistic Hamiltonian were performed on the $(SmF_8Ca_{12})^{19+}$ cluster under the effects of an AIMP embedding potential of CaF_2, using the following Gaussian basis sets: All-electron atomic natural orbital relativistic basis sets ANO-RCC for Sm $(25s22p15d11f4g2h)[9s8p5d4f3g2h]$ (quadruple-zeta with polarization quality) and F $(14s9p4d)[5s4p3d]$ (quadruple-zeta with polarization quality, without f and g; additional $(13s7p)[1s1p]$ outermost $3s, 3p$ orbital shells for second-neighbor Ca^{2+} ions for a better cluster-host orthogonality; and further $(5s)[5s]$ Gaussian primitives in the next six neighboring interstitial sites for an adequate representation of possible impurity-trapped-exciton states. The D_{2h} abelian point group was used. A RAS space was used resulting from the distribution of 5 electrons in 20 molecular orbitals with main characters of Sm $4f$, $5d$, $6s$, and $5f$ (double f-shell), which belong to the O_h irreducible representations: a_{2u} (2), t_{1u} (6), t_{2u} (6), e_g (2), t_{2g} (3), and a_{1g} (1). All possible occupations were allowed in the Sm $4f$ shell and a maximum of four electrons were allowed in the Sm $5d$, $6s$, and $5f$ shells. The numbers of configuration state functions (CSF) generated in each spin and D_{2h} symmetry block range between 1729 and 9096. The results obtained in the so defined RAS space should be fairly close to those of a CAS space of 5 electrons in 20 orbitals. The SA-RASSCF calculations with this RAS space are referred to as SA-RASSCF($4f/5d6s5f$-$4e$). In order to optimize the molecular orbitals, the following average energies were minimized in SA-RASSCF calculations: one $^6A_{2u}$ and one ^6E_u; four $^6T_{1u}$ and two $^6T_{2u}$; ten $^4A_{1u}$, ten $^4A_{2u}$, and eighteen ^4E_u; and twenty-eight $^4T_{1u}$ and twenty-eight $^4T_{2u}$ terms. All valence electrons were correlated except those with main character Sm $4d$. A standard IPEA value (0.25 au) was used and an imaginary level shift of 0.15 au to avoid intruder states. Symmetry analyses of the SA-RASSCF multiconfigurational reference wave functions allowed to perform MS-RASPT2 calculations over actual O_h $^{2S+1}\Gamma$ symmetry blocks, which avoids symmetry breakings at the MS-RASPT2 level and their transfer to the RASSI-SO step. All spin-orbit free states of spin multiplicity $2S + 1 = 6$ and 4 obtained in the MS-RASPT2 calculations were allowed to couple in the RASSI-SO calculations.

The results are shown in Fig. 6.2 and Table 6.3 (only the first and last states of each band are given here). As in all studied $4f^N$ cases, all states of $4f^5$ configuration have basically the same bond length and vibrational frequency. The Sm-F bond length is 2.265±0.002 Å and the breathing mode harmonic vibrational frequency 499±1 cm^{-1}. The Sm^{3+}-F$^-$ equilibrium distance implies a contraction of –0.10 Å of the Ca^{2+}-F$^-$ distance in CaF$_2$ (2.366 Å), which means additional –0.06 Å over the

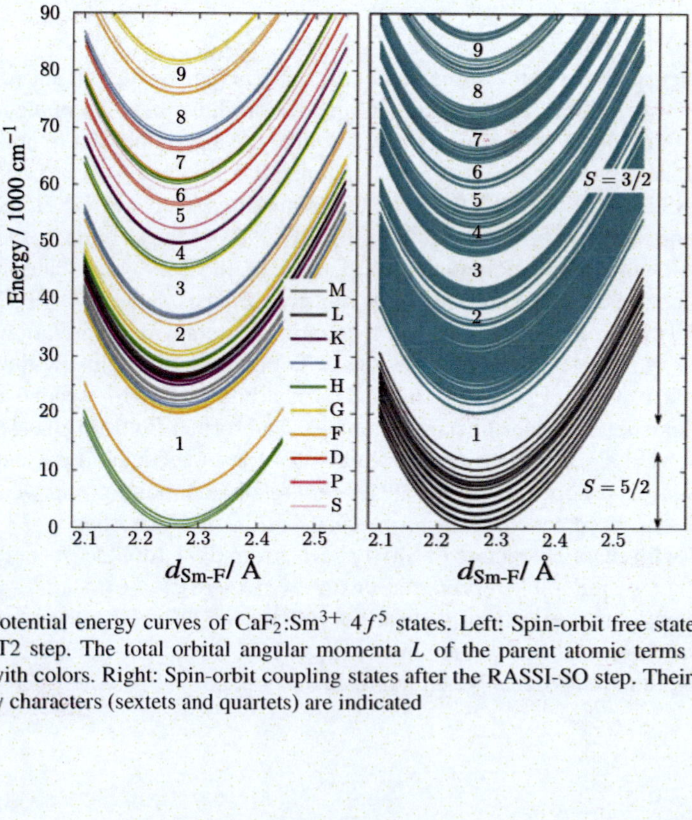

Fig. 6.2 Potential energy curves of CaF$_2$:Sm^{3+} $4f^5$ states. Left: Spin-orbit free states after the MS-RASPT2 step. The total orbital angular momenta L of the parent atomic terms ^{2S+1}L are indicated with colors. Right: Spin-orbit coupling states after the RASSI-SO step. Their main spin multiplicity characters (sextets and quartets) are indicated

Table 6.3 Calculated $4f^5$ energy levels of Sm^{3+} in CaF_2: Spectroscopic constants $d_{e,Sm\text{-}F}$, $\bar{\nu}_{a_{1g}}$, and T_e; and spin-orbit-free state characters of the wave functions. Only selected states are shown

	State	$d_{e,Sm\text{-}F}$[a]	$\bar{\nu}_{a_{1g}}$[b]	T_e[c]	Spin-orbit-free character (%)[d]							
$4f^5$ configuration: $\langle d_{e,Sm\text{-}F}\rangle = 2.265 \pm 0.002$ Å; $\langle \bar{\nu}_{a_{1g}}\rangle = 499 \pm 1$ cm^{-1}												
Derived from $^6H_J, ^6F_J$												
	1 Γ_{8u}	2.266	499	0	42.47	1 $^6T_{2u}$	33.36	1 $^6T_{1u}$	12.29	2 $^6T_{1u}$		
up to	9 Γ_{6u}	2.266	500	11725	49.52	2 $^6T_{2u}$	33.68	1 $^6A_{2u}$	14.46	3 $^6T_{1u}$		
				gap 1								
Derived from 1 ^4F_J, 1 ^4G_J, 1 ^4I_J, 1 ^4M_J; 1 ^4K_J, 1 ^4L_J, 1 ^4D_J; 1 ^4H_J; 2 ^4G_J, 1 ^4P_J, and ^6P_J												
	19 Γ_{8u}	2.266	498	19984	41.63	1 $^4T_{1u}$	15.62	1 $^4T_{2u}$	14.55	2 $^4T_{2u}$		
up to	93 Γ_{8u}	2.266	500	36268	53.18	14 $^4T_{1u}$	18.62	13 $^4T_{1u}$	12.63	4 $^6T_{1u}$		
				gap 2								
Derived from 2 ^4F_J, 2 ^4I_J												
	47 Γ_{6u}	2.265	499	38205	80.27	15 $^4T_{1u}$	14.35	14 $^4T_{2u}$				
up to	107 Γ_{8u}	2.266	500	41946	46.42	6 $^4A_{1u}$	41.32	16 $^4T_{1u}$				
				gap 3								
Derived from 3 ^4G_J, 2 ^4H_J												
	54 Γ_{7u}	2.264	499	47480	41.80	17 $^4T_{1u}$	27.99	17 $^4T_{2u}$	20.60	11 ^4E_u		
up to	120 Γ_{8u}	2.264	500	51142	32.07	19 $^4T_{1u}$	21.92	7 $^4A_{1u}$	21.10	18 $^4T_{1u}$	10.95	18 $^4T_{2u}$
				gap 4								
Derived from 2 ^4K_J, 2 ^4P_J												
	60 Γ_{6u}	2.264	499	52431	78.15	20 $^4T_{1u}$	11.99	13 ^4E_u				
up to	132 Γ_{8u}	2.265	499	55747	70.16	22 $^4T_{1u}$	19.48	21 $^4T_{2u}$				
				gap 5								

(continued)

Table 6.3 (continued)

	State	$d_{e,\text{Sm-F}}$[a]	$\bar{\nu}_{a_{1g}}$[b]	T_e[c]	Spin-orbit-free character (%)[d]							
Derived from $2\,^4D_J$												
	66 Γ_{6u}	2.264	499	59296	52.82	14 ^4E_u	37.90	21 $^4T_{2u}$				
up to	68 Γ_{7u}	2.264	499	60684	36.96	14 ^4E_u	27.14	22 $^4T_{1u}$	12.16	23 $^4T_{1u}$	12.07	21 $^4T_{2u}$
				gap 6								
Derived from $1\,^4S_J, 3\,^4H_J, 3\,^4F_J$												
	136 Γ_{8u}	2.264	499	63689	53.46	8 $^4A_{1u}$	20.99	23 $^4T_{1u}$	10.89	23 $^4T_{2u}$		
up to	74 Γ_{7u}	2.265	499	66389	67.11	23 $^4T_{2u}$	18.17	21 $^4T_{2u}$				
				gap 7								
Derived from $3\,^4D_J, 3\,^4I_J$												
	75 Γ_{7u}	2.264	499	69664	85.47	24 $^4T_{2u}$						
up to	79 Γ_{6u}	2.264	500	73436	54.48	17 ^4E_u	32.47	26 $^4T_{2u}$	12.19	25 $^4T_{2u}$		
				gap 8								
Derived from $4\,^4F_J$												
	161 Γ_{8u}	2.262	499	78837	48.07	10 $^4A_{2u}$	44.19	27 $^4T_{2u}$				
up to	165 Γ_{8u}	2.264	500	81811	77.96	27 $^4T_{1u}$	17.72	27 $^4T_{2u}$				
				gap 9								
Derived from $4\,^4G_J$												
	166 Γ_{8u}	2.262	499	85027	69.22	10 $^4A_{1u}$	21.66	28 $^4T_{1u}$				
up to	84 Γ_{6u}	2.264	500	86309	74.67	28 $^4T_{1u}$	15.77	18 ^4E_u				

[a] Sm-F equilibrium distance, in Å
[b] Breathing mode harmonic vibrational frequency, in cm^{-1}
[c] Adiabatic (minimum-to-minimum) transition energy, in cm^{-1}
[d] Weights calculated at $d_{\text{Sm-F}} = 2.250$ Å; only those larger than 10% are shown

ionic radii mismatch (–0.04 Å = 1.08 Å(Sm^{3+}) – 1.12 Å(Ca^{2+})) [19]. This amount is the same found for the Pr^{3+} ion (Sect. 6.1.2) and can be attributed to the electrostatic effect of the defect positive charge.

6.1.4 $4f^6$ Manifold of Sm^{2+}

Sm^{2+} in CaF_2

The $4f^6$ energy levels of Sm^{2+} in CaF_2 have been calculated as in Sect. 6.1.3, except: the $(SmF_8Ca_{12})^{18+}$ embedded-cluster was used; 6 electrons were distributed among the 20 molecular orbitals of main character Sm $4f$, $5d$, $6s$, and $5f$ in the SA-RASSCF($4f/5d6s5f$-$4e$) calculations, the number of CSFs ranging between 3460 and 24093; the state-averages for molecular orbital optimization were: one $^7A_{2g}$; one $^7T_{1g}$ and one $^7T_{2g}$; seven $^5A_{1g}$, five $^5A_{2g}$, and thirteen ^5E_g; and sixteen $^5T_{1g}$ and eighteen $^5T_{2g}$ terms; and all spin-orbit free states of spin multiplicity $2S+1=7$ and 5 were used in the RASSI-SO calculations. The results are shown in Table 6.4 and Fig. 6.5(left). The common features of all $4f^N$ cases are observed: all $4f^N$ states show basically the same bond length and breathing mode harmonic vibrational frequency; and the impurity-ligand bond length corresponds to a distortion of the host bond length which is a fraction of the ionic radii mismatch [22] plus the electrostatic effect of the defect charge. For Sm^{2+} $4f^6$ states in CaF_2, the common bond length and vibrational frequency are 2.387 Å and 427 cm^{-1}; the Sm^{2+}-F^- bond length means a +0.02 Å elongation of the 2.366 Å Ca^{2+}-F^- one, smaller than the +0.15 Å elongation estimated from ionic radii: 1.27 Å (Sm^{2+}) – 1.12 Å (Ca^{2+}) [19]; and electrostatic effects due to defect charge are non-existing.

6.1.5 $4f^6$ Manifold of Eu^{3+}

Eu^{3+} is a lanthanide ion with a rich history in scientific and technological applications. E.g., its characteristic red or orange $^5D_0 \rightarrow {}^7F_{1,2}$ emission grounds its successful application in fluorescent lighting [23] and the low J-values of the multiplets involved in the luminescence make it suitable as a local symmetry probe [24–26].

Eu^{3+} in CaF_2, Cubic Site

The $4f^6$ energy levels of the cubic defect of Eu^{3+} in CaF_2, where it substitutes for a Ca^{2+} ion with a long-range charge compensation, have been computed in [27], together with those of the substitutional defects in the hosts SrF_2, BaF_2, CaS, SrS, and BaS. Besides their own value, these results are also interesting to analyze the levels of Eu^{2+} that result from the $4f^6 \times 5d$ coupling.

SA-RASSCF/MS-RASPT2/RASSI-SO calculations with the DKH relativistic Hamiltonian were performed on the $(EuF_8Ca_{12})^{19+}$ cluster under the effects of an AIMP embedding potential of the CaF_2 host (cf. Sect. 2). In the other Eu^{3+}-

Table 6.4 Calculated $4f^6$ energy levels of Sm^{2+} in CaF_2: Spectroscopic constants $d_{e,Sm\text{-}F}$, $\bar{\nu}_{a_{1g}}$, and T_e; emission oscillator strengths; and spin-orbit-free state characters of the wave functions. Only selected states are shown

	State	$d_{e,Sm\text{-}F}$ [a]	$\bar{\nu}_{a_{1g}}$ [b]	T_e [c]	f^{emi}/f_{ref} [d]	Spin-orbit-free character (%)[e]					
$4f^6$ configuration: $\langle d_{e,Sm\text{-}F}\rangle = 2.387 \pm 0.003$ Å; $\langle \bar{\nu}_{a_{1g}}\rangle = 427 \pm 2$ cm^{-1}											
Derived from $^7F_{0-6}$											
	1 A_{1g}	2.388	425	0		37.05	1 $^7A_{2g}$	34.23	1 $^7T_{1g}$	23.59	1 $^7T_{2g}$
	1 T_{1g}	2.388	425	244	1.00	44.16	1 $^7A_{2g}$	33.00	1 $^7T_{1g}$	18.57	1 $^7T_{2g}$
	2 T_{1g}	2.389	427	1629	0.02	46.93	1 $^7T_{2g}$	25.67	1 $^7T_{1g}$	24.81	1 $^7A_{2g}$
	4 T_{1g}	2.389	427	3591	0.03	82.23	1 $^7T_{1g}$				
up to	4 E_g	2.390	427	4846		67.44	1 $^7T_{2g}$	29.50	1 $^7T_{1g}$		
Derived from 1 $^5D_{0-3}$											
	4 A_{1g}	2.388	426	16497		54.25	1 $^5T_{2g}$	40.12	1 ^5E_g		
	7 T_{1g}	2.388	426	17147		54.02	1 $^5T_{2g}$	40.96	1 ^5E_g		
	5 E_g	2.387	426	18379		55.35	1 ^5E_g	40.65	1 $^5T_{2g}$		
	7 T_{2g}	2.388	425	18580		61.92	1 $^5T_{2g}$	33.77	1 ^5E_g		
	3 A_{2g}	2.387	426	20518		96.32	1 ^5E_g				
	8 T_{2g}	2.387	426	20653		55.14	1 $^5T_{2g}$	41.03	1 ^5E_g		
	8 T_{1g}	2.387	425	20679		88.68	1 $^5T_{2g}$				
Derived from 1 $^5D_4$, $^5L_{6-10}$, 1 $^5G_{2-6}$											
	9 T_{2g}	2.385	426	23054		40.16	2 $^5T_{2g}$	32.88	1 $^5T_{2g}$	10.02	1 ^5E_g
up to	26 T_{2g}	2.389	426	27400		70.58	4 $^5T_{2g}$	16.70	4 ^5E_g		
Derived from 1 $^5H_{3-7}$, 1 $^5F_{1-5}$, 1 $^5I_{4-8}$, 1 $^5K_{5-9}$, 2 $^5G_{2-6}$, 2 $^5D_{1-4}$, 1 $^5P_{1-3}$, 2 $^5H_{3-7}$											
	10 A_{2g}	2.387	426	28027		56.98	4 $^5T_{1g}$	33.06	5 ^5E_g		
up to	71 T_{1g}	2.387	426	44495		86.11	13 $^5T_{1g}$				

(continued)

Table 6.4 (continued)

	State	$d_{e,\text{Sm-F}}$[a]	$\bar{\nu}_{a_{1g}}$[b]	T_e[c]	f^{emi}/f_{ref}[d]	Spin-orbit-free character (%)[e]					
Derived from $1\,^5S_2$											
	$73\ T_{2g}$	2.387	426	50207		90.96	$5\,^5A_{1g}$				
	$48\ E_g$	2.387	426	50208		90.99	$5\,^5A_{1g}$				
Derived from $2\,^5I_{4-8}$											
	$27\ A_{1g}$	2.386	426	52442		55.36	$14\,^5T_{2g}$	41.28	$11\,^5E_g$		
up to	$81\ T_{2g}$	2.387	426	55134		50.95	$14\,^5T_{1g}$	24.36	$15\,^5T_{2g}$	21.18	$6\,^5A_{1g}$
Derived from $2\,^5F_{1-5}$											
	$80\ T_{1g}$	2.386	426	55889		74.76	$15\,^5T_{1g}$	14.87	$16\,^5T_{2g}$		
up to	$84\ T_{1g}$	2.386	426	57023		58.11	$16\,^5T_{2g}$	33.75	$15\,^5T_{1g}$		
Derived from $3\,^5G_{2-6}$, $3\,^5D_{0-4}$											
	$31\ A_{1g}$	2.384	426	66391		39.33	$12\,^5E_g$	37.83	$17\,^5T_{2g}$	19.38	$13\,^5E_g$
up to	$92\ T_{1g}$	2.387	427	71271		73.25	$18\,^5T_{2g}$	21.54	$13\,^5E_g$		

[a] Sm-F equilibrium distance, in Å

[b] Breathing mode harmonic vibrational frequency, in cm^{-1}

[c] Adiabatic (minimum-to-minimum) transition energy, in cm^{-1}

[d] Relative emission oscillator strength from the $4f^5 5de_g\ 1A_{1u}$ state, calculated at $d_{\text{Sm-F}} = 2.350$ Å; the reference value is $f_{ref} = 4.851 \times 10^{-3}$

[e] Weights calculated at $d_{\text{Sm-F}} = 2.400$ Å; only those larger than 10% are shown

doped alkaline earth fluorides, equivalent clusters and corresponding host embedding potentials were used. The $(EuS_6Ca_6)^{3+}$ cluster embedded in the AIMP embedding potential of CaS and corresponding clusters and embedding potentials were used for the alkaline earth sulfides. The following Gaussian basis sets were used: All-electron atomic natural orbital relativistic basis sets ANO-RCC for Eu $(25s22p15d11f4g2h)[9s8p5d4f3g2h]$ (quadruple-zeta with polarization quality) and F $(14s9p4d)[5s4p3d]$ and S $(17s12p5d)[6s5p3d]$ (quadruple-zeta with polarization quality, without f and g. Additional $[1s1p]$ contractions of the outermost $3s, 3p$ orbital shells for second-neighbor frozen Ca^{2+} ions, $4s, 4p$ shells for Sr^{2+} ions, and $5s, 5p$ shells for Ba^{2+} ions, were used for a better cluster-host orthogonality in the MF_2 alkaline earth difluoride hosts. In the MS alkaline sulfide hosts, the description of the six M^{2+} ions located next to the sulfur ligands in the $\langle 100 \rangle$ axes was improved further: only the inner shells of the M^{2+} ions were frozen ([Mg] core for Ca^{2+}, [Zn] core for Sr^{2+}, and [Cd] core for Ba^{2+}); the corresponding atomic natural orbitals of the ANO-RCC set were used to build the core AIMP operators; [28] the 6-electron valence of the M^{2+} were treated explicitly using the following valence basis sets, adapted from the ANO-RCC primitive sets: Ca$(20s16p6d)[3s4p1d]$, Sr$(23s19p12d)[3s4p1d]$, Ba$(26s22p15d)[3s4p1d]$. Further $(5s)[5s]$ Gaussian primitives were added in the next six neighboring interstitial sites of the MF_2 hosts [29] for an adequate representation of possible impurity-trapped-exciton states. The D_{2h} abelian point group was used. A RAS space was used resulting from the distribution of 6 electrons in 20 molecular orbitals with main characters of Eu $4f$, $5d$, $6s$, and $5f$ (double f-shell), which belong to the O_h irreducible representations: a_{2u} (2), t_{1u} (6), t_{2u} (6), e_g (2), t_{2g} (3), and a_{1g} (1). All possible occupations were allowed in the Eu $4f$ shell and a maximum of four electrons were allowed in the Eu $5d$, $6s$, and $5f$ shells. The numbers of configuration state functions (CSF) generated in each spin and D_{2h} symmetry block range between 3400 and 24000. The results obtained in the so defined RAS space should be fairly close to those of a CAS space of 6 electrons in 20 orbitals. The SA-RASSCF calculations with this RAS space are referred to as SA-RASSCF($4f/5d6s5f$-$4e$). In order to optimize the molecular orbitals, the following average energies were minimized in SA-RASSCF calculations: one $^7A_{2g}$; one $^7T_{1g}$ and one $^7T_{2g}$; seven $^5A_{1g}$, five $^5A_{2g}$, and thirteen ^5E_g; and sixteen $^5T_{1g}$ and eighteen $^5T_{2g}$ terms. All valence electrons were correlated except those with main character Eu $4d$. A standard IPEA value (0.25 au) was used and an imaginary level shift of 0.15 au to avoid intruder states. Symmetry analyses of the SA-RASSCF multiconfigurational reference wave functions allowed to perform MS-RASPT2 calculations over actual O_h $^{2S+1}\Gamma$ symmetry blocks, which avoids symmetry breakings at the MS-RASPT2 level and their transfer to the RASSI-SO step. All spin-orbit free states of spin multiplicity $2S + 1 = 7$ and 5 obtained in the MS-RASPT2 calculations were allowed to couple in the RASSI-SO calculations.

The potential energy curves of the $4f^6$ manifold of the cubic defect of Eu^{3+} in CaF_2 are shown in Fig. 6.3, both for the final spin-orbit coupling results (right pannel) and the intermediate spin-orbit-free results (left panel). Selected details of the potential energy curves are summarized in Table 6.5.

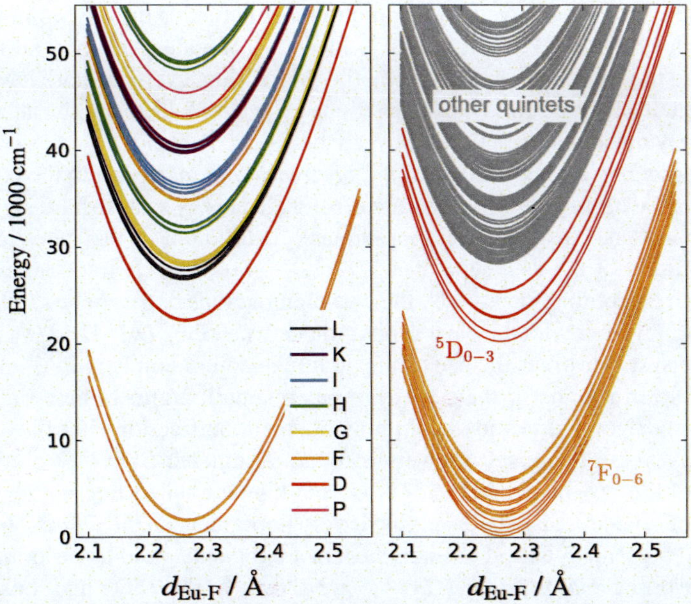

Fig. 6.3 Potential energy curves of CaF_2:Eu^{3+} $4f^6$ states. Left: Spin-orbit free states after the MS-RASPT2 step. The total orbital angular momenta L of the parent atomic terms ^{2S+1}L are indicated with colors. Right: Spin-orbit coupling states after the RASSI-SO step

Table 6.5 Calculated $4f^6$ energy levels of Eu^{3+} in CaF_2: Spectroscopic constants $d_{e,\text{Eu-F}}$, $\bar{\nu}_{a_{1g}}$, and T_e; emission oscillator strengths; and spin-orbit-free state characters of the wave functions. Only selected states are shown

	State	$d_{e,\text{Eu-F}}$ [a]	$\bar{\nu}_{a_{1g}}$ [b]	T_e [c]	Spin-orbit-free character (%) [d]					
$4f^6$ configuration: $\langle d_{e,\text{Eu-F}} \rangle = 2.260 \pm 0.002$ Å; $\langle \bar{\nu}_{a_{1g}} \rangle = 495 \pm 3$ cm^{-1}										
Derived from $^7F_{0-6}$										
	1 A_{1g}	2.261	495	0	35.10	1 $^7T_{2g}$	30.28	1 $^7A_{2g}$	28.94	1 $^7T_{1g}$
	1 T_{1g}	2.260	495	345	35.77	1 $^7A_{2g}$	34.53	1 $^7T_{2g}$	24.93	1 $^7T_{1g}$
	1 T_{2g}	2.259	497	803	55.37	1 $^7A_{2g}$	32.46	1 $^7T_{2g}$		
up to	2 A_{2g}	2.261	494	5827	50.05	1 $^7T_{2g}$	46.56	1 $^7T_{1g}$		
Derived from 1 $^5D_{0-3}$										
	4 A_{1g}	2.260	494	20023	55.09	1 $^5T_{2g}$	38.95	1 ^5E_g		
up to	8 T_{1g}	2.260	494	25180	87.42	1 $^5T_{2g}$				
Derived from 1 $^5D_4$, $^5L_{6-10}$, 1 $^5G_{2-6}$										
	9 T_{2g}	2.258	492	27978	36.81	1 $^5T_{1g}$	29.22	2 ^5E_g		
up to	11 A_{1g}	2.262	493	32458	52.98	4 $^5T_{2g}$	32.54	4 ^5E_g		
	...									

[a] Eu-F equilibrium distance, in Å

[b] Breathing mode harmonic vibrational frequency, in cm^{-1}

[c] Adiabatic (minimum-to-minimum) transition energy, in cm^{-1}

[d] Weights calculated at $d_{\text{Sm-F}} = 2.250$ Å; only those larger than 10% are shown

Figure 6.3 illustrates the well known multiplet structure of the $4f^6$ configuration under the effect of a weak crystal-field. A lower set of states spanning 6000 cm^{-1} correspond to the splitting of the ^6F_J $(J = 0 - 6)$ free-ion levels, or Stark levels. The analysis of their many-electron wave functions shown in Table 6.5 indicate they are almost entirely made of interactions between the $^7A_{2g}$, $^7T_{1g}$, and $^7T_{2g}$ spin-orbit-free states corresponding to the splitting of the ^6F free-ion term, shown in the left panel of the Figure; the nearest spin-quintets are too high in energy so as to have significant contributions. Next, after a gap of approximately 12000 cm^{-1}, the states resulting from the splitting of the well-separated $^5D_0$, $^5D_1$, $^5D_2$ and $^5D_3$ free-ion levels are found. The small splittings caused by the weak interactions of the Eu^{3+} $4f$ electrons with the host, together with the clear separation between the $^5D_0$, $^5D_1$, $^5D_2$ and $^5D_3$ free-ion levels, still justifies the use of the atomic quantum numbers L, S and J for these states, even if none of them are rigorous quantum numbers here (in fact, L and S were not rigorously valid quantum numbers in the free-ion either). At higher energies, however, a dense set of levels is found, originated from the splitting and interaction of the $^5D_4$ level and the ^5L, ^5G, and higher excited free-ion terms. For these states of Eu^{3+} in CaF_2, the J-mixing is very important and the O_h' double-group irreducible representations and ordinal integers are the only valid labels or quantum numbers. Two of them, 9 T_{2g} and 11 A_{1g} are shown in the Table and those up to 55000 cm^{-1} are shown in gray in the Figure. The ones in the upper part are expected to have lower quality in this calculation, because they are in an energy region where interactions with spin-triplets, not considered in the calculation, can be important.

A comparison with experimental results on the Stark levels of the $^7F_{0-6}$, $^5D_0$, and $^5D_1$ levels of CaF_2:Eu^{3+} cubic centers, obtained after a thorough analysis of combined excitation and emission spectroscopy [30], is done in Sect. 9.4.1.

The structural results are not an exception in that all $4f^6$ states have the same Eu-F bond length (2.260±0.002 Å) and breathing mode harmonic vibrational frequency (495±3 cm^{-1}). These values mean a very small lanthanide contraction of –0.005 Å with respect to Sm^{3+} in the same host, which is accompanied with a very small increase of the vibrational frequency of +5 cm^{-1} (cf. Sect. 6.1.3).

6.1.6 $4f^7$ Manifold of Eu^{2+}

Eu^{2+} in CaF_2

The $4f^7$ energy levels of Eu^{2+} in CaF_2 have been calculated as in Sect. 6.1.5, except: the $(EuF_8Ca_{12})^{18+}$ embedded-cluster was used; 7 electrons were distributed among the 20 molecular orbitals of main character Eu $4f$, $5d$, $6s$, and $5f$ in the SA-RASSCF($4f/5d6s5f$-$4e$) calculations; the state-averages for molecular orbital optimization were: one $^8A_{1u}$; two $^6A_{1u}$, two $^6A_{2u}$, and four ^6E_u; and six $^6T_{1u}$ and six $^6T_{2u}$ terms; and all spin-orbit-free states of spin multiplicity $2S + 1 = 8$ and 6 were used in the RASSI-SO calculations. The particular choice of state-averaging was made in an attempt to calculate with even quality the states derived from the lowest $4f^7$ free-ion terms ^8S, ^6P, ^6D, ^6F, ^6G, ^6H, and ^6I, whose O_h splittings are shown in Table 6.6. In it, an accounting of the Γ_{6u}, Γ_{7u}, and Γ_{8u} many-electron states (and degeneracies) that will result in the RASSI-SO calculations is also made; it facilitates the identification of the RASSI-SO O_h' wave functions and their free-ion parentage.

The results are shown in Fig. 6.6 (black lines) and in Table 6.7. Comparison with two-photon absorption spectroscopy experiments on $CaF_2:Eu^{2+}$, $SrF_2:Eu^{2+}$, and $BaF_2:Eu^{2+}$ [31] is done in Sect. 9.4.2.

The lowest state of the $4f^7$ configuration, which is the ground state of Eu^{2+}, derives from the atomic term ^8S and level $^8S_{7/2}$. In O_h symmetry, spin-orbit coupling splits this level into three Kramers doublets, Γ_{6u}, Γ_{7u}, and Γ_{8u}, the latter with an additional double-degeneracy and total degeneracy 4. The splitting, of at most 0.2 cm^{-1} in the calculations, has no measurable influence on optical and luminescence properties. The rest of the $4f^7$ manifold is immersed in the dense $4f^65d$ manifold. The states derived from the atomic terms ^6P, ^6I, and ^6D are the first in energy; they start at 30000 cm^{-1} and share the energy region of $4f^6(^7F)5de_g$ states. They are followed by the states derived from ^6G, ^6F, and ^6H, which start at 48000 cm^{-1}, right above the onset of states with main configurational characters $4f^6(^7F)5dt_{2g}$ and $4f^6(^7F)\phi_{ITE,a_{1g}}$.

Eu^{2+}-ligand bond lengths or equilibrium distances in the $4f^7$ ground state are shown in Table 6.15 for the host series CaF_2-SrF_2-BaF_2 and CaS-SrS-BaS, together with ground state bond lengths of Eu^{3+} and $4f^65d$ excited state bond lengths of Eu^{2+}. Although very few experimental data exist on ground state quantitative geometries and bond lengths of lanthanide dopants in luminescent materials, the Eu-F distance in $CaF_2:Eu^{2+}$ is a happy exception to this shortage: $d_{e,\text{Eu-F}} = (2.413 \pm 0.027)$ Å have been reported from extended X-ray absorption fine structure (EXAFS) [32], which agrees with the calculated value of 2.388 Å. The Eu^{2+}-ligand bond lengths increase with the size of the alkaline earth ion within a series and they are larger in sulfides than in fluorides, irrespective of their smaller coordination number (6 vs. 8). In all cases, the bond length is slightly smaller in the ground state than in the higher-lying $4f^7$ spin-sextets (only shown for CaF_2, in Table 6.7) about 0.002 Å in fluorides and 0.006 Å in sulfides.

At this point, it is interesting to note that the difference between Eu-ligand bond lengths in the Eu^{2+} and Eu^{3+} oxidation states, $\Delta d_e(Eu^{2+}/Eu^{3+}) = d_{e,\text{Eu-L}}(Eu^{2+}) - d_{e,\text{Eu-L}}(Eu^{3+})$, is a parameter of paramount importance in charge transfer phenomena involving a (photo)reduction of Eu^{3+} or (photo-)oxidation of Eu^{2+}. A notable example is the electron transfer between Eu^{2+} and Eu^{3+} in a mixed-valence material (inter-valence charge transfer, IVCT). The bond length difference, $\Delta d_e(Eu^{2+}/Eu^{3+})$, increases across the CaF_2-SrF_2-BaF_2 series of hosts (from 0.13 Å to 0.16 Å to 0.20Å) and the CaS-SrS-BaS series of hosts (from 0.15 Å to 0.18 Å to 0.26Å). As it will be discussed in Sect. 11.2.5, it is this increase of bond length mismatch towards Ba compounds that makes the Eu^{2+} $4f^65d \rightarrow 4f^7$ emission to be completely quenched in BaF_2:Eu in case some (unintended) Eu^{3+} is found in addition to Eu^{2+}, while the $4f^65d \rightarrow 4f^7$ emission is still present in CaF_2:Eu and SrF_2:Eu, even with Eu^{3+} being present [33].

Table 6.6 Numbers of spin-orbit-free states of the spin × O_h *ungerade* irreducible representation blocks, and of spin-orbit coupling states of the O'_h double-group *ungerade* irreducible representations Γ_{6u}, Γ_{7u}, and Γ_{8u}, derived from the lowest $4f^7$ terms and levels

	Term	Spin × O_h irrep						Level	2J									O'_h irrep		
		$^8A_{1u}$	$^6A_{1u}$	$^6A_{2u}$	^6E_u	$^6T_{1u}$	$^6T_{2u}$											Γ_{6u}	Γ_{7u}	Γ_{8u}
degen.:		8	6	6	12	18	18											2	2	4
	^8S	1						^8S_J				7						1	1	1
	^6P					1		^6P_J		3	5	7						1	2	3
	^6D				1		1	^6D_J	1	3	5	7	9					3	2	5
	^6F			1		1	1	^6F_J	1	3	5	7	9	11				4	3	7
	^6G		1		1	1	1	^6G_J		3	5	7	9	11	13			4	5	9
	^6H				1	2	1	^6H_J			5	7	9	11	13	15		5	6	11
	^6I		1	1	1	1	2	^6I_J				7	9	11	13	15	17	7	6	13

Table 6.7 Calculated $4f^7$ energy levels of Eu^{2+} in CaF_2: Spectroscopic constants $d_{e,\text{Eu-F}}$, $\bar{\nu}_{a_{1g}}$, and T_e; emission oscillator strengths; and spin-orbit-free state characters of the wave functions. Only selected states are shown

	State	$d_{e,\text{Eu-F}}$ [a]	$\bar{\nu}_{a_{1g}}$ [b]	T_e [c]	f^{emi}/f_{ref} [d]	Spin-orbit-free character (%) [e]					
$4f^7$ configuration: $\langle d_{e,\text{Eu-F}}\rangle = 2.387 \pm 0.002$ Å; $\langle \bar{\nu}_{a_{1g}}\rangle = 422 \pm 3$ cm^{-1}											
Derived from $^8S_{7/2}$											
	$1\,\Gamma_{6u}$	2.388	423	0	1.00	97.76	$1\,^8A_{1u}$				
	$1\,\Gamma_{8u}$	2.388	423	0	0.86	97.76	$1\,^8A_{1u}$				
	$1\,\Gamma_{7u}$	2.388	423	0	0.04	97.76	$1\,^8A_{1u}$				
Derived from $^6P_{3/2-7/2}$											
	$2\,\Gamma_{8u}$	2.386	422	30442		89.24	$1\,^6T_{1u}$				
up to	$3\,\Gamma_{7u}$	2.386	422	31087		85.28	$1\,^6T_{1u}$	10.76	$2\,^6E_u$		
Derived from $^6I_{7/2-17/2}$											
	$3\,\Gamma_{6u}$	2.385	422	34758		77.14	$1\,^6T_{2u}$	19.36	$1\,^6E_u$		
up to	$9\,\Gamma_{7u}$	2.386	422	35510		56.03	$2\,^6T_{1u}$	32.90	$1\,^6A_{1u}$		
Derived from $^6D_{1/2-7/2}$											
	$18\,\Gamma_{8u}$	2.385	422	37485		78.62	$3\,^6T_{2u}$	17.76	$2\,^6E_u$		
up to	$11\,\Gamma_{7u}$	2.386	422	38679		84.08	$3\,^6T_{2u}$	10.50	$1\,^6T_{1u}$		
Derived from $^6G_{3/2-13/2}$											
	$23\,\Gamma_{8u}$	2.384	422	47769		62.96	$4\,^6T_{2u}$	21.55	$2\,^6A_{2u}$		
up to	$31\,\Gamma_{8u}$	2.385	422	50076		41.99	$3\,^6T_{1u}$	22.15	$3\,^6E_u$	21.60	$4\,^6T_{1u}$
Derived from $^6F_{1/2-11/2}$											
	$17\,\Gamma_{6u}$	2.386	422	50640		44.37	$5\,^6T_{2u}$	24.72	$2\,^6A_{2u}$	18.88	$4\,^6T_{1u}$
up to	$38\,\Gamma_{8u}$	2.386	422	52909		47.86	$5\,^6T_{2u}$	31.75	$4\,^6T_{1u}$		
Derived from $^6H_{5/2-15/2}$											
	$20\,\Gamma_{7u}$	2.385	423	55493		57.10	$5\,^6T_{1u}$	27.00	$4\,^6E_u$		
up to	$25\,\Gamma_{7u}$	2.387	423	56875		50.14	$6\,^6T_{1u}$	39.91	$6\,^6T_{2u}$		

[a] Eu-F equilibrium distance, in Å

[b] Breathing mode harmonic vibrational frequency, in cm^{-1}

[c] Adiabatic (minimum-to-minimum) transition energy, in cm^{-1}

[d] Relative emission oscillator strength from the $4f^6(^7F_J)5de_g$ $1\Gamma_{8g}$ excited state (radiative lifetime of 0.553 μs), calculated at $d_{\text{Eu-F}} = 2.350$ Å; the reference value is $f_{ref} = 2.233 \times 10^{-3}$

[e] Weights calculated at $d_{\text{Eu-F}} = 2.400$ Å; only those larger than 10% are shown

6.1.7 $4f^{13}$ Manifold

The $4f^{13}$ manifold of Tm^{2+} and Yb^{3+} lanthanide ions in solid hosts is well described as that of the corresponding free-ion, perturbed by crystal-field and bonding host effects. Spin-orbit coupling is important in heavy lanthanides (second-half of the series) and it produces an important splitting of the ^2F atomic term into $^2F_{7/2}$ and $^2F_{5/2}$ energy levels, which grows with the atomic number from 8774 cm^{-1} in Tm^{2+} to 10214 cm^{-1} in Yb^{3+} [34]. These levels have degeneracy 4 and 3, respectively, on top of the twofold Kramer's degeneracy inherent to the odd number of electrons. In solids, bonding interactions slightly change the reference $^2F_{7/2}$–$^2F_{5/2}$ spin-orbit coupling splitting from host to host and the degeneracies are broken according to crystal-field symmetry rules. In an O_h environment, the lowest $^2F_{7/2}$ level splits into Γ_{6u}, Γ_{8u}, and Γ_{7u} Kramer's doublets of respective degeneracies 2, 4, and 2, and the highest $^2F_{5/2}$ level splits into Γ_{8u} and Γ_{7u} Kramer's doublets of respective degeneracies 4 and 2. C_{4v} and lower site symmetries totally break the spatial degeneracies, given four lower and three upper levels, all of them with Kramer's double degeneracy.

Tm^{2+} in $CsCaCl_3$ and $CsCaBr_3$

The $4f^{13}$ manifold of Tm^{2+} has been calculated in six-fold coordinated octahedral sites of the $CsCaCl_3$ and $CsCaBr_3$ perovskites [35], as part of a study on several $4f^{12}5d \rightarrow 4f^{13}$ transitions of these multiple emission materials [36–38]. The results on $CsCaCl_3$:Tm^{2+}, obtained in SA-RASSCF/MS-RASPT2/RASSI-SO multiconfigurational ab initio calculations on the $(TmCl_6Cs_8)^{4+}$ cluster under the effect of the quantum mechanical embedding potential of $CsCaCl_3$, are summarized in Table 6.8. All the $4f^{13}$ levels show the same activator-ligand bond length and bond force constant (breathing mode vibrational frequency); the O_h splittings are clear and the analysis of the spin-orbit coupling wave functions in terms of spin-orbit-free levels $^2T_{1u}$, $^2T_{2u}$, and $^2A_{2u}$ derived from the atomic term ^2F, are included in the table. In $CsCaBr_3$:Tm^{2+}, the calculated values are $d_{e,\text{Tm-Br}} = 2.692$ Å, $\bar{\nu}_{a_{1g}} = 218$ cm^{-1}, and $T_e = 0$ cm^{-1} ($1\Gamma_{6u}$), 128 cm^{-1} ($1\Gamma_{8u}$), 248 cm^{-1} ($1\Gamma_{7u}$), 9072 cm^{-1} ($2\Gamma_{8u}$), and 9241 cm^{-1} ($2\Gamma_{7u}$). A hint on the accuracies of the theoretical spin-orbit coupling splittings is given by the $2\Gamma_{8u} \rightarrow 1\Gamma_{6u}$ spin-orbit coupling emission, which is experimentally found to be of 8810 and 8796 cm^{-1}, respectively, in $CsCaCl_3$:Tm^{2+} and $CsCaBr_3$:Tm^{2+} [38], the corresponding theoretical values being 9102 and 9072 cm^{-1}, with a 3.3% overestimation. Oscillator strengths for the multiple emissions of these materials from the $1\Gamma_{8g}$, $3\Gamma_{8g}$, and $8\Gamma_{6g}$ excited states of the $4f^{12}5dt_{2g}$ electronic configuration to the $4f^{13}$ manifold can be calculated with the many-electron wave functions; the ones of $CsCaCl_3$:Tm^{2+} are included in Table 6.8. These emissions are discussed in Sect. 9.5.

Yb^{2+} in Cubic Sites of Fluorite-Type Hosts

The $4f^{13}$ manifold of Yb^{3+} in the CaF_2, SrF_2, BaF_2, and $SrCl_2$ series of fluorite-type of hosts have been calculated in [29] with the method described in Sect. 6.2.6 for the $4f^{12}5d$ manifolds. The state averages are done over all states in the same

Table 6.8 Calculated $4f^{13}$ energy levels of Tm^{2+} in $CsCaCl_3$: Spectroscopic constants $d_{e,\text{Tm-Cl}}$, $\bar{\nu}_{a_{1g}}$, and T_e; emission oscillator strengths; and spin-orbit-free state characters of the wave functions

	State	$d_{e,\text{Tm-Cl}}$[a]	$\bar{\nu}_{a_{1g}}$[b]	T_e[c]	f^{emi}/f_{ref}[d]			Spin-orbit-free character (%)[e]			
$4f^{13}$ configuration: $\langle d_{e,\text{Tm-Cl}} \rangle = 2.557 \pm 0.001$ Å; $\langle \bar{\nu}_{a_{1g}} \rangle = 349 \pm 1$ cm^{-1}											
					Emitting state						
					$1\Gamma_{8g}$	$3\Gamma_{8g}$	$8\Gamma_{6g}$				
derived from $^2F_{7/2}$											
	$1\,\Gamma_{6u}$	2.556	350	0		0.47	0.23	100	$1\,^2T_{1u}$		
	$1\,\Gamma_{8u}$	2.557	349	176	0.03	1.00	1.00	63.15	$1\,^2T_{2u}$	36.85	$1\,^2T_{1u}$
	$1\,\Gamma_{7u}$	2.557	349	330	1.00	0.02		58.28	$1\,^2A_{2u}$	41.74	$1\,^2T_{2u}$
derived from $^2F_{5/2}$											
	$2\,\Gamma_{8u}$	2.556	349	9102			0.01	63.14	$1\,^2T_{1u}$	36.85	$1\,^2T_{2u}$
	$2\,\Gamma_{7u}$	2.557	349	9318				58.26	$1\,^2T_{2u}$	41.73	$1\,^2A_{2u}$

[a] Tm-Cl equilibrium distance, in Å
[b] Breathing mode harmonic vibrational frequency, in cm^{-1}
[c] Adiabatic (minimum-to-minimum) transition energy, in cm^{-1}
[d] Relative emission oscillator strengths from the $4f^{12}5dt_{2g}$ $1\Gamma_{8g}$, $3\Gamma_{8g}$, and $8\Gamma_{6g}$ excited states, calculated at $d_{\text{Tm-Cl}} = 2.540$ Å; reference values are $f_{ref}(1\Gamma_{8g}) = 2.678 \times 10^{-6}$, $f_{ref}(3\Gamma_{8g}) = 4.184 \times 10^{-5}$, and $f_{ref}(8\Gamma_{6g}) = 6.892 \times 10^{-4}$; calculated radiative emission lifetimes are 704 μs for $1\Gamma_{8g}$, 27.8 μs for $3\Gamma_{8g}$, and 1.41 μs for $8\Gamma_{6g}$ excited states
[e] Weights calculated at $d_{\text{Tm-Cl}} = 2.540$ Å

Table 6.9 Calculated $4f^{13}$ energy levels of Yb^{2+} in fluorite-type hosts: Spectroscopic constants $d_{e,\text{Yb-L}}$, $\bar{\nu}_{a_{1g}}$, and T_e; emission oscillator strengths; and spin-orbit-free state characters of the wave functions

	State	$d_{e,\text{Yb-L}}$ [a]	$\bar{\nu}_{a_{1g}}$ [b]	T_e [c]	Experiment[d]	Spin-orbit-free character (%)[e]			
CaF_2:Yb^{3+}, cubic site									
$4f^{13}$ configuration: $\langle d_{e,\text{Yb-F}} \rangle = 2.201 \pm 0.001$ Å; $\langle \bar{\nu}_{a_{1g}} \rangle = 493 \pm 0$ cm^{-1}									
derived from $^2F_{7/2}$									
	$1\,\Gamma_{7u}$	2.200	493	0		62.74	$1\,^2A_{2u}$	37.26	$1\,^2T_{2u}$
	$1\,\Gamma_{6u}$	2.202	493	686		99.99	$1\,^2T_{1u}$		
	$1\,\Gamma_{8u}$	2.202	494	729		63.93	$1\,^2T_{2u}$	36.06	$1\,^2T_{1u}$
derived from $^2F_{5/2}$									
	$2\,\Gamma_{7u}$	2.201	494	10764	10384	62.74	$1\,^2T_{2u}$	37.26	$1\,^2A_{2u}$
	$2\,\Gamma_{8u}$	2.202	493	11196	10794	63.94	$1\,^2T_{1u}$	36.06	$1\,^2T_{2u}$
SrF_2:Yb^{3+}, cubic site									
$4f^{13}$ configuration: $\langle d_{e,\text{Yb-F}} \rangle = 2.252 \pm 0.001$ Å; $\langle \bar{\nu}_{a_{1g}} \rangle = 442 \pm 0$ cm^{-1}									
derived from $^2F_{7/2}$									
	$1\,\Gamma_{7u}$	2.251	442	0		61.97	$1\,^2A_{2u}$	38.02	$1\,^2T_{2u}$
	$1\,\Gamma_{6u}$	2.252	442	581		100.00	$1\,^2T_{1u}$		
	$1\,\Gamma_{8u}$	2.253	442	615		64.01	$1\,^2T_{2u}$	36.00	$1\,^2T_{1u}$
derived from $^2F_{5/2}$									
	$2\,\Gamma_{7u}$	2.251	442	10720		61.97	$1\,^2T_{2u}$	38.03	$1\,^2A_{2u}$
	$2\,\Gamma_{8u}$	2.252	442	11093		64.01	$1\,^2T_{1u}$	35.99	$1\,^2T_{2u}$

(continued)

Table 6.9 (continued)

	State	$d_{e,\text{Yb-L}}$[a]	$\bar{\nu}_{a_{1g}}$[b]	T_e[c]	Experiment[d]	Spin-orbit-free character (%)[e]			
BaF_2:Yb^{3+}, cubic site									
$4f^{13}$ configuration: $\langle d_{e,\text{Yb-F}}\rangle = 2.297 \pm 0.001$ Å; $\langle \bar{\nu}_{a_{1g}}\rangle = 400 \pm 0$ cm^{-1}									
derived from $^2F_{7/2}$									
	1 Γ_{7u}	2.296	399	0		61.32	1 $^2A_{2u}$	38.66	1 $^2T_{2u}$
	1 Γ_{6u}	2.298	400	500		99.99	1 $^2T_{1u}$		
	1 Γ_{8u}	2.298	400	529		64.05	1 $^2T_{2u}$	35.95	1 $^2T_{1u}$
derived from $^2F_{5/2}$									
	2 Γ_{7u}	2.297	400	10690		61.32	1 $^2T_{2u}$	38.68	1 $^2A_{2u}$
	2 Γ_{8u}	2.298	400	11016		64.05	1 $^2T_{1u}$	35.95	1 $^2T_{2u}$
$SrCl_2$:Yb^{3+}, cubic site									
$4f^{13}$ configuration: $\langle d_{e,\text{Yb-Cl}}\rangle = 2.727 \pm 0.001$ Å; $\langle \bar{\nu}_{a_{1g}}\rangle = 266 \pm 0$ cm^{-1}									
derived from $^2F_{7/2}$									
	1 Γ_{7u}	2.726	266	0		60.76	1 $^2A_{2u}$	39.23	1 $^2T_{2u}$
	1 Γ_{6u}	2.728	266	368		100.00	1 $^2T_{1u}$		
	1 Γ_{8u}	2.728	266	416		63.92	1 $^2T_{2u}$	36.08	1 $^2T_{1u}$
derived from $^2F_{5/2}$									
	2 Γ_{7u}	2.727	266	10641		60.78	1 $^2T_{2u}$	39.24	1 $^2A_{2u}$
	2 Γ_{8u}	2.728	266	10896		63.92	1 $^2T_{1u}$	36.08	1 $^2T_{2u}$

[a]Yb-ligand equilibrium distance, in Å

[b]Breathing mode harmonic vibrational frequency, in cm^{-1}

[c]Adiabatic (minimum-to-minimum) transition energy, in cm^{-1}

[d]Magnetic dipole transition, in cm^{-1}

[e]Weights calculated at $d_{\text{Yb-F}} = 2.200$ Å (CaF_2), $d_{\text{Yb-F}} = 2.246$ Å (SrF_2), $d_{\text{Yb-F}} = 2.292$ Å (BaF_2), and $d_{\text{Yb-Cl}} = 2.704$ Å ($SrCl_2$)

block of spin and O_h irrep and all spin states are included in the spin-orbit coupling calculation step. The results are shown in Table 6.9.

Only the energies of the Kramers doublets derived from the atomic level $^2F_{5/2}$ in the cubic site of CaF_2:Yb^{3+} have been compared with experiments, with a good agreement. Identifying the spectral lines of the cubic sites of Yb^{2+} in fluorite-type hosts, where the charge compensation is local most of the times and lowers the symmetry of the O_h site, is a difficult task, but a thorough combined electron paramagnetic resonance (EPR) and optical absorption spectroscopy study on CaF_2:Eu^{3+} was able to identify the absorption lines of the cubic site [39]. Identification of the lines attributed to cubic sites in the emission bands of CaF_2:Eu^{3+} both γ-irradiated and annealed in hydrogen was not so clear [40]. The same is true for the other hosts.

Table 6.9 shows the $4f^{13}$ levels of Yb^{2+} split in a cubic O_h site, different than their splitting in the octahedral O_h site of Tm^{3+} in perovskites (Table 6.8). The limited effect of the host embedding, which is expected due to the internal nature of the unfilled $4f$ shell, protected from the ligands and the rest of the host by the more external $5s^25p^6$ filled shells, is expressed minimally in the separation between the lower states, derived from the atomic level $^2F_{7/2}$, and the upper states, derived from $^2F_{5/2}$, which is governed by the spin-orbit coupling of strongly atomic character. Their effects on the Stark splittings of the mentioned atomic levels are more significant; they follow the expected trend of crystal-field strength: decreasing monotonically across the CaF_2-SrF_2-BaF_2 series following the increase of the Yb-F distance, and with a larger drop in $SrCl_2$, where the Yb-Cl distance is much longer and the ligand ionicity smaller. The contributions of the many-electron spin-orbit-free wave functions to the spin-orbit coupling many-electron wave functions are pretty constant in the four hosts. They even are very similar to those of Tm^{2+} in perovskites, where only the Γ_{7u} states are prone to a slightly more significant change.

6.2 $4f^{N-1}5d$ Manifolds

In addition to the $4f^N$ configuration, all lanthanide ions feature a $4f^{n-1}5d$ configuration towards which electronic transitions from the $4f^N$ ground state are electric dipole allowed. Strong absorptions and efficient luminescence can originate as a consequence. For all stable Ln^{4+} and most Ln^{3+} ions, this $4f^{n-1}5d$ configuration lies at high energies, yielding transitions in the UVC region. However, in some cases, transitions in the visible (e.g. Ce^{3+}) or near-UV (e.g. Tb^{3+}) ranges are possible. For Ln^{2+} ions, the $4f^{N-1}5d$ configuration lies at significantly lower energies, often corresponding to (near-)infrared absorption bands. For the divalent ions, the largest $4f^N - 4f^{N-1}5d$ energies are found in the visible range for ions such as Eu^{2+} or Yb^{2+}. Due to the spatially extensive nature of the $5d$ orbitals, the energy range and structure of the $4f^{N-1}5d$ manifolds are highly affected by the chemical environment. The combination of high transition probabilities and extreme tune-ability make the $4f^{N-1}5d$ configurations, and the associated $4f^N$-$4f^{N-1}5d$ transitions, highly suitable for various applications and therefore among the most studied and intriguing electronic structures in lanthanide photophysics. This section summarizes the most important findings and systematics of the various $4f^{N-1}5d$ configurations resulting from quantum chemical ab initio calculations.

In a lanthanide activator, a $4f \rightarrow 5d$ excitation originates a large number of excited states with a complex electronic structure and associated rich luminescence properties. These are more important in divalent than in trivalent ions because they have much lower excitation energies and the $4f^{N-1}5d$ manifolds are more available.[2]

The electronic structure of the $4f^{N-1}5d$ manifold of a lanthanide ion in a solid host results from the coupling between the $4f^{N-1}$ and the $5d$ sub-shells.

The $4f^{N-1}$ sub-shell has, basically, the electronic structure of the oxidized ion, due to the inner character of the $4f$ orbitals. As discussed in Sect. 6.1, the energy levels of the free-ion are determined by the couplings between $4f$ electrons and by strong $4f$ spin-orbit coupling interactions; in a host, these levels are split into Stark levels by interactions with the solid, but the splittings are small due to the very effective screening of the $5s^25p^6$ shell.

The $5d$ sub-shell, on the contrary, experiences very small spin-orbit coupling interactions and its electronic structure is controlled by the ligand-field splitting of the free-ion ^2D term. In other words, in an O_h site for instance, the $5d$ sub-shell is well described with two well separated sub-levels $5de_g(^2E_g)$ and $5dt_{2g}(^2T_{2g})$, the latter slightly split by spin-orbit coupling.

Accordingly, e.g. for lanthanide ion activators in O_h sites, besides the double-group O_h' irreducible representations and ordinal numbers, which are the only truly good quantum numbers of the excited states, it is insightful to refer to their main configurational and spin characters in forms like $4f^{N-1}(^{2S_{4f}+1}L_J)5de_g$ main configuration, and $S = S_{4f} + 1/2$ (High Spin, HS) or $S = S_{4f} - 1/2$ (Low Spin, LS) main spins, and alike, always having in mind that the states also have contributions from other configurations and spins,

An important additional question on $4f^{N-1}5d$ manifolds is that, whatever complex they are, they are not alone. They expand over broad energy regions and, at some energies, they may coincide and interact with other impurity states, like impurity-trapped-excitons of $4f^{N-1}\phi_{ITE}$ configuration, or states of $4f^{N-1}6s$ configuration. At sufficiently high energies, they will always overlap excited states of the host, like anion-to-cation charge-transfer states (i.e. the many-electron equivalent of the conduction band states used in one-electron formulations), self-trapped excitons, and others.

---

[2] The reason for this is the same behind the ionization potential series: The fourth ionization potential of a lanthanide element Ln (i.e. the ionization potential of Ln^{3+}) is smaller than the third ionization potential (i.e. the ionization potential of Ln^{2+}) due to the increase in positive charge. This is also applicable to a $4f \rightarrow 5d$ excitation because it is a first step towards ionization, an incomplete ionization, where the inner $4f$ electron has to cross the more external $5s^25p^6$ shell ending up in an even more external $5d$ orbital shell. As a matter of fact, only a $5d$ ionization remains for a full ionization, and this is small and almost independent of the lanthanide [41]. Hence, $4f \rightarrow 5d$ excitations and $4f$ ionization potentials follow the same trends.

6.2.1 $5d^1$ Manifold of Ce^{3+}

The $5d^1$ manifold of Ce^{3+} in hosts is made by a set of five Kramers doublets, which may have additional degeneracies in high symmetry sites. In the octahedral sites of elpasolites like Cs_2NaYCl_6, the strong-field splitting of the free-ion ^2D term into $^2T_{2g}$ and ^2E_g is partly split further by spin-orbit coupling, which results into two lower Γ_{8g}, Γ_{7g} levels, and an upper Γ_{8g} level, with respective additional degeneracies 2, 1, and 2 [2, 3]. The $5dt_{2g}$-Γ_{8g} and $5dt_{2g}$-Γ_{7g} levels share a common value of the Ce-ligand bond length, shorter than in the $4f^1$ states; in $5de_g$-Γ_{8g} the bond length is not only longer than in the states of the $5dt_{2g}$ configuration, but also than in the states in the $4f^1$ manifold. In $Cs_2LiLuCl_6$:Ce^{3+} [3] the Ce-Cl distances are: 2.660 Å ($4f$-$1\Gamma_{7u}$), 2.618 Å ($5dt_{2g}$-Γ_{8g}, Γ_{7g}), and 2.669 Å ($5de_g$-Γ_{8g}). In Cs_2NaYCl_6:Ce^{3+} [2]: 2.723 Å ($4f$-$1\Gamma_{7u}$), 2.676 Å ($5dt_{2g}$-Γ_{8g}, Γ_{7g}), and 2.740 Å ($5de_g$-Γ_{8g}). The theoretical –0.047 Å bond length contraction in the $5dt_{2g}$ configuration over the $4f$ configuration in the latter material agrees very well with the 0.046-0.051 Å estimations for the absolute value of the bond-length change deduced experimentally from the analyses of the $4f \rightarrow 5dt_{2g}$ absorption and $5dt_{2g} \rightarrow 4f$ emission vibrational progressions [2]. Such analyses are unable to give information on the sign of the change. The pattern of bond length change upon $4f \rightarrow 5d$ excitations is found to be general for lanthanide ions and it is discussed in detail in Sect. 10.1. The breathing mode vibrational frequency is very similar in all $5d^1$ states. They range between 319 cm^{-1} and 322 cm^{-1} in $Cs_2LiLuCl_6$:Ce^{3+} and between 293 cm^{-1} and 297 cm^{-1} in Cs_2NaYCl_6:Ce^{3+}. In the latter material, the vibrational progression of the $4f \rightarrow 5dt_{2g}$ absorption gives an experimental value of 299 cm^{-1} [2].

In cubic sites, like CaF_2 with non-local charge compensation [4], the order of the ligand-field splitting of ^2D is reversed, but not the order of the spin-orbit coupling splitting of $^2T_{2g}$, which results into a lower Γ_{8g} level and two upper Γ_{8g}, Γ_{7g} levels. In lower symmetry sites, like the D_2 sites of garnets and other sites with even lower symmetry (either in low-symmetry hosts or in high-symmetry hosts with local charge compensations) [4, 7, 42] the $5d^1$ levels split further into five Kramers doublets with no additional degeneracies (see Fig. 8.3 in Sect. 8.2.1).

In Sect. 9.1 the spectroscopic implications of the states of the $5d^1$ manifold of Ce^{3+} in a number of representative hosts is discussed.

6.2.2 $4f5d$ Manifold of Pr^{3+}

Pr^{3+} Free-Ion

The results of SA-CASSCF/MS-CASPT2/RASSI-SO calculations on the $4f5d$ manifold of Pr^{3+} free-ion [9] are shown and compared with experiments [14] in Table 6.10 (Cf. Sect. 6.1.2 for calculation details). The energy levels are underestimated between –1900 cm^{-1} and –2900 cm^{-1}, with a largest relative error of –4.5%. The deviations are uniform and the order of states agrees very well with experiments; order alter-

Table 6.10 Energy levels of the $4f5d$ configuration of the Pr^{3+} free-ion: Experiments [14] and SA-CASSCF/MS-CASPT2/RASSI-SO calculations (deviations from experiment are given in parentheses)

Level	Experiment	Calculation
$^1G^o_4$	61171	58742 (−2429)
$^3F^o_2$	61457	58900 (−2557)
$^3G^o_3$	63356	60483 (−2873)
$^3H^o_4$	63581	61415 (−2166)
$^3F^o_3$	64124	61435 (−2689)
$^3H^o_5$	65240	63270 (−1970)
$^1D^o_2$	65322	63050 (−2272)
$^3G^o_4$	65640	62816 (−2824)
$^3F^o_4$	66518	64115 (−2403)
$^3D^o_1$	66968	64613 (−2355)
$^3G^o_5$	67899	65135 (−2765)
$^3H^o_6$	68078	66192 (−1886)
$^3D^o_2$	68412	66133 (−2278)
$^3D^o_3$	68496	66155 (−2341)
$^3P^o_1$	70755	68703 (−2053)
$^3P^o_0$	70843	68824 (−2019)
$^1F^o_3$	71725	69436 (−2288)
$^3P^o_2$	72185	70168 (−2017)
$^1H^o_5$	75266	72964 (−2301)
$^1P^o_1$	78776	76739 (−2037)

ations are observed only in two narrow ranges: $^3H^o_5$, $^1D^o_2$, and $^3G^o_4$, within a 400 cm^{-1} range, and $^3H^o_6$, $^3D^o_2$, and $^3D^o_3$, within a 420 cm^{-1} range.

There is something which is interesting to remark here, in the $4f5d$ manifold of Pr^{3+}: The lowest level of this two-electron system is a spin singlet, $^1G^o_4$, rather than a spin triplet. We make this remark because it is not uncommon to see the wrong assumption that Hund's rule applies to any system of electrons and expect a spin triplet to be the lowest state of a two-electron open-shell configuration. The rule applies only for two electrons in an equivalent shell (e.g. $4f^2$) but not for two electrons in non-equivalent shells, like $4f5d$. Examples of this can be found in atomic spectroscopy (e.g. the Condon-Shortley rule for an atomic df configuration that states that the singlet-triplet energy order is determined by the parity of the total orbital angular momentum L [43]: positive for odd L, giving $E(^1P) > E(^3P)$, $E(^1F) > E(^3F)$, ..., and negative for even L, giving $E(^1D) < E(^3D)$, $E(^1G) < E(^3G)$, ...); but also in other fields, like magnetic interactions, where two electrons in two orbitals of separated atoms can either have ferromagnetic and antiferromagnetic lowest states. In general, and in active centers of luminescent materials in particular, no rule applies that can state if a spin singlet or a spin triplet is the lowest state of two electrons

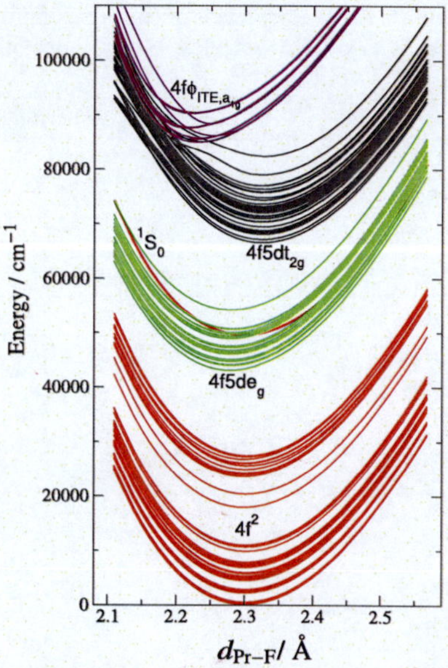

Fig. 6.4 Energy levels of the $(PrF_8)^{5-}$ cubic cluster embedded in CaF_2 corresponding to the electronic configurations $4f^2$ (red), $4f5de_g$ (green), $4f5dt_{2g}$ (black), and $4f\phi_{ITE,a_{1g}}$ (violet), as a function of the Pr-F distance, $d_{\mathrm{Pr-F}}$

in non-equivalent shells (nor if a spin-orbit state with main spin singlet or spin triplet character is). Electron correlation and spin-orbit coupling can play key roles at determining this and only direct calculations or direct measurements of the energy levels can provide reliable information about their actual order.

Pr^{3+} in CaF_2, Cubic Site

The levels of the $4f5d$ configuration (cf. Sect. 6.1.2 for calculation details) are split into the $4f5de_g$ and $4f5dt_{2g}$ manifolds, respectively shown in green and black in Fig. 6.4, as expected from the $5d$ crystal field splitting. Spectroscopic constants and wave function analyses are given in Table 6.11.

The energy curves within each set are basically parallel. The Pr-F bond length in the $4f5de_g$ manifold (2.282±0.001 Å) is 0.02 Å shorter than in the $4f^2$ manifold. On the contrary, the Pr-F bond length in the $4f5dt_{2g}$ manifold (2.327±0.004 Å) is 0.03 Å longer than in the $4f^2$ manifold. This bond length pattern is common to all studied lanthanide ions in O_h sites with eight-fold coordination and it is discussed in Sect. 10.1. In brief, three factors contribute to the Pr-F bond contracting upon $4f \rightarrow 5de_g$ excitation and expanding upon $4f \rightarrow 5dt_{2g}$ excitation: First, the bond length is controlled by the size of the lanthanide $5p^6$ shell (rather than by the inner $4f$ orbitals), which does not change its occupation in the excitation. This means that a one electron excitation from inner $4f$ orbitals to outer $5d$ orbitals, crossing the full $5p$ shell, only experience small orbital size effects on the bond length, much smaller than a simple estimate made from differences between $5d$ and $4f$ orbital radii. Second, a $4f \rightarrow 5d$ excitation creates an inner hole inside the $5p^6$ shell that enhances ligand-to-$4f$ charge transfer; this charge transfer effect tends to shorten the bond length, its contribution being similar to the orbital size effect. Orbital size and

Table 6.11 Calculated $4f5de_g$ and $4f5dt_{2g}$ energy levels of Pr^{3+} in CaF_2: Spectroscopic constants $d_{e,\text{Pr-F}}$, $\bar{\nu}_{a_{1g}}$, and T_e; percentage spin-triplet characters; absorption oscillator strengths; and spin-orbit-free state characters of the wave functions

	State	$d_{e,\text{Pr-F}}$ [a]	$\bar{\nu}_{a_{1g}}$ [b]	T_e [c]	%($S = 1$)	f^{abs}/f_{ref} [d]	Spin-orbit-free character (%)[e]							
$4f5de_g$ dominant configuration: $\langle d_{e,\text{Pr-F}} \rangle = 2.282 \pm 0.001$ Å; $\langle \bar{\nu}_{a_{1g}} \rangle = 508 \pm 2\ \text{cm}^{-1}$														
	$1\ T_{2u}$	2.282	507	43042	20.2	0.190	79.7	1^1T_{2u}	12.8	$1\ ^3T_{2u}$				
	$1\ E_u$	2.281	508	43922	97.3	0.018	83.0	1^3T_{1u}						
	$2\ T_{2u}$	2.281	508	44063	96.2	0.038	77.1	1^3T_{1u}	17.0	$1\ ^3T_{2u}$				
	$1\ A_{2u}$	2.281	508	44661	99.9	0.727	97.6	1^3T_{2u}						
	$1T_{1u}$	2.281	508	45007	90.1	0.053	65.2	1^3T_{1u}						
	$3\ T_{2u}$	2.281	507	45996	81.6	0.436	51.8	1^3T_{2u}						
	$2\ E_u$	2.282	506	46088	78.7	0.031	64.6	2^3T_{2u}	21.3	$1\ ^1E_u$				
	$1\ A_{1u}$	2.282	508	46104	100.0		77.6	2^3T_{1u}	22.0	$1\ ^3T_{1u}$				
	$2\ T_{1u}$	2.281	507	46310	99.7	0.015	46.3	2^3T_{2u}	36.5	$2\ ^3T_{1u}$				
	$2A_{1u}$	2.281	507	46641	99.7		77.7	1^3T_{1u}	22.0	$2\ ^3T_{1u}$				
	$3T_{1u}$	2.282	508	47179	97.6	0.015	39.0	1^3T_{2u}	21.8	2^3T_{1u}	20.2	2^3T_{2u}	16.3	1^3T_{1u}
	$4T_{2u}$	2.282	507	47340	77.8	0.007	30.3	$2\ ^3T_{2u}$	18.0	$1\ ^3T_{2u}$				
	$4T_{1u}$	2.281	508	47423	96.3	0.028	36.3	$2\ ^3T_{1u}$	34.5	$1\ ^3T_{2u}$				
	$3E_u$	2.281	508	47456	99.9	0.000	88.1	$1\ ^3T_{2u}$						
	$2A_{2u}$	2.281	507	48841	100.0	0.021	97.5	$2\ ^3T_{2u}$						
	$5T_{2u}$	2.281	507	49185	98.0	0.003	46.8	$2\ ^3T_{1u}$	43.4	$2\ ^3T_{2u}$				
	$4E_u$	2.283	504	49267	64.4	0.000	35.6	$1\ ^1E_u$	28.1	$2\ ^3T_{1u}$	31.0	$2\ ^3T_{2u}$		
	$5T_{1u}$	2.282	506	49281	35.5	0.000	64.4	$1\ ^1T_{1u}$	17.6	$1\ ^3E_u$				
	$5E_u$	2.284	513	49657	59.8	0.002	40.2	$1\ ^1E_u$	58.5	$2\ ^3T_{1u}$				
	$6T_{2u}$	2.282	507	49911	37.4	0.000	62.4	$2\ ^1T_{2u}$	24.1	$2\ ^3T_{1u}$				
	$7T_{2u}$	2.285	509	50523	89.2	0.000	76.3	$1\ ^3E_u$						
	$6T_{1u}$	2.285	509	50632	76.3	0.000	67.8	$1\ ^3E_u$	23.5	$1\ ^1T_{1u}$				
	$7T_{1u}$	2.282	506	54127	5.0	0.000	95.0	$2\ ^1T_{1u}$						

(continued)

Table 6.11 (continued)

State	$d_{e,\text{Pr-F}}$[a]	$\bar{\nu}_{a_{1g}}$[b]	T_e[c]	%($S = 1$)	f^{abs}/f_{ref}[d]	Spin-orbit-free character (%)[e]							
$4f5dt_{2g}$ dominant configuration: $\langle d_{e,\text{Pr-F}} \rangle = 2.327 \pm 0.004$ Å; $\langle \bar{\nu}_{a_{1g}} \rangle = 479 \pm 10$ cm^{-1}													
$3A_{1u}$	2.327	490	66695	12.9		87.0	$1\,^1A_{1u}$						
$6E_u$	2.331	472	67593	23.9	0.009	76.2	$2\,^1E_u$	15.1	$3\,^3T_{2u}$				
$8T_{2u}$	2.332	468	68596	94.4	0.228	47.1	$3\,^3T_{2u}$	38.3	$1\,^3A_{2u}$				
$8T_{1u}$	2.329	480	68773	96.7	0.294	53.3	$3\,^3T_{1u}$	29.0	$2\,^3E_u$				
$9T_{1u}$	2.332	467	68929	85.3	1.000	37.2	$3\,^3T_{2u}$	22.8	$2\,^3E_u$	20.6	$3\,^3T_{1u}$		
$10T_{1u}$	2.329	481	69053	69.6	0.062	32.2	$3\,^1T_{1u}$	26.0	$3\,^3T_{1u}$	16.3	$1\,^3A_{1u}$		
$7E_u$	2.330	476	69103	85.2	0.092	41.2	$3\,^3T_{2u}$	36.9	$4\,^3T_{1u}$				
$3A_{2u}$	2.332	468	69147	94.1	0.309	90.3	$3\,^3T_{2u}$						
$9T_{2u}$	2.333	462	70004	81.3	0.014	24.3	4^3T_{1u}	23.8	1^3A_{2u}	21.9	3^3T_{2u}	18.1	3^1T_{2u}
$8E_u$	2.326	485	70493	91.6	0.054	80.0	$3\,^3T_{1u}$						
$11T_{1u}$	2.327	491	70525	75.1	0.005	41.1	$3\,^3T_{2u}$	22.1	$3\,^1T_{1u}$				
$4A_{1u}$	2.329	482	70621	90.0		78.5	$3\,^3T_{1u}$						
$12T_{1u}$	2.327	490	71054	86.6	0.006	32.8	$4\,^3T_{2u}$	23.2	$2\,^3E_u$	22.0	$3\,^3E_u$		
…													

[a]Pr-F equilibrium distance, in Å

[b]Breathing mode harmonic vibrational frequency, in cm^{-1}

[c]Adiabatic (minimum-to-minimum) transition energy, in cm^{-1}

[d]Relative absorption oscillator strength from the $4f^2$ $1T_{2g}$ ground state, calculated at $d_{\text{Pr-F}} = 2.291$ Å; the reference value is $f_{ref} = 0.0810$ [e]Weights larger than 15% are given. All weights are calculated at $d_{\text{Pr-F}} = 2.291$ Å

charge transfer effects together either cancel each other or give a small contraction of the bond length. Finally, the crystal or ligand field stabilizes the $5de_g$ shell, which tends to shorten the bond length; this ligand field effect is larger than the other two effects (although not very much), with an overall result of bond length contraction upon $4f \rightarrow 5de_g$ excitation. The ligand field effect is opposite for excitation to the highest $5dt_{2g}$ shell, with the overall result of a bond length expansion upon $4f \rightarrow 5dt_{2g}$ excitation.

As for the $4f^2$ manifold, all the energy levels of the $4f5de_g$ and $4f5dt_{2g}$ manifolds show basically the same breathing mode harmonic vibrational frequency (508±2 cm^{-1} in $4f5de_g$ and 479±10 cm^{-1} in $4f5dt_{2g}$). A 420 cm^{-1} replica found in excitation spectra of CaF_2:Pr^{3+} was attributed to the cubic defect [44]. If this has to be compared with the theoretical 508 cm^{-1}, the overestimation would be much larger than expected from comparisons between theory and experiments in other materials, which questions the 420 cm^{-1} assignment. Unassigned vibronic replica have been found of 460 cm^{-1} [45] and 485 cm^{-1} [46], which would agree better with the theoretical vibrational frequency of the cubic defect.

The $4f5de_g$ manifold is made of 23 levels starting at 43000 cm^{-1} and spanning an energy window of 11000 cm^{-1}. The largest gap has 3500 cm^{-1} and separates the last level, $7T_{1u}$, from the others. Additional, smaller gaps are found below the second level ($1E_u$, 900 cm^{-1}), sixth level ($3T_{2u}$, 1000 cm^{-1}) and fifteenth level ($2A_{2u}$, 1400 cm^{-1}). The lowest level of this manifold, $1T_{2u}$, has a main character of spin singlet, 80% 1^1T_{2u}, in agreement with $^1G_4^{\circ}$ being the lowest state of the free-ion's $4f5d$ configuration; and the first spin-triplet is around 900 cm^{-1} higher, $1E_u(1^3T_{1u})$. As discussed in the case of the free-ion, this is normal and does not contradict Hund's rule, which does not apply to electrons in non-equivalent shells.

The $4f^2\ ^1S_0$ level is embedded in this manifold. This means that no $4f^2 \rightarrow 4f^2$ emission can result from this state and, consequently, no quantum cutting can occur in the Pr cubic site of CaF_2:Pr^{3+}, but a strong $4f5de_g \rightarrow 4f^2$ emission must be expected. These are the emission properties assumed in [47] for the cubic defect.

The $4f5dt_{2g}$ manifold starts 23500 cm^{-1} above the beginning of the $4f5de_g$ manifold and 12500 cm^{-1} after its end, and it spans an energy window of 15000 cm^{-1}. Its lowest level also has a main character of spin singlet, $3A_{1u}(1^1A_{1u})$, in correspondence with the O_h splitting of the free-ion's $^1G_4^{\circ}$ and the lowest $4f5de_g$ level: In an O_h field, ^1G splits into $^1T_{2u} +^1 A_{1u} +^1 E_u +^1 T_{1u}$ and $J = 4$ splits into $T_{2u} + A_{1u} + E_u + T_{1u}$; and accordingly, the lowest $4f5de_g$ state is $1T_{2u}(1^1T_{2u})$, the lowest $4f5dt_{2g}$ state is $3A_{1u}(1^1A_{1u})$ and the second lowest is $6E_u(2^1E_u)$, and $10T_{2u}(3^1T_{1u})$ lies in the low range of this manifold.

6.2.3 *$4f^5 5d$ Manifold of Sm^{2+}*

Sm^{2+} in CaF_2

A number of $4f^5 5de_g$ energy levels of Sm^{2+} in CaF_2 have been calculated with the same procedures used for the $4f^6$ levels in Sect. 6.1.4. Given the large number of excited states in the $4f^5 5d$ manifold and its high density of states, criteria to limit the number of roots to be calculated at the SA-RASSCF level of theory were discussed in Sect. 2.3.2. The roots chosen to be calculated and to define the state-averages for molecular orbital optimization were: one $^7A_{1u}$, one $^7A_{2u}$, and two ^7E_u; five $^7T_{1u}$ and five $^7T_{2u}$; one $^5A_{1u}$, one $^5A_{2u}$, and two ^5E_u; and five $^5T_{1u}$ and five $^5T_{2u}$. Exploratory calculations showed these spin-septet and quintet states of main $4f^5(^6H + ^6F)5de_g$ character span an energy region of around 20000 cm^{-1} below an energy gap that is followed by a continuum of excited states (cf. Sect. 2.3.2).

Once the choices of states to calculate and state-averages to optimize molecular orbitals have been made, SA-RASSCF($4f/5d6s5f$-$4e$) calculations were done; this RAS space is very flexible and the results are expected to be close to CASSCF calculations of 6 electrons in 20 orbitals, the $4f$, $5d$, $6s$, and $5f$ orbital shells. When these calculations are finished and the wave functions are analyzed, they are followed by MS-CASPT2 calculations correlating all valence electrons except those with main character Sm $4d$. The previous analyses, which identify the O_h irrep of all the SA-RASSCF wave functions, allow to perform the subsequent MS-CASPT2 calculations over actual O_h symmetry blocks, rather than over D_{2h} blocks. This procedure increases accuracy and decreases computational demands: It reduces symmetry breakings associated with spurious interactions among wave functions of different irreps due to limited accuracy and increases the natural parallelism of site symmetry. As an example, MS-CASPT2 calculations of wave functions of the O_h blocks $^7A_{1u}$, $^7A_{2u}$, and ^7E_u are done separately, in parallel, rather than all together within the D_{2h} ^7A_u block that might contain them. Finally, RASSI-SO calculations are done where all calculated spin-septets and spin-quintets are allowed to couple via spin-orbit-coupling, as indicated in Sect. 6.1.4. The potential energy curves and the underlying data are summarized in Fig. 6.5 (right) and Table 6.12.

All potential energy curves are basically parallel, as they all correspond to the same $4f^5 5de_g$ configuration, with only small contributions from $4f^5 5dt_{2g}$ and $4f^5\phi_{ITE}$ configurations. States of a dominant character $4f^5 5dt_{2g}$ and $4f^5\phi_{ITE}$ configurations appear at higher energies. The $4f^6$ wave functions (Table 6.4) and the $4f^5 5de_g$ wave functions (Table 6.12) are used to calculate $4f^5 5de_g \rightarrow 4f^6$ emission and $4f^6 \rightarrow 4f^5 5de_g$ absorption oscillator strengths (cf. both Tables), which, together with the minimum-to-minimum transition energies T_e, the horizontal offsets (changes of $d_{\mathrm{e,Sm\text{-}F}}$ in the transition), and the breathing mode vibrational frequencies $\bar{\nu}_{a_{1g}}$, allow to calculate emission and absorption spectra profiles. These are used to discuss the spectroscopy of the $CaF_2{:}Sm^{2+}$ doped optical materials (Sect. 9.3). It is this discussion that assesses the reliability of the theoretical calculations. Reliability is only gained when there is an agreement between theory and experiment in the known facts. Once the reliability is confirmed, the results of the calculations provide further detailed interpretation of the measurements and are in a position to shed light into troublesome issues, if necessary.

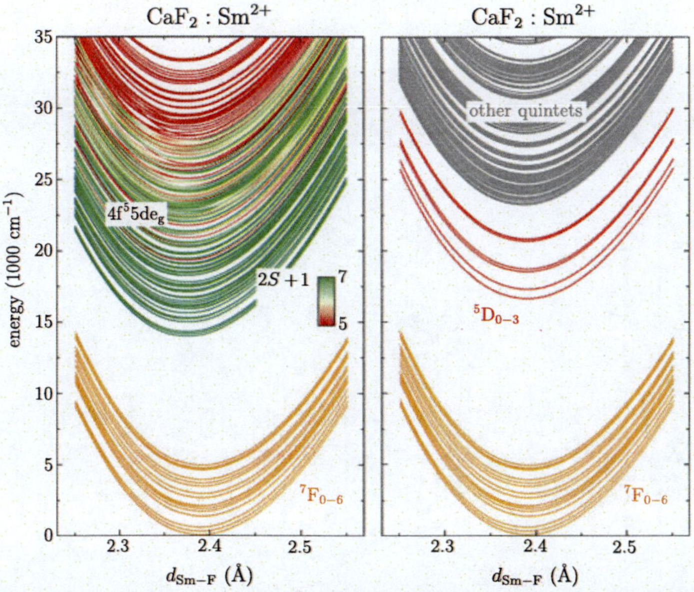

Fig. 6.5 Calculated potential energy curves of $4f^6$ and $4f^5 5de_g$ states of Sm^{2+} in CaF_2 after the RASSI-SO step. Left: $4f^6(^7F_{0-6})$ states (orange) and $4f^5(^6H+^6F)5de_g$ states (color depending on their spin character as indicated). Right: $4f^6$ states of main spin septet and quintet characters

6.2.4 $4f^6 5d$ Manifold of Eu^{2+}

If the $4f^{N-1}5d$ manifolds of lanthanide ions in solid hosts are difficult to study in general, the $4f^6 5d$ manifold of Eu^{2+} is particularly complex due to the huge number of states: 30030, or 15015 Kramers doublets. Yet, a good understanding of the $4f^6 5d$ configuration is very important because many modern technologies rely on it.

It is empirically known that the color of the Eu^{2+} $5d \rightarrow 4f$ luminescence, i.e. the onset of the Eu^{2+} $4f^6 5d$ manifold above the $4f^7$ ground state, is very sensitive to the chemical nature of the ligands. Fluorides and other ionic-bond forming ligands typically have high onsets and violet to near-UV emissions. Less ionic, more covalent-bond forming ligands like sulfides or nitrides lower the $4f^6 5d$ onset and red-shift the emission, allowing a color tuning across the visible spectral range: E.g. CaS:Eu^{2+} and SrS:Eu^{2+} feature red and orange emissions [48]. These two examples illustrate that not only the ligands are relevant here, but host cations also play a relevant role by means of the structural and host embedding changes that they produce.

A comprehensive multiconfigurational ab initio study aimed at providing a quantitative understanding of the $4f^6 5d$ manifold of Eu^{2+} in alkaline earth difluorides CaF_2, SrF_2, and BaF_2, and in alkaline earth sulfides CaS, SrS, and CaS was made in [27], Next in this book, the results on CaF_2:Eu^{2+} are summarized and the differences

Table 6.12 Calculated $4f^5 5de_g$ energy levels of Sm^{2+} in CaF_2: Spectroscopic constants $d_{e,\text{Sm-F}}$, $\bar{\nu}_{a_{1g}}$, and T_e; absorption oscillator strengths; and spin-orbit-free state characters of the wave functions. Only states with significant absorption oscillator strength and a few selected states are shown

	State	$d_{e,\text{Sm-F}}$[a]	$\bar{\nu}_{a_{1g}}$[b]	T_e[c]	f^{abs}/f_{ref}[d]	Spin-orbit-free character (%)[e]							
$4f^5 5de_g$ dominant configuration: $\langle d_{e,\text{Sm-F}}\rangle = 2.374 \pm 0.001$ Å; $\langle \bar{\nu}_{a_{1g}}\rangle = 432 \pm 3$ cm^{-1}													
$4f^5(^6H)5de_g$ main character													
High-spin states													
	1 A_{1u}	2.375	432	13855		54.28	1 $^7T_{1u}$	34.92	1 $^7T_{2u}$				
	1 T_{1u}	2.375	432	14020	0.12	50.96	1 $^7T_{1u}$	37.39	1 $^7T_{2u}$				
	2 T_{1u}	2.374	432	15061	1.00	58.62	1 $^7T_{2u}$	14.45	2 $^7T_{2u}$	13.25	1 $^7T_{1u}$		
	3 T_{1u}	2.377	432	15754	0.13	33.75	2 $^7T_{2u}$	18.32	1 ^7E_u	17.98	2 $^7T_{1u}$	12.67	1 $^7A_{2u}$
	4 T_{1u}	2.376	431	16516	0.07	23.69	1 ^7E_u	23.03	2 $^7T_{2u}$	16.51	3 $^7T_{2u}$	12.04	2 $^7T_{1u}$
						10.21	1 $^7T_{2u}$						
	7 T_{1u}	2.375	432	17615	0.01	24.67	2 $^7T_{2u}$	20.14	3 $^7T_{2u}$	16.45	1 ^7E_u	12.34	1 $^7T_{1u}$
	8 T_{1u}	2.375	433	18097	0.01	25.61	1 $^7T_{2u}$	20.59	2 $^7T_{2u}$	18.39	1 $^7T_{1u}$	14.81	3 $^7T_{1u}$
	9 T_{1u}	2.373	432	18355	0.02	18.36	3 $^7T_{2u}$	16.70	1 $^7T_{1u}$	12.13	2 $^7T_{1u}$	11.86	3 $^7T_{1u}$
						11.74	1 $^7T_{2u}$	10.06	1 ^7E_u				
up to	7 E_u	2.375	433	18883		48.45	2 $^7T_{2u}$	12.86	1 ^7E_u	11.77	2 $^7T_{1u}$		
Intercalated high, low, and mixed-spin states													
	11 T_{1u}	2.374	432	18956	0.01	49.97	1 $^5T_{1u}$	38.52	1 $^5T_{2u}$				
	12 T_{1u}	2.374	432	19284	0.03	44.46	3 $^7T_{2u}$	17.64	3 $^7T_{1u}$				
	13 T_{1u}	2.374	433	19681	0.14	51.85	1 $^7T_{2u}$	17.54	3 $^7T_{2u}$	11.93	1 $^7T_{1u}$		
up to	19 T_{2u}	2.374	431	21599		44.63	1 $^5T_{1u}$	24.90	2 $^5T_{2u}$	11.18	2 $^5T_{1u}$		

(continued)

Table 6.12 (continued)

	State	$d_{e,\text{Sm-F}}$[a]	$\bar{\nu}_{a_{1g}}$[b]	T_e[c]	f^{abs}/f_{ref}[d]	Spin-orbit-free character (%)[e]							
$4f^5(^6H+^6F)5de_g$ mixed character													
Intercalated high, low, and mixed-spin states													
	18 T_{1u}	2.374	428	21771	0.01	42.51	1 $^5T_{2u}$	20.65	4 $^7T_{2u}$	13.19	1 $^5T_{1u}$	12.69	2 $^5T_{2u}$
	21 T_{1u}	2.374	433	22122	0.07	20.93	2 $^7T_{2u}$	16.76	1 ^7E_u	16.68	2 $^7T_{1u}$		
	23 T_{1u}	2.374	417	22628	0.18	22.26	4 $^7T_{2u}$	18.35	3 $^7T_{2u}$	15.30	3 $^7T_{1u}$	12.65	5 $^7T_{1u}$
	25 T_{1u}	2.374	440	22851	0.04	25.58	3 $^7T_{2u}$	25.19	4 $^7T_{2u}$	20.70	3 $^7T_{1u}$		
	29 T_{1u}	2.374	431	23592	0.25	23.09	2 $^5T_{2u}$	20.80	1 ^5E_u	14.69	1 $^5T_{1u}$	12.97	1 $^5A_{1u}$
	30 T_{1u}	2.374	433	23825	0.63	31.67	2 ^7E_u	20.79	1 $^5A_{1u}$	13.91	4 $^7T_{2u}$		
	32 T_{1u}	2.374	430	24325	0.02	21.44	2 $^5T_{1u}$	18.04	3 $^5T_{2u}$	15.80	1 $^5T_{2u}$	13.22	1 $^5A_{2u}$
	33 T_{1u}	2.374	435	24345	0.01	34.79	5 $^7T_{1u}$	16.26	1 $^5A_{1u}$	15.24	4 $^7T_{1u}$		
	36 T_{1u}	2.375	434	25194	0.01	53.18	4 $^7T_{2u}$	23.75	1 $^5A_{1u}$	13.37	5 $^7T_{1u}$		
	38 T_{1u}	2.373	440	25577	0.01	30.69	3 $^5T_{1u}$	13.17	2 $^5T_{2u}$	13.03	3 $^5T_{2u}$	12.53	1 $^5T_{1u}$
	42 T_{1u}	2.374	445	26601	0.02	32.21	2 ^7E_u	19.63	1 $^5A_{1u}$	18.07	4 $^7T_{2u}$		
up to	46 T_{1u}	2.373	449	27580		26.82	2 $^5T_{1u}$	13.54	1 ^5E_u	12.92	2 $^5T_{2u}$	11.87	4 $^5T_{1u}$
Low-spin states													
	15 A_{2u}	2.373	431	27813		37.26	4 $^5T_{1u}$	18.10	2 ^5E_u	16.96	3 $^5T_{1u}$	15.26	2 $^5T_{1u}$
up to	47 T_{2u}	2.373	433	29258		24.97	3 $^5T_{2u}$	23.39	2 ^5E_u	21.91	4 $^5T_{1u}$	13.41	3 $^5T_{1u}$
						11.37	4 $^5T_{2u}$						
$4f^5(^6F)5de_g$ main character													
Low-spin states													
	49 T_{1u}	2.373	432	29364		51.27	4 $^5T_{2u}$	23.91	5 $^5T_{2u}$				
	...												

[a] Sm-F equilibrium distance, in Å

[b] Breathing mode harmonic vibrational frequency, in cm^{-1}

[c] Adiabatic (minimum-to-minimum) transition energy, in cm^{-1}

[d] Relative absorption oscillator strength from the $4f^6$ 1 A_{1g} ground state, calculated at $d_{\text{Sm-F}} = 2.400$ Å; the reference value is $f_{ref} = 9.724 \times 10^{-3}$ [e] Weights larger than 10% are given. All weights are calculated at $d_{\text{Sm-F}} = 2.350$ Å

with CaS:Eu^{2+} pinpointed. The spectroscopic implications of the ab initio results on the six hosts are discussed in Sect. 9.4.3. Section 10.5 puts a special focus on the insights provided by such a set of complex calculations.

Eu^{2+} in CaF_2

A number of $4f^65de_g$ and $4f^6\phi_{ITE,a_{1g}}$ energy levels of Eu^{2+} in CaF_2 have been calculated in [27] with the same procedures used for the $4f^7$ levels in Sect. 6.1.6. The state-averages used for molecular orbital optimization were one $^8A_{1g}$, two $^8A_{2g}$, and three ^8E_g; six $^8T_{1g}$ and five $^8T_{2g}$; thirty-nine $^6A_{1g}$, thirty-six $^6A_{2g}$, and seventy-five ^6E_g; one hundred-eight $^6T_{1g}$ and one hundred-eleven $^6T_{2g}$ terms. The distribution of 7 electrons among the 20 molecular orbitals of main Eu $4f$, $5d$, $6s$, and $5f$ character in the SA-RASSCF($4f/5d6s5f$-$4e$) double f-shell calculations led to a number of CSFs ranging between 4500 and 41500. As in the $4f^7$ manifold, all valence electrons were correlated in the MS-RASPT2 calculations except those with main Eu $4d$character. Analyses of the MS-RASPT2 potential energy surfaces revealed small spikes or irregularities coinciding with sharp avoided crossings between some high-energy spin-sextet excited states of $4f^65d$ and $4f^6\phi_{ITE,a_{1g}}$ main configurational character; these did not disappear using different reasonable values of MS-RASPT2 parameters. As a consequence, and bearing in mind that the high energy range where these problems were found is not relevant for the spectroscopic discussion of these materials, the high energy spin-sextet states were excluded from the subsequent RASSI-SO calculations, which included all spin-orbit free states of spin multiplicities $2S+1=8$ and 6 computed at the MS-RASPT2 level up to the indicated energies.

Table 6.13 shows the maximum numbers of states of main configurational characters $4f^65d$ and $4f^6\phi_{ITE,a_{1g}}$ (and main high-spin $S=7/2$ and low-spin $S=5/2$ characters) with a ^7F_J many-electron state for the $4f^6$ subshell. This is important for the analysis of the many-electron wave functions resulting from the calculations, having in mind that the energy difference between the ^7F_J and ^5D_J states of the $4f^6$ subshell is high (of around 20000 cm^{-1}, cf. Table 6.5) and $4f^6(^5D_J)5d$ states with main spin-sextet character are expected to lie more or less at that energy difference above the corresponding $4f^6(^7F_J)5d$ states. Such states, besides being high in energy, correspond to double and higher electronic excitations from the $4f^7$ ground state (an intershell $4f$-$5d$ excitation plus intrashell $4f$-$4f$ excitations) and they are not expected to play a relevant role in regular spectroscopic absorptions and emissions.

The results on CaF_2:Eu^{2+} are summarized in Fig. 6.6, together with corresponding results for CaS:Eu^{2+}, and Table 6.14.

Four crowded sub-manifolds constitute the Eu^{2+} $4f^65d$ manifold in CaF_2:Eu^{2+}. This is so also in the other difluorides and in the sulfides.[3] The two sub-manifolds of lower energy have clear configurational and spin characters and the two of higher

[3] Any reference in the following discussion to the $5de_g$ shell in difluorides, where Eu^{2+} occupies an 8-fold coordinated cubic O_h site, must be changed to $5dt_{2g}$ in sulfides, where Eu^{2+} occupies a 6-fold coordinated octahedral O_h site, meaning the lowest of the ligand-field split $5d$ shells. Equivalently, $5dt_{2g}$ in difluorides corresponds to $5de_g$ in sulfides, meaning the highest of the $5d$ shells.

Table 6.13 O_h Γ_{6g}, Γ_{7g}, and Γ_{8g} states derived from the ${}^7F_J \times ({}^2E_g + {}^2T_{2g} + {}^2A_{1g})$ coupling: corresponding to the $4f^6({}^7F_J)5d$ and $4f^6({}^7F_J)\phi_{ITE,a_{1g}}$ configurations

	Total			High-spin			Low-spin		
J	Γ_{6g}	Γ_{7g}	Γ_{8g}	Γ_{6g}	Γ_{7g}	Γ_{8g}	Γ_{6g}	Γ_{7g}	Γ_{8g}
$4f^6({}^7F_J) \times 5de_g({}^2E_g)$									
0	0	0	1	0	0	1	0	0	0
1	1	1	2	1	1	1	0	0	1
2	2	2	3	1	1	2	1	1	1
3	2	2	5	1	1	3	1	1	2
4	3	3	6	2	2	3	1	1	3
5	4	4	7	2	2	4	2	2	3
6	4	4	9	2	2	5	2	2	4
${}^7F_{0-6}$	16	16	33	9	9	19	7	7	14
$4f^6({}^7F_J) \times 5dt_{2g}({}^2T_{2g})$									
0	0	1	1	0	1	1	0	0	0
1	1	2	3	1	1	2	0	1	1
2	3	2	5	2	1	3	1	1	2
3	4	3	7	2	2	4	2	1	3
4	4	5	9	2	3	5	2	2	4
5	5	6	11	3	3	6	2	3	5
6	7	6	13	4	3	7	3	3	6
${}^7F_{0-6}$	24	25	49	14	14	28	10	11	21
$4f^6({}^7F_J) \times \phi_{ITE,a_{1g}}({}^2A_{1g})$									
0	1	0	0	1	0	0	0	0	0
1	1	0	1	0	0	1	1	0	0
2	0	1	2	0	1	1	0	0	1
3	1	2	2	1	1	1	0	1	1
4	2	1	3	1	0	2	1	1	1
5	2	1	4	1	1	2	1	0	2
6	2	3	4	1	2	2	1	1	2
${}^7F_{0-6}$	9	8	16	5	5	9	4	3	7

energy overlap and interact significantly, which gives them either mixed configurational or mixed spin character, or both.

The two lowest sub-manifolds result from the coupling between the $4f^6$ inner electron subshell in all states derived from the ${}^7F$ atomic term and the $5de_g$ outer electron subshell; hence, they have a $4f^6({}^7F_J)5de_g$ configuration. The $4f^6$ inner subshell is well shielded from external fields by the more external $5s^25p^6$ closed-shells and it is weakly split by the crystal-field but strongly split by spin-orbit coupling, very much like the $4f^6$ shell of Eu^{3+}. In consequence, a first set of states derived from the ${}^7F$ atomic term is very well separated from the next set, derived from ${}^5D$

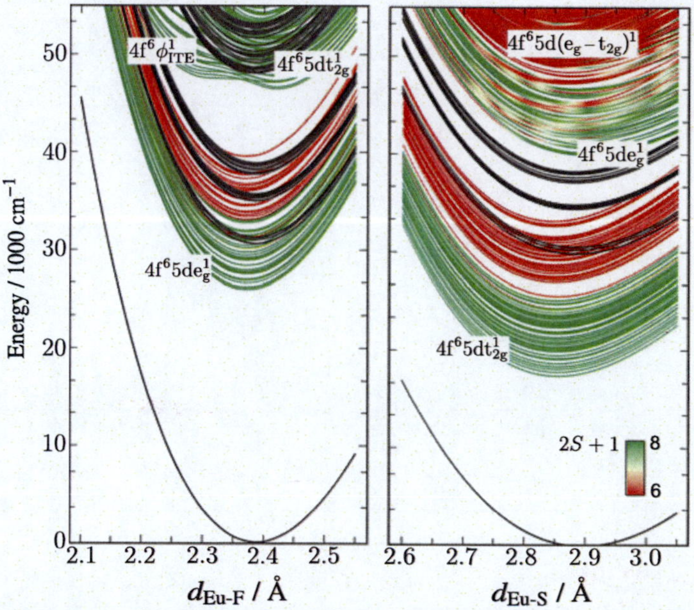

Fig. 6.6 Potential energy curves of Eu^{2+} in CaF_2 (left) and CaS (right). Levels of the $4f^7$ configuration are shown in black. Colored curves indicate excited $4f^6(5d{+}\phi_{ITE,a_{1g}})$ configurations; Their colors represent the spin-character of the eigenstates, ranging from green for pure spin-octets to red for pure spin-sextets

(Table 6.5). On the other hand, the states of the $5de_g$ outer subshell experience a small spin-orbit coupling and are well separated from the next $5dt_{2g}$ subshell due to strong ligand-field effects, so that they are well described as $5de_g\ {}^2E_g$. The states resulting from the coupling $4f^6({}^7F_J) \times 5de_g({}^2E_g)$ split, in turn, into those with main high-spin character (HS, spin-octets, $S = 7/2$) and main low-spin character (LS, spin-sextets, $S = 5/2$), always having in mind that this is only a main character, because spin-orbit coupling prevents S from being a true good quantum number. These two sets of states constitute, respectively, the first and the second Eu^{2+} $4f^65d$ sub-manifolds: High-spin $4f^6({}^7F_J)5de_g$ and low-spin $4f^6({}^7F_J)5de_g$, respectively shown in green and red in Fig. 6.6. How many of them are produced in each case is shown in Table 6.13. At energies where both sub-manifolds overlap, the high- and low-spin characters are not so clearly dominant and the states are better described as mixed-spin states; in fact, the spin-orbit coupling between the two sub-manifolds turns out to be very small and this only happens for a very small number of states. All this is shown in Table 6.14, In it, is is also clear that all these states have very similar bond lengths and harmonic vibrational frequencies, which is a consequence of their common configurational character $4f^65de_g$.

Table 6.15 shows the Eu-ligand bond length in the two lowest sub-manifolds in alkaline-earth difluorides and sulfides. The bond length is shorter in these states of

Table 6.14 Calculated $4f^65d$ and $4f^6\phi_{ITE,a_{1g}}$ energy levels of Eu^{2+} in CaF_2: Spectroscopic constants $d_{e,\text{Eu-F}}$, $\bar{\nu}_{a_{1g}}$, and T_e; and spin-orbit-free state characters of the wave functions. Only a few selected states are shown

	State	$d_{e,\text{Eu-F}}$[a]	$\bar{\nu}_{a_{1g}}$[b]	T_e[c]	Spin-orbit-free character (%)[d]							
$4f^65de_g$ dominant configuration: $\langle d_{e,\text{Eu-F}}\rangle = 2.371 \pm 0.002$ Å; $\langle\bar{\nu}_{a_{1g}}\rangle = 428 \pm 2$ cm^{-1}												
$4f^6(^7F_J)5de_g$ main character												
High-spin states												
	$1\,\Gamma_{8g}$	2.373	428	25720	51.00	$1\,^8T_{2g}$	31.49	$1\,^8T_{1g}$	11.47	$1\,^8E_g$		
up to	$10\,\Gamma_{7g}$	2.372	429	32742	80.31	$2\,^8T_{2g}$	18.12	$2\,^8T_{1g}$				
Mixed-spin states												
	$12\,\Gamma_{8g}$	2.373	427	29855	50.18	$1\,^6T_{1g}$	13.53	$1\,^8T_{2g}$	12.57	$2\,^8T_{1g}$	10.93	$2\,^8T_{2g}$
	$13\,\Gamma_{8g}$	2.372	428	29950	42.22	$1\,^6T_{1g}$	15.14	$1\,^8E_g$	12.74	$2\,^8T_{2g}$	12.61	$2\,^8T_{1g}$
						10.38	$1\,^8T_{2g}$					
Low-spin states												
	$8\,\Gamma_{7g}$	2.372	427	31106	83.67	$1\,^6T_{1g}$	12.47	$1\,^6T_{2g}$				
up to	$16\,\Gamma_{7g}$	2.370	429	39354	90.02	$2\,^6T_{2g}$						
$4f^6(^7F_J)(5dt_{2g}+\phi_{ITE,a_{1g}})$ main character												
High-spin states												
	$34\,\Gamma_{8g}$	2.419	401	46209	39.85	$3\,^8T_{2g}$	32.64	$3\,^8T_{1g}$				
	...											
	$35\,\Gamma_{6g}$	2.400	1074	55022	48.20	$6\,^8T_{1g}$	14.55	$3\,^8E_g$				
	...											

[a]Eu-F equilibrium distance, in Å

[b]Breathing mode harmonic vibrational frequency, in cm^{-1}

[c]Adiabatic (minimum-to-minimum) transition energy, in cm^{-1}

[d]Weights larger than 10% are given. All weights are calculated at $d_{\text{Sm-F}} = 2.350$ Å

Table 6.15 Calculated Eu-F and Eu-S bond-lengths $d_{e,\text{Eu-F}}$ and $d_{e,\text{Eu-S}}$ (in Å) and breathing mode harmonic vibrational frequencies $\bar{\nu}_{a_{1g}}$ (in cm^{-1}) of Eu^{3+} in its $4f^6$ ground state and Eu^{2+} in its $4f^7$ ground state and lowest $4f^65de_g$ and $4f^65dt_{2g}$ states, in alkaline earth difluorides and sulfides

			CaF_2	SrF_2	BaF_2			CaS	SrS	BaS
	Config.	State	$d_{e,\text{Eu-F}}$			Config.	State	$d_{e,\text{Eu-S}}$		
Eu^{3+}	$4f^6(^7F_0)$	$1\ A_{1g}$	2.261	2.313	2.363			2.729	2.800	2.829
Eu^{2+}	$4f^7(^8S_{7/2})$	$1\ \Gamma_{6u}$	2.388	2.470	2.558			2.876	2.975	3.087
Eu^{2+}	$4f^65de_g$	$1\ \Gamma_{8g}$	2.373	2.451	2.535	$4f^65dt_{2g}$	$1\ \Gamma_{8g}$	2.849	2.945	3.037
Eu^{2+}	$4f^65dt_{2g}$ [a]	$34\ \Gamma_{8g}$	2.419	2.502	2.575	$4f^65de_g$ [a]	$50\ \Gamma_{8g}$	2.882	2.983	3.054
			$\bar{\nu}_{a_{1g}}$					$\bar{\nu}_{a_{1g}}$		
Eu^{3+}	$4f^6(^7F_0)$	$1\ A_{1g}$	495	445	402			308	286	248
Eu^{2+}	$4f^7(^8S_{7/2})$	$1\ \Gamma_{6u}$	423	370	321			286	273	204
Eu^{2+}	$4f^65de_g$	$1\ \Gamma_{8g}$	428	370	328	$4f^65dt_{2g}$	$1\ \Gamma_{8g}$	283	263	196

[a] Besides the indicated dominant configurational character, this state has significant $4f^6\phi_{ITE,a_{1g}}$ character. The mixed configurational character makes its potential energy curve far from parabolic and its harmonic vibrational frequency meaningless

$4f^65de_g$ (difluorides) and $4f^65dt_{2g}$ (sulfides) configurations than in the states of $4f^7$ configuration, without exception. This confirms the general observation that an excitation from the $4f$ shell to the lowest of the ligand-field split $5d$ shells contracts the impurity-ligand bond length (Sect. 10.1). An associated systematic increment of the vibrational frequency does not exist and it must not be expected, because there is a significant change in the nature of the bonding interactions in $4f^7$ states and $4f^65d$ states and bond lengths and force constants are only correlated in series of molecules or solids with the same chemical bond.

In a similar way to the $4f^6(^7F_J) \times 5de_g(^2E_g)$ case, the coupling between the states of the $4f^6(^7F_J)$ inner subshell and the states of the ligand-field excited $5dt_{2g}$ outer subshell are well described as $4f^6(^7F_J) \times 5dt_{2g}(^2T_{2g})$ and it leads to two separated sets of high-spin and low-spin states, which respectively constitute the third and fourth sub-manifolds. In CaF_2:Eu^{2+} their onset is at around 46000 cm^{-1} above the ground state; in CaS:Eu^{2+} they start at around 40000 cm^{-1}. However, in contrast with the first and second sub-manifolds, the high-spin $4f^6(^7F_J)5dt_{2g}$states of the third sub-manifold loose their clear spin and configurational character. This is so because they interact with ITE states of $4f^6(^7F_J)\phi_{ITE,a_{1g}}$ configuration and similar energies and, besides, the ^5F_J-^5D_J excitation energies within the $4f^6$ inner subshell are similar to the $5de_g$-$5dt_{2g}$ excitation energies within the $5d$ outer subshell ($5d$ ligand-field splitting), which results into $4f^6(^7F_J)5dt_{2g} + 4f^6(^7F_J)5de_g$ configurational mixing. In the results of CaF_2:Eu^{2+} this configurational mixing is important only above to 60000 cm^{-1}, but in CaS:Eu^{2+} a strong interaction between high-spin $4f^6(^7F_J)5de_g$ states, with their $5d$ subshell excited, and low-spin $4f^6(^5D_J)5dt_{2g}$ states of similar energy, with their $4f^6$ subshell excited, exists right above 40000 cm^{-1}. Here, it has to be brought under consideration that all spin-sextets have been included in the RASSI-SO spin-orbit coupling calculations on CaS:Eu^{2+}, but only a limited number of them could be included in the calculations on CaF_2:Eu^{2+}; this suggests that a larger spin and configurational mixing than the calculated could really exist in the third sub-manifold of the difluoride. It will be later shown that, fortunately, the impact on the absorption spectrum of the spin-sextets in the third sub-manifold is negligible.

The configurational mixing that extends over the third and fourth $4f^65d$ sub-manifolds result into more complex patterns of bond length and harmonic vibrational frequency. States with a main $4f^65dt_{2g}$ character show a longer bond length that $4f^65de_g$ states (in CaF_2 an the other difluorides; the opposite is true in sulfides), those with a main $4f^6\phi_{ITE}$ character show a shorter bond length that $4f^7$ states, and those with important $4f^65de_g + 4f^65dt_{2g}$ configurational mixing show an intermediate bond length, between those of both individual configurations. Again, the lack of a common chemical bond in the discussed states makes the harmonic vibrational frequencies not to follow a corresponding pattern. The Eu-ligand bond length pattern associated with the configurational character, when this is pure or dominant, is, in the cubic site of the difluorides: $d_{\text{e,Eu-F}}(\text{Eu}^{3+}\ 4f^6) < d_{\text{e,Eu-F}}(\text{Eu}^{2+}\ 4f^6\phi_{ITE}) < d_{\text{e,Eu-F}}(\text{Eu}^{2+}\ 4f^65de_g) < d_{\text{e,Eu-F}}(\text{Eu}^{2+}\ 4f^7) < d_{\text{e,Eu-F}}(\text{Eu}^{2+}\ 4f^65dt_{2g})$, and in the octahedral site of the sulfides: $d_{\text{e,Eu-S}}(\text{Eu}^{3+}\ 4f^6) < d_{\text{e,Eu-S}}(\text{Eu}^{2+}\ 4f^6\phi_{ITE}) < d_{\text{e,Eu-S}}(\text{Eu}^{2+}\ 4f^65dt_{2g}) < d_{\text{e,Eu-S}}(\text{Eu}^{2+}\ 4f^7) < d_{\text{e,Eu-S}}(\text{Eu}^{2+}\ 4f^65de_g)$. This is partly shown in Table 6.15. It is a confirmation of the general rule (Sect. 10.1) for bond length distances between Ln^{3+} and Ln^{2+} lanthanide ions and their L ligands in sites where the lanthanide $5d$ shell splits into lower $5d_{\text{LOW}}$ and higher $5d_{\text{HIGH}}$ ligand-field components: $d_{\text{e,Ln-L}}(\text{Ln}^{3+}\ 4f^{N-1}) < d_{\text{e,Ln-L}}(\text{Ln}^{2+}\ 4f^{N-1}\phi_{ITE}) < d_{\text{e,Ln-L}}(\text{Ln}^{2+}\ 4f^{N-1}5d_{\text{LOW}}) < d_{\text{e,Ln-L}}(\text{Ln}^{2+}\ 4f^N) < d_{\text{e,Ln-L}}(\text{Ln}^{2+}\ 4f^{N-1}5d_{\text{HIGH}})$.

In the present calculations of the $\text{Eu}^{2+}\ 4f^65d$ manifold, like in any sophisticated many-electron descriptions of complex manifolds, it is not an easy task to make simple analyses based on one-electron concepts, such as ligand-field splittings and exchange interactions. Nevertheless, the organization of the complex and crowded $4f^65d$ manifold into sub-manifolds combining main spin and configurational characters, i.e. HS and LS, $4f^6(^7F_J)5de_g$ and $4f^6(^7F_J)5dt_{2g}$, make it possible to introduce some parameters to characterize ligand-field and exchange interactions which are meaningful both experimentally and theoretically, namely a high-spin ligand-field splitting parameter $\varepsilon_{\text{lfs}}^{\text{HS}}$, and e_g and t_{2g} exchange splitting parameters $\varepsilon_{\text{exch}}^{e_g}$ and $\varepsilon_{\text{exch}}^{t_{2g}}$. This largely facilitates a way to compare theory and experiments which can provide important insights into the physical nature of the interactions governing the complex electronic structures, in parallel with and apart from the central, necessary, direct contrast between, e.g., experimental and calculated absorption spectra. Direct comparison between theoretical and experimental spectroscopy results is shown in Sect. 9.4.3. Physical insights gained by the mentioned analyses are discussed in Sect. 10.5.

6.2.5 $4f^{12}5d$ Manifold of Tm^{2+}

Tm^{2+} in $CsCaCl_3$ and $CsCaBr_3$

As the $4f^{13}$ manifolds of Tm^{2+} in six-fold coordinated octahedral sites of the $CsCaCl_3$ and $CsCaBr_3$ perovskites (Sect. 6.1.7), the $4f^{12}5d$ manifolds have also been obtained in [35] by means of SA-RASSCF/MS-RASPT2/RASSI-SO multiconfigurational ab initio calculations on the $(TmCl_6Cs_8)^{4+}$ and $(TmBr_6Cs_8)^{4+}$ clusters respectively embedded in $CsCaCl_3$ and $CsCaBr_3$, With the two manifolds of each material, the theoretical spectroscopic study of the multiple emissions of these materials [36–38] has been carried out (Sect. 9.5). The main features of the potential energy curves of the $4f^{12}5d$ manifold of $CsCaCl_3:Tm^{2+}$ are summarized in Table 6.16 and Fig. 6.7. Emphasis has been put on the energy gaps found between dense bands of local states because the multiple emissions require their existence: only large enough gaps can create metastable, emitting electronic states and prevent from fast non-radiative decays that lead to emission quenching. They are graphically shown in the bottom panel of the figure and more detailedly identified in the table.

Having included a Tm 6*s* molecular orbital in the active space, states of impurity-trapped-exciton character $4f^{13}\phi_{ITE,a_{1g}}$ may also appear in the calculations and, in fact, they do. As we will see, they start in the calculations at around 36000 cm^{-1}. Probably, their quality is inferior to that of $4f^{13}5d$ states because a similar quality is expected to demand the use of interstitial basis set functions that were omitted in the study, but the extra computational efforts required for a better accounting of them was not considered to be worthy because their high energy suggests they are not involved in any interesting photon absorption and emission processes in these materials.

The top panels in Fig. 6.7 show the spin-orbit-free states calculated at the MS-RASPT2 step, which are key intermediate results of the whole calculation. They are classified according the D_{2h} irreducible representations (A_g, B_{1g}, B_{2g}, and B_{3g} -the last three are degenerate in real octahedral symmetry) and total spin quantum numbers ($S = 1/2$ and $S = 3/2$). This is so because the relativistic many-electron DKH Hamiltonian used in MS-RASPT2 step is spin-orbit-free, which makes S a good quantum number, and the spatial symmetry restrictions of the abelian point-symmetry group D_{2h} have been imposed by choice. The conditions of the real O_h symmetry of the embedded-clusters could not be imposed because the algorithms used are bound to abelian groups [15] and the real calculations are made in the D_{2h} blocks ${}^4A_g$, ${}^4B_{1g}$, ${}^4B_{2g}$, ${}^4B_{3g}$, ${}^2A_g$, ${}^2B_{1g}$, ${}^2B_{2g}$, and ${}^2B_{3g}$. Analysis of the resulting many-electron wave functions allow to identify their O_h irreps, and these are indicated in the Figure. The O_h characterizations (quantum numbers) of the spin-orbit-free wave functions ${}^4A_{1g}$, ${}^4A_{2g}$, ${}^4E_g$, ${}^4T_{1g}$, ${}^4T_{2g}$, ${}^2A_{1g}$, ${}^2A_{2g}$, ${}^2E_g$, ${}^2T_{1g}$, and ${}^2T_{2g}$, are used to help identifying the spin-orbit coupling wave functions according to their irreducible representations of the O'_h double-group: The RASSI-SO calculations are done in the D'_{2h} double-group and all states of the $4f^{12}5d$ manifold belong to its Γ_{5g} irrep; those with degeneracy 4 are necessarily O'_h Γ_{8g} states; those with degeneracy 2 can be either O'_h Γ_{6g} or Γ_{7g} states; among these, those having contributions from ${}^2A_{1g}$ spin-orbit-free states

Table 6.16 Calculated $4f^{12}5d$ energy levels of Tm^{2+} in $CsCaCl_3$: Spectroscopic constants $d_{e,\text{Tm-Cl}}$, $\bar{\nu}_{a_{1g}}$, and T_e; and spin-orbit-free state characters of the wave functions

	State	$d_{e,\text{Tm-Cl}}$[a]	$\bar{\nu}_{a_{1g}}$[b]	T_e[c]	Spin-orbit-free character (%)[d]					
$4f^{12}5dt_{2g}$ dominant configuration: $\langle d_{e,\text{Tm-Cl}}\rangle = 2.537 \pm 0.001$ Å; $\langle \bar{\nu}_{a_{1g}}\rangle = 352 \pm 4$ cm^{-1}										
	above upper $4f^{13}$ state, gap: 4264 cm^{-1}									
$4f^{12}(^3H_6)5dt_{2g}$ main character										
	$1\,\Gamma_{8g}$	2.540	354	13582	46.28	$1\,^4T_{2g}$	28.62	$1\,^4A_{2g}$	13.45	$1\,^4T_{1g}$
	$2\,\Gamma_{8g}$	2.540	355	13700	41.94	$1\,^4T_{1g}$	41.63	$1\,^4T_{2g}$		
	$1\,\Gamma_{6g}$	2.540	353	13786	67.30	$1\,^4T_{1g}$	15.15	$1\,^4T_{2g}$		
				gap: 1454 cm^{-1}						
	$3\,\Gamma_{8g}$	2.538	351	15240	53.17	$2\,^4T_{2g}$	21.15	$2\,^4T_{1g}$		
up to	$13\,\Gamma_{8g}$	2.537	351	18578	21.20	$4\,^2T_{1g}$	17.31	$3\,^2E_g$	10.27	$3\,^4T_{1g}$
				gap: 2597 cm^{-1}						
$4f^{12}(^3F_4,^3H_5)5dt_{2g}$ main character										
	$8\,\Gamma_{6g}$	2.538	352	21175	33.86	$5\,^4T_{2g}$	25.10	$4\,^4E_g$	11.53	$5\,^2T_{1g}$
up to	$33\,\Gamma_{8g}$	2.537	350	26542	33.15	$4\,^4T_{2g}$	12.99	$3\,^4E_g$	10.75	$4\,^2T_{2g}$
				gap: 1200 cm^{-1}						
$4f^{12}(^3H_4,^3F_{3,2})5dt_{2g}$ main character										
	$34\,\Gamma_{8g}$	2.538	352	27742	23.82	$1\,^4T_{1g}$	22.58	$1\,^4T_{2g}$	11.74	$7\,^2T_{1g}$
up to	$54\,\Gamma_{8g}$	2.536	358	34144	13.50	$5\,^2T_{1g}$	10.09	$6\,^2T_{2g}$		
Interacting $4f^{12}5d$, $4f^{12}\phi_{ITE,a_{1g}}$ character										
	$28\,\Gamma_{6g}$	2.520	309	35433	85.11	$8\,^4T_{1g}$				
	...									

[a] Tm-Cl equilibrium distance, in Å
[b] Breathing mode harmonic vibrational frequency, in cm^{-1}
[c] Adiabatic (minimum-to-minimum) transition energy, in cm^{-1}
[d] Weights larger than 10% are given. All weights are calculated at $d_{\text{Tm-Cl}} = 2.540$ Å

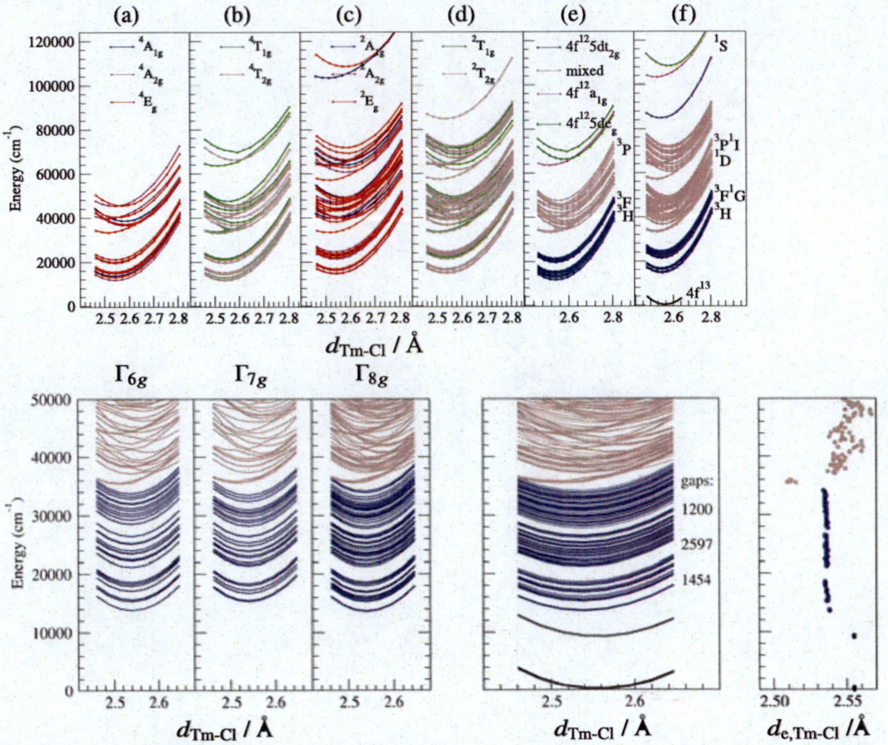

Fig. 6.7 Breathing mode potential energy curves of the $4f^{13}$, $4f^{12}5dt_{2g}$, $4f^{12}5de_g$, and $4f^{12}\phi_{ITE,a_{1g}}$ manifolds of Tm^{2+} in $CsCaCl_3$. Top panels: Spin-orbit-free MS-RASPT2 calculations; **a** $^4A_{1g}$, $^4A_{2g}$, and ^4E_g O_h states identified in a calculation on the D_{2h} ^4A_g block; **b** $^4T_{1g}$ and $^4T_{2g}$ O_h states identified in a calculation on the D_{2h} $^4B_{1g}$ block (identical results are obtained in calculations on the D_{2h} $^4B_{2g}$ and $^4B_{3g}$ blocks); **c** $^2A_{1g}$, $^2A_{2g}$, and ^2E_g O_h states identified in a calculation on the D_{2h} ^2A_g block; **d** $^2T_{1g}$ and $^2T_{2g}$ O_h states identified in a calculation on the D_{2h} $^2B_{1g}$ block (identical results are obtained in calculations on the D_{2h} $^2B_{2g}$ and $^2B_{3g}$ blocks); **e** spin-quartets; **f** spin-doublets. Parentages with atomic $4f^{12}$ terms are indicated in panels **e** and **f**. Bottom panels: Spin-orbit coupling RASSI-SO calculations. In the first three panels, Γ_{6g}, Γ_{7g}, and Γ_{8g} O'_h states identified in a calculation on the Γ_{5g} irreducible representation of the D'_{2h} double-group are respectively represented. All of them are represented in the fourth panel, together with the $4f^{13}$ Γ_{6u}, Γ_{8u}, Γ_{7u} ($^2F_{7/2}$), and Γ_{8u}, Γ_{7u} ($^2F_{5/2}$) states (in black). Bottom right panel: Tm-Cl equilibrium distance $d_{\text{e,Tm-Cl}}$ of all states. In top panels **e** and **f**, and all the bottom panels, colors are used to indicate the main configurational character of the states: $4f^{13}$ (black), $4f^{12}5dt_{2g}$ (blue), $4f^{12}5de_g$ (green), $4f^{12}\phi_{ITE,a_{1g}}$ (magenta), and mixed configurational character (brown)

are necessarily Γ_{6g} states, whereas those with $^2A_{2g}$ spin-orbit-free contributions are necessarily Γ_{7g} states.

As in other $4f^{N-1}5d$ cases, all states with a clearly dominant character of a given electronic configuration share the values of activator-ligand bond length and breathing mode vibrational frequency. In fact, the bond length (equilibrium bond distance) of a state, gives a fast, correct indication of its configurational character: a slightly shorter value than the $4f^{13}$ bond length indicates a dominant $4f^{12}5dt_{2g}$ configurational character (in O_h six-fold coordination), a slightly longer value indicates a main character $4f^{12}5de_g$, a much shorter value, similar to and slightly longer than the bond length of the ionized cluster, indicates a main impurity-trap-exciton character $4f^{12}\phi_{ITE}$, and states with a mixed configurational character have intermediate bond lengths. Panels (e) and (f) of Fig. 6.7 illustrate all this. In the bottom-right panel of the figure, the Tm-Cl bond lengths of all states up to 50000 cm^{-1} are represented and a simple inspection allows to identify $4f^{13}$ states, $4f^{12}5dt_{2g}$ states, $4f^{12}\phi_{ITE}$ states, and states of mixed $4f^{12}5dt_{2g}$+ $4f^{12}5de_g$+ $4f^{12}\phi_{ITE}$ character. This property was instrumental to understand and interpret excited state excitation spectroscopy experiments on Tm^{2+}-doped $CsCaCl_3$ and $CsCaBr_3$ [49], to assign its narrow and broad bands, and to identify one- and two-photon excitations. In energy regions where there is a interaction between states of different configurations, not only the bond lengths are intermediate, but also the parabolic character is strongly broken, the curvatures (vibrational frequencies) show strong anomalies and states with multiple energy minima can be found.

6.2.6 $4f^{13}5d$ Manifold of Yb^{2+}

The basic electronic structure of the $4f^{13}5d$ manifold of Yb^{2+} in solids results from the coupling between the $^2F_{7/2}$ and $^2F_{5/2}$ $4f^{13}$ sub-levels, slightly split by interactions of the inner $4f^{13}$ sub-shell with the host (Sect. 6.1.7), and the ligand-field split 5d sub-levels. In O_h sites, the main configurational characters will be $4f^{13}(^2F_{7/2})5de_g$, $4f^{13}(^2F_{5/2})5de_g$, $4f^{13}(^2F_{7/2})5dt_{2g}$, and $4f^{13}(^2F_{5/2})5dt_{2g}$, and the main spin characters will be spin-triplets ($S = 1/2 + 1/2 = 1$) and spin-singlets ($S = 1/2 - 1/2 = 0$).

Besides the inherent complexity of the $4f^{13}5d$ manifold, its states may also interact, at certain energies, with impurity-trapped-excitons of configurational character $4f^{13}\phi_{ITE}$ and states resulting from Yb^{2+} $4f \rightarrow 6s$ excitations, of the $4f^{13}6s$ configuration [50, 51].

Next, multiconfigurational ab initio results from [29] on the structure of the different $4f^{13}5d$ sub-manifolds and their interplay with $4f^{13}\phi_{ITE}$ and $4f^{13}6s$ states are discussed for Yb^{2+}-doped fluorite-type crystals CaF_2, SrF_2, BaF_2, and $SrCl_2$, more detailedly in $SrCl_2$. The analysis of the spectroscopic implications of the whole set of excited states is discussed in Sect. 9.6 for $SrCl_2$:Yb^{2+}. The electronic structure of $SrCl_2$:Yb^{2+} has also been studied in detail in [52] using a slightly less developed version of the computational choices discussed here and, although some numerical results are quantitatively different, they are qualitatively equivalent and provide the same general image of the electronic structure and interpretation of the spectroscopy.

Yb^{2+} in CaF_2, SrF_2, BaF_2, and $SrCl_2$

SA-RASSCF/MS-RASPT2/RASSI-SO calculations with the DKH relativistic Hamiltonian were performed on the clusters $(YbF_8Ca_{12})^{18+}$, $(YbF_8Sr_{12})^{18+}$, $(YbF_8Ba_{12})^{18+}$, and $(YbCl_8Sr_{12})^{18+}$, under the effects of respective AIMP embedding potentials of the CaF_2, SrF_2, BaF_2, and $SrCl_2$ hosts (cf. Sect. 2). The following Gaussian basis sets were used: All-electron atomic natural orbital relativistic basis sets ANO-RCC for Yb $(25s22p15d11f4g2h)[9s8p5d4f3g2h]$ (quadruple-zeta with polarization quality) and F $(14s9p4d)[5s4p3d]$ and Cl $(17s12p5d)[6s5p3d]$ (quadruple-zeta with polarization quality, without f and g. Additional $[1s1p]$ contractions of the outermost $3s, 3p$ orbital shells for second-neighbor frozen Ca^{2+} ions $(20s15p)[1s1p]$, $4s, 4p$ shells for Sr^{2+} ions $(23s19p)[1s1p]$, and $5s, 5p$ shells for Ba^{2+} ions $(27s23p)[1s1p]$, were used for a better cluster-host orthogonality, as well as $(5s)[5s]$ Gaussian primitives in the next six neighboring interstitial sites of the fluorite-like hosts for an adequate representation of possible impurity-trapped-exciton states. The D_{2h} abelian point group was used. A RAS space was used resulting from the distribution of 14 electrons in 20 molecular orbitals with main characters of Yb $4f$, $5d$, $6s$, and $5f$ (double f-shell), which belong to the O_h irreducible representations: a_{2u} (2), t_{1u} (6), t_{2u} (6), e_g (2), t_{2g} (3), and a_{1g} (1). All possible occupations were allowed in the Yb $4f$ shell and a maximum of 2 electrons were allowed in the Yb $5f$ shell (to account for the large change of radial correlation upon $4f \rightarrow 5d$ excitations and ionization) and 1 in the Yb $5d$ and $6s$ shells (to provide the minimal configurational space required for the spectroscopy). Taking into account the restrictions imposed by the inversion symmetry, these conditions are achieved if the $4f$ shell is chosen to be the RAS1 orbital space and a maximum of 3 holes are allowed in it, the $5f$ shell is the RAS2 orbital space, and the $5d$ and $6s$ shells together are the RAS3 orbital space and a maximum of 1 electron is allowed in it. I.e. SA-RASSCF($4f$-$3h$/$5f$/$5d6s$-$1e$). In order to optimize the molecular orbitals, state averages are done over all states in the same block of spin and O_h irrep. All valence electrons were correlated except those with main character Yb $4d$. A standard IPEA value (0.25 au) was used and an imaginary level shift of 0.15 au to avoid intruder states. Symmetry analyses of the SA-RASSCF multiconfigurational reference wave functions allowed to perform MS-RASPT2 calculations over actual O_h $^{2S+1}\Gamma$ symmetry blocks, which avoids symmetry breakings at the MS-RASPT2 level and their transfer to the RASSI-SO step. All spin-orbit free states of spin multiplicity $2S+1=3$ and 1 obtained in the MS-RASPT2 calculations were allowed to interact in the RASSI-SO calculations.

The potential energy curves of Yb^{2+} in the four fluorite-type hosts are shown in Fig. 6.8. More detailed data are given for $SrCl_2$:Yb^{2+} in Table 6.17; the spectroscopy of this material will be later discussed in Sect. 9.6.

Figure 6.8 shows in a glimpse several interesting features. Firstly, prior to including spin-orbit coupling interactions (upper panel), the states show very well defined configurational characters in all hosts, illustrated by the associated geometrical properties: the $4f^{14}$ ground state (in black); a set of spin-triplets and spin-singlets of basically pure $4f^{13}5de_g$ configurational character (in blue), sharing a common bond length $d_{e,\text{Yb-L}}$, shorter than in the $4f^{14}$ state; a higher set of spin-triplets and spin-

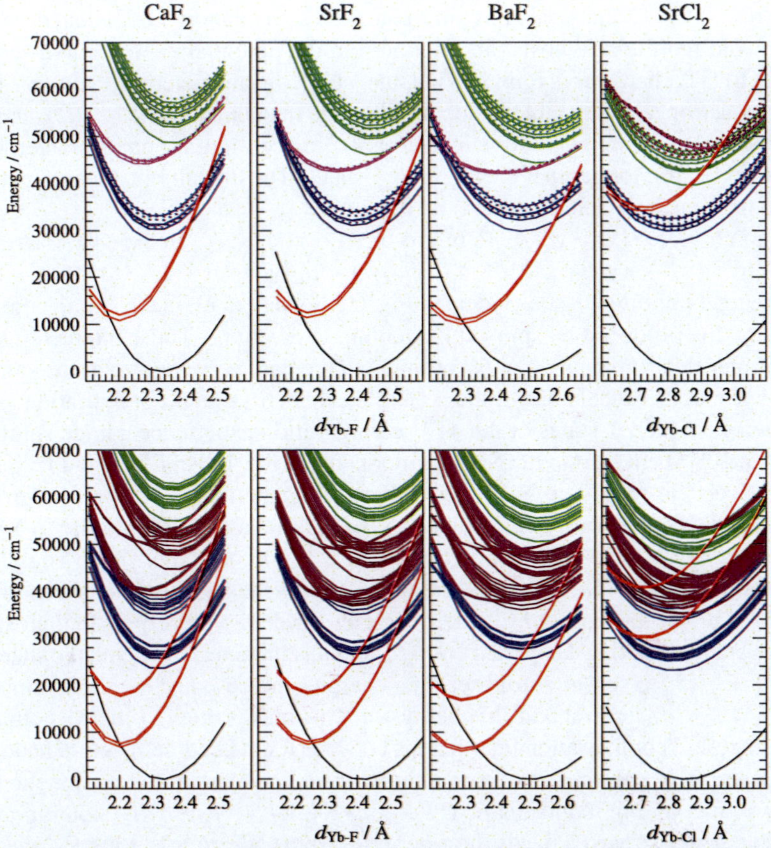

Fig. 6.8 Potential energy curves of Yb^{2+} and Yb^{3+} in fluorite-type hosts CaF_2, SrF_2, BaF_2, and $SrCl_2$. Upper panel: RASPT2 spin-orbit-free calculations; full lines are used for spin-triplet states and dotted lines for spin-singlet states. Lower panel: RASSI-SO spin-orbit coupling calculations. Colors are used to identify main configurational characters in both panels: black for Yb^{2+} $4f^{14}$ ground state, blue for Yb^{2+} $4f^{13}5de_g$ states, green for Yb^{2+} $4f^{13}5dt_{2g}$ states, magenta for Yb^{2+} $4f^{13}\phi_{ITE,a_{1g}}$ states, and maroon for Yb^{2+} states with strongly mixed configurational characters. Yb^{3+} $4f^{13}$ states are shown in red. In $SrCl_2$:Yb^{2+}, the states shown in magenta have $4f^{13}\phi_{ITE,a_{1g}}$ ITE character only at very low Yb-Cl distances ($d_{\text{Yb-Cl}} < 2.7$ Å); at longer distances their configurational character is $4f^{13}6s$

singlets of basically pure $4f^{13}5dt_{2g}$ configurational character (in green), also sharing a common bond length, longer than in the $4f^{13}5de_g$ states and, in this case, than in the $4f^{14}$ state; a set of spin-triplet and spin-singlet ITE states of $4f^{13}\phi_{ITE,a_{1g}}$ configurational character in CaF_2:Yb^{2+}, SrF_2:Yb^{2+}, and BaF_2:Yb^{2+} (in magenta), with a shorter bond length than in the $4f^{14}$ configuration, and slightly longer than the Yb^{3+}-ligand equilibrium distance after ionization (the potential energy curves of Yb^{3+}, corresponding to ionization, are given for a reference and shown in red). In

$SrCl_2$:Yb^{2+}, the ITE states only have this character at very short Yb-Cl distances, but in contrast to the fluoride hosts, no minimum in the potential energy curves corresponding to ITE character is found. When the Yb-Cl distance increases above 2.7 Å, the outer electron is more stable inside the Cl_8 cage than outside it (Fig. 2.2), and the states, rather than being ITE's, obtain a clear $4f^{13}6s$ configurational character, with a long bond length, longer that the $4f^{13}5dt_{2g}$ states. The overall bond length pattern found in the fourth fluorite-like hosts is:

$d_{\text{e,Yb-L}}(\text{Yb}^{3+}\ 4f^{13}) < d_{\text{e,Yb-L}}(\text{Yb}^{2+}\ 4f^{13}\phi_{ITE}) < d_{\text{e,Yb-L}}(\text{Yb}^{2+}\ 4f^{13}5de_g) < d_{\text{e,Yb-L}}(\text{Yb}^{2+}\ 4f^{14}) < d_{\text{e,Yb-L}}(\text{Yb}^{2+}\ 4f^{13}5dt_{2g}) < d_{\text{e,Yb-L}}(\text{Yb}^{2+}\ 4f^{13}6s)$.

The second important issue evident in Fig. 6.8 (lower panel) is the strong configurational mixing produced by spin-orbit coupling interactions. These interactions are small for one electron in an outer $5d$ orbital, but they are very important for unfilled inner $4f$ shells. It cancels out in the $4f^{14}$ closed-shell ground configuration. As a consequence, the Yb^{3+} states of the $4f^{13}$ configuration experience a large splitting, larger than 10000 cm^{-1}, into the Stark components of the $4f^{13}\ {}^2F_{7/2}$ and $4f^{13}\ {}^2F_{5/2}$ free-ion levels (in red). The Yb^{2+} ground state remains intact, and all its groups of excited states experience spin-orbit coupling splittings of a size similar to Yb^{3+}, because they are basically produced by their $4f^{13}$ inner sub-shell, with negligible contributions from the $5d$, $\phi_{ITE,a_{1g}}$, and $6s$ outer sub-shells. Then, the states of the $4f^{13}({}^2F_{7/2})5de_g$ sub-manifold see their energies lowered with respect to their spin-orbit-free values, and they keep their configurational character. However, the states of the $4f^{13}({}^2F_{5/2})5de_g$ sub-manifold see their energies increased, at the same time that the $4f^{13}({}^2F_{7/2})5dt_{2g}$ sub-manifold goes down. This makes them to share a common energy range and configuration interactions between them become strong. The lower the $5d$ ligand-field splitting, the more important the configurational mixing, so that the spin-orbit coupling induced $4f^{13}({}^2F_{5/2})5de_g + 4f^{13}({}^2F_{7/2})5dt_{2g}$ configuration interaction increases across the di-fluoride series from CaF_2 to SrF_2 to BaF_2, and it is even larger in $SrCl_2$. This is shown in the maroon lines in the lower panel of Fig. 6.8. $4f^{13}\phi_{ITE}$ and $4f^{13}6s$ states also experience the same spin-orbit coupling splitting and contribute to an even more extended and complex interaction between multiple configurations. An obvious consequence of all this, is a loose of the harmonic nature of the potential energy curves of the mixed-configuration states, which now show a variety of bond lengths intermediate between those of the pure configurations, and even multiple minima. Finally, the $4f^{13}({}^2F_{5/2})5dt_{2g}$ sub-manifold, at higher energies, remains free from configurational mixing, except in $SrCl_2$, where it interacts with $4f^{13}({}^2F_{5/2})6s$ states.

The overall multiconfigurational ab initio result for the electronic structure of the $4f^{13}5d$ excited manifold of Yb^{2+} given in the lower panel of Fig. 6.8 is the following: There is a first sub-manifold of almost pure configurational character $4f^{13}({}^2F_{7/2})5de_g$ in the four hosts. In the di-fluorides, a second sub-manifold of main mixed configurational character $4f^{13}({}^2F_{5/2})5de_g$ plus $4f^{13}({}^2F_{7/2})\phi_{ITE,a_{1g}}$, and a third sub-manifold of main mixed configurational character $4f^{13}({}^2F_{7/2})5dt_{2g}$ plus $4f^{13}({}^2F_{5/2})\phi_{ITE,a_{1g}}$ follow, with mutual interactions that are small in CaF_2, become large in SrF_2, and are very large in BaF_2. In $SrCl_2$, the second sub-manifold is of $4f^{13}({}^2F_{5/2})5de_g$ character, the third sub-manifold is of mixed $4f^{13}({}^2F_{7/2})5dt_{2g}$ plus

Table 6.17 Calculated $4f^{14}$, $4f^{13}5d$ and $4f^{12}6s$ energy levels of Yb^{2+} in $SrCl_2$: Spectroscopic constants $d_{e,\text{Yb-Cl}}$, $\bar{\nu}_{a_{1g}}$, and T_e; absorption oscillator strengths; and and spin-orbit-free state characters of the wave functions. Only the first and last states of the sub-manifolds are shown, as well as those with significant absorption oscillator strengths

	State	$d_{e,\text{Yb-Cl}}$ [a]	$\bar{\nu}_{a_{1g}}$ [b]	T_e [c]	f^{abs}/f_{ref} [d]	Spin-orbit-free character (%) [e]					
$4f^{14}$	1 A_{1g}	2.853	226	0		100.0	1^1A_{1g}				
$4f^{13}5de_g$ dominant configuration: $\langle d_{e,\text{Yb-Cl}}\rangle = 2.811 \pm 0.002$ Å; $\langle \bar{\nu}_{a_{1g}}\rangle = 225 \pm 1$ cm^{-1}											
$4f^{13}(^2F_{7/2})5de_g$ main character											
	1 E_u	2.815	225	23189		88.99	$1\,^3T_{1u}$				
	1 T_{2u}	2.815	225	23227		89.81	$1\,^3T_{1u}$				
	1 T_{1u}	2.811	224	25201	1.000	36.58	$1\,^1T_{1u}$	40.82	$1\,^3T_{1u}$	18.31	$1\,^3T_{2u}$
	2 T_{2u}	2.812	223	25436		63.77	$1\,^3T_{2u}$	34.34	$1\,^1T_{2u}$		
	1 A_{2u}	2.811	226	25504		99.10	$1\,^3T_{2u}$				
	2 T_{1u}	2.810	226	26257	0.127	50.03	$1\,^3E_u$	21.32	$2\,^3T_{1u}$	16.53	$1\,^3T_{2u}$
	3 T_{1u}	2.812	225	26589	0.028	55.56	$2\,^3T_{1u}$	27.08	$2\,^1T_{1u}$		
	4 T_{1u}	2.810	225	27800	0.006	77.98	$2\,^3T_{2u}$	17.66	$2\,^1T_{1u}$		
up to	4 T_{2u}	2.808	222	27870		51.47	$2\,^1T_{2u}$	26.47	$2\,^3T_{2u}$	15.42	$2\,^3T_{1u}$
$4f^{13}(^2F_{5/2})5de_g$, $4f^{13}(^2F_{7/2})5dt_{2g}$, $4f^{13}(^2F_{7/2})6s$ mixed characters [f]											
	2 A_{1u}	2.814	225	33745		94.99	$1\,^3T_{1u}$				
	5 T_{1u}	2.818	219	34433	0.382	53.20	$1\,^3T_{1u}$	22.34	$1\,^1T_{1u}$		
	6 T_{1u}	2.813	221	36288	0.694	49.38	$1\,^3T_{2u}$	24.47	$1\,^1T_{1u}$	17.18	$1\,^3E_u$
	7 T_{1u}	2.808	210	37692	0.001	39.61	$2\,^1T_{1u}$	21.60	$1\,^3E_u$	16.18	$2\,^3T_{1u}$
						14.49	$2\,^3T_{2u}$				
	8 T_{1u}	2.851	246	38513	0.079	61.14	$4\,^3T_{1u}$	14.04	$2\,^3E_u$		
	9 T_{1u}	2.855	234	38788	3.673	31.14	$3\,^3T_{1u}$	17.57	$2\,^3E_u$	15.21	$4\,^1T_{1u}$
	10 T_{1u}	2.860	262	38899	0.353	41.50	$2\,^3E_u$	32.44	$3\,^1T_{1u}$	10.67	$3\,^3T_{2u}$
	11 T_{1u}	2.861	224	40576	0.005	42.62	$4\,^3T_{2u}$	38.24	$3\,^3E_u$	11.46	$3\,^1T_{1u}$
	12 T_{1u}	2.862	225	41238	0.434	38.80	$5\,^1T_{1u}$	15.43	$3\,^3T_{2u}$	12.90	$4\,^3T_{2u}$
	13 T_{1u}	2.859	189	42713	0.921	36.99	$5\,^3T_{1u}$	30.44	$1\,^3A_{1u}$	14.32	$4\,^1T_{1u}$
						10.43	$3\,^3E_u$				
	14 T_{1u}	2.894	210	42950	0.051	41.02	$6\,^1T_{1u}$	39.19	$6\,^3T_{1u}$		
up to	15 T_{1u}	2.889	254	43219	0.028	53.52	$5\,^3T_{2u}$	37.70	$6\,^3T_{1u}$		

(continued)

Table 6.17 (continued)

	State	$d_{e,\text{Yb-Cl}}$[a]	$\bar{\nu}_{a_{1g}}$[b]	T_e[c]	f^{abs}/f_{ref}[d]	Spin-orbit-free character (%)[e]					
$4f^{13}5dt_{2g}$ dominant configuration: $\langle d_{e,\text{Yb-Cl}}\rangle = 2.864 \pm 0.004$ Å; $\langle \bar{\nu}_{a_{1g}}\rangle = 224 \pm 2$ cm^{-1}											
$4f^{13}(^2F_{5/2})5dt_{2g}$ main character											
	6 A_{1u}	2.865	225	45751		88.08	3 $^3T_{1u}$				
	16 T_{1u}	2.870	226	47820	2.399	32.01	3 $^3T_{1u}$	24.74	3 $^3T_{2u}$	16.93	5 $^1T_{1u}$
						13.08	2 ^3E_u				
	17 T_{1u}	2.875	230	49471	4.909	41.00	4 $^1T_{1u}$	19.47	3 $^3T_{2u}$	13.95	2 ^3E_u
	18 T_{1u}	2.862	224	50240	0.017	39.34	3 $^1T_{1u}$	12.58	3 $^3T_{2u}$	11.51	4 $^3T_{1u}$
						11.27	2 ^3E_u				
	19 T_{1u}	2.863	220	51935	0.011	29.73	5 $^3T_{1u}$	20.54	3 ^3E_u	19.53	4 $^3T_{2u}$
						13.52	1 $^3A_{1u}$				
	20 T_{1u}	2.862	226	52532	0.153	29.24	5 $^1T_{1u}$	27.45	3 ^3E_u	18.10	1 $^3A_{1u}$
						11.80	4 $^3T_{2u}$				
up to	7 A_{1u}	2.862	223	52663		53.95	5 $^3T_{1u}$	40.54	1 $^1A_{1u}$		
$4f^{13}6s$ dominant configuration: $\langle d_{e,\text{Yb-Cl}}\rangle = 2.909 \pm 0.002$ Å; $\langle \bar{\nu}_{a_{1g}}\rangle = 224 \pm 9$ cm^{-1}											
$4f^{13}(^2F_{5/2})6s$ main character											
	14 E_u	2.908	236	53396		62.54	6^3T_{1u}	37.29	5 $^3T_{2u}$		
	21 T_{1u}	2.909	234	53559	0.149	37.83	5 $^3T_{2u}$	37.53	6 $^1T_{1u}$	21.09	6 $^3T_{1u}$
up to	7 A_{2u}	2.911	214	53636		55.31	5 $^3T_{2u}$	39.10	2 $^1A_{2u}$		

[a] Yb-Cl equilibrium distance, in Å

[b] Breathing mode harmonic vibrational frequency, in cm^{-1}

[c] Adiabatic (minimum-to-minimum) transition energy, in cm^{-1}

[d] Relative absorption oscillator strength from the $4f^{14}$ 1 A_{1g} ground state, calculated at $d_{\text{Yb-Cl}} = 2.841$ Å; the reference value is $f_{ref} = 1.972 \times 10^{-2}$ [e] Weights larger than 10% are given. All weights are calculated at $d_{\text{Yb-Cl}} = 2.841$ Å

[f] Avoided crossings in these states make the energy curves strongly anharmonic in many states; in them, the second derivatives obtained from fitting and their associated vibrational frequencies are uncertain

$4f^{13}(^2F_{7/2})6s$ character, and the two sub-manifolds have a strong mutual configuration interaction. Finally, the four hosts show a fourth sub-manifold of higher energy with almost pure $4f^{13}(^2F_{5/2})5dt_{2g}$ character, except $SrCl_2$, where it interacts with $4f^{13}(^2F_{5/2})6s$ states.

Detailed data of the potential energy curves and absorption oscillator strengths are given for $SrCl_2$:Yb^{2+} in Table 6.17, where only the first and last state of each sub-manifold is shown, together with T_{1u} states with allowed $A_{1g}\rightarrow T_{1u}$ absorptions. These data will be used to discuss the spectroscopy of this material (Sect. 9.6).

6.3 $4f^{N-1}6s$ and Impurity-Trapped-Exciton Manifolds

In 1985, the existence of impurity-trapped exciton (ITE) states was hypothesized by McClure and Pedrini in order to explain the so-called anomalous emission in various Eu^{2+} and Yb^{2+}-doped crystals [50]. The authors conjectured that a valence electron of the lanthanide gets partially delocalized over the nearest neighbor cations and remains attracted by the hole that is left localized in the $4f$ shell of the lanthanide. No direct experimental evidence or theoretical support for these states was given, but the ITE concept became very popular as a hallmark for multiple unexplained phenomena in luminescent materials such as anomalous emissions and absorption bands, quenching pathways etc. Ab initio quantum chemical studies confirm the existence of ITE states, having the geometric and electronic structure as hypothesized by McClure and Pedrini. Interestingly, these states are well-hidden from experiments and do not give rise to the phenomenology that is usually associated with them. Alternative explanations are hence required. Calculations reveal an intimate connection between ITE states and states of the excited $4f^{N-1}6s^1$ configurations, justifying their simultaneous discussion in this Section.

Archetypal for impurity-trapped-exciton states is that electron charge density gets partly delocalized to neighboring sites of the lanthanide ion. The hole left behind in the $4f$ shell attracts the remote electron, preventing a complete ionization. As a consequence of this partial ionization, the equilibrium lanthanide-ligand bond length of this state decreases appreciably compared to the ground state and approaches the ligand-bond length for the fully oxidized ion. As shown in detail in the specific examples below, a connection between the presence of ITE states and the crystal structure has also been found: the remote electron tends to localize in voids in the structure, e.g. the interstices of fluorite-type structures. As a consequence, no ITE states have been found for lanthanide ions in rocksalt-type crystals that have a close-packing structure.

Computationally, ITE states show up while attempting variational calculations of $4f^{N-1}6s$ states. Adding a basis set on the nearest neighbor cation and interstitial sites yields a significant gain in energy minimization by allowing the electron density to leak towards these more distant regions. Details and technicalities are described in Sect. 2.1.2. The optimized natural orbitals as displayed in Fig. 6.9 illustrate the

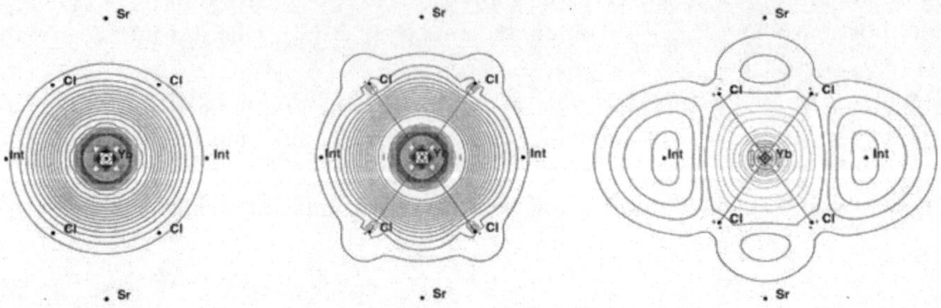

Fig. 6.9 Natural orbitals calculated for the $4f^{13}6s$ and $4f^{13}\phi_{ITE}$ configurations of Yb^{2+}. **a** Atomic 6*s* orbital of Yb^{2+} *in vacuo*, displayed on top of the atomic positions of Sr, Cl and the interstices (Int) in $SrCl_2$ to allow comparison with **b** and **c**. **b** and **c** a_{1g} molecular orbital of the $4f^{13}\phi(a_{1g})\ 2\,^3A_{2u}$ state of the $YbCl_8Sr_{12}$ cluster in $SrCl_2$ at $d_{Yb-Cl} = 2.75$ Å, having a pronounced 6*s* character (**b**) and at $d_{Yb-Cl} = 3.18$ Å, having a pronounced impurity-trapped-exciton (ITE) character (**c**)

spatially extended nature of the ϕ_{ITE} orbital, compared to the 6*s* atomic or molecular orbital. Below, specific cases where ITE states were found are discussed in detail.

6.3.1 Impurity-Trapped-Exciton of Ce^{3+}

In $Cs_2LiLuCl_6$:Ce^{3+}, SA-CASSCF/MS-CASPT2/RASSI-SO calculations [3] find a state of $\phi_{ITE,a_{1g}}$ configuration at higher energies than the $5de_g$ states (55810 cm^{-1} vs. 45860 cm^{-1}). In this impurity-trapped-exciton state, the $\phi_{ITE,a_{1g}}$ electron is outside the $CeCl_6$ moiety, localized in its neighborhood, although much more delocalized than in $4f$ and $5d$ states, and the hole is localized in the inner $4f$ shell, hence the name of impurity-trapped-exciton [50].

Its structural properties, which are common to all ITE states found in luminescent materials activated with lanthanide ions, are defined by an impurity-ligand distance slightly longer than in the ionized state (Ce^{4+}) and significantly shorter than in the $4f$ manifold (2.555 Å for Ce^{3+} ITE vs. 2.542 Å for Ce^{4+} and 2.660 Å for Ce^{3+} $4f$ in $Cs_2LiLuCl_6$ [3]) and an intermediate value of the breathing mode vibrational frequency (325 cm^{-1} for Ce^{3+} ITE vs. 355 cm^{-1} for Ce^{4+} and 313 cm^{-1} for Ce^{3+} $4f$ in $Cs_2LiLuCl_6$). The calculated $4f \rightarrow \phi_{ITE}$ absorption band is a very broad low intensity band (3000 cm^{-1} FWHM in $Cs_2LiLuCl_6$).

6.3.2 $4f6s$ and $4f\phi_{ITE}$ Manifolds of Pr^{3+}

Pr^{3+} Free-Ion

The results of SA-CASSCF/MS-CASPT2/RASSI-SO calculations on the $4f6s$ manifold of Pr^{3+} free-ion [9] are shown and compared with experiments [14] in Table 6.18 (cf. Sect.6.1.2 for calculation details). These states, which appear at very high energies, above 100000 cm^{-1}, show positive deviations around +2%, lower than 2000 cm^{-1}.

Pr^{3+} in CaF_2, Cubic Site

Right above the $4f5dt_{2g}$ manifold of the cubic site of CaF_2:Pr^{3+}, a new set of states localized around the Pr^{3+} impurity appears, as it is shown in Fig. 6.4 (violet curves). These states do not correspond to the free-ion's $4f6s$ configuration, but rather to a $4f\phi_{ITE,a_{1g}}$ configuration, in which the occupied a_{1g} active molecular orbital $\phi_{ITE,a_{1g}}$ clearly shows Pr ITE character. This orbital has significant electron density outside the $(PrF_8)^{5-}$ moiety, in the interstitial sites found along the 100 directions. The electronic charge in the large 6*s* atomic orbital cannot stay confined within the fluorine cage when the Pr^{3+} ion enters it, and it spreads away, although it remains in the vicinities of the impurity, trapped by the attraction of the $4f$ hole created by the excitation. Hence the impurity-trapped-exciton name [50].

A structural effect brought about by the ITE formation is an important shrinkage of the Pr-F bond length, originated by the Pr hole, and an associated increment of the harmonic vibrational frequency.

A good description of an ITE is known to be very demanding in terms of basis set, requiring the use of Gaussian orbitals at the interstices [52], and the present description of these ITE states must definitely be poorer than the $4f5d$ states; nevertheless, the improvements would not lower their energies enough to perturbe the Pr^{3+} $4f5dt_{2g}$ and $4f5de_g$ states. It is interesting, however, that Loh [45] assigned absorptions at 76000 cm^{-1} to excitations to Pr 6*s* orbitals and, although the multiconfigurational ab initio calculations do not support such interpretation in a strict sense, a more careful study of the ITE states could lower their energies from 85000 cm^{-1} to this range, which would support a variant of the assignment since the ITE orbitals can be described as a (strong) modification of the 6*s* orbitals.

Table 6.18 Energy levels of the $4f6s$ configuration of the Pr^{3+} free-ion: Experiments [14] and SA-CASSCF/MS-CASPT2/RASSI-SO calculations (deviations from experiment are given in parentheses)

Level	Experiment	Calculation
$^3F_2$	100258	101856 (+1598)
$^3F_3$	100544	102228 (+1684)
$^3F_4$	103171	104949 (+1678)
$^1F_3$	103754	105660 (+1906)

6.3.3 $4f^6\phi_{ITE}$ Manifold of Eu^{2+}

A double-well structure is found in the $4f^65d$ electronic states of Eu^{2+} in the alkaline earth difluorides CaF_2, SrF_2, and BaF_2, immediately above the gap after the states of main configurational character $4f^6(^7F_J)5de_g$, where high-spin states of the $4f^6(^7F_J)5dt_{2g}$ configuration are expected. It can be seen in Fig. 6.6 for CaF_2:Eu^{2+}. Such anomalous structure is not found in any of the alkaline earth sulfides.

Analyses of the many-electron wave functions at Eu-F distances corresponding to both wells reveal a main $4f^65dt_{2g}$ configurational character at the long-distance well and a main $4f^6\phi_{ITE,a_{1g}}$ configurational character at the short-distance well (cf. Table 6.14 for CaF_2:Eu^{2+}). In the latter, the molecular orbital $\phi_{ITE,a_{1g}}$ describes an electron strongly delocalized towards the nearby interstitial sites of the fluorite structure, almost ionized but still trapped outside the EuF_6 moiety; the combination of this trapped delocalized electron with the hole left in the $4f$ orbital shell after the excitation, is an Eu-trapped exciton or impurity-trapped exciton. The spill of electron charge out of the EuF_6 moiety causes a contraction of the fluoride cage and results into an Eu-F equilibrium distance in the inner well much smaller than in regular Eu^{2+} states (of $4f^7$ and $4f^65d$ configurations) and only slightly larger than in Eu^{3+} $4f^6$ states. The interaction between $4f^6\phi_{ITE,a_{1g}}$ and $4f^65dt_{2g}$ states of equal symmetry (i.e. spin and irreducible representation in spin-orbit-free calculations and double-group irreducible representation in spin-orbit coupling calculations) and respective inner and outer parabolas, according to the Wigner-Von Neumann rule, explains the double-well structure of the potential energy curves.

The energies at which the ITE states appear in these materials makes them not to play important qualitative roles in the spectroscopic properties. However, they must have an effect on the shapes of the $4f \rightarrow 5dt_{2g}$ absorption bands of the Eu^{2+}-doped difluorides.

6.3.4 $4f^{13}6s$ and $4f^{13}\phi_{ITE}$ Manifolds of Yb^{2+}

Excitation of one $4f$ electron from the full $4f^{14}$ shell of Yb^{2+} to orbitals other than $5d$, but still belonging to the Yb^{2+} impurity or at least attached to it, have traditionally been assumed in order to explain singular absorption and emission features of Yb^{2+}-doped solids. E.g., peaks in the absorption spectrum of $SrCl_2$:Yb^{2+} have been interpreted as due to absorptions from the ground state of Yb^{2+} to excited states of the $4f^{13}6s$ configuration [51]. This interpretation was based on the observation that the peaks' energies were similar to those of $4f^{13}6s$ states in the free-ion and on the fact that root-mean-square deviations between experimental peak energies and calculations were reduced upon inclusion of $4f^{13}6s$ configurations in empirical CFT calculations. Also, impurity-trapped exciton states, consisting in "a bound electron-hole pair with the hole localized on the impurity and the electron on nearby lattice

Table 6.19 Calculated Yb-F and Yb-Cl bond-lentghs $d_{e,\text{Yb-F}}$ and $d_{e,\text{Yb-Cl}}$ (in Å) in the fluorite-like hosts CaF_2, SrF_2, BaF_2, and $SrCl_2$, for Yb^{3+} in its $4f^{13}$ ground state and Yb^{2+} in its $4f^{14}$ ground state and selected states with clear configurational characters $4f^{13}5de_g$, $4f^{13}5dt_{2g}$, $4f^{13}\phi_{ITE}$, and $4f^{13}6s$

			CaF_2	SrF_2	BaF_2	$SrCl_2$
	Configuration	State	$d_{e,\text{Yb-F}}$			$d_{e,\text{Yb-Cl}}$
Yb^{3+}	$4f^{13}$	$1\,\Gamma_{7u}$	2.200	2.251	2.296	2.726
Yb^{2+}	$4f^{13}\phi_{ITE}$	[a]	2.270 ($9\,T_{1u}$)	2.334 ($3\,A_{1u}$)	2.414 ($2\,A_{2u}$)	–
Yb^{2+}	$4f^{13}5de_g$	$1\,E_u$	2.317	2.396	2.478	2.815
Yb^{2+}	$4f^{14}$	$1\,A_{1g}$	2.329	2.413	2.501	2.853
Yb^{2+}	$4f^{13}5dt_{2g}$	$6\,A_{1u}$	2.363	2.445	2.535	2.865
Yb^{2+}	$4f^{13}6s$	$7\,A_{2u}$	–	–	–	2.911

[a] ITE states are strongly mixed with $4f^{13}5d$ states (cf. Fig. 6.8) in these materials and the many-electron states with clearly dominant $4f^{13}\phi_{ITE}$ configurational character change from host to thost

sites", hence of the $4f^{13}\phi_{ITE}$ configuration, have been proposed to be responsible for the anomalous emissions of Yb^{2+} in CaF_2 and SrF_2 [50, 53],

The multiconfigurational ab initio calculations of the excited manifolds of Yb^{2+} in CaF_2, SrF_2, BaF_2, and $SrCl_2$ summarized in Fig. 6.8 do indeed support the formation of both types of excited states. They indicate that $4f^{13}\phi_{ITE}$ and $4f^{13}6s$ states are somehow related and that the formation of one or the other types of states depends on the host.

In $SrCl_2$:Yb^{2+} [29, 52], a big Cl_8 ligand cage allows the stabilization of the excited electron in a molecular orbital with main character Yb^{2+} $6s$ (cf. Fig. 2.2, bottom-right panel), hence the use of the $4f^{13}6s$ name for the excited configuration, which contains a number of many-electron states. This molecular orbital is very diffuse, much more than those with main character Yb^{2+} $5d$, but all its electronic charge density rests inside the $YbCl_8$ moiety. The Yb-Cl bond length is in these states longer than in the $4f^{13}5d$ states (cf. Table 6.19 and Fig. 6.8, right panels) and the energies of these states lie in the upper part of the energy window of $4f^{13}5dt_{2g}$ states. Nevertheless, as discussed in Sect. 9.6, no $4f^{13}6s$ states are found to be responsible for important peaks in the absorption spectrum, in opposition with the mentioned assumptions on their involvement.

It is interesting to observe that, at sufficiently short Yb-Cl distances, when the Cl_8 ligand cage is small enough, the squeezing of the $6s$ electron becomes so intense that the molecular orbital cannot retain all the electronic charge inside anymore and it spills off the cage, stabilizing outside it but in its neighborhood (cf. Fig. 2.2, bottom-left panel), in the next interstitial sites.

The same behavior is found in $CsCaBr_3$:Yb^{2+} [54]. In this bromide host, it is shown that $4f^{13}6s$ states have their energies very significantly shifted upwards with respect to their gas-phase values and that this phenomenon is a direct effect of host embedding interactions between $CsCaBr_3$ and Yb^{2+} [54] (cf. Sect. 10.4).

In the CaF_2, SrF_2, and BaF_2 difluorides, in opposition with iso-structural $SrCl_2$, the small F_8 ligand cage cannot retain inside the $6s$ electronic charge at any reasonable values of the Yb-F distance (at very long $d_{\text{Yb-F}}$, where the $6s$ electron would stabilize, the F ligands get too close to lattice cations and get very much destabilized) and the ITE states with $4f^{13}\phi_{ITE}$ configuration are formed and stabilized (cf. Fig. 6.8). All these states have bond lengths shorter than any other Yb^{2+} impurity state and slightly longer than Yb^{3+} states, where the excited electron is fully ionized (cf. Table 6.19). The energies of the ITE states in CaF_2, SrF_2, and BaF_2, lie between $4f^{13}5de_g$ and $4f^{13}5dt_{2g}$; upon including spin-orbit coupling interactions, $4f^{13}(^2F_{7/2})\phi_{ITE}$ states are similar in energy with $4f^{13}(^2F_{5/2})5de_g$ states and both of them strongly mix their configurational characters, something that also happens with $4f^{13}(^2F_{5/2})\phi_{ITE}$ states and $4f^{13}(^2F_{7/2})5dt_{2g}$ states.

The assumed involvement of these ITE states on the anomalous emissions of Yb^{2+}-doped CaF_2 and SrF_2 has, however, not been supported by the multiconfigurational ab initio calculations. The calculations fully support the existence of the hypothesized states, but they do not appear at the right energies, nor their potential energy curves have a sufficiently large horizontal offset, so as to be able to explain the extremely broad anomalous band emissions of these materials. The calculations indicate that these anomalous emissions are, instead, inter-valence charge-transfer (IVCT) emissions between Yb^{2+} and Yb^{3+} [29]. These transitions and their complex interplay with regular $5d \rightarrow 4f$ emission and emission quenching are the subject of Sect. 11.2.1.1.

References

1. G.H. Dieke, *Spectra and Energy Levels of Rare Earth Ions in Crystals* (Interscience Publishers, New York, 1968)
2. P.A. Tanner, C.S.K. Mak, N.M. Edelstein, K.M. Murdoch, G. Liu, J. Huang, L. Seijo, Z. Barandiarán, J. Amer. Chem. Soc. **125**, 13225 (2003)
3. L. Seijo, Z. Barandiarán, J. Chem. Phys. **141**, 214706/1 (2014)
4. J. Wen, L. Ning, C.K. Duan, Y. Chen, Y. Zhang, M. Yin, J. Phys. Chem. C **116**, 20513 (2012)
5. L. Ning, C. Wu, L. Li, L. Lin, C. Duan, Y. Zhang, L. Seijo, J. Phys. Chem. C **116**, 18419 (2012)
6. L. Ning, F. Yang, C. Duan, Y. Zhang, J. Liang, Z. Cui, J. Phys. Condens. Matter **24**, 05502 (2012)
7. J. Gracia, L. Seijo, Z. Barandiarán, D. Curulla, H. Niemansverdriet, W. van Gennip, J. Lumin. **128**, 1248 (2008)
8. L. Seijo, Z. Barandiarán, Phys. Chem. Chem. Phys. **16**, 3830 (2014)
9. M. Krośnicki, A. Kedziorski, L. Seijo, Z. Barandiarán, J. Phys. Chem. A **118**, 358 (2014)
10. L. Seijo, J. Chem. Phys. **102**, 8078 (1995)
11. L. Seijo, Z. Barandiarán, B. Ordejón, Mol. Phys. **101**, 73 (2003)
12. Z. Barandiarán, L. Seijo, Can. J. Chem. **70**, 409 (1992)
13. J.L. Pascual, L. Seijo, J. Chem. Phys. **102**, 5368 (1995)
14. W.C. Martin, R. Zalubas, L. Hagan, *Atomic Energy Levels - The Rare Earth Elements*, Natl. Stand. Ref. Data Ser., Natl. Bur. Stand. No. 60 (U. S. GPO, Washington, D. C., 1978)
15. G. Karlström, R. Lindh, P.A. Malmqvist, B.O. Roos, U. Ryde, V. Veryazov, P.O. Widmark, M. Cossi, B. Schimmelpfennig, P. Neogrady, L. Seijo, Comput. Mater. Sci. **28**, 222 (2003)
16. Z. Cai, V. Meiser-Uma, C. Froese-Fischer, Phys. Rev. Lett. **68**, 297 (1992)
17. E. Eliav, U. Kaldor, Y. Ishikawa, Phys. Rev. A **51**, 225 (1995)
18. R.W.G. Wyckoff, *Crystal Structures* (Wiley, New York, 1963)

19. R.D. Shannon, Acta Crystallogr. A **32**, 751 (1976)
20. L. van Pieterson, R.T. Wegh, A. Meijerink, M.F. Reid, J. Chem. Phys. **115**, 9382 (2001)
21. J.J. Joos, D.V. der Heggen, L. Amidani, L. Seijo, Z. Barandiarán, Phys. Rev. B **104**, L201108 (2021)
22. L. Seijo, Z. Barandiarán, Int. J. Quantum Chem. **60**, 617 (1996)
23. G. Blasse, B. Grabmaier, *Luminescent Materials* (Springer, 1994), pp. 1–9
24. P.A. Tanner, Chem. Soc. Rev. **42**, 5090 (2013)
25. C. Cascales, J. Fernández, R. Balda, Opt. Express **13**, 2141 (2005)
26. V.A. Morozov, A. Bertha, K.W. Meert, S. Van Rompaey, D. Batuk, G.T. Martinez, S. Van Aert, P.F. Smet, M.V. Raskina, D. Poelman, A.M. Abakumov, J. Hadermann, Chem. Mater. **25**, 4387 (2013)
27. J.J. Joos, P.F. Smet, L. Seijo, Z. Barandiarán, Inorg. Chem. Front. **7**, 871 (2020)
28. L. Seijo, Z. Barandiarán, *Computational Chemistry: Reviews of Current Trends*, vol. 4, ed. by J. Leszczyński (World Scientific, Singapore, 1999), pp. 55–152
29. Z. Barandiarán, L. Seijo, J. Chem. Phys. **141**, 234704 (2014)
30. S.V. Gastev, J.K. Choi, R.J. Reeves, Phys. Solid State **51**(1), 44 (2009)
31. M.C. Downer, C.D. Corderomontalvo, H. Crosswhite, Phys. Rev. B **28**(9), 4931 (1983)
32. V. Möllmann, J. Selling, B. Henke, S. Schweizer, P. Keil, V. Lavin, G. Wortmann, Desy annual report (2007)
33. J. Joos, L. Seijo, Z. Barandiarán, J. Phys. Chem. Lett. **10**, 1581 (2019)
34. A. Kramida, Yu. Ralchenko, J. Reader, and NIST ASD Team. NIST Atomic Spectra Database (ver. 5.8), [Online]. Available: https://physics.nist.gov/asd [2021, January 25]. National Institute of Standards and Technology, Gaithersburg, MD (2020)
35. M. de Jong, A. Meijerink, L. Seijo, Z. Barandiarán, J. Phys. Chem. C **121**, 10095 (2017)
36. J. Grimm, H.U. Güdel, Chem. Phys. Lett. **404**, 40 (2005)
37. J. Grimm, E. Beurer, H.U. Güdel, Inorg. Chem. **45**, 10905 (2006)
38. J. Grimm, J.F. Suyver, E. Beurer, G. Carver, H.U. Güdel, J. Phys. Chem. B **110**, 2093 (2006)
39. J. Kirton, S.D. McLaughlan, Phys. Rev. **155**, 279 (1967)
40. S.M. Kaczmarek, T. Tsuboi, M. Ito, G. Boulon, G. Leniec, J. Phys. Condens. Matter **17**, 3771 (2005)
41. F. Ruipérez, Z. Barandiarán, L. Seijo, J. Chem. Phys. **123**, 244703 (2005)
42. L. Seijo, Z. Barandiarán, Opt. Mater. **35**, 1932 (2013)
43. E.U. Condon, G.H. Shortley, *The Theory of Atomic Spectra* (Cambridge University Press, Cambridge, 1935)
44. L. van Pieterson, M.F. Reid, R.T. Wegh, S. Soverna, A. Meijerink, Phys. Rev. B **65**, 045113 (2002)
45. E. Loh, Phys. Rev. **158**, 273 (1967)
46. T. Szczurek, M. Schlesinger, in *Rare Earth Spectroscopy*. ed. by J. Legendziewicz, W. Strek (World Scientific, Singapore, 1985), p. 309
47. K.D. Oskam, A.J. Houtepen, A. Meijerink, J. Lumin. **97**, 107 (2002)
48. N. Yamashita, O. Harada, K. Nakamura, Jpn. J. Appl. Phys. **34**(10), 5539 (1995)
49. M. de Jong, D. Biner, K. Krämer, Z. Barandiarán, L. Seijo, A. Meijerink, J. Phys. Chem. Lett. **7**, 2730 (2016)
50. D.S. McClure, C. Pédrini, Phys. Rev. B **32**, 8465 (1985)
51. Z. Pan, C. Duan, P.A. Tanner, Phys. Rev. B **77**, 085114 (2008)
52. G. Sánchez-Sanz, L. Seijo, Z. Barandiarán, J. Chem. Phys. **133**, 114509 (2010)
53. B. Moine, B. Courtois, C. Pédrini, J. Phys. France **50**, 2105 (1989)
54. G. Sánchez-Sanz, L. Seijo, Z. Barandiarán, Spec. Lett. **43**, 393 (2010)

Chapter 7
Charge Transfer States

Abstract Up to now, the electronic transitions that have been discussed are intra-atomic, i.e. the involved electrons reorganize in orbitals that can be associated with the luminescence activator itself. However, some important processes in luminescent materials involve the interatomic transfer of an electron. From the computational point of view, this is a nontrivial complication, involving enlarged active spaces, lower symmetries and a significantly larger number of excited states. In this chapter, the basics of several types of charge transfer (CT) states and the corresponding transitions are discussed. Special attention is paid to the construction of diabatic electron transfer configurational coordinate diagrams from single-center ab initio calculations or empirical data. The diabatic approximation circumvents computational demands that can be enormous and eventually make the calculations unfeasible and, although it comes together with some quantitative and qualitative limitations, it provides much significant information on luminescence related processes. The application of the diabatic approximation recipes to luminescence is discussed in Chap. 11.

7.1 Ligand-to-Metal Charge Transfer LMCT

For an optically active center made of a lanthanide ion, or a transition metal ion, surrounded by a first coordination shell of ligands and by the rest of the host, the states that result from an electronic transition from a molecular orbital with dominant ligand character to another one with dominant lanthanide (or transition metal) character are called Ligand-to-Metal Charge Transfer (LMCT) states and the electronic transition is referred to as a LMCT transition. In these transitions, an electron is partially transferred from the anionic ligands to the metallic cation and the oxidation state of the latter is reduced in the excited state.

LMCT states are also formed in undoped or unactivated hosts when the electron is transferred from the valence band (the filled shell of anionic orbitals) to the conduction band (the either empty or partially occupied shell of cationic orbitals). In this case, the electronic transitions are usually seen as excitations from delocalized, periodic orbitals (Bloch orbitals) of the anions to also periodic and delocalized orbitals of the cations and they are mostly referred to as interband excitations or valence-band

Z. Barandiarán et al., *Luminescent Materials*, Springer Series in Materials Science 322,
https://doi.org/10.1007/978-3-030-94984-6_7

to conduction-band excitations. These delocalized transitions are equivalent to transitions between localized orbitals of the anions and the cations (Wannier orbitals), in which the initial and final orbitals of the transition are obtained from the Bloch orbitals by appropriate unitary transformations and they are usually localized on a few anions and one cation, respectively. Under this localized description equivalent to the delocalized one, it is normal to see the missing valence electron as a hole and the new conduction electron as an electron, and call the excitation an electron-hole creation (and the de-excitation an electron-hole recombination). Also, the molecular language results natural in the localized formalism and the terms LMCT states and LMCT transitions initially developed in the field of transition metal complexes are applied as well. Beyond language issues, the localized description of these electron transfers in periodic insulators opens the way to their theoretical quantitative study with the same multiconfigurational ab initio methods used for the local LMCT states of active sites made by a lanthanide activator and the surrounding ligands.

A structural consequence of the ligand-to-metal electron transfer, which is derived from the weakening of the electrostatic attraction due to the partial oxidation of ligands and reduction of the cation, is the elongation of the metal-ligand bond length. An associated lowering of the vibrational frequencies should not be expected, because the correlation between longer bond lengths and lower vibrational frequencies stands for states with the same bonding interactions, but these experience a strong change from the ground state to LMCT states.

7.1.1 LMCT States of $CaTiO_3$ and $CaZrO_3$

In the following example, the importance of hosts' LMCT states is illustrated. Host effects have been found to favor one or another of two competing $4f \rightarrow 4f$ emissions of Pr^{3+}-doped materials: The so called $^3P_0 \rightarrow ^3H_4$ and $^1D_2 \rightarrow ^3H_4$ emissions. The former is a faster, greenish-blue emission and the latter is a slower, red emission, so that gaining control over them offers a variety of practical applications for Pr^{3+} based phosphors. E.g., in the search for energy efficient red phosphors for solid-state lighting, it has been found that Pr^{3+}-doped $CaTiO_3$ exhibits red $^1D_2$ luminescence whereas the luminescence of Pr^{3+}-doped $CaZrO_3$ is dominated by the $^3P_0$ greenish-blue emission. As it is discussed in Sect. 8.2.3, multiconfigurational ab initio calculations revealed that the ultimate reason for this is the presence of LMCT states at intermediate energies between the $^3P_0$ and $^1D_2$ emitting states of Pr^{3+} in $CaTiO_3$:Pr^{3+} (O-to-Ti states) but not in $CaZrO_3$:Pr^{3+}, where they (O-to-Zr states) appear at higher energies. Let us then discuss the LMCT states of $CaTiO_3$ and $CaZrO_3$.

The O-to-Ti and O-to-Zr LMCT states of $CaTiO_3$ and $CaZrO_3$ have been calculated at a SA-CASSCF/MS-CASPT2 level in [1] using, respectively, the $(TiO_6Ca_8Ti_6)^{32+}$ and $(ZrO_6Ca_8Zr_6)^{32+}$ clusters of O_h point symmetry (Fig 8.11d) properly embedded in their respective $CaTiO_3$ and $CaZrO_3$ hosts by means of AIMP embedding potentials. In these calculations, Gaussian atomic natural orbital relativistic basis sets ANO-RCC for Titanium [2], Zirconium [2], Calcium [3], and Oxygen [4] were used. For the innermost Ti atoms the contraction $(21s15p10d6f4g)[7s6p4d3f2g]$ (quadruple-zeta with polarization quality) was used and for the outermost Ti atoms $(21s15p10d)[5s4p2d]$ (double-zeta quality except for f functions); equivalently, for the innermost Zr atoms $(21s18p13d6f4g)[8s7p5d3f2g]$ (quadruple-zeta with polarization quality) and for the outermost Zr atoms $(21s18p13d)[6s5p3d]$ (double-zeta quality except for f functions); for O $(14s9p4d)[5s4p3d]$ (quadruple-zeta with polarization quality except for f functions); and for Ca $(20s16p6d)[5s3p2d]$ (double-zeta with polarization quality). The calculations were performed in actual D_{2h} symmetry with a total of 630 symmetry adapted basis functions for the $(TiO_6Ca_8Ti_6)^{32+}$ cluster and 693 symmetry adapted basis functions for the $(ZrO_6Ca_8Zr_6)^{32+}$ cluster.

In order to decide what active orbitals should be used in the CASSCF calculations, preliminary Hartree-Fock calculation were done on the simpler, embedded-cluster $(TiO_6Ti_6)^{16+}$ at d_{Ti-O}=1.95 Å. They gave the following frontier orbital energies (in hartree): occupied MOs with dominant character O_{2p}: $\varepsilon(a_{1g}(O_{2p\sigma})) = -0.436$, $\varepsilon(e_g(O_{2p\sigma})) = -0.416$, $\varepsilon(t_{1u}(O_{2p\sigma})) = -0.379$, $\varepsilon(t_{2g}(O_{2p\pi})) = -0.378$, $\varepsilon(t_{2u}(O_{2p\pi})) = -0.312$, $\varepsilon(t_{1u}(O_{2p\pi})) = -0.305$, $\varepsilon(t_{1g}(O_{2p\pi})) = -0.281$; and unoccupied MOs with dominant character of Ti: $\varepsilon(t_{2g}(Ti_{3d})) = 0.301$, $\varepsilon(e_g(Ti_{3d})) = 0.401$, $\varepsilon(a_{1g}(Ti_{4s})) = 0.553$, $\varepsilon(t_{1u}(Ti_{4p})) = 0.661$. This suggest that the lowest O_{2p}→Ti LMCT excitations will be $t_{1g}(O_{2p\pi})\rightarrow t_{2g}(Ti_{3d})$, $t_{1u}(O_{2p\pi})\rightarrow t_{2g}(Ti_{3d})$, and $t_{2u}(O_{2p\pi})\rightarrow t_{2g}(Ti_{3d})$. This was confirmed with MS-RASPT2 calculations with a restricted-active-space RAS in which one hole only was allowed in the 18 MOs with dominant character O_{2p} and one electron only was allowed in the 3 MOs t_{2g} with dominant character Ti_{3d}. In such calculations, the mentioned LMCT excitations lie in 40600–49800 cm^{-1} whereas all others are in the range 57500–78300 cm^{-1}.

According to the results of the exploratory HF calculation, the decided configurational complete-active-spaces correspond to the distribution of 6 electrons in the six t_{1g} and t_{2g} MOs of main character $t_{1g}(O_{2p\pi})$ and $t_{2g}(Ti_{3d}/Zr_{4d})$ for spin singlet ant spin triplet *gerade* LMCT states (as well as for the $^1A_{1g}$ ground state), and of 12 electrons in the nine t_{1u}, t_{2u}, and t_{2g} MOs of main characters $t_{1u}(O_{2p\pi})$, $t_{2u}(O_{2p\pi})$, and $t_{2g}(Ti_{3d}/Zr_{4d})$ for spin singlet ant spin triplet *ungerade* LMCT tates. The following state-averages were used for the MO optimizations in SA-CASSCF calculations: one $^1A_{1g}$, one $^1A_{2g}$, and one ^1E_g levels; one $^1T_{1g}$ and one $^1T_{2g}$; one $^1A_{1u}$, one $^1A_{2u}$, and two ^1E_u; two $^1T_{1u}$ and two $^1T_{2u}$; one $^3A_{2g}$ and one ^3E_g; one $^3T_{1g}$ and one $^3T_{2g}$; one $^3A_{1u}$, one $^3A_{2u}$, and two ^3E_u; and two $^1T_{1u}$ and two $^1T_{2u}$. In the MS-CASPT2 calculations all valence electrons were correlated. Spin-orbit coupling was not included in these calculations.

The results are shown in the configuration coordiate energy diagram of Fig. 7.1 and in Table 7.1.

The calculated LMCT states, of type $O_{2p\pi}(t_{1g}, t_{1u}, t_{2u})\rightarrow Ti_{3d}/Zr_{4d}(t_{2g})$, span an energy window of around 12000 cm^{-1} in both materials and the remaining LMCT states are expected well above that. The O-to-Ti states start at around 20700 cm^{-1} whereas the O-to-Zr LMCT states appear 22600 cm^{-1} higher. This is a manifestation of the larger electron affinity of Ti^{4+} over Zr^{4+} (i.e. the fourth ionization potential

Table 7.1 Spectroscopic constants of O-to-Ti and O-to-Zr LMCT states of $CaTiO_3$ and $CaZrO_3$, respectively, according to MS-CASPT2 calculations on the $(TiO_6Ca_8Ti_6)^{32+}$ and $(ZrO_6Ca_8Zr_6)^{32+}$ embedded-clusters. Ti–O and Zr–O equilibrium distances $d_{Ti-O,e}$ and $d_{Zr-O,e}$, in Angstrom, breathing mode harmonic vibrational frequencies $\omega_{a_{1g}}$, in cm^{-1}, and minimum-to-minimum transition energies T_e, in cm^{-1}. The allowed LMCT absorptions ($^1A_{1g} \rightarrow ^1T_{1u}$) are highlighted

Level	$d_{Ti-O,e}$	$\omega_{a_{1g}}$	T_e	$d_{Zr-O,e}$	$\omega_{a_{1g}}$	T_e
	$CaTiO_3$			$CaZrO_3$		
Ground state						
$1\,^1A_{1g}$	1.957	794	0	2.105	775	0
LMCT states						
$1\,^1A_{2g}$	2.001	775	20720	2.143	744	43360
$1\,^1E_g$	2.000	786	21110	2.139	732	44010
$1\,^3E_g$	1.999	777	23490	2.145	763	44960
$1\,^3A_{2g}$	1.999	775	23790	2.145	761	45680
$1\,^1T_{1g}$	1.995	793	26650	2.140	765	48910
$1\,^1T_{2g}$	1.995	793	26720	2.139	764	48720
$1\,^1E_u$	2.006	817	27240	2.147	505	48430
$1\,^3E_u$	2.006	800	27790	2.150	830	48810
$1\,^3T_{2g}$	1.996	791	27830	2.141	767	50550
$1\,^3T_{1g}$	1.996	792	28030	2.141	769	50020
$1\,^1A_{1u}$	2.006	801	28620	2.152	767	50530
$2\,^3E_u$	2.007	803	29200	2.150	965	50050
$1\,^3T_{1u}$	2.005	807	29320	2.147	736	49410
$1\,^3A_{1u}$	2.006	801	29630	2.152	771	51290
$1\,^1A_{2u}$	2.008	801	29800	2.150	771	49650
$2\,^1E_u$	2.010	777	28900	2.155	711	49450
$1\,^1T_{1u}$	**2.004**	**783**	**29470**	**2.145**	**782**	**49290**
$2\,^1T_{2u}$	2.003	746	30410	2.147	898	51130
$2\,^3T_{2u}$	2.001	724	30500	2.148	763	50620
$3\,^1T_{1u}$	**2.006**	**812**	**30710**	**2.147**	**732**	**52800**
$1\,^3A_{2u}$	2.008	801	30890	2.151	771	50910
$3\,^3T_{1u}$	2.009	803	31110	2.145	668	53130
$4\,^1T_{2u}$	2.004	789	31140	2.149	688	55140
$4\,^3T_{2u}$	2.008	799	31690	2.162	755	54540

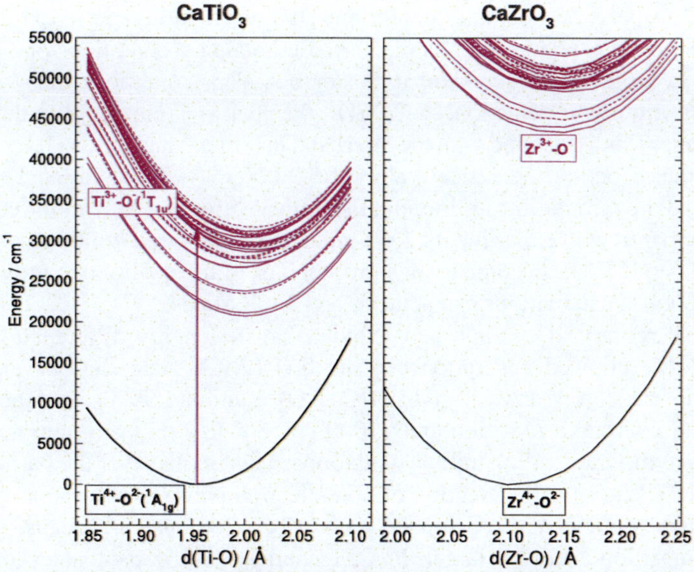

Fig. 7.1 Potential energy curves of the $^1A_{1g}$ ground state (in black) and O^{2-}-to-Ti^{4+} LMCT states of $CaTiO_3$ (left) and O^{2-}-to-Zr^{4+} LMCT states of $CaZrO_3$ (right) after MS-CASPT2 calculations on the $(TiO_6Ca_8Ti_6)^{32+}$ and $(ZrO_6Ca_8Zr_6)^{32+}$ clusters embedded in $CaTiO_3$ and $CaZrO_3$ respectively. The first allowed $^1A_{1g}\rightarrow{}^1T_{1u}$ LMCT absorption of $CaTiO_3$ is indicated (cf. Table 7.1)

being larger in Ti than in Zr: IP(Ti^{3+}) > IP(Zr^{3+}) or EA(Ti^{4+}) > EA(Zr^{4+})). The energies of the lowest LMCT states of $CaTiO_3$ lie between those of the $^3P_0$ and $^1D_2$ states of Pr^{3+} and favor non-radiative decays to $^1D_2$, so enhancing its red emission. The equivalent states of $CaZrO_3$ cannot play this role because they have much higher energies.

In order to evaluate the quantitative reliability of these results, comparisons can be made with experimental spectroscopic results and, for this, the selection rules have to be considered. The electric dipole allowed LMCT absorptions have a $^1A_{1g}\rightarrow{}^1T_{1u}$ many-electron character, which restricts the LMCT states that play relevant roles in this respect, in contrast with non-radiative decay mechanisms that can selectively quench or enhance particular emissions, where all LMCT states are relevant.

The allowed minimum-to-minimum $^1A_{1g}\rightarrow{}^1T_{1u}$ transitions are shown in bold font in Table 7.1. The first corresponding, allowed Frank-Condon transition of $CaTiO_3$ is signaled with a vertical magenta arrow in Fig. 7.1. Orbital analysis indicates their main one-electron character is $O_{2p\pi}(t_{1u}, t_{2u})\rightarrow Ti_{3d}/Zr_{4d}(t_{2g})$. The origin of the interband transition was found at around 28000 cm^{-1} in the experimental excitation spectrum of $CaTiO_3$:Pr^{3+} [5]; it should be compared with the theoretical mininum-to-minimum allowed LMCT transition of $CaTiO_3$ at 29500 cm^{-1}. On the other hand, the experimental interband transition maximum, which was found at 30000 cm^{-1}, should be compared with the theoretical Frank-Condon allowed LMCT transition at

31400 cm^{-1}. A periodic-boundary-conditions DFT calculation reported 31500 cm^{-1} for the direct band gap [6], which is the periodic-boundary-conditions one-electron equivalent to molecular or embedded-cluster boundary-conditions many-electron minimum-to-minimum transition. In $CaZrO_3$, the peak at 47600 cm^{-1} in the excitation spectrum [7] is associated with the interband maximum and it must be compared with the Frank-Condon theoretical value of 50600 cm^{-1}. The interband origin is unknown and the minimum-to-minimum theoretical value of 49300 cm^{-1} cannot be directly compared with experiments. DFT calculations reported 46000 cm^{-1} for the direct band gap [8]. These comparisons give an idea of the quality that multiconfigurational ab initio calculations can achieve on LMCT states.

As a final remark, it is important to note that the energy difference between the first LMCT allowed absorption and the first LMCT state can be very large. For example, the lowest LMCT state of $CaTiO_3$ is around 9000 cm^{-1} below the origin of the interband transition and 11000 cm^{-1} below its maximum according to the multiconfigurational ab initio calculations. This implies that the participation of host LMCT states (or the creation of hole-electron pairs) in non-radiative decay mechanisms of optically active dopants cannot be ruled out just because the host interband transition energy is larger than the energy of the dopant states above the dopant ground state. The case of $CaTiO_3$:Pr^{3+} is paradigmatic: Although the $4f^2$ $^1D_2$ and $^3P_0$ states of Pr^{3+} are around 20000–25000 cm^{-1} above its $^3H_4$ ground state and the $CaTiO_3$ interband transition is clearly higher, of around 30000 cm^{-1}, a number of LMCT states of this host play a key role in the $^3P_0$-to-$^1D_2$ non-radiative decay because they are well below 30000 cm^{-1}, intercalated between Pr^{3+} $^1D_2$ and $^3P_0$.

7.2 Inter-Valence Charge Transfer IVCT

Electron transfer between two metal sites differing only in oxidation state is conventionally called inter-valence charge transfer (IVCT) [9, 10]. This means that IVCT is a particular case of metal-to-metal charge transfer (MMCT) in which two metal ions of the same element are involved in the redox process, or, in other words, IVCT is the homonuclear symmetric MMCT. To avoid confusion, it is important to mention that the term IVCT has also been used in the literature for MMCT between non-equivalent metal ions, as well as for electron transfer processes not involving metals. In this section, it is the conventional meaning which is followed. Electron transfer between ions of different metallic elements is discussed in Sect. 7.3 under its conventional acronym MMCT.

IVCT is important in so-called mixed-valence compounds, i.e. compounds or materials where two or more oxidation states of the same element coexist. An old example is Prussian Blue, $Fe_4^{III}[Fe^{II}(CN)_6]_3$, where the +3 and +2 oxidations states of Fe are present. Analogously, the same importance must be expected in luminescent materials activated with ions of lanthanide elements that have two stable oxidation states, because electron transfers in active redox pairs can play relevant roles in

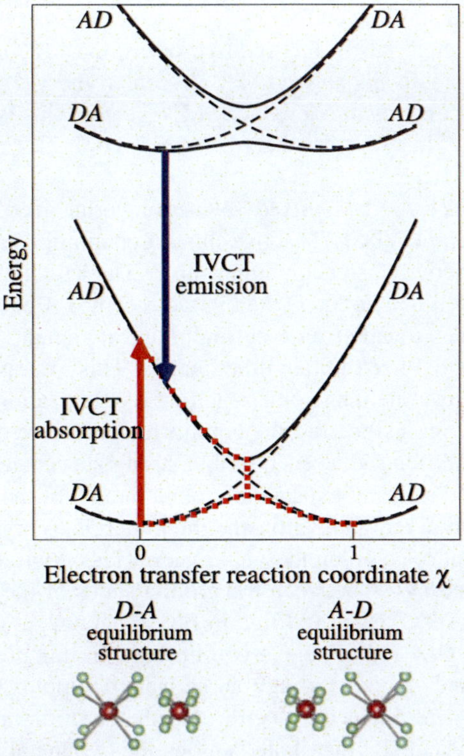

Fig. 7.2 Schematic configuration coordinate diagram of the lowest IVCT states of a $M^{n+}/M^{(n+1)+}$ mixed-valence active pair. M^{n+} is the donor *D* and $M^{(n+1)+}$ is the acceptor *A*. The chosen electron transfer reaction coordinate χ takes the values 0 and 1 at the respective equilibrium structures of the *D-A* and *A-D* configurations of the *D*/*A* pair, i.e. M^{n+}–$M^{(n+1)+}$ and $M^{(n+1)+}$–M^{n+}. Photon induced IVCT absorption and emission are shown in red and blue arrows respectively. Non-radiative decay after IVCT transitions is shown in dashed red lines with two branching channels: one leading to the initial configuration *DA* (back-transfer IVCT) and one to the IVCT configuration *AD*. The lowest dashed red line represents ground state thermally induced IVCT. Adiabatic (full lines) and diabatic (dashed lines) potential energy curves are shown

absorptions, non-radiative decays, and luminescence. This is the case of phosphor materials (consciously or involuntarily) co-doped with Ce^{3+} and Ce^{4+}, Eu^{2+} and Eu^{3+}, or Yb^{2+} and Yb^{3+}, among other mixed-valence pairs.

The basic theory for electron transfer was formulated by Marcus [11]. It was very successfully applied to the understanding of electron transfer reactions, supported on systematic experimental studies of mixed-valence compounds, mostly involving their ground states [12, 13]. A schematic configuration coordinate diagram of the lowest states of a $M^{n+}/M^{(n+1)+}$ mixed-valence active pair is shown in Fig. 7.2. Here, M^{n+} and $M^{(n+1)+}$ do not represent single ions, but two active centers that contain them. Each center involves, at least, the respective ion and its first-neighbors, i.e. the ion and its ligands, but it can eventually include more distant lattice components. A

M^{n+} center can donate an electron to a $M^{(n+1)+}$ center present in its neighborhood in the mixed-valence material. In this redox process, the M^{n+} center is called the donor D and it converts into an oxidized $M^{(n+1)+}$ center; the $M^{(n+1)+}$ center is called the acceptor A and it converts into a reduced M^{n+} center. This is to say: a DA pair converts into a AD pair. In a reversed electron transfer, or back-transfer, the AD pair can convert into DA.

For instance, in a Yb^{2+}/Yb^{3+} mixed-valence material, like Yb^{2+},Yb^{3+}-doped CaF_2, the donor D can be a $(YbCl_8)^{6-}$ active cluster containing Yb^{2+} and the accpetor A can be a $(YbCl_8)^{5-}$ active cluster containing Yb^{3+}. The Yb-Cl equilibrium distance $d_{e,\text{Yb-Cl}}$ is much longer in the donor D than in the acceptor A because the oxidation of Yb^{2+} to Yb^{3+} comes together with an important increment of the electrostatic attraction between the oxidized ion and the ligands. This example is chosen in the bottom of Fig. 7.2 to illustrate the strong structural reorganization that accompanies an inter-valence charge transfer from the equilibrium structure of DA to the equilibrium structure of AD, and viceversa: both a strong shrinkage around the Yb of the donor center D and a strong expansion around the Yb of the acceptor center A take place. This structural reorganization and the important reorganization energy it involves play important roles in luminescent materials, as it will be discussed.

A thermally induced IVCT between the ground states of the two metal sites is illustrated with the lowest dashed red line in Fig. 7.2 connecting the DA and AD configurations of the D/A redox pair at equilibrium. In this process, the electron transfer between D and A goes through an activated complex with an activation energy barrier, which determines the electron transfer reaction rate. In the activated complex the electron is equally distributed among the two metallic centers.

The Figure also illustrates the key importance of electronic interactions between D and A for the thermal electron transfer reaction rate. In effect, it shows in dashed lines *diabatic* potential energy curves of the DA and AD configurations of the pair; these are calculated including electrostatic interactions between D and A, but not bonding interactions between them. When such bonding interactions are included, the *adiabatic* potential energy curves result, shown in full lines. This is equivalent to consider resonant interactions between DA and AD configurations or to allow their valence bond electronic structures to couple. As it is clear in the Figure, the thermal electron transfer energy barrier totally depends on DA bonding interactions at the crossing point between the diabatic DA and AD curves, with structures of D and A that make DA and AD degenerate. The larger the bonding interactions are, the larger the splitting and the lower the electron transfer activation energy. They are very important when the transferred electron is not well localized on D, but equally distributed between D and A. Moreover, the electron transfer probability itself is null within the diabatic approximation. The DA bonding interactions decrease with the distance between the oxidized and reduced forms, which produce an accompanying increase of the barrier and lowering of the electron transfer probability. It is noticeable that electron transfers between redox centers as distant as 25 Å have been proven in binuclear mixed-valence metal complexes [14].

However, when the transferred electron is well localized on D, i.e. at the minima of the potential energy curves, the resonance between DA and AD electronic configurations is lower or much lower, the difference between dashed and full potential energy curves is small, and the diabatic approximation works well in many cases. This implies that energy differences between potential energy curves calculated near the equilibrium structures of the DA and AD configurations of the D/A redox pair have a good chance to be correct within the diabatic approximation. This is important because diabatic potential energy surfaces are far less demanding than adiabatic ones from the computational point of view. All this means that vertical energy differences at the minima in an IVCT configuration coordinate diagram like the one in Fig. 7.2 may be expected to be calculated with sufficient accuracy within the diabatic approximation. Transition probabilities, though, still require the far more demanding adiabatic calculations. The IVCT and MMCT studies shown in the book make use of the diabatic approximation.

Vertical energy differences between DA and AD curves in an IVCT diagram like Fig. 7.2 represent photo-induced electron transfer energies. The red upwards arrow represents the so called IVCT absorption. In it, the absorption of a photon produces the transfer of an electron from D to A at fixed nuclei positions (also known as vertical or Frank-Condon transition). The strong slope of the AD curve at the position of the red arrow, which is associated with a very large horizontal offset between the minimum of this AD curve (final state of the transition) and the minimum of the DA curve (initial state of the transition), implies that the IVCT absorption band is very very broad. The IVCT absorption cannot be followed by any emission. Instead, it is followed by a two-step non-radiative decay involving a strong nuclear reorganization: In the first step (dashed red line down to the minimum of the second adiabatic potential energy curve), the initial D, after having donated one electron and become an actual A, strongly diminishes its M-L distance, and, simultaneously, the initial A, after having accepted one electron and become an actual D, strongly increments its M-L distance; the first non-radiative decay step ends at the structure of the activated complex, when both M-L distances are equal and the transferred electron is equally distributed between the two metallic centers. At this point, there is branching and two second non-radiative steps are possible; either the electron is transferred back and the original DA configuration of the pair is reestablished (left branching of the dashed red line) or it is definitely transferred and the final AD configuration acquires its equilibrium structure (right branching of the dashed red line). In the first case, the M-L distance in the left site increases again up to its initial value at the time that the M-L distance in the right site decreases to its initial value. In the second case, the M-L distance in the right increases and the one in the left decreases. In both cases, the structural reorganization is the same, although symmetrical, and the reorganization energy is identical. The Figure lets clear that this reorganization energy is very large and equal to the IVCT absorption energy. Observations of dark color in substances containing platinum in two oxidation states [15] were explained as due to very broad IVCT absorption and the black, dark brown, dark blue, and dark color in general were found in a large number of mixed-valence compounds [12, 13]. In fact, the mere observation of dark color in a material is often interpreted as a possible first indication of mixed-valence. The Marcus theory of thermally induced and photo-induced electron transfers shown in red in Fig. 7.2 and involving the two lowest states was complemented by Hush [16, 17] with a theory of heteronuclear MMCT and homonuclear IVCT considering a two-state problem. Piepho et al. [18] formulated a vibronic model for the IVCT absorption of the two-state problem. However, development of extensions of these theories to higher excited states of the mixed-valence compounds were missing.

It is clear in Fig. 7.2 that not only IVCT absorption but also IVCT *emissions* like the one represented with the vertical downwards blue arrow must also be expected and that the diabatic approximation may perform well for them. However, extensions of the above theories to higher excited states of the mixed-valence compounds was not considered a priority and, as a matter of fact, IVCT emissions had not been reported, identified, or predicted prior to the works discussed in this book, which conclude that IVCT emissions have in fact been observed in Ce-doped elpasolites and in Yb-doped fluorites, even though they have not been identified as such.

In the book, both IVCT absorptions, IVCT emissions, and IVCT nonradiate decays leading either to emission quenching or to the activation of new luminescence, are considered with equal interest and shown to play relevant roles in luminescence. The multiconfigurational ab initio calculation of IVCT potential energy surfaces and configuration coordinate diagrams is discussed in Sect. 7.4. Several applications to the understanding of complex luminescence phenomena are discussed in Chap. 11, including the complex interplay between regular and anomalous emissions of Yb^{2+} in fluorite-type hosts, where the luminescence in similar materials ranges from regular, to anomalous, to a mix of both, and to total absence of it; the anomalous emission of Ce^{3+} in several hosts; the laser-induced white-light emission of Sr_2CeO_4; the direct evidence of IVCT states found in Eu-doped materials; and the invariance of the IVCT absorption onset across the lanthanide series.

7.3 Metal-to-Metal Charge Transfer MMCT

When the two metallic ions involved in an electron transfer belong to different elements, the process is conventionally called metal-to-metal charge transfer (MMCT). The importance of MMCT transitions in optically active doped solids has long been recognized [19], mostly for transition metal dopants, but, as in the case of IVCT transitions, they haven't been the subject of extensive research. In the case of lanthanide doped phosphors, the number of MMCT studies is very small. A traditional example is Pr-to-Ti MMCT, which has long been considered responsible for the quenching of the green-blue luminescence and simultaneous enhancement of the red emission in Pr^{3+}-doped compounds [20], although this interpretation was later unsupported by ab initio calculations (cf. Sects. 8.2.3 and 11.1). More recently, Eu-to-Tb MMCT IR luminescence has been experimentally proven in the doubly doped material CaS:Eu^{2+},Tb^{3+} and multiconfigurational ab initio calculations have attributed it to the radiative decay of a metastable Eu^{3+}-Tb^{2+} configuration of the Eu/Tb pair, or in other words a metastable MMCT state of a Eu^{2+}-Tb^{3+} pair [21].

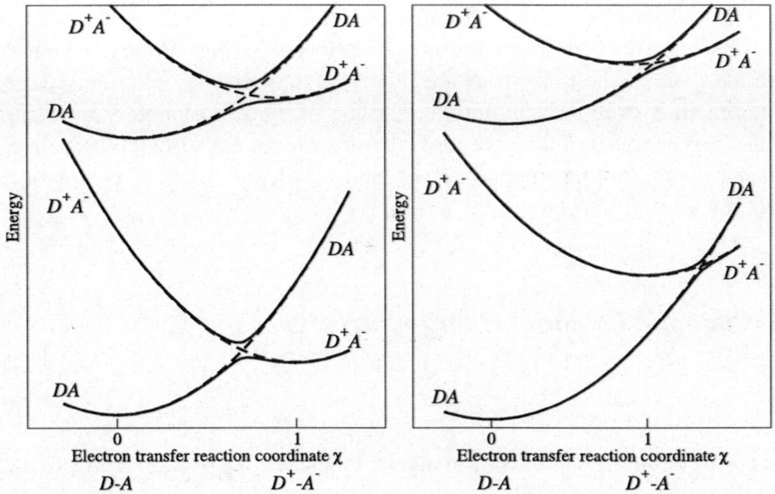

Fig. 7.3 Two cases of configuration coordinate diagrams of the lowest MMCT states of a $M_1^{n_1+}/M_2^{n_2+}$ active pair. $M_1^{n_1+}$ is the donor D and $M_2^{n_2+}$ is the acceptor A. $M_1^{(n_1+1)+}$ is D^+ and $M_2^{(n_2-1)+}$ is A^-. Dashed lines: diabatic potential energy curves. Full lines: adiabatic potential energy curves

The calculations also explained why the similar material SrS:Eu^{2+},Tb^{3+} does not have an equivalent MMCT emission (Sect. 11.3).

The basic structure of a configuration coordinate diagram of the MMCT states of a $M_1^{n_1+}/M_2^{n_2+}$ pair, in which $M_1^{n_1+}$ acts as a donor D and $M_2^{n_2+}$ as an acceptor A, is shown in Fig. 7.3. It resembles the diagram IVCT states in Fig. 7.2, with the important differences that the left-right symmetry is lost and the left and right minima are not equivalent. This is because the acceptor A does not coincide with the ionized donor D^+ anymore, like in the IVCT case. The case shown in Fig. 7.3 (left) corresponds to not very different values of the electron affinity of A and ionization potential of D; hence, to a diagram not very different to the IVCT diagram. In Fig. 7.3 (right), however, very different values of these two energies produce a quite different diagram from the quantitative point of view, with the minima of the DA and D^+A^- curves lying at very different energies and their avoided crossing not lying between them. These diagrams are further discussed in Sect. 7.4.

7.4 Ab Initio Diabatic IVCT and MMCT Configuration Coordinate Diagrams

In this section, a presentation is made of the adiabatic potential energy surfaces and the configuration coordinate diagrams derived from them that would be required for a basic understanding of optically relevant electron transfers between metal ions in solids. Their multiconfigurational ab initio calculations involve direct treatment of metallic dimers and they are in general very highly demanding. When two lanthanide ions with many open-shell electrons are involved, they may be extremely demanding and unpractical from the computational point of view. Fortunately, as discussed above, basic key features of the potential energy surfaces are sufficiently well reproduced at the much simpler, diabatic level, and the details of their calculation are discussed in a second step.

7.4.1 Adiabatic Potential Energy Surfaces

MMCT

Taking a specific luminescent doped material like $CaF_2{:}Yb^{2+}$ as a descriptive case, an electron transfer from the Yb^{2+} lanthanide dopant to the CaF_2 host, either photoinduced or thermally induced, is an example of a MMCT process, which can be represented by the equation

$$Yb^{2+} + Ca^{2+} \rightarrow Yb^{3+} + Ca^{+} \ .$$

But the electron transfer produces strong changes in the surroundings of the Yb and Ca sites and a better representation of the MMCT process involves, at least, their ligands:

$$(YbF_8)^{6-} + (CaF_8)^{6-} \rightarrow (YbF_8)^{5-} + (CaF_8)^{7-} \ ,$$

where some of the F ligands of Yb and Ca can eventually be shared if the dopant and the host cation are very close to each other. In this representation, $(YbF_8)^{6-}$ is the electron *donor* D and $(CaF_8)^{6-}$ is the *acceptor* A. Improved representations would involve second and more distant neighbors of Yb and Ca. In correspondence, $(YbF_8)^{5-}$ is the ionized donor D^+ and $(CaF_8)^{7-}$ is the electron attached acceptor A^-, and the above MMCT process can be represented as $DA \rightarrow D^+A^-$. This representation is more realistic because it lets more clear that donor and acceptor might not be separated entities but, rather, a single chemical moiety where the electron is transferred from one part to another. Moreover, the electron transfers may involve not only the ground states of DA and D^+A^-, but also many of their excited states. If these can be linked with the excited states of the separated donor and acceptor, which is definitely possible for long D-A distances d_{DA}, a better and more complete representation of the possible MMCT processes is

$$D_i A_j \rightarrow D_k^+ A_l^- \ , \tag{7.1}$$

where the subindexes i, j, k, l indicate the electronic states of donor, acceptor, ionized donor, and electron attached acceptor, respectively.

The many-electron wave functions and total energies of DA and D^+A^- embedded in the solid host can be calculated for any of their ground and excited states with multiconfigurational ab initio methods as functions of their internal nuclear coordinates. The result of such calculations is a (large) set of (complex) adiabatic potential energy surfaces. Fortunately, photon absorptions and emissions are dominated by totally symmetric vibrational coordinates and the key features of optically relevant electron transfers are expected to be well represented by considering only the breathing modes of the simplest clusters involving the active metal sites and their ligands, i.e. $(YbF_8)^{6-}$, $(CaF_8)^{6-}$, $(YbF_8)^{5-}$, and $(CaF_8)^{7-}$ in the example, which limits the relevant vibrational degrees of freedom to the Yb-F and Ca-F distances, $d_{\text{Yb-F}}$ and $d_{\text{Ca-F}}$. In a general case these distances will be called d_D and d_A, respectively. I.e. d_D is the distance between the active donor metal site and its ligands (assuming a highly symmetric site) and d_A is the distance between the active acceptor metal site and its ligands. Then, the relevant adiabatic potential energy surfaces are $E_{D_i A_j}(d_D, d_A)$ and $E_{D_k^+ A_l^-}(d_D, d_A)$. In general, they result from direct quantum mechanical calculations on metallic dimers, which are highly demanding and, depending on the number of open-shell electrons of the lanthanide, can be extremely demanding.

IVCT

IVCT is a particular case of MMCT where the acceptor is an oxidized form of the donor and some specific notation conventions are convenient. As an example, in the co-doped material CaF_2:Yb^{2+},Yb^{3+}, an electron transfer from the reduced form of the lanthanide dopant to the oxidized form, either photoinduced or thermally induced, is an IVCT process that can be represented with the equation

$$Yb^{2+} + Yb^{3+} \rightarrow Yb^{3+} + Yb^{2+} \ .$$

Here two specific Yb ions are involved: one can be called the left Yb and the other the right Yb. In the represented IVCT process, the left Yb (Yb^{2+}) transfers one electron to the right Yb (Yb^{3+}) and both of them change oxidation state: the left Yb becomes Yb^{3+} and the right Yb becomes Yb^{2+}. As before, a better representation is

$$(YbF_8)^{6-} + (YbF_8)^{5-} \rightarrow (YbF_8)^{5-} + (YbF_8)^{6-} \ ,$$

where $(YbF_8)^{6-}$ and $(YbF_8)^{5-}$ in the left-hand side of the equation are, respectively, the *donor D* and the *acceptor A* sites, corresponding to the left Yb and the right Yb, respectively. Now, in the right-hand side of the equation, after the electron transfer, the ionized donor D^+ is $(YbF_8)^{5-}$, which is the left Yb converted into an acceptor A, and the electron attached acceptor A^- is $(YbF_8)^{6-}$, which is the right Yb converted into a donor D. Then, the IVCT process can be represented as $DA \rightarrow AD$, and, considering all possible excited states,

$$D_i A_j \rightarrow A_k D_l \ . \tag{7.2}$$

The adiabatic potential energy surfaces of interest are $E_{D_i A_j}(d_L, d_R)$ and $E_{A_k D_l}$ (d_L, d_R), where d_L is the distance between the left metal site and its ligands and d_R is the distance between the right metal site and its ligands. In the example, d_L is $d_{\text{Yb-F}}$ in the left cluster, which is $(YbF_8)^{6-}$ before the electron transfer and $(YbF_8)^{5-}$ after it took place, and d_R is $d_{\text{Yb-F}}$ in the right cluster, which is $(YbF_8)^{5-}$ before the electron transfer and $(YbF_8)^{6-}$ after it. The symmetry of the problem implies $E_{D_iA_j}(d_L, d_R) = E_{A_jD_i}(d_R, d_L)$.

7.4.2 *Diabatic Potential Energy Surfaces*

The adiabatic potential energy surfaces can be calculated in several ways. Regardless of the way they are calculated, they can be considered to result from the diagonalization of an interaction matrix between diabatic many-electron states. This is a very convenient point of view that will be discussed here, following [22–24], because it leads to the much simpler but fruitful diabatic approximation, which, in spite of its intrinsic limitations, captures the basics of the photoinduced electron transfers.

Definition

The adiabatic energies of the states of a donor/acceptor pair prior and after an electron transfer, $E_{D_iA_j}$, $E_{D_k^+A_l^-}$ (full lines in Figs. 7.2 and 7.3), are the eigenvalues of the many-electron Hamiltonian of the pair embedded in the host, $\hat{H}_{\text{EC}}$ (1.3). The adiabatic states are the corresponding eigenfunctions Φ_{EC} (1.1). Regardless of the way the adiabatic energies are obtained, they can be calculated as the eigenvalues of an interaction matrix between diabatic many-electron wave functions, the diabatic basis $\{\Phi^{\text{diab}}\}$. The diagonal elements of this matrix are the diabatic energies E^{diab} (dashed lines in Figs. 7.2 and 7.3) and the off-diagonal elements are the electronic couplings between diabatic states. When the matrix is calculated for a number of nuclear configurations, the eigenvalues map the adiabatic potential energy surfaces and the diagonal elements map the diabatic potential energy surfaces.

The diabatic basis $\{\Phi^{\text{diab}}\}$ is arbitrary. In principle, the full solution of the vibronic problem of the donor/acceptor pair can be equally obtained using the full set of adiabatic electronic wave functions $\{\Phi_{\text{EC}}\}$ or any of the diabatic basis of choice, because all these alternatives lead to the same vibronic energies and wave functions. Choosing one or another set of electronic wave functions is then a matter of convenience. However, when the focus is on features of the electronic spectra (like zero-phonon lines, absorption peaks and emission peaks, band widths, etc.) or on properties of the electronic states of the donor/acceptor pair (like ground and excited state equilibrium structures, electron transfer activation energy barriers, etc.), the adiabatic potential energy surfaces are the valid ones and the diabatic ones are an approximation, better or worse. Then, if one is willing to use the diabatic potential energy surfaces for MMCT absorption and emission peaks, for instance, the choice of diabatic states is crucial.

A very convenient choice of diabatic states is the set of generalized antisymmetric products [25] of donor and acceptor wave functions $\{\Phi_D\}$, $\{\Phi_A\}$, $\{\Phi_{D^+}\}$, $\{\Phi_{A^-}\}$, i.e.

$$\Phi^{\text{diab}}_{D_iA_j} = M\hat{A}(\Phi_{Di}\Phi_{Aj}), \quad \Phi^{\text{diab}}_{D_k^+A_l^-} = M\hat{A}(\Phi_{D_k^+}\Phi_{A_l^-}). \tag{7.3}$$

Here, Φ_{D_i}, Φ_{A_j}, $\Phi_{D_k^+}$, and $\Phi_{A_l^-}$ are the many-electron wave functions of the ground and excited states of the D, A, D^+ and A^- moieties of the metal-metal pair, M is a normalization constant and $\hat{A}$ is the inter-group antisymmetrization operator [25].

The diabatic energies are

$$E^{\text{diab}}_{D_i A_j} = \langle \Phi^{\text{diab}}_{D_i A_j} | \hat{H}_{\text{EC}} | \Phi^{\text{diab}}_{D_i A_j} \rangle \,, \quad E^{\text{diab}}_{D_k^+ A_l^-} = \langle \Phi^{\text{diab}}_{D_k^+ A_l^-} | \hat{H}_{\text{EC}} | \Phi^{\text{diab}}_{D_k^+ A_l^-} \rangle \,. \tag{7.4}$$

The diabatic energy surfaces can cross each other and the electronic characters of the diabatic states remain unchanged at both sides of the crossings. In contrast, the crossings are avoided in the adiabatic case due to the electronic couplings. The electronic character of an adiabatic state is different at both sides of an avoided crossing, where it suffers a drastic change, passing through a superposition of both characters. E.g. the adiabatic ground state in Fig. 7.3 changes from DA to D^+A^- character across the electron transfer energy barrier at the avoided crossing. In it, its character is a strong mixture of DA and D^+A^-, so that the transferred electron is not in D nor in A, but equally distributed in both of them.

On the contrary, the diabatic ground state DA remains as such across the crossing and becomes the first excited state, at the time that the diabatic state of D^+A^- character becomes the ground state. The differences are a consequence of the electronic couplings $V^{\text{diab}}_{D_i A_j, D_k^+ A_l^-} = \langle \Phi^{\text{diab}}_{D_i A_j} | \hat{H}_{\text{EC}} | \Phi^{\text{diab}}_{D_k^+ A_l^-} \rangle$. These have their maximum impact at the crossings between diabatic states. This is why they determine the electron transfer energy barriers and the kinetics of the thermal electron transfer. As a matter of fact, the electronic couplings are responsible for the electron transfer itself, which is strictly forbidden for null couplings.

However, around the minima of DA and D^+A^- potential energy surfaces, away from the crossing points, the electronic couplings are far less important and the diabatic potential energy surfaces calculated with the choice of diabatic wave functions (7.3) can be very close to the adiabatic ones. This opens the door to an understanding of the radiative and non-radiative processes involved in many IVCT and MMCT processes at a first order level, without resorting to the expensive adiabatic level. IVCT luminescence of Ce^{3+}/Ce^{4+} and Yb^{2+}/Yb^{3+} mixed-valence pairs [22, 23] and the role of MMCT in Yb^{2+}/Ca^{2+} pairs on the photoconductivity of CaF_2:Yb^{2+} [24] are good examples of this.

Approximations

Although the calculations of the diabatic energies (7.4) are much simpler than the calculations of the adiabatic ones, they are still cumbersome and the use of additional, sound, simplifying approximations is advisable. Two alternative strategies are discussed in [23]: one in which the symmetry lowering around one entity in the solid host due to the presence of the other (around D by the presence of A and viceversa, and around D^+ by the presence of A^- and viceversa) is considered from the beginning; and another one in which this symmetry lowering is neglected in a fist step and considered later, at the same time that the electronic couplings, when the adiabatic surfaces are eventually calculated. The fist strategy has the advantage of giving symmetry driven splittings already at the diabatic level, which are dependent on the D-A distance and orientation. The advantage of the second strategy is to capture the basics of the potential energy surfaces by means of independent calculations on the embedded D, A, D^+, and A^- entities.

The second strategy is discussed here. The IVCT case is detailed first and the necessary extensions for the MMCT case are presented later.

The diabatic energies correspond to group function theory electronic wave functions [26] and they are the sums of donor and acceptor energies plus their mutual Coulomb and exchange interactions, whose state dependence can be neglected:

$$E^{\mathrm{diab}}_{D_iA_j} = E^*_{D_i} + E^*_{A_j} + E^{cx}_{D_iA_j} \approx E^*_{D_i} + E^*_{A_j} + E^{cx}_{DA} \,. \qquad (7.5)$$

Here, $E^*_{D_i}$ and $E^*_{A_j}$ include the embedding interactions of D and A with the crystalline environment *of the DA pair*. Similar equations can be written for $E^{\mathrm{diab}}_{A_kD_l}$ in the IVCT case and for $E^{\mathrm{diab}}_{D^+_kA^-_l}$ in the MMCT case.

Considering the descriptive example of IVCT in the double doped CaF_2:Yb^{2+}, Yb^{3+} material, a calculation on the $(YbF_8)^{6-}$ cluster, the donor, embedded in CaF_2 gives the embedded-cluster energy $E_{D_i} = E^*_{D_i} + E^{cx}_{DC_A}$, where $E^{cx}_{DC_A}$ stands for the Coulomb and exchange interaction energy between the donor cluster $(YbF_8)^{6-}$ and a cluster C_A located where the acceptor is, but containing the original host cation that has been substituted by the acceptor dopant, i.e. $(CaF_8)^{6-}$. Another calculation on the $(YbF_8)^{5-}$ cluster, the acceptor, embedded in CaF_2, gives $E_{A_j} = E^*_{A_j} + E^{cx}_{AC_D}$, where, in parallel with the above, $E^{cx}_{AC_D}$ stands for the Coulomb and exchange interaction energy between the acceptor cluster $(YbF_8)^{5-}$ and a cluster C_D located where the donor is, but containing the original host cation that has been substituted by the donor dopant, i.e. $(CaF_8)^{6-}$. In the example, C_D and C_A are identical, but the difference is maintained here in order to consider cases where donor and acceptor substitute different lattice ions. Then, $E^{\mathrm{diab}}_{D_iA_j} = E_{D_i} + E_{A_j} - E^{cx}_{DC_A} - E^{cx}_{AC_D} + E^{cx}_{DA}$. In this approximate strategy, E_{D_i} and E_{A_j} are independent of the relative positions of donor and acceptor. Besides, except for short cation-cation distances, the Coulomb and exchange correction term can be approximated, $-E^{cx}_{DC_A} - E^{cx}_{AC_D} + E^{cx}_{DA} \approx (-q_Dq_{C_A} - q_Aq_{C_D} + q_Dq_A)e^2/d_{DA}$, where d_{DA} is the donor-acceptor distance, and q_D, q_A, q_{C_D}, and q_{C_A} are the charges of the donor, the acceptor, and the host ions they substitute, i.e. +2, +3, +2 and +2, respectively in the example.

In summary, the adopted approximations lead to the DA diabatic potential energy surfaces:

$$E^{\mathrm{diab}}_{D_iA_j}(d_L, d_R) = E_{D_i}(d_L) + E_{A_j}(d_R) + \Delta E^{cx}_{DA}(d_{DA}) \,, \qquad (7.6)$$

with $\Delta E^{cx}_{DA}(d_{DA}) = (q_Dq_A - q_Dq_{C_A} - q_Aq_{C_D})e^2/d_{DA}$. In the example, d_L and d_R are the $d_{\text{Yb-F}}$ distances in the left Yb and in the right Yb, respectively. $E_{D_i}(d_L)$ and $E_{A_j}(d_R)$ are, then, curves of the E verses $d_{\text{Yb-F}}$ energy diagrams of $(YbF_8)^{6-}$ and $(YbF_8)^{5-}$, respectively.

The diabatic potential energy surfaces of the states of the AD configuration of the mixed-valence D/A pair in the IVCT problem, obtained after donor-to-acceptor electron transfer, are:

$$E^{\text{diab}}_{A_k D_l}(d_L, d_R) = E_{A_k}(d_L) + E_{D_l}(d_R) + \Delta E^{cx}_{AD}(d_{DA}), \tag{7.7}$$

with $\Delta E^{cx}_{AD}(d_{DA}) = (q_D q_A - q_D q_{C_D} - q_A q_{C_A})e^2/d_{DA}$.

In the most common IVCT cases, like the CaF_2:Yb^{2+},Yb^{3+} example, where both donor and acceptor substitute for the same host cation, the Coulomb and exchange correction terms $\Delta E^{cx}_{DA}(d_{DA})$ and $\Delta E^{cx}_{AD}(d_{DA})$ are identical for all states of the DA and AD configurations of the D/A pair; they produce a common shift of all states; and they do not have any relevant effects on the IVCT diabatic potential energy surface diagram. As a result, the IVCT energy diagram is independent of the donor-acceptor distance because the Coulomb and exchange correction terms are the only ones that depend on d_{DA}.

For convenience, the quantities represented in IVCT diagrams are the state energies referred to the minimum of the DA ground state,

$$E^{\text{diab}}_{D_i A_j}(d_L, d_R) - E^{\text{diab}}_{D_0 A_0, e} = [E_{D_i}(d_L) - E_{D_0,e}] + [E_{A_j}(d_R) - E_{A_0,e}], \tag{7.8}$$

and

$$E^{\text{diab}}_{A_k D_l}(d_L, d_R) - E^{\text{diab}}_{D_0 A_0, e} = [E_{A_k}(d_L) - E_{A_0,e}] + [E_{D_l}(d_R) - E_{D_0,e}]. \tag{7.9}$$

The components of (7.8) and (7.9) in square brackets are the potential energy curves of donor and acceptor referred to their respective ground state minima. An example of these surfaces for the ground state of a mixed-valence D/A pair $(YbF_8)^{6-}$/$(YbF_8)^{5-}$ in CaF_2:Yb^{2+},Yb^{3+} is shown in Fig. 7.4.

Let us now discuss the calculation of diabatic potential energy surfaces in a MMCT case, like Yb-to-Ca electron transfer in CaF_2:Yb^{2+}. The DA diabatic potential energy surfaces are equally described by (7.6), like in the IVCT case, but the fact that the donor and acceptor ions are now different, not just two oxidation states of the same element, makes it more convenient to update the labels of the metal-ligand distances: the left cation is always Yb in the example, Yb^{2+} in the donor D and Yb^{3+} in the ionized donor D^+, and the right cation is always Ca, Ca^{2+} in the acceptor A and Ca^+ in the electron attached acceptor A^-. Then, d_D will be used instead of d_L and d_A instead of d_R. In the example d_D is $d_{\text{Yb-F}}$ and d_A is $d_{\text{Ca-F}}$.

Then, the DA diabatic potential energy surfaces are

$$E^{\text{diab}}_{D_i A_j}(d_D, d_A) = E_{D_i}(d_D) + E_{A_j}(d_A) + \Delta E^{cx}_{DA}(d_{DA}), \tag{7.10}$$

with $\Delta E^{cx}_{DA}(d_{DA}) = (q_D q_A - q_D q_{C_A} - q_A q_{C_D})e^2/d_{DA}$. In the example, $E_{D_i}(d_D)$ is an E verses $d_{\text{Yb-F}}$ curve of $(YbF_8)^{6-}$, $E_{A_j}(d_A)$ is an E verses $d_{\text{Ca-F}}$ curve of $(CaF_8)^{6-}$, $q_D = q_{Yb^{2+}} = +2$, and $q_A = q_{C_A} = q_{C_D} = q_{Ca^{2+}} = +2$.

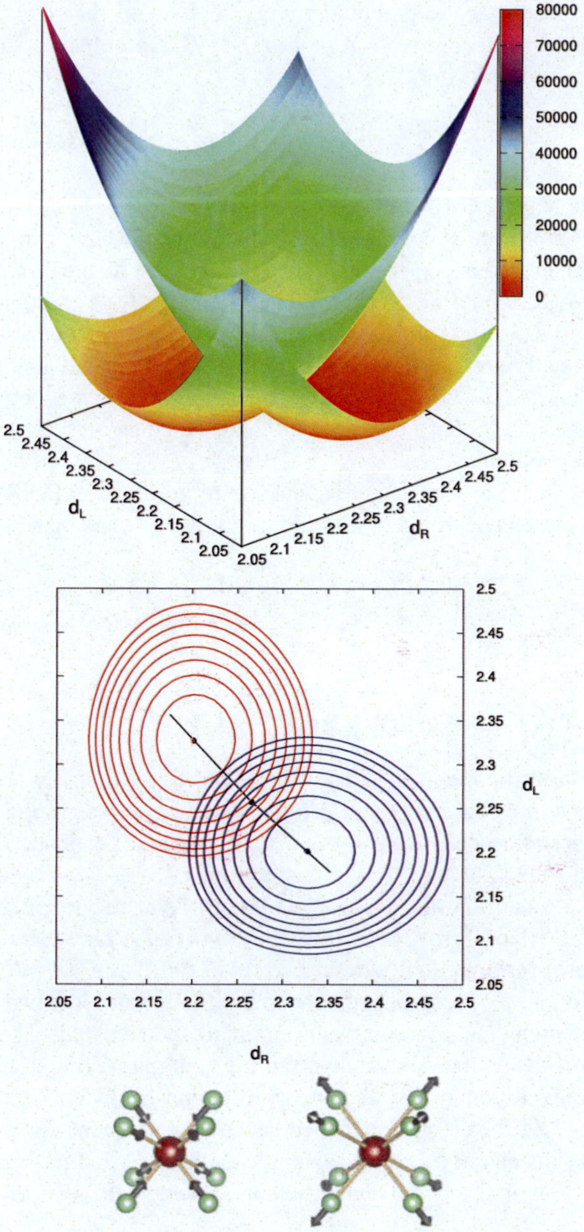

Fig. 7.4 Ground state diabatic potential energy surfaces (in cm^{-1}) of the mixed-valence D/A pair $(YbF_8)^{6-}/(YbF_8)^{5-}$ in CaF_2, shown in color code (top graph) and with isolines (middle graph). The surface with the minimum in the left (red isolines) corresponds to the D-A configuration of the pair, $(Yb_LF_8)^{6-} - (Yb_RF_8)^{5-}$; the surface with the minimum in the right (blue isolines) corresponds to the A-D configuration of the pair, $(Yb_LF_8)^{5-} - (Yb_RF_8)^{6-}$. d_L and d_R stand respectively for d_{Yb_L-F} and d_{Yb_R-F}, in Å. The IVCT or electron transfer reaction coordinate Q_{et} (minimal energy reaction path connecting the two equivalent minima) is indicated in the middle graph with a black line. The two minima and the activated complex are indicated with full black circles. Coordinated atomic displacements of the IVCT reaction coordinate are shown in the bottom graph. The variations of the energies of all states with this electron transfer reaction coordinate are the configuration coordinate energy diagrams

The D^+A^- diabatic potential energy surfaces are:

$$E^{\text{diab}}_{D_k^+A_l^-}(d_D, d_A) = E_{D_k^+}(d_D) + E_{A_l^-}(d_A) + \Delta E^{cx}_{D^+A^-}(d_{DA}), \quad (7.11)$$

with $\Delta E^{cx}_{D^+A^-}(d_{DA}) = [(q_D + 1)(q_A - 1) - (q_D + 1)q_{C_A} - (q_A - 1)q_{C_D}]e^2/d_{DA}$.

The MMCT energy diagrams are usually referred to the minimum of the DA ground state, as in the IVCT case. Then, the DA surfaces are

$$E^{\text{diab}}_{D_iA_j}(d_D, d_A) - E^{\text{diab}}_{D_0A_0,e} = [E_{D_i}(d_D) - E_{D_0,e}] + [E_{A_j}(d_A) - E_{A_0,e}], \quad (7.12)$$

where the usual potential energy curves of donor and acceptor referred to their respective ground state minima appear in square brackets. The D^+A^- surfaces are

$$\begin{aligned} E^{\text{diab}}_{D_k^+A_l^-}(d_D, d_A) - E^{\text{diab}}_{D_0A_0,e} &= E^{\text{diab}}_{D_k^+A_l^-}(d_D, d_A) - E^{\text{diab}}_{D_0^+A_0^-,e} + E^{\text{diab}}_{D_0^+A_0^-,e} - E^{\text{diab}}_{D_0A_0,e} \\ &= [E_{D_k^+}(d_D) - E_{D_0^+,e}] + [E_{A_l^-}(d_A) - E_{A_0^-,e}] \\ &\quad + \text{IP}_D - \text{EA}_A \\ &\quad -[(q_D - q_{C_D}) - (q_A - q_{C_A}) + 1]e^2/d_{DA}, \end{aligned} \quad (7.13)$$

which contain the minimum-to-minimum ionization potential of the donor $\text{IP}_D = E_{D_0^+,e} - E_{D_0,e}$ and electron affinity of the acceptor $\text{EA}_A = E_{A_0,e} - E_{A_0^-,e}$. In (7.13), the components in square brackets are the potential energy curves of the ionized donor and the electron attached acceptor ($(YbF_8)^{5-}$ and $(CaF_8)^{7-}$, respectively, in the example) referred to their respective ground state minima. In the MMCT case, in opposition to the IVCT case, the donor ionization potential and the acceptor electron affinity are different in general and their difference shifts the D^+A^- surfaces with respect to the DA ones. This shift is uniform and independent of the donor-acceptor distance. There is another shift, however, that depends on this distance: it is the Coulomb and exchange correction represented by the last term in the equation. It depends on the charge defects of donor and acceptor, $q_D - q_{C_D}$ and $q_A - q_{C_A}$, and it is the only contribution that induces a donor-acceptor distance, d_{DA}, dependence in the MMCT diagrams.

An example of two MMCT diabatic potential energy surfaces of the Yb-to-Ca electron transfer in $CaF_2{:}Yb^{2+}$ is shown in Fig. 7.5: The ground state of the D/A pair $(YbF_8)^{6-}/(CaF_8)^{6-}$, which corresponds to a $(YbF_8)^{6-}$-$(CaF_8)^{6-}$ configuration of the pair with both Yb^{2+} and Ca^{2+} in their ground states of respective $4f^{14}$ and $3p^6$ closed-shell electronic configurations; and an excited state corresponding to a D^+-A^- configuration of the pair, $(YbF_8)^{5-}$-$(CaF_8)^{7-}$, with Yb^{3+} in its $4f^{13}\,{}^2F_{7/2}$ ground state and Ca^+ in the lowest state of its $4s^1$ configuration.

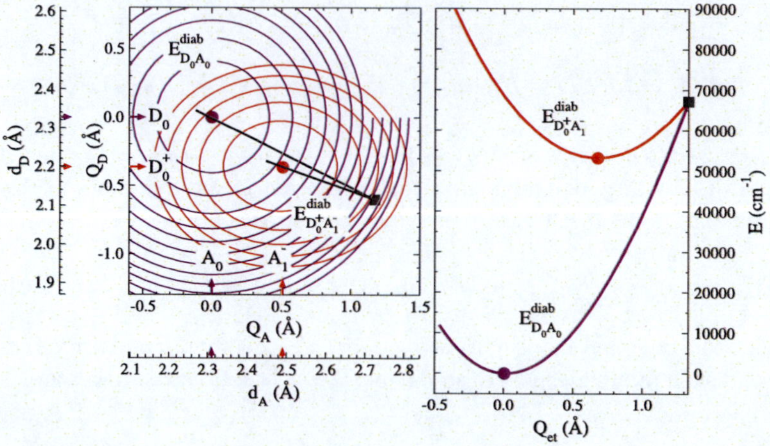

Fig. 7.5 Left panel: Two diabatic potential energy surfaces of the D/A MMCT pair $(YbF_8)^{6-}/(CaF_8)^{6-}$ in CaF_2:Yb^{2+}: Ground state, corresponding to the D-A configuration of the pair, $(YbF_8)^{6-}$-$(CaF_8)^{6-}$ (violet isolines) and an excited state corresponding to a D^+-A^- configuration of the pair, $(YbF_8)^{5-}$-$(CaF_8)^{7-}$, with Yb^{3+} in its $4f^{13}\ {}^2F_{7/2}$ ground state and Ca^+ in the lowest state of its $4s^1$ configuration (red isolines). d_D and d_A stand respectively for $d_{\text{Yb-F}}$ and $d_{\text{Ca-F}}$. Corresponding breathing normal vibrational modes are Q_D and Q_A. The MMCT or electron transfer reaction coordinate Q_{et} (minimal energy reaction path connecting the two equivalent minima) is indicated with a black line. The two minima are shown with violet and red full circles and the activated complex with a black full circle. Right panel: Partial configuration coordinate energy diagram: Energies of the two states along the electron transfer reaction coordinate. (Note that the state of DA is evaluated along the line connecting the violet and black circles in the left panel, whereas the state of D^+A^- is evaluated along the line connecting the black and red circles.)

Topology

IVCT

Let us start with the topology of IVCT diabatic potential energy surfaces. Figure 7.4 shows those of the ground state of the mixed-valence D/A pair $(YbF_8)^{6-}/(YbF_8)^{5-}$ in CaF_2 [22] and it is a good example of a general case. They are a couple of symmetry related surfaces: one of them, $E^{\text{diab}}_{D_0A_0}$, corresponds to the D-A configuration of the pair, $(Yb_LF_8)^{6-}$-$(Yb_RF_8)^{5-}$ and the other, $E^{\text{diab}}_{A_0D_0}$, corresponds to the A-D configuration of the pair, $(Yb_LF_8)^{5-}$-$(Yb_RF_8)^{6-}$. They fulfill $E^{\text{diab}}_{D_0A_0}(d_L, d_R) = E^{\text{diab}}_{A_0D_0}(d_R, d_L)$. This is a general result for any pair of excited state surfaces $E^{\text{diab}}_{D_iA_j}$ and $E^{\text{diab}}_{A_jD_i}$: $E^{\text{diab}}_{D_iA_j}(d_L, d_R) = E^{\text{diab}}_{A_jD_i}(d_R, d_L)$. $E^{\text{diab}}_{D_iA_j}$ has its minimum at left and right moiety distances equal to the respective equilibrium metal-ligand distances of donor in its i state and acceptor in its j state, i.e. $d_L = d_{e,D_i}, d_R = d_{e,A_j}$. Correspondingly, $E^{\text{diab}}_{A_jD_i}$ has its minimum at $d_L = d_{e,A_j}, d_R = d_{e,D_i}$ and exactly the same energy. For example, in the reference example, the minimum of $E^{\text{diab}}_{D_0A_0}$ is at $d_L = 2.329$ Å, $d_R = 2.200$ Å, which are the respective Yb-F equilibrium distances of $(YbF_8)^{6-}$ in its ground state

$4f^{14}\,{}^1A_{1g}$ and $(YbF_8)^{5-}$ in its ground state $4f^{13}\,{}^2F_{7/2}\,1\Gamma_{7u}$. The minimum of $E^{\rm diab}_{A_0D_0}$ is at d_L=2.200 Å, d_R=2.329 Å,

Each pair of DA and AD diabatic surfaces is degenerate along the diagonal d_L=d_R. In consequence, finding the minimum of any of them along the diagonal gives the diabatic activated complex of the $D + A \rightarrow A + D$ thermal electron transfer reaction, which is the lowest energy pass from one surface to the other. In the diabatic activated complex, where the electron is equally distributed between D and A, the left and right metal-ligand distances are equal $d_{L,ac} = d_{R,ac} = d_{ac}$. Although diabatic and adiabatic activated complexes should have significant energy differences, they are expected to have similar locations. In the example, the diabatic is at the Yb-F distance $d_{ac} = 2.261$ Å.

A good estimate of d_{ac} can be obtained from the harmonic approximation of the diabatic surfaces as the ω^2-weighted average of donor and acceptor equilibrium distances: $d_{ac} = (\omega_D^2 d_{eD} + \omega_A^2 d_{eA})/(\omega_D^2 + \omega_A^2)$, ω_D and ω_A being the vibrational frequencies of D and A respectively. If equal values for the vibrational frequencies of donor and acceptor are assumed d_{ac} is the simple average, $d_{ac} = (d_{eD} + d_{eA})/2$. Since the acceptor has a higher force constant than the donor, d_{ac} should be close to the average of d_{eD} and d_{eA} but closer to d_{eA} than to d_{eD}. In the ground state IVCT surfaces of the example, the ab initio values $\omega_D = 417$ cm^{-1} and $\omega_A = 493$ cm^{-1} give $d_{ac} = 2.253$ Å for the different-frequencies harmonic approximation and $d_{ac} = 2.265$ Å for the equal-frequencies harmonic approximation.

As represented in Fig. 7.4, the ground state reaction coordinate Q_{et}, i.e. the reaction path of minimum energy from one minimum to the other, can be approximated with the straight lines that connect the activated complex (d_{ac}, d_{ac}) with the two minima (d_{e,D_0}, d_{e,A_0}) and (d_{e,A_0}, d_{e,D_0}) in the (d_L, d_R) plane. Similar reaction coordinates can be defined for any pair of IVCT states. Obeying to the donor having a larger equilibrium metal-ligand distance than the acceptor, the DA configuration of the pair has the lowest diabatic energy in the left side of the activation complex $(d_L > d_R)$ and the AD pair in the right side $(d_L < d_R)$. The reaction coordinate contains the most relevant information of the diabatic energy surfaces and, accordingly, instead of cumbersome surfaces, it is convenient to make simplified plots of the energies of the DA and AD configurations of a D/A pair along the reaction coordinate: $E^{\rm diab}_{D_iA_j}(Q_{et})$ and $E^{\rm diab}_{A_jD_i}(Q_{et})$, which are called configuration coordinate energy diagrams. Hence the importance of a detailed knowledge of Q_{et}.

A detailed definition of the coordinated nuclear displacements involved in Q_{et} can be made in the Yb^{2+}/Yb^{3+} IVCT example of CaF_2:Yb^{2+},Yb^{3+} without loss of generality:

$$Q_{et} = \frac{1}{\sqrt{1+m^2}}\,(Q_R + m Q_L)\ . \tag{7.14}$$

In (7.14), m is the slope of Q_{et} in the (d_L, d_R) plane:

$$d_L - d_{ac} = m(d_R - d_{ac}) \begin{cases} m = \frac{d_{e,D}-d_{ac}}{d_{e,A}-d_{ac}} : d_L \geq d_R \\ m = \frac{d_{e,A}-d_{ac}}{d_{e,D}-d_{ac}} : d_L \leq d_R \end{cases}, \tag{7.15}$$

and Q_L and Q_R are the normal breathing modes of the left and right YbF_8 moieties with respect to their structures *in the activated complex*:

$$\begin{aligned} Q_L &= \frac{1}{\sqrt{8}} \left(\delta_{F_{L1}} + \delta_{F_{L2}} + \cdots + \delta_{F_{L8}} \right) , \\ Q_R &= \frac{1}{\sqrt{8}} \left(\delta_{F_{R1}} + \delta_{F_{R2}} + \cdots + \delta_{F_{R8}} \right) , \end{aligned} \tag{7.16}$$

where and $\delta_{F_{Lk}}$ and $\delta_{F_{Rk}}$ stand for ligand displacements in the left and right YbF_8 moieties with respect to the positions they occupy in the activated complex (cf. bottom panel in Fig. 7.4). Q_L and Q_R can be further simplified having in mind that the ligand breathings imply $\delta_{F_{L1}} = \delta_{F_{L2}} = \cdots = d_L - d_{ac}$ and $\delta_{F_{R1}} = \delta_{F_{R2}} = \cdots = d_R - d_{ac}$, in which case

$$Q_L = \sqrt{8}(d_L - d_{ac}), \quad Q_R = \sqrt{8}(d_R - d_{ac}), \tag{7.17}$$

and

$$\begin{aligned} Q_{et} &= \sqrt{\frac{8}{1+m^2}} \left[(d_R - d_{ac}) + m(d_L - d_{ac}) \right] \\ &= \sqrt{8(1+m^2)}(d_R - d_{ac}) \\ &= \sqrt{8(1+m^2)}(d_L - d_{ac})/m . \end{aligned} \tag{7.18}$$

The ground state diabatic electron transfer activation energy is $E^{\ddagger,\text{diab}}_{D_0A_0} = E^{\text{diab}}_{D_0A_0}(d_{ac}, d_{ac}) - E_{e,D_0A_0}$. It is independent of the distance between donor and acceptor and it is an upper bound of the adiabatic activation energy, which is d_{DA}-dependent through the values of the adiabatic couplings between $\Phi^{\text{diab}}_{D_0A_0}$ and $\Phi^{\text{diab}}_{A_0D_0}$ group wave functions.

An IVCT configuration coordinate energy diagram is the representaion of the energies of all DA and AD states along the IVCT reaction coordinate. See, for instance, Fig. 2 in [27] for a series of Ln^{2+}/Ln^{3+} multiconfigurational ab initio IVCT configuration coordinate energy diagrams in CaF_2:Ln^{2+}, Ln^{3+} doubly-doped phosphor materials.

MMCT

The MMCT diabatic potential energy surfaces loose the symmetry of the IVCT ones and they do not group in pairs. As it was mentioned in (7.13), the D^+A^- surfaces have their energy shifted above the DA ones due to the non-equality of the donor's ionization potential and the acceptor's electron affinity and to the different ionic charges interacting in D-A and D^+-A^-.

Regarding the positions of the minima, they are an extension of the IVCT case: $E^{\text{diab}}_{D_k^+A_l^-}$ has its minimum when d_D and d_A equal the metal-ligand equilibrium distances of the ionized donor in its k state and of the electron attached acceptor in its l state, respectively, i.e. at d_D=d_{e,D_k^+}, d_A=d_{e,A_l^-}. In the example of Yb-to-Ca MMCT in CaF_2:Yb^{2+}, the minimum of $E^{\text{diab}}_{D_0^+A_0^-}$ is at d_D=2.200 Å, d_A=2.459 Å, which are the respective Yb-F equilibrium distance of $(YbF_8)^{5-}$ in its ground state $4f^{13}\,{}^2F_{7/2}$ $1\Gamma_{7u}$ and Ca-F equilibrium distance of $(CaF_8)^{7-}$ in its ground state $3de_g^1$ $1\Gamma_{8g}$. The corresponding minimum of $E^{\text{diab}}_{D_0A_0}$ is at d_D=2.329 Å, d_A=2.309 Å, which are the respective Yb-F equilibrium distance of $(YbF_8)^{6-}$ in its ground state $4f^{14}\,{}^1A_{1g}$ and Ca-F equilibrium distance of $(CaF_8)^{6-}$ in its ground state $3p^64s^0\,{}^1A_{1g}$.

The definition of the MMCT electron transfer reaction coordinate Q_{et} in, e.g. a process $D_0A_0 \rightarrow ac \rightarrow D_k^+A_l^-$, starts with the identification of the diabatic activated complex ac, i.e. the lowest energy pass between the D_0A_0 and the $D_k^+A_l^-$ surfaces, or in other words the minimum of the intersection between $E^{\text{diab}}_{D_0A_0}(d_D, d_A)$ and $E^{\text{diab}}_{D_k^+A_l^-}(d_D, d_A)$ in the (d_D, d_A) plane. This is the point $(d_{ac,D}, d_{ac,A})$ (black square box in Fig. 7.5).

Besides, it is useful to define the vibrational breathing modes of the donor and acceptor moieties YbF_8 and CaF_8 in terms of the changes of the Yb-F and Ca-F distances with respect to the initial equilibrium structure, which is the D_0A_0 equilibrium structure in this case (the equilibrium structures of $(YbF_8)^{6-}$ and $(CaF_8)^{6-}$ in their respective ground states). These are $\delta_{F_1}, \ldots, \delta_{F_8}$ and $\delta_{F_9}, \ldots, \delta_{F_{16}}$ respectively, where the subindexes 1–8 identify ligands of Yb and the subindexes 9–16 identify ligands of Ca:

$$Q_D = \frac{1}{\sqrt{8}}\left(\delta_{F_1} + \delta_{F_2} + \cdots + \delta_{F_8}\right) = \sqrt{8}(d_D - d_{e,D_0}),$$
$$Q_A = \frac{1}{\sqrt{8}}\left(\delta_{F_9} + \delta_{F_{10}} + \ldots + \delta_{F_{16}}\right) = \sqrt{8}(d_A - d_{e,A_0}). \tag{7.19}$$

According to this choice, the values of the vibrational modes at the D_0A_0 equilibrium structure are null: $Q_{e,D_0} = 0$, $Q_{e,A_0} = 0$. At the activated complex of the $D_0A_0 \rightarrow ac \rightarrow D_k^+A_l^-$ thermal MMCT process they take the values $Q_{ac,D} = \sqrt{8}(d_{ac,D} - d_{e,D_0})$, $Q_{ac,A} = \sqrt{8}(d_{ac,A} - d_{e,A_0})$.

The electron transfer reaction coordinate is defined as

$$Q_{et} = \frac{1}{\sqrt{1+m^2}}\left[(Q_A - Q_{e,A_0}) + m(Q_D - Q_{e,D_0})\right], \tag{7.20}$$

if it is chosen to be referred to the D_0A_0 minimum. The reference values Q_{e,D_0} and Q_{e,A_0} are made explicit, even if they are zero with the choice (7.19). This way, the expression is still valid if a different reference is adopted. In (7.20), m is the slope of the straight line connecting either the D_0A_0 minimum and the activated complex ac or the activated complex ac and the $D_k^+A_l^-$ minimum (cf. Fig. 7.5):

$$\begin{aligned} m &= (Q_{ac,D} - Q_{e,D_0})/(Q_{ac,A} - Q_{e,A_0}) \quad \text{for the } D_0A_0 \rightarrow ac \text{ part,} \\ m &= (Q_{e,D_k^+} - Q_{ac,D})/(Q_{e,A_l^-} - Q_{ac,A}) \quad \text{for the } ac \rightarrow D_k^+A_l^- \text{ part.} \end{aligned} \tag{7.21}$$

If donor and acceptor have the same ligand coordination, (7.21) becomes

$$\begin{aligned} m &= (d_{ac,D} - d_{e,D_0})/(d_{ac,A} - d_{e,A_0}) \quad \text{for the } D_0A_0 \rightarrow ac \text{ part,} \\ m &= (d_{e,D_k^+} - d_{ac,D})/(d_{e,A_l^-} - d_{ac,A}) \quad \text{for the } ac \rightarrow D_k^+A_l^- \text{ part.} \end{aligned} \tag{7.22}$$

In parametric form, the MMCT reaction coordinate reads:

$$\begin{cases} Q_D - Q_{e,D_0} = \dfrac{m}{\sqrt{1+m^2}} Q_{et} \,, \\ Q_A - Q_{e,A_0} = \dfrac{1}{\sqrt{1+m^2}} Q_{et} \,. \end{cases} \tag{7.23}$$

This form is more convenient to evaluate MMCT trajectory, i.e. the points (Q_D, Q_A) followed by the MMCT process.

Having seen the MMCT diabatic potential energy surfaces and the MMCT diabatic reaction coordinate, a MMCT configuration coordinate energy diagram is the representaion of the energies of all DA states along the $D_0A_0 \rightarrow ac$ part and all D^+A^- states along the $ac \rightarrow D_k^+A_l^-$ part. Figure 7.5 shows, for the Yb-to-Ca MMCT in CaF_2:Yb^{2+}, the D_0A_0 and $D_0^+A_1^-$ potential energy surfaces, the $D_0A_0 \rightarrow ac \rightarrow D_0^+A_1^-$ thermal MMCT process, including its activated complex ac, and a partial configuration coordinate energy diagram where only those two states are represented. It must be remarked that the electron transfer reaction coordinate is specific for a pair of DA and D^+A^- states. This implies that the energies of all other states in a particular configuration coordinate diagram may be slightly larger than their respective values along their specific reaction coordinates. In other words, configuration coordinate diagrams have to be read with care: they provide qualitative information, but the potential energy surfaces themselves have to be used if more quantitative data are required.

A series of Eu^{2+}-to-Ln^{3+} multiconfigurational ab initio MMCT configuration coordinate energy diagrams in CaF_2:Eu^{2+}, Ln^{3+} doubly-doped phosphor materials shown in Fig. 1 of [28] may serve as a good illustration of the diversity of MMCT diagrams.

7.5 Empirical IVCT and MMCT Diagrams

In addition to their ab initio calculations, it is also possible to make diabatic IVCT and MMCT configuration coordinate energy diagrams using energetic and structural data of the individual active centers that participate in these electron transfer processes [29]. This is the issue discussed in this section. The empirical, quantitative IVCT and MMCT diagrams obtained using measured or estimated data are, then, an alternative to schematic configuration coordinate energy level diagrams traditionally used to address the participation of MMCT and IVCT states in energy transfer, non-radiative decay, and radiative processes [19, 30]

Configuration Coordinate

Firstly, it is important to quantitatively define the configuration coordinate Q_{cc} of a single-ion active center, which is the abscissa of its individual configuration coordinate energy diagram and will be later used to define the abscissa of IVCT and MMCT diagrams of metal ion pairs. The configuration coordinate energy diagram of a single-ion center is a very useful simplified representation of the energy variation of its electronic levels with the nuclear displacements in which only one vibrational coordinate is used, Q_{cc}, in order to give a gross description of relative positions and crossings of the electronic levels. The use of additional vibrational coordinates is required only to provide finer details like Jahn-Teller distortions and couplings.

Having in mind that, in the absence of Jahn-Teller or vibronic couplings between degenerate states, the excited state potential energy surfaces have null horizontal offsets with respect to the ground state's along non-totally symmetric normal vibrational coordinates, and that the contributions to electronic energy differences due to different force constants or curvatures are much smaller than those due to vibrational equilibrium offsets, the excess energy of an excited state i with respect to the ground state 0 can be approximately written as a sum of quadratic contributions of totally symmetric vibrational modes

$$E_i - E_0 = E_{i,e} - E_{0,e} + \frac{1}{2}\sum_{\mu}^{N_{TS}} k_{\mu i}(Q_\mu - Q_{\mu i,e})^2 - \frac{1}{2}\sum_{\mu}^{N_{TS}} k_{\mu 0}Q_\mu^2 + \ldots . \quad (7.24)$$

Here, if a unitary transformation (rotation) is made of the N_{TS} Q_μ into a new set such that one of the transformed coordinates, Q_{cc}, connects directly the minima of the two states, then, only this coordinate has a first-order contribution to the energy difference (the contribution of all other coordinates is a second-order one, via differences in force constants),

$$E_i - E_0 = E_{i,e} - E_{0,e} + \frac{1}{2}k_{cc,i}(Q_{cc} - Q_{cc\,i,e})^2 - \frac{1}{2}k_{cc,0}Q_{cc}^2 + \cdots . \quad (7.25)$$

Q_{cc} is, then, *the configuration coordinate* of the single ion active center, because it is the only one with horizontal offset and the only one that contributes to the leading term of the electronic energy differences. In a highly symmetric center with n-fold coordination ML_n the breathing mode is the only totally-symmetric mode and, hence, the configuration coordinate,

$$Q_{cc} = \frac{1}{\sqrt{n}}\left(\delta_{L_1} + \delta_{L_2} + \cdots + \delta_{L_n}\right) , \quad (7.26)$$

where δ_{L_i} stands for the displacement of ligand L_i from its equilibrium position along the breathing.

IVCT

The simplest IVCT diabatic configuration coordinate diagram of a D/A mixed-valence pair is

$$E_{D_iA_j}(Q_{et}) = E_{D_iA_j,e} + \frac{1}{2}k(Q_{et} - Q_{et,D_0A_0})^2 ,$$
$$E_{A_jD_i}(Q_{et}) = E_{D_iA_j,e} + \frac{1}{2}k(Q_{et} - Q_{et,A_0D_0})^2 . \tag{7.27}$$

In (7.27), Q_{et} is the electron transfer reaction coordinate or IVCT configuration coordinate; the vertical offsets $E_{D_iA_j,e}$ are the sum of the minimum-to-minimum excitation energies from the ground state to the i excited state of D and from the ground state to the j excited state of A: $E_{D_iA_j,e} = (E_{D_i,e} - E_{D_0,e}) + (E_{A_j,e} - E_{A_0,e})$; the horizontal offsets are $-Q_{et,D_0A_0} = Q_{et,A_0D_0} = (Q_{D_0,e} - Q_{A_0,e})/\sqrt{2}$, and $-Q_{et,D_0A_0} = Q_{et,A_0D_0} = \sqrt{n/2}(d_{e,D_0} - d_{e,A_0})$ for breathing vibrational modes (7.26); and k is an effective force constant (the breathing mode force constant if (7.26) holds) common for donor and acceptor, $k = 0.029660\,\mu\,\bar{\nu}^2$, when k is in cm^{-1}/Å^2, the effective vibrational mass μ is given in amu and the effective vibrational frequency in cm^{-1}.

The values of the configuration coordinates of the left and right moieties of the D/A pair that correspond to value of Q_{et} are $Q_L = Q_{ac} - Q_{et}/\sqrt{2}$, $Q_R = Q_{ac} + Q_{et}/\sqrt{2}$, and their metal-ligand distances for breathing vibrational modes are $d_L = d_{ac} - Q_{et}/\sqrt{2n}$, $d_R = d_{ac} + Q_{et}/\sqrt{2n}$.

IVCT Configuration Coordinate Diagram of Yb^{2+}/Yb^{3+} Pairs in YAG

Figure 7.6 shows an empirical IVCT configuration coordinate diagram of the $4f^N$ levels of an Yb^{2+}/Yb^{3+} mixed-valence pair in YAG resulting from the use of the above model and the following empirical data, as discussed in [29]: coordination number $n = 8$; experimental $4f \rightarrow 4f$ excitation energies of YAG:Yb^{3+}, the acceptor [31] (the donor Yb^{2+} only has the $4f^{14}$ ground state); $d_{e,D_0} - d_{e,A_0} = 0.14$ Å, which is 90% of the difference between the ionic radii of Yb^{2+} and Yb^{3+} in coordination 8 (1.14 Å and 0.985 Å respectively [32]); and a force constant $k = 50434$ cm^{-1} A^{-2}, which is 90% of the one corresponding to μ=m(O)=15.999 amu and $\bar{\nu} = 326$ cm^{-1} resulting from an analysis of resonant vibronic effects in YAG:Yb^{3+} [33].

The seven levels of the Yb_L^{2+}-Yb_R^{3+} and Yb_L^{3+}-Yb_R^{2+} configurations of the Yb^{2+}/Yb^{3+} pair are shown in red and blue respectively, grouped in lower sets of four levels corresponding to $Yb_R^{3+}(^2F_{7/2})$ and $Yb_L^{3+}(^2F_{7/2})$ and upper sets of three levels about 10000 cm^{-1} above corresponding to $Yb_R^{3+}(^2F_{5/2})$ and $Yb_L^{3+}(^2F_{5/2})$. Full lines are used for configuration coordinate values around the minima and dashes lines for values corresponding to stressed structures. For instance, Yb_L^{2+}-Yb_R^{3+} is in equilibrium at Q_{et}=–0.28 Å with a long Yb_L-O distance and a short Yb_R-O distance; on the contrary, an Yb_L^{3+}-Yb_R^{2+} state is under strong stress in such structure, at around 7500 cm^{-1} higher energy, which corresponds to the Yb^{2+}-to-Yb^{3+} IVCT absorption peak. This

Fig. 7.6 Empirical IVCT configuration coordiante diagram of a Yb^{2+}/Yb^{3+} mixed-valence pair in YAG

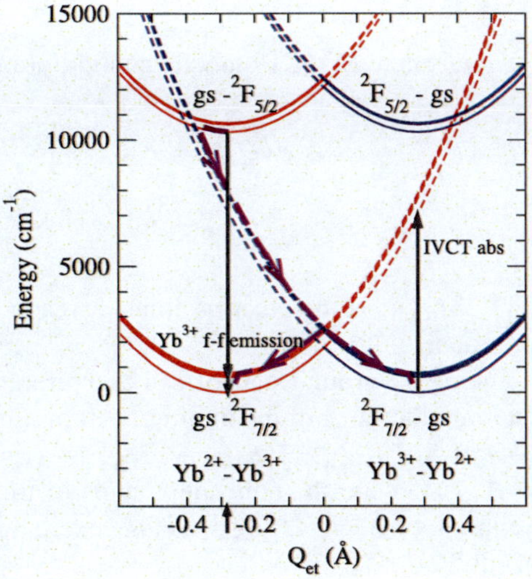

energy will be released after an eventual photoexcitation, in a non-radiative relaxation towards a short Yb_L-O distance and a long Yb_R-O distance, in which Q_{et} is increased up to Q_{et}=+0.28 Å. This is why such energy is also called reorganization energy.

The diagram shows that thermal IVCT can produce non-radiative decays via crossings between the excited Yb_L^{2+}-$Yb_R^{3+}(^2F_{5/2})$ and the ground $Yb_L^{3+}(^2F_{7/2})$-Yb_R^{2+} manifolds (and their symmetrical counterparts), where it is possible to transfer an electron between an Yb_L^{2+} and an excited Yb_R^{3+} at a cost of only 200 cm^{-1}, leaving the newly formed Yb_L^{3+} and Yb_R^{2+} in their respective ground states (violet arrows in the Figure). This result indicates that non-radiative $Yb^{2+}(4f_{7/2}) \rightarrow Yb^{3+}(4f_{5/2})$ electron transfer in Yb^{2+}/Yb^{3+} pairs is a possible mechanism responsible for Yb^{3+} emission quenching, meaning that Yb^{2+}/Yb^{3+} pairs can be the active quenchers in the observed concentration quenching of the Yb^{3+} emission [34]. Also, they can explain the temperature quenching of the Yb^{3+} emission excited with Ce^{3+} $4f \rightarrow 5d$ absorption in Ce,Yb-codoped YAG if Yb^{2+} is present or is formed in the excitation process [30], as it will be discussed below.

The qualitative value of diagrams like this one is to be remarked, but care must be taken not to make quantitative over-interpretations because simplifications of the model and uncertainties of the empirical data limit their quantitative value. E.g. taking 95% or 85% of the ionic radii difference instead of 90% changes the discussed 200 cm^{-1} energy barrier to 75 cm^{-1} and 400 cm^{-1} respectively, and adiabatic couplings would lower these values in all cases, besides other considerations. What seems to be solid in any case is that a reasonable configuration coordinate diagram indicates a potential crossing between an emitting level Yb_L^{2+}-$Yb_R^{3+}(^2F_{5/2})$ and a structurally stressed IVCT ground state level $Yb_L^{3+}(^2F_{7/2})$-Yb_R^{2+}, and an associated Yb^{3+} emission quenching.

MMCT

The equivalent MMCT diabatic configuration coordinate diagram of a D/A pair is

$$E_{D_iA_j}(Q_{et}) = E_{D_iA_j,e} + \frac{1}{2}k_{et}(Q_{et} - Q_{et,D_0A_0})^2$$
$$E_{D_k^+A_l^-}(Q_{et}) = E_{D_k^+A_l^-,e} + \frac{1}{2}k_{et}(Q_{et} - Q_{et,D_0^+A_0^-})^2 , \qquad (7.28)$$

In (7.28), Q_{et} is the electron transfer reaction coordinate or MMCT configuration coordinate.

Taking the ground state of DA as a reference, the vertical offsets of all its excited states are the sums of the minimum-to-minimum excitation energies $0 \rightarrow i$ of D and $0 \rightarrow j$ of A, $E_{D_iA_j,e} = (E_{D_i,e} - E_{D_0,e}) + (E_{A_j,e} - E_{A_0,e})$. The vertical offsets of the D^+A^- states are the equivalent sums of the $0 \rightarrow k$ excitation of D^+ and the $0 \rightarrow l$ excitation of A^- in D^+A^-, plus the excitation energy between the ground states of DA and D^+A^-,

$$E_{D_k^+A_l^-,e} = (E_{D_k^+,e} - E_{D_0^+,e}) + (E_{A_l^-,e} - E_{A_0^+,e}) + (E_{D_0^+A_0^-,e} - E_{D_0A_0,e}) . \quad (7.29)$$

The latter is the addition of (1) the adiabatic ionization potential of D in the host, IP_D (energy difference between the ground states of D^+ and D at their relaxed structures), (2) the negative adiabatic electron affinity of A in the host, $-EA_A$ (energy difference between the ground states of A^- and A at their relaxed structures), and (3) the interaction energy change due to the creation of a hole in D and an electron in A; for long D-A separations, this can be approximated by $(q_A - q_D - 1)/d_{DA}$, d_{DA} being the distance between donor and acceptor, and q_D and q_A their respective defect charges:

$$E_{D_0^+A_0^-,e} - E_{D_0A_0,e} = IP_D - EA_A + (q_A - q_D - 1)e^2/d_{DA} . \qquad (7.30)$$

The horizontal offsets are $Q_{et,D_0A_0} = 0$ and $Q_{et,D_0^+A_0^-} = [(Q_{D_0,e} - Q_{D_0^+,e})^2 + (Q_{A_0^-,e} - Q_{A_0,e})^2]^{1/2}$. And k_{et} is an effective curvature along the electron transfer reaction coordinate, which can be written in terms of the donor and acceptor force constants as $k_{et} = (m^2k_D + k_A)/(1 + m^2)$ with $m = (Q_{D_0^+,e} - Q_{D_0,e})/(Q_{A_0^-,e} - Q_{A_0,e})$.

If Q_D and Q_A are the breathing modes of the donor and acceptor, with n_D and n_A equal ligands respectively (7.26), then the horizontal offsets are $Q_{et,D_0A_0} = 0$ and $Q_{et,D_0^+A_0^-} = (n_D(d_{e,D_0} - d_{e,D_0^+})^2 + n_A(d_{e,A_0^-} - d_{e,A_0})^2)^{1/2}$. And the curvature along the electron transfer reaction coordinate is $k_{et} = (m^2k_D + k_A)/(1 + m^2)$, with donor and acceptor force constants $k_D = 0.029660\, \mu_D\, \bar{\nu}_D^2$ and $k_A = 0.029660\, \mu_A\, \bar{\nu}_A^2$, in $cm^{-1}/Å^2$ when the donor and acceptor effective vibrational masses μ_D and μ_A are given in amu and their effective vibrational frequencies ν_D and ν_A are given in cm^{-1}, and $m = \sqrt{n_D}(d_{e,D_0^+} - d_{e,D_0})/\sqrt{n_A}(d_{e,A_0^-} - d_{e,A_0})$.

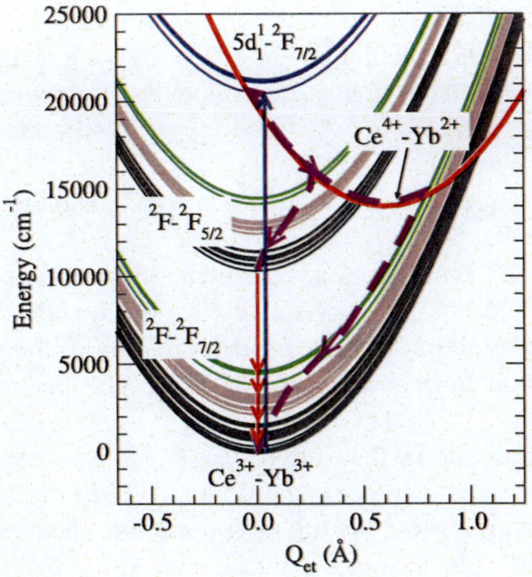

Fig. 7.7 Empirical MMCT configuration coordiante diagram of a Ce^{3+}/Yb^{3+} pair in YAG. The lowest $4f \rightarrow 5d$ excitation of Ce^{3+} is shown in a vertical blue arrow. Dashed violet lines indicate two decay mechanisms mediated by a non-radiative electron transfer to a Ce^{4+}-Yb^{2+} MMCT state (red curve): a non-radiative electron back-transfer to an excited $Yb^{3+}(^2F_{5/2})$, which is followed by Yb^{3+} $4f \rightarrow 4f$ emission (red arrows), and another non-radiative electron back-transfer to the ground states of Ce^{3+} and Yb^{3+} that quench all emisisons

The values of the configuration coordinates of donor and acceptor that correspond to value of Q_{et} are $Q_D = Q_{D_0,e} + mQ_{et}/\sqrt{1+m^2}$, $Q_A = Q_{A_0,e} + Q_{et}/\sqrt{1+m^2}$, and their metal-ligand distances for breathing vibrational modes are $d_D = d_{e,D_0} + mQ_{et}/\sqrt{n_D(1+m^2)}$, $d_A = d_{e,A_0} + Q_{et}/\sqrt{n_A(1+m^2)}$.

MMCT Configuration Coordinate Diagram of Ce^{3+}/Yb^{2+} Pairs in YAG

Figure 7.7 shows an empirical MMCT configuration coordinate diagram of a Ce^{3+}/Yb^{3+} pair in YAG, drawn along the ground state electron transfer reaction coordinate. It results from using the above model with the following parameters: A common coordination number 8 for donor and acceptor, $n_D = n_A = 8$. Experimental $4f \rightarrow 4f$ excitation energies of YAG:Ce^{3+} for the donor [35] and YAG:Yb^{3+} for the acceptor [31], for the $E_{D_iA_j,e}$ energy offsets $(E(Ce_i^{3+}\text{-}Yb_j^{3+})_e)$. In the lack of empirical data on $IP_D - EA_A$ and the minimum-to-minimum Ce^{3+}-to-Yb^{3+} charge transfer excitation energy, $E_{D_0^+A_0^-,e} - E_{D_0A_0,e}$ was used as an empirical parameter able to provide an energy diagram consistent with the experiments: $E_{D_0^+A_0^-,e} - E_{D_0A_0,e} = E_e(Ce^{4+} - Yb^{2+}) - E_e(Ce^{3+} - Yb^{3+}) = 14000$ cm^{-1} was used. The following horizontal offsets: $d_{e,D_0} - d_{e,D_0^+} = d_{e,Ce^{3+}-O} - d_{e,Ce^{4+}-O} = 0.16$ Å, $d_{e,A_0^-} - d_{e,A_0} = d_{e,Yb^{2+}-O} - d_{e,Yb^{3+}-O} = 0.14$ Å, in correspondence with the procedure used above for the IVCT diagram (for Ce^{3+} and Ce^{4+}, 90% of the difference between their ionic radii in coordination 8, 1.143 Å and 0.97 Å respectively [32], as it was done above for Yb^{2+} and Yb^{3+}). The same Ce-O distance offset was used in all $4f^1$ states of

Ce^{3+}; for the lowest $5d^1$ state, a -0.02 Å offset was used, slightly larger than the -0.014 Å found in ab initio calculations in YAG:Ce^{3+}, which has been considered to be underestimated [36]. And, finally, the following effective donor and acceptor force constants: $k_D = 20928$ cm^{-1} A^{-2} and $k_A = 50434$ cm^{-1} A^{-2}, which correspond to $\mu_D = \mu_A = m(\mathrm{O}) = 15.999$ amu, $\bar{\nu}_D = 210$ cm^{-1}, 5% larger than the 200 cm^{-1} vibrational sequence found in YAG:Ce^{3+} [37], and the Yb data used in the above IVCT case.

The empirical MMCT configuration coordinate diagram is basically consistent with the experiments and interpretations given in [30]. It shows the state manifold of the Ce^{3+}-Yb^{3+} configuration of the pair and, crossing it, the ground state of the Ce^{4+}-Yb^{2+} configuration, horizontally shifted 0.60 Å. The locations of the crossings are interpretative keys.

In effect, the diagram shows that, after a Ce^{3+} $4f\rightarrow 5d$ excitation, the lowest $Ce^{3+}(5d_1^1)$-$Yb^{3+}(^2F_{7/2})$ state can decay non-radiatively to the Ce^{4+}-Yb^{2+} MMCT state with a small energy barrier. At the crossing point between these two states, where the Ce^{3+}-to-Yb^{3+} electron transfer takes place, the Ce^{4+}-Yb^{2+} structure is very much stressed and it undergoes a strong structural relaxation with the release of a large reorganization energy (violet dashed arrows). In this process, a number of pairs branch at the crossing points between $Ce^{3+}(4f^1)$-$Yb^{3+}(^2F_{7/2})$ and Ce^{4+}-Yb^{2+}, and the Ce^{4+}-Yb^{2+} MMCT state decays non-radiatively directly to the $Ce^{3+}(4f^1)$-$Yb^{3+}(^2F_{5/2})$ manifold with a Yb^{2+}-to-Ce^{4+} electron back transfer; the excited Yb^{3+} finally yields $4f\rightarrow 4f$ emission (red arrows). This is consistent with the observations of Ce^{3+} to Yb^{3+} energy transfer in co-doped YAG:Ce,Yb, and its temperature dependence, and with the Yb^{3+} emission not showing rise time; it supports the interpretations that attributed it to a thermally activated decay through an intermediate Ce^{4+}-Yb^{2+} charge transfer state [30] and it gives a more detailed description of such an interpretation. It must be pointed out that the empirical parameters used here are not the only ones valid: E.g., a value of $E_e(Ce^{4+}$-$Yb^{2+})$ - $E_e(Ce^{3+}$-$Yb^{3+})$ around 2000 cm^{-1} higher is also consistent with the experiments. Also, it is to be noted that several $E_e(Ce^{4+}$-$Yb^{2+})$ - $E_e(Ce^{3+}$-$Yb^{3+})$ values associated with different Ce-Yb distances can coexist in the material (7.30).

It is also interesting to realize the interpretative opportunities offered by empirical quantitative diagrams as the ones discussed here. For instance, the intensity of the Yb^{3+} $4f\rightarrow 4f$ emission excited with Ce^{3+} $4f\rightarrow 5d$ absorption was found to strongly decrease above 110 K, after an initial increase with temperature due to the thermally activated crossing to the Ce^{4+}-Yb^{2+} MMCT state, and the crossing between the Ce^{4+}-Yb^{2+} MMCT state and the lowest $Ce^{3+}(4f^1)$-$Yb^{3+}(^2F_{7/2})$ manifold was suggested as a possible explanation. The present MMCT diagram does not rule out this explanation because such a crossing exists, although it is predicted to produce a relatively high energy barrier. But, in addition to this, the combination of the Ce^{3+}/Yb^{3+} MMCT and the Yb^{3+}/Yb^{2+} IVCT diagrams suggest an alternative explanation, because the intermediate MMCT state implies the formation of Yb^{2+} and the rising of the probability of formation of Yb^{2+}-Yb^{3+} pairs, which, as it was discussed in the analysis of the IVCT diagram, quench the Yb^{3+} $4f\rightarrow 4f$ emission via IVCT non-radiative decay.

References

1. Z. Barandiarán, M. Bettinelli, L. Seijo, J. Phys. Chem. Lett. 8, 3095 (2017)
2. B.O. Roos, R. Lindh, P.A. Malmqvist, V. Veryazov, P.O. Widmark, J. Phys. Chem. A 109, 6575 (2005)
3. B.O. Roos, V. Veryazov, P.O. Widmark, Theor. Chim. Acta 111, 345 (2003)
4. B.O. Roos, R. Lindh, P.A. Malmqvist, V. Veryazov, P.O. Widmark, J. Phys. Chem. A 108, 2851 (2004)
5. E. Pinel, P. Boutinaud, R. Mahiou, J. Alloys Compd. 380, 225 (2004)
6. M.L. Moreira, E.C. Paris, G.S. do Nascimento, V.M. Longo, J.R. Sambrano, V.R. Mastelaro, M.I.B. Bernardi, J. André, J.A. Varela, E. Longo, Acta Mater. **57**, 5174 (2009)
7. M. Kaneyoshi, J. Lumin. 121, 102 (2006)
8. I.L.V. Rosa, M.C. Oliveira, M. Assis, M. Ferrer, R.S. Andre, E. Longo, M.F.C. Gurgel, Ceram. Int. 41, 3069 (2015)
9. J.W. Verhoeven, Pure Appl. Chem. 68, 2223 (1996)
10. IUPAC, Compendium of Chemical Terminology, 2nd ed. (the "Gold Book"). Compiled by A.D. McNaught, A. Wilkinson. Blackwell Scientific Publications, Oxford (1997). Online version (2019-) created by S. J. Chalk. ISBN 0-9678550-9-8. https://doi.org/10.1351/goldbook.I03125. Accessed on 2021-06-08
11. R.A. Marcus, Annu. Rev. Phys. Chem. 15, 155 (1964)
12. G.C. Allen, N.S. Hush, Prog. Inorg. Chem. 8, 357 (2007)
13. M. Robin, P. Day, Adv. Inorg. Chem. Radiochem. 10, 247 (1968)
14. J.P. Launay, Chem. Soc. Rev. 30, 386 (2001)
15. A. Werner, Z. Anorg. Chem. 12, 46 (1896)
16. N.S. Hush, Prog. Inorg. Chem. 8, 391 (1967)
17. N.S. Hush, Coord. Chem. Rev. 64, 135 (1985)
18. S.B. Piepho, E.R. Krausz, P.N. Schatz, J. Am. Chem. Soc. **100**, 2996 (1978)
19. G. Blasse, Struct. Bond. 76, 153 (1991)
20. P. Boutinaud, R. Mahiou, E. Cavalli, M. Bettinelli, J. Lumin. 122–123, 430 (2007)
21. J.J. Joos, D.V. der Heggen, L.I.D.J. Martin, L. Amidani, P.F. Smet, Z. Barandiarán, L. Seijo, Nat. Commun. 11, 3647 (2020)
22. Z. Barandiarán, L. Seijo, J. Chem. Phys. 141, 234704 (2014)
23. L. Seijo, Z. Barandiarán, J. Chem. Phys. **141**, 214706/1 (2014)
24. Z. Barandiarán, L. Seijo, J. Chem. Phys. 143, 144702 (2015)
25. R. McWeeny, Proc. R. Soc. Lond. A 253, 242 (1959)
26. R. McWeeny, *Methods of Molecular Quantum Mechanics*, 2nd edn. (Academic, London, 1989). (Ch. 14)
27. I. Neefjes, J.J. Joos, Z. Barandiarán, L. Seijo, J. Phys. Chem. C 124, 2619 (2020)
28. J.J. Joos, I. Neefjes, L. Seijo, Z. Barandiarán, J. Chem. Phys. 154, 064704 (2021)
29. Z. Barandiarán, A. Meijerink, L. Seijo, Phys. Chem. Chem. Phys. 17, 19874 (2015)
30. D.C. Yu, F.T. Rabouw, W.Q. Boon, T. Kieboom, S. Ye, Q.Y. Zhang, A. Meijerink, Phys. Rev. B 90, 165126 (2014)
31. R.A. Buchanan, K.A. Wickersheim, J.J. Pearson, G.F. Herrmann, Phys. Rev. 159, 245 (1967)
32. R.D. Shannon, Acta Crystallogr. A 32, 751 (1976)
33. A. Lupei, V. Enaki, V. Lupei, C. Presura, A. Petraru, J. Alloys Compd. 275–277, 196 (1998)
34. P. Yang, P. Deng, Z. Yin, J. Lumin. 97, 51 (2002)
35. H. Przybylińska, C.G. Ma, M.G. Brik, A. Kamińska, P. Sybilski, A. Wittlin, M. Berkowski, Y. Zorenko, V. Gorbenko, H. Wrzesinski, A. Suchocki, Appl. Phys. Lett. 102, 241112 (2013)
36. J. Gracia, L. Seijo, Z. Barandiarán, D. Curulla, H. Niemansverdriet, W. van Gennip, J. Lumin. 128, 1248 (2008)
37. V. Bachmann, C. Ronda, A. Meijerink, Chem. Mater. 21, 2077 (2009)

Part IV
Fundamental Studies on Luminescence

Results of typical multiconfigurational *ab initio* studies on luminescence with a fundamental character are shown in this Part. The understanding of archetype solid-state lighting phosphors and the search for technologically important red-phosphors are tackled in a first chapter. A second chapter is dedicated to fundamental spectroscopic studies.

Chapter 8
Solid-State Lighting Phosphors

Abstract The invention of efficient blue light-emitting diodes (LED) based on Gallium Nitride (GaN) by Nakamura, Akasaki, Amano and their coworkers opened the door to a revolution in illumination. The promise of efficient white phosphor-converted LEDs (pc-LEDs) surpassing the efficiency and versatility of traditional incandescent and fluorescent lamp lighting technologies, sparked the interest for phosphors which can be excited by blue light. This made urgent the need of accurate structure-property relations that allow to control the phosphors optical and luminescent properties by engineering their chemical composition and structure. The potentialities and limitations of multiconfigurational ab initio calculations in this respect are illustrated in this Chapter with summaries of detailed studies on the seminal blue-excited yellow phosphor $Y_3Al_5O_{12}$:Ce^{3+}, which was and still is central in solid-state lighting, and on several alternatives in the search for new red phosphors, which, among other uses, can control the color rendering index of the lighting devices in order to improve the quality of the illumination.

8.1 Ce-Doped Yttrium Aluminum Garnet

The appearance of an efficient blue light-emitting diode (LED) based on Gallium Nitride (GaN) [1] and the resulting promise of efficient white phosphor-converted LEDs (pc-LEDs) surpassing the efficiency and versatility of traditional incandescent and fluorescent lamp technologies, sparked the interest for phosphors which can be excited by blue light. This requirement disqualified the established (fluorescent) lamp phosphors, typically based on trivalent lanthanide (Eu^{3+} or Tb^{3+}) ions, which have been selected for their efficient UVC excitability, but that are typically poorly excited by blue light.

The nature of pc-LED technology enables a far-reaching color engineering where the properties of the individual materials, as well as their mixtures, determine the output spectrum of the white LED, as well as its energetic and luminous efficiency [2].

Z. Barandiarán et al., *Luminescent Materials*, Springer Series in Materials Science 322,
https://doi.org/10.1007/978-3-030-94984-6_8

Color quality considerations translate into requirements for the luminescent materials. In view of lighting applications, a high color rendering requires the emission spectrum of the white LED to extend over a large part of the visible range. Therefore, one ends up in a trade-off between color quality and luminous efficiency [3]. Once a luminescent material and a blue pumping LED are selected, the correlated color temperature (CCT) of the white lamp can be engineered by tweaking the intensity ratios of the different emissions.

Next to the requirements for the blue excitability and those related to the color of the phosphors and the resulting LED lamp, additional requirements on the performance of the luminescent materials have to be fulfilled simultaneously [3]. This includes a high conversion (or quantum) efficiency and the absence of thermal quenching at the typical working temperatures that can reach up to 150 °C [4]. In other words, the quantum efficiency should be high and remain high at elevated temperatures. Often, high-power blue pumping LEDs are used to increase the luminous flux. In those cases, care should be taken to avoid a sub-linear power-dependence of the conversion efficiency due to the saturation of luminescent centers inside the phosphor.

As for LED lighting, other applications of phosphors impose their own requirements [5]. It is hence desirable to have access to accurate structure-property relations that allow to control the phosphors optical and luminescent properties by engineering their chemical composition and structure.

It is obvious that the control of excitations, non-radiative decays, and emissions of the phosphors must be rooted in a fundamental knowledge of the materials. And, while this knowledge is necessarily experimental, theoretical ab initio calculations should speed it up significantly. Furthermore, the calculations are also expected to shed light into corners that remain hidden to experiments or where reasonable hypotheses formulated at one point are very hard to check in the long term. For this, the theoretical methods should be able to provide, with sufficient accuracy and in a reasonable time, reliable descriptions of the bonding, the electronic structures, and the geometrical structures of the active defects in their ground and excited states, their absorption and emission energies and profiles, and their radiative and non-radiative decay mechanisms.

Ce^{3+}-doped YAG is not only a good, but a necessary entry point for an ab initio approach to enter the field of luminescent materials and show its potentialities and limitations. This is what has been done in [6] and is summarized in this chapter. Can the multiconfigurational ab initio calculations handle the $4f$–$5d$ absorption and emission transitions in this reference luminescent material? How accurately?

In YAG:Ce^{3+}, Ce substitutes for Y in an eight-fold oxygen-coordinated site with D_2 point symmetry (Fig. 8.1) [7]. SA-CASSCF/MS-CASPT2/RASSI-SO DKH or equivalent calculations[1] have been performed on $(CeO_8)^{13-}$ and $(CeO_8Al_2O_4)^{15-}$

[1] In earlier times, due to the different stages of the methodological developments that finally stabilized after software development and accumulation of experience, other methods of equivalent quality were used for some purposes. For the sake of simplicity, the details of these methods are omitted here. Nevertheless, cross-checking their results, for selected properties, with SA-CASSCF/MS-

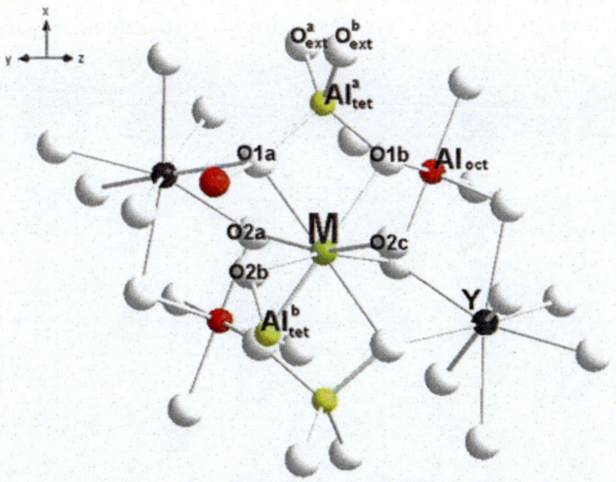

Fig. 8.1 Coordination of a Ce_Y substitutional defect in YAG:Ce^{3+}. D_2 Ce_Y site (M, green), with 3 C_2 perpendicular symmetry axes. First shell of 8 oxygens: 4 O1 in a quasi-planar axial twisted cross and 4 O2 in a quasi-planar equatorial twisted cross. Second shell of 2 axial tetrahedral Al^a_{tet} (yellow), plus 4 tetrahedral Al^b_{tet} (yellow), 4 octahedral Al_{oct} (red), and 4 dodecahedral Y (black). Oxygens are shared as follows: 2 O1 with axial Al^a_{tet}, 1 O1 and 1 O2 with Al_{oct}, 1 O1 and 1 O2 with Y, and 1 O2 with Al^b_{tet}. Each first shell O is linked to Ce, 1 Al_{tet}, 1 Al_{oct}, and 1 Y

embedded-clusters, with basically equivalent results and only small, irrelevant differences between them [6, 10].

Structural results are shown in Table 8.1 for the ground and excited states of the Ce^{3+} dopant. Local expansions are calculated in the ground state that are very similar to EXAFS[2] results [13] for axial oxygens and somehow smaller for the equatorial oxygens. This might be due to allowing relaxation of the first coordination shell only, although the agreement with DFT calculations in which all coordination shells were relaxed [14] is remarkable. The local expansion around Ce^{3+} is in qualitative agreement with the ionic radii mismatch between Ce^{3+} (1.143 Å) and Y^{3+} (1.019 Å) [15];

CASPT2/RASSI-SO DKH results, when these became available, was a standard procedure. Having this in mind, some of the results reported here might not exactly correspond to SA-CASSCF/MS-CASPT2/RASSI-SO DKH calculations, but to equivalent calculations whose details are reported for reproducibility in the original references. The differences with SA-CASSCF/MS-CASPT2/RASSI-SO DKH results are, in any case, not significant. The equivalent but slightly different methodologies refer to the consideration of relativistic effects and of the simultaneous treatment of spin-orbit coupling and correlation. In the first case, a Wood-Boring relativistic Hamiltonian with effective core potentials (core AIMPs) [8] was used to consider relativity instead of the all-electron DKH Hamiltonian. In the second case, *spin-free-state-shifted* spin-orbit configuration interaction (*sfss*-SOCI) [9] calculations were done instead of RASSI-SO calculations. A CI space defined as the CAS multireference plus all single-excitations from the orbitals that are dynamically correlated was the standard choice in *sfss*-SOCI calculations.

[2] Extended X-ray Absorption Fine Structure, an X-ray spectroscopic technique that allows to obtain element-specific local structural information such as geometries, coordination numbers and bond lengths [11, 12].

Table 8.1 Local structure (Ce-O distances only; see [6] for bond and torsion angles) and $4f$–$5d$ transitions of the Ce_Y subsitutional defect in YAG:Ce^{3+}

Local structure			YAG:Ce^{3+} absorptions and emissions		
	Distances (Å)				
	Ce-O_1	Ce-O_2	Experiment	Calculation	
	YAG				
Experiment [17]	2.303	2.432	$4f \rightarrow 5d$ absorptions		
	YAG:Ce^{3+}		21740^a, 22000^b	$1\Gamma_5 \rightarrow 8\Gamma_5(2^2A)$	24040
EXAFS experiment [13]	2.38(1)	2.52(2)	29400^{a,b}	$\rightarrow 9\Gamma_5(3^2B_3)$	31660
DFT calculation [14]	2.373	2.468	44000^b, 44800^a	$\rightarrow 10\Gamma_5(3^2A)$	48070
Multiconfigurational calculation [6]				$\rightarrow 11\Gamma_5(3^2B_2)$	52540
$4f^1 - 1^2B_2$	2.370	2.441		$\rightarrow 12\Gamma_5(3^2B_1)$	54190
$5d^1$ excited states					
2^2A	2.356	2.427	$5d \rightarrow 4f$ emissions		
3^2B_3	2.364	2.436	18700^c	$8\Gamma_5(2^2A) \rightarrow 1, 2, 3\Gamma_5$	23350
3^2A	2.383	2.454	17400^c	$\rightarrow 4, 5, 6\Gamma_5$	21130
3^2B_2	2.322	2.510	29200^d	$9\Gamma_5(3^2B_3) \rightarrow 1, 2, 3\Gamma_5$	<31600
3^2B_1	2.441	2.403	27800^d	$\rightarrow 4, 5, 6\Gamma_5$	<29380

[a]Absorption peak [18] [b]Excitation peak [7] [c]Local maximum of a double peak emission band [19] [d]Local maximum of a double peak emission band [7]

however, from the quantitative point of view, they are significantly smaller. This behavior is systematically found in substitutional defects and its quantitative effect (reduction factor from ionic radii mismatch to local distortion) is host dependent [16].

Absorptions and emissions, also summarized in Table 8.1, show a reasonable agreement with experimental data extracted from absorption, excitation, and emission spectra. The spin-orbit coupling assignments are simple here, because all many-electron states belong to the Γ_5 irreducible representation of the D_2' double group, but the analyses of the wave functions also allows to know, besides the main configurational character $4f^1$ or $5d^1$, which is not strictly pure here, the main spin-orbit coupling free character of the states, i.e. the irreducible representations of the point group D_2. These characters are almost pure, due to the small spin-orbit coupling between them, and they are shown in the table. The first three $4f \rightarrow 5d$ absorptions are identified as transitions from the ground state to Γ_5 states of almost pure character 2^2A, 3^2B_3 and 3^2A, respectively, and are overestimated 2300 and 3300 cm^{-1} with respect to experiment. The fourth and fifth absorptions are hidden in the experiments by the absorptions of the host crystal itself. They are found in the calculations right above 50000 cm^{-1} and are identified as Γ_5 states of dominant character 3^2B_2 and 3^2B_1. Together with the many-electron state energies, the calculation of oscillator strengths and fundamental vibrational frequencies allows to calculate absorption and emission band profiles, which also agree reasonably well with experiments [6].

The lowest $5d \rightarrow 4f$ emission takes place from the Γ_5 (2^2A) state and is made of two overlapping bands, each of them built on three electronic origins. The calculated emission energies overestimate the experiments more than the absorption energies, which gives smaller Stokes shifts than the experimental ones. Although a careful analysis of concentration and temperature effects resulted into a lower Stokes shift value (2400 cm^{-1}) than the simple energy difference between experimental absorption/excitation and emission bands, regardless of the doping concentration or temperature [20]. Still, the theoretical value is too small. This result is somehow surprising, because an equivalent calculation on the Stokes shift of the actinide isoelectronic ion Pa^{4+} in the structurally simpler Cs_2ZrCl_6 host [21], gave an excellent agreement with the experimental result obtained from high-resolution absorption and emission spectra [22, 23].

In addition to the visible emission in YAG:Ce^{3+}, a similar double emission band can be found in the near-UV upon 254 nm excitation [7]. This additional emission, just under 30000 cm^{-1} , was tentatively attributed to $5d \rightarrow 4f$ transitions from higher $5d^1$ levels [7]. The calculations support this assignment. In particular, this double peak band is interpreted as the emission from the second $5d^1$ level, $\Gamma_5(3^2B_3)$.

Here, it is particularly interesting to recall that the double peak in the $5d \rightarrow 4f$ emission was commonly attributed to the split of the seven $4f^1$ levels of Ce^{3+} into two groups of levels separated by about 2000–2500 cm^{-1} and associated with small crystal field splittings of three $^2F_{5/2}$ and four $^2F_{7/2}$ atomic levels [7, 24, 25]. The pioneering work of Herrmann et al. [26] pointed out otherwise: The splitting should follow a pattern of: three levels of predominantly $^2F_{5/2}$ character spanning ~400 cm^{-1}; 2000 cm^{-1} above, three additional levels of predominantly $^2F_{7/2}$ character spanning ~200 cm^{-1}; and 3500–4000 cm^{-1} above the first set, a single level of mixed character. This work was basically ignored, but the multiconfigurational ab initio calculations in YAG:Ce^{3+} confirmed its prediction. Additional systematic analyses of theoretical calculations of the same kind in the doped garnets $Lu_3Al_5O_{12}$:Ce^{3+}, $Gd_3Al_5O_{12}$:Ce^{3+}, $Y_3Ga_5O_{12}$:Ce^{3+}, $Lu_3Ga_5O_{12}$:Ce^{3+}, and $Gd_3Ga_5O_{12}$:Ce^{3+} [27], further confirmed the general case and explained the reason for the wrong assumption to have been working so well (Fig. 8.2): the intensity of the emission to the seventh $4f^1$ level is only one hundredth of the integrated intensity of the emissions to the other six levels grouped into two sets of three. The calculations showed that effective field splittings and spin-orbit coupling splittings are of the same order of magnitude in these materials. They confirmed the weak field point of view of the pattern [26], i.e. the first three levels result from the splitting of $^2F_{5/2}$, the second three levels from the splitting of $^2F_{7/2}$, and the seventh level from a strong, cubic field driven interaction between $^2F_{7/2}$ and $^2F_{5/2}$ components. They also explain the pattern from the strong field point of view: the first six levels result from the (D_2 field and spin-orbit coupling) interactions between $^2T_{2u}$ and $^2T_{1u}$ cubic levels; and the higher, isolated seventh level comes directly from the cubic $^2A_{2u}$.

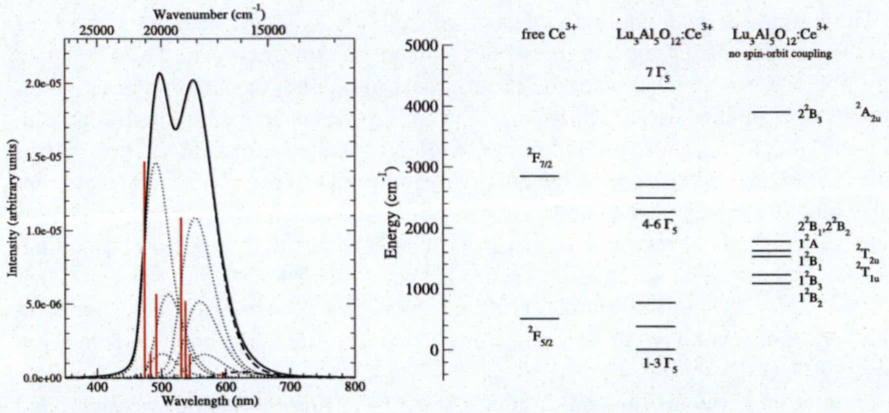

Fig. 8.2 Left box: Calculated $5d \rightarrow 4f$ double peak emission band of $Lu_3Al_5O_{12}$:Ce^{3+}. Full line: superposition of seven individual transitions $8\,\Gamma_5 \rightarrow 1\text{–}7\,\Gamma_5$. Dashed line: superposition of the first six individual transitions only $8\,\Gamma_5 \rightarrow 1\text{–}6\,\Gamma_5$. The individual contribution of $8\,\Gamma_5 \rightarrow 7\,\Gamma_5$ is shown in dashed line; those of the other six emissions are shown in dotted lines. Vertical bars with heights proportional to the oscillator strengths are shown in the zero-phonon positions of the individual $5d \rightarrow 4f$ transitions. Right box: $4f^1$ manifold of Ce^{3+} as a free-ion and in $Lu_3Al_5O_{12}$, with and without spin-orbit coupling included

8.2 Search for Red Phosphors

In case of a phosphor-converted white LED solely based on YAG:Ce^{3+}, the CCT has a lower limit in the range of 5000-6000 K. Further increasing the yellow:blue ratio will lead to an unacceptable deviation from the back-body locus (a high D_{uv} value) and ultimately a yellow LED. Furthermore, due to the limited red spectral contribution the resulting color rendering and the overall color quality are low. The use of this first generation of dichromatic LEDs is hence limited to lower-end applications where color quality is less important, such as bicycle headlights.

Trichromatic LEDs are targeted to fabricate a white pc-LED with improved color quality and a low CCT, respresenting warm-white light. These pc-LEDs feature a blue pumping LED with a yellow+red or green+red phosphor blend. Especially the former solution, where YAG:Ce^{3+} is used to render a broad green-yellow spectral band is a suitable approach to achieve LED lamps for lighting with a high color rendering. In any case, a red phosphor, excitable by the blue pumping LED, is required. During the quest for efficient red phosphors, various avenues have been explored where novel hosts were developed and unconventional optical dopants and oxidation states have been examined.

The numerous combinations of available hosts, dopants, syntheses and optimization strategies indicate that also in this case a computationally-aided design approach is highly desirable.

The promising results of multiconfigurational ab initio calculations on the blue-to-yellow converter phosphor YAG:Ce^{3+} (Sect. 8.1), pointed out the strong predictive and interpretative potential of these theoretical methods. The search for novel red phosphors is, then, a challenging area for them.

One of the avenues followed in this search is to use the same activator, Ce^{3+}, in hosts of the same family of YAG, i.e. garnets. This way, one should be able to take advantage of one important but not well understood fact, which is that the $5d \rightarrow 4f$ emission of Ce^{3+} is remarkably red-shifted in YAG in comparison to other types of hosts. Along this line, different garnets have been used to host Ce^{3+}, in the understanding that changes in the local D_2 field shift its emission. But the detailed structure-spectroscopy relations in these phosphors could not be well understood and a wise color control of the Ce^{3+} emission was elusive. E.g. to the question: why has the $5d \rightarrow 4f$ emission of YAG:Ce^{3+} so long wavelength?, one could add: why is the wavelength slightly shorter in $Lu_3Al_5O_{12}$:Ce^{3+}(LuAG:Ce^{3+}) than in YAG:Ce^{3+}? Why isn't it slightly longer? A related line of work is co-doping YAG:Ce^{3+} and other Ce^{3+}-doped garnets with optically inactive ions that shift the Ce^{3+} emission. Color control along this line was also elusive and more questions were added to the list of unanswered questions, because different co-dopants produced similar overall structural effects but opposite emission shifts.

The use of multiconfigurational ab initio calculations to tackle the interpretative and predictive challenges posed by Ce^{3+}-doped and co-doped garnets in the search for red phosphors is summarized in Sect. 8.2.1.

Another line of work in the search for new, efficient, tailored red phosphors, is to use appropriate emissions of other activators, like Eu^{2+}, Bi^{2+}, Pr^{3+}, and others. Further sections are dedicated to a better, more complete understanding of their red luminescence provided by the ab initio calculations.

8.2.1 *Ce^{3+}-Doped Garnets*

8.2.1.1 Red and Blue Shifts Induced by Lanthanum and Gallium Co-doping

From the very beginning of the discovery of YAG:Ce^{3+} and related phosphors, it was found that co-doping the material with optically inactive ions could shift the emission wavelength [7]. This observation opened a promising research avenue towards the control of the phosphor's color by means of co-doping [28, 29]. Color control was considered a key issue for the success of solid-state lighting technologies [30] and using co-dopants that didn't play a role as co-activators, but only as wavelength shifters, was not complicated. However, a detailed understanding of the reasons behind the wavelength shifts remained elusive for a long while, until systematic

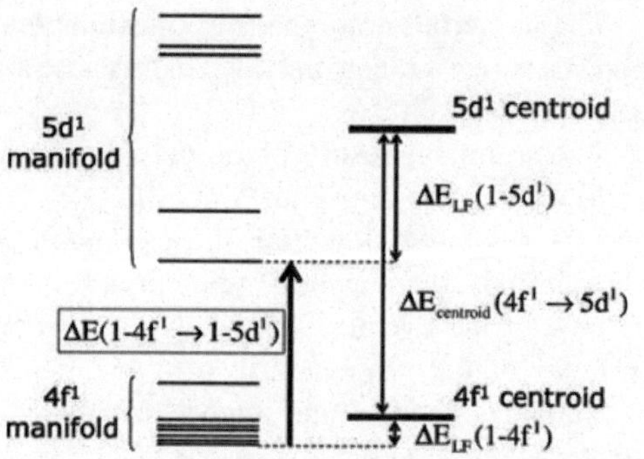

Fig. 8.3 Schematic representation of the $4f^1$ and $5d^1$ manifolds of the Ce_Y defect

multiconfigurational ab initio calculations could be performed and their analyses provided an answer [10, 14, 31].

In the mean time, it was known that Gd^{3+} and La^{3+} co-dopants red shifted the Ce^{3+} $5d \rightarrow 4f$ emisson of YAG:Ce^{3+} and LuAG:Ce^{3+} (to longer wavelengths and lower energies), whereas Ga^{3+} and In^{3+} co-dopants blue shifted it (to shorter wavelengths and higher energies). The corresponding, first $4f \rightarrow 5d$ absorption of Ce^{3+}, experienced the same red and blue shifts. Remarkably, the Ce^{3+} second $4f \rightarrow 5d$ absorption showed the opposite shifts, i.e. red shifts upon Ga^{3+} and In^{3+} co-doping and blue shifts upon Gd^{3+} and La^{3+} co-doping [7]. However, a sound knowledge about the structure-spectroscopy relationship governing these shifts was missing. It was limited to an empirical rule stating that, substitutions of dodecahedral Y^{3+} by larger ions produces red shift whereas substitutions of octahedral Al^{3+} by larger ions produces blue shift of the YAG:Ce^{3+} $5d \rightarrow 4f$ luminescence (and the first $4f \rightarrow 5d$ absorption) [7, 32–35]. But why is it so? The answer to this question was puzzling.

The fact that both types of substitutions by larger ions, of dodecahedral Y^{3+} and octahedral Al^{3+}, produce clear lattice expansions, from low concentrations to full substitution [33], creates the interpretative puzzle. In effect, the lattice expansion due to Ga^{3+} substituting for octahedral Al^{3+} should diminish the crystal field splitting of the Ce^{3+} $5d^1$ manifold; a similar effect on the $4f^1$ manifold should be expected, but of a much lower magnitude, due to the inner character of the $4f$ shell, which is very efficiently shielded from the external crystal field by the full $5s^2$ and $5p^6$ shells. The overall result should be a rising of the first state of the $5d^1$ configuration with respect to the first state of the $4f^1$ configuration, hence a blue shift of the emission and the first absorption (Fig. 8.3). But, if La^{3+} substituting for dodecahedral Y^{3+} also produces a lattice expansion [29, 33], the same kind of blue shift should be expected. This not being the case, other factors than an homogeneous lattice expansion lowering the crystal field splitting must be playing a key role on the red shift.

A hypothesis was formulated according to which the La^{3+} co-doping would create local compressions around Ce^{3+} in spite of them producing lattice expansions [29, 34]. The local compressions would enlarge the effective crystal field on the Ce^{3+} site and so lower the energy difference between the ground state and the first $5d^1$ state, hence producing an emission red shift. What would produce the local compressions was unexplained. And the effects of La^{3+} co-doping on the local structure around Ce^{3+} was unknown, due to the lack of experiments, like EXAFS, able to unveil the local structure around Ce^{3+} before and after La^{3+} co-doping. Nevertheless, the local compression was the only reasonable assumption for decades. In the lack of experimental confirmation or rejection of the hypothesis, first-principles DFT calculations were performed on the co-doped material YAG:Ce^{3+},La^{3+} [31] that, a priori, could either support or unsupport it. They concluded that the structural effects of La^{3+} co-doping YAG:Ce^{3+} predicted by the theoretical method used are: an overall lattice expansion, in agreement with the experimental observations, and a local expansion around Ce^{3+}. This finding cannot disproof the local compression hypothesis, but seriously questions whether it is a reasonable one.

Then, if the La^{3+} co-doping induced red shift is not due to a local compression around the Ce^{3+} active center, what is the reason for it? Why does La^{3+} co-doping produces a red shift of the Ce^{3+} $5d \rightarrow 4f$ emission in YAG:Ce^{3+} instead of a blue shift?

A series of systematic multiconfigurational ab initio calculations of the Ce^{3+} $4f^1$ and $5d^1$ manifolds in YAG:Ce^{3+}, YAG:Ce^{3+},Ga^{3+}, and YAG:Ce^{3+},La^{3+}, combined with structural first-principles DFT calculations on the ground states of the same materials, were designed to find an answer to these questions and, so, helping to establish a recipe for color control of YAG:Ce^{3+} and Ce^{3+}-doped garnets in general via co-doping [10, 14, 31]. They are summarized here.

Even though the effect of Ga^{3+} co-doping was basically known and no serious questions remained unanswered in this respect, ab initio calculations that could reproduce the known facts were considered necessary. More for endorsing the theoretical approach that would also be applied to the puzzling case of La^{3+} co-doping than for shedding light on the Ga^{3+} co-doping issue. The calculations [10] studied substitutional defects in which Ga^{3+} substitutes for Al^{3+} in YAG and in YAG:Ce^{3+}, in octahedral sites, Ga_{Al}^{oct}, and in tetrahedral sites, Ga_{Al}^{tet}. They found slightly more favorable formation energies for Ga_{Al}^{oct} than for Ga_{Al}^{tet} (4.1 kJ/mol more favorable with unrelaxed lattice and 4.8 kJ/mol with relaxed lattice), in an overall picture in which Ga_{Al} substitutions at octahedral sites are only slightly preferred over substitutions at tetrahedral sites under no relaxation constraints, although any hindering of relaxation strongly favors the formation of the octahedral substitutional defects. Since increasing defect concentration tends to hinder relaxation, Ga_{Al}^{oct} substitutions are expected to be dominant over Ga_{Al}^{tet} substitutions at increasing defect concentrations. This conclusion qualitatively agrees with the common assumption, based on the evolution of Ce^{3+} luminescence upon monotonous increasing of Ga_{Al} concentration, that, between 10 and 80 at.%, substitutions at octahedral sites are made before substitutions at tetrahedral sites take place [33, 34]. The multiconfigurational ab initio calculations performed at the DFT structures give blue shifts experienced by the first Ce^{3+} $4f \rightarrow 5d$ transition upon Ga substitutions for Al^{oct} (74 cm^{-1}) and for Al^{tet} (211 cm^{-1}) at Ga doping concentrations of 2.5 at.%. This result is in qualitative agreement with experiments [7, 32, 33, 35]. An extrapolation of the experimental data to the 2.5 at.% defect concentration used in the calculations, gives a 50 cm^{-1} blue shift, which should be compared with the 74 cm^{-1} blue shift induced by Ga_{Al}^{oct}

defects. The quantitative agreement is also remarkable. The analysis of the reasons behind the blue shift reveal that it is basically due to a reduction of the effective ligand field splitting of the $5d^1$ manifold. The effect of Ga^{3+} co-doping on the centroids of the Ce^{3+} $4f^1$ and $5d^1$ manifolds is negligible. Direct electronic effects of Ga^{3+} on the Ce^{3+} transitions are totally irrelevant. The only role of Ga^{3+} co-doping is to produce an anisotropic structural expansion of the first Oxygen coordination shell around Ce^{3+}.

An equivalent study on the effect of La^{3+} co-doping on the luminescence of YAG:Ce^{3+}, started with a structural analysis of local defects in the doubly-doped YAG:Ce^{3+},La^{3+} material by means of DFT periodic-boundary-conditions calculations [31]. Both Ce_Y and La_Y substitutional defects produced overall lattice expansions, with lattice constant increments never larger than +0.25%. They also produced local expansions around them, which in the case of Ce_Y confirmed the results of previous multiconfigurational ab initio and DFT calculations. The case of La_Y is not an exception and it reaffirms the observation that local distortions are qualitatively governed by ionic radii mismatches, although they are quantitatively much smaller [16]. In all cases, the local expansions are limited within the first and second coordination spheres and atoms beyond that limit do not experience significant displacements. An interesting conclusion is that La_Y defects induce local expansions around Ce_Y defects in line with the overall lattice expansion. This result is in opposition to the hypothesis that La^{3+} co-doping could perhaps induce local compressions around Ce_Y in spite of it producing overall lattice expansions, and rules it out as the explanation of the La^{3+} induced red shift of the Ce^{3+} emission. The expansions around Ce_Y induced by La_Y are anisotropic and, as expected, their magnitude decreases with the Ce-La distance.

But the most interesting conclusion of the structural study, because it will play a key role in the explanation of the La^{3+} induced red shift of the Ce^{3+} emission, is that La_Y defects tend to be as close as possible to the optically active Ce_Y centers (It corresponds to a substitution of any of the two Y atoms shown in black in Fig. 8.1 by a La atom.) The analyses of the atomistic distortions produced by the La^{3+} co-doping and the electronic structure of the doubly-doped material do not indicate the presence of important bonding disruptions or changes, so that electronic interactions cannot be responsible for the attraction between Ce_Y and La_Y defects. The Ce_Y-La_Y attraction is rather due to simple mechanical reasons, namely that the distortions are minimized by the clustering of the defects. It is the same attraction force that puts together two marbles on a drum. This issue is further discussed in Sect. 10.2.

The multiconfigurational ab initio calculations on the doubly doped material YAG:Ce^{3+},La^{3+} that followed the structural study, focused on the excited manifold of the Ce_Y active center and how it was impacted by the presence of close La_Y defects [14]. Here, the defect structures of the DFT study were used and CASSCF/CASPT2 calculations performed on the states of main $4f^1$ and $5d^1$ configurational characters of the embedded-cluster $(CeO_8Al_2O_4)^{15-}$. Since the effects of spin-orbit couplings on the $4f \rightarrow 5d$ transitions were found to be small and constant over the $5d^1$ manifold [6] they were not included to calculate the La^{3+}-induced shifts. The lowest Ce^{3+} $4f \rightarrow 5d$ absorption, which corresponds to the $5d \rightarrow 4f$ emission, was found to experience a red shift by the presence of the La_Y defect nearby. This defect, however, blue shifted the second Ce^{3+} $4f \rightarrow 5d$ absorption. Both results were in agreement with experimental observations of La^{3+} and Gd^{3+} induced spectral shifts [7, 29, 33].

This result means that the ab initio calculations correctly reproduce the opposite shifts of the Ce^{3+} first $4f \rightarrow 5d$ absorption and corresponding $5d \rightarrow 4f$ emission induced by Ga^{3+} (blue shift) and La^{3+} (red shift) co-doping, as well as the lattice expansions produced by both co-dopings, together with the local expansions that one would expect but have been elusive to experimental quantification. This is the most important observation at this point, because simple interpretative models have failed to put all these points together.

So, ab initio calculations can be done and they reproduce the experimental results, no matter how disconcerting these are, which enables them as predictive tools for color control. But, additionally, it is convenient to use the ab initio calculations as interpretative tools as well, i.e. as instruments able to provide simple descriptions that can help fast and correct thinking. In other words, if the calculations reproduce the experiments, not just once but systematically, it is because they consider the governing effects in the right proportions, together with many other minor effects, so that extracting the roles of the governing effects should help to expect results of new calculations and to make predictions.

In order to identify the governing effects on the shifts, the contributions of the $4f^1$ and $5d^1$ energy centroids and ligand-field splitting energies to the $4f \rightarrow 5d$ transitions (Fig 8.3) were calculated and analyzed. Comparisons between calculations with different combinations of global and local structures (without and with La_Y defects) and embedding potentials (La^{3+} and Y^{3+} embedding potentials), allowed to establish the roles of first-shell distortions, long-range distortions, and co-dopant electronic effects on the Ce_Y luminescence color, via their effects on the energy centroids and ligand-field splittings. Well-established understanding of ligand-field splittings ascertains that they increase with more negative ligand charges and shorter Ce-ligand distances [36, 37]. Regarding the difference between energy centroids, the only available model for it [38, 39] relates it to the ligand polarizability, because it associates it with the different ligand polarization produced by the Ce^{3+} electron when it occupies $4f$ and $5d$ orbitals. According to this model, the centroid difference should increase with the Ce–O distance.

The analysis of the multiconfigurational ab initio calculations indicate that the La^{3+}-induced first-shell expansion around Ce^{3+} produces a negligible effect, which results from the compensation of two opposite effects of almost equal magnitude: There is a small blue shift (+86 cm^{-1}) due to a decrease of the $5d$ ligand-field splitting (significantly larger than that of the $4f$ ligand-field splitting), as expected from the knowledge of the ligand-field splittings. But the first-shell expansion also produces a small red shift (–77 cm^{-1}) due to a decrease of the energy difference between $5d^1$ and $4f^1$ centroids. Such a decrease is however in opposition to the expectations from the ligand polarizability model. This result (also found in many other analyses of multiconfigurational ab initio results) seriously questions this phenomenological model for the energy centroid difference and suggests than other effects than the ligand polarization with the active electron must also be considered in a reliable model for the centroid difference.

Long-range distortions produce a small red shift (–58 cm^{-1}), with similar contributions from ligand-field splitting (–34 cm^{-1}) and centroids (–24 cm^{-1}); both of them are hard to predict from any simple model and hard to attribute to specific distortions.

Finally, co-dopant electronic effects produce a significant red shift (–171 cm^{-1}), with a large contribution from ligand-field splitting (–149 cm^{-1}) and a minor one from energy centroids (–22 cm^{-1}). The direct effect of co-dopants means the main contribution to the total La^{3+}-induced red shift (–220 cm^{-1}). Such electronic effect of La^{3+} is basically a steric effect known as Pauli repulsion (cf. Sect. 10.3). It has its origin in the fact that two electrons of equal spin occupying two orbitals fulfill the Pauli antisymmetry principle only when the two orbitals in the Slater determinant are orthogonal. If the orbitals are initially non-orthogonal, they have to mutually orthogonalize. Such orthogonalization creates nodes and slope changes that increase the energy (hence Pauli repulsion) and it also creates some delocalization. In AIMP embedded-cluster calculations, Pauli repulsion is taken into account by means of a projection operator on each environmental ion (Sect. 1.2) that enforces orthogonality between active center orbitals and environmental orbitals. Its effect is, in part, the same effect created by a repulsive, additional negative charge on the environmental ion and, in part, producing a delocalization of the active center orbitals. Then, the presence of a La^{3+} ion instead of a smaller Y^{3+} ion in the vicinities of Ce^{3+} brings a larger Pauli repulsion and forces the Ce^{3+} $5d$ orbitals to orthogonalize themselves to larger orbitals (La^{3+} $5p$ orbitals instead of Y^{3+} $4p$ orbitals). The part of the effect equivalent to an additional environmental negative charge increases the $5d$ ligand-field splitting and red shifts the $5d \rightarrow 4f$ emission; the part that creates a Ce^{3+} $5d$ delocalization lowers the $5d^1$ centroid and enhances the $5d \rightarrow 4f$ red shift.

In summary, the fact that a La_Y defect tends to get as close as possible to a Ce_Y defect and the fact that La^{3+} is larger than Y^{3+} and, because of it, has a stronger Pauli repulsion effect on Ce^{3+}, together explain why La^{3+} co-doping YAG:Ce^{3+} produces a red shift of the Ce^{3+} $5d \rightarrow 4f$ emission in spite of the local and global lattice expansions it generates. On the contrary, in the case of interaction between Ga_{Al} and Ce_Y defects in YAG:Ce^{3+},Ga^{3+}, both clustering of substitutional defects and mutual electronic interactions are basically insignificant, which leaves the impact of lattice expansion on the crystal field splittings, and its associated blue shift of the Ce^{3+} $5d \rightarrow 4f$ emission, as the only relevant effect of Ga^{3+} co-doping. Hence the difference between La^{3+} and Ga^{3+} co-doping.

8.2.1.2 Role of Antisite Defects

We have seen how the roles of Ga^{3+} and La^{3+} types of co-doping on the $5d \rightarrow 4f$ emission of YAG:Ce^{3+} are elucidated by means of a combined use of DFT and multiconfigurational ab initio calculations: A DFT study of the luminescent material with multiple defects shows the structural impact on the Ce_Y active site of Ga_{Al} and

La_Y substitutional defects. Then, a multiconfigurational ab initio study reveals the effects on the excited manifold of the Ce_Y active center of the structural modifications and electronic interactions. A similar procedure can be used to pinpoint the influence of other defects on the luminescence of YAG:Ce^{3+}.

YAG:Ce^{3+} is known to have different optical behaviors in its different forms, like single-crystal, single-crystalline film, nanopowder, and transparent optical ceramics [19, 40–42], which are believed to be due to the interactions of the Ce^{3+} ions with different defects. Defects are known to be present in all forms of YAG and their types and concentrations strongly depend on the preparation method. Some defects are intentional, like co-dopants, but many of them are formed unintentionally. Among these, antisite defects, interstitials, and vacancies are common and antisites (AD) are dominant among the intrinsic defects [43, 44]. In them, octahedral Al^{3+} and dodecahedral Y^{3+} ions exchange positions to create a doubly substitutional defect Y_{Al}-Al_Y. In the single crystal form of YAG, these ADs are known to lower the actual symmetry of the pure garnet from cubic to trigonal [43], but these defects are hard to measure directly, their presence is often deduced indirectly and detailed experimental knowledge on their properties is limited. E.g., they have tentatively been considered responsible for the observation of 1650 cm^{-1} below the zero-phonon line excitation of the $5d \rightarrow 4f$ YAG:Ce^{3+} emission, where phonon-assisted non-radiative energy transfer between different distorted Ce^{3+} centers was assumed [41], but a direct proof was elusive.

A combined DFT and multiconfigurational ab initio study of the interplay between the Ce_Y active site of the luminescent material YAG:Ce^{3+} and Y_{Al}-Al_Y ADs was performed in [45]. Taking advantage of a previous study that provided a detailed picture of the local structures and the distribution of ADs in un-doped YAG [46], the analysis of all possible distributions of the Ce_Y defect around single and double Y_{Al}-Al_Y defects (Fig. 8.4a, b) provided a map of the energetics of defects inter-conversions (Fig. 8.4c) and structures of the most stable Ce_Y-Al_Y-Y_{Al} and Ce_Y-Al_Y-Y_{Al}-Al_Y-Y_{Al} defects (Fig. 8.4d, e). The multiconfigurational ab initio study was performed at the CASSCF/CASPT2 level of theory on the $(CeO_8Al_2O_4)^{15-}$ cluster under the effects of embedding potentials of YAG with one and two antisite defects.

Interestingly, it was found that Ce_Y substitutional defects and Al_Y-Y_{Al} antisite defects attract each other. As in the case of attraction between Ce_Y and La_Y defects in the doubly doped material $Y_3Al_5O_{12}$:Ce^{3+},La^{3+} (Sect. 8.2.1.1), there is no indication (neither in the structures nor in the wave functions) of relevant bond changes and electronic interactions when the defects are together, which makes the minimization of lattice distortions the most likely reason for their mutual attraction: the local lattice distortions brought about by two independent Ce_Y and Al_Y-Y_{Al} defects is larger than the distortions induced by a single Ce_Y-Al_Y-Y_{Al} defect.

It was also found that Ce_Y defects tend to lower the concentration of ADs in YAG, because the formation of one AD is energetically more favorable in YAG:Ce^{3+} than in YAG, whereas the formation of a second AD is less favorable.

The spectroscopic effects of ADs is revealed by an analysis of the $4f^1$, $5d^1$, and $6s^1$ electronic manifolds of the active site Ce_Y. In the first two Ce^{3+} $4f \rightarrow 5d$ transitions, they produce a small blue shift. It implies that, in order to maximize the red shift

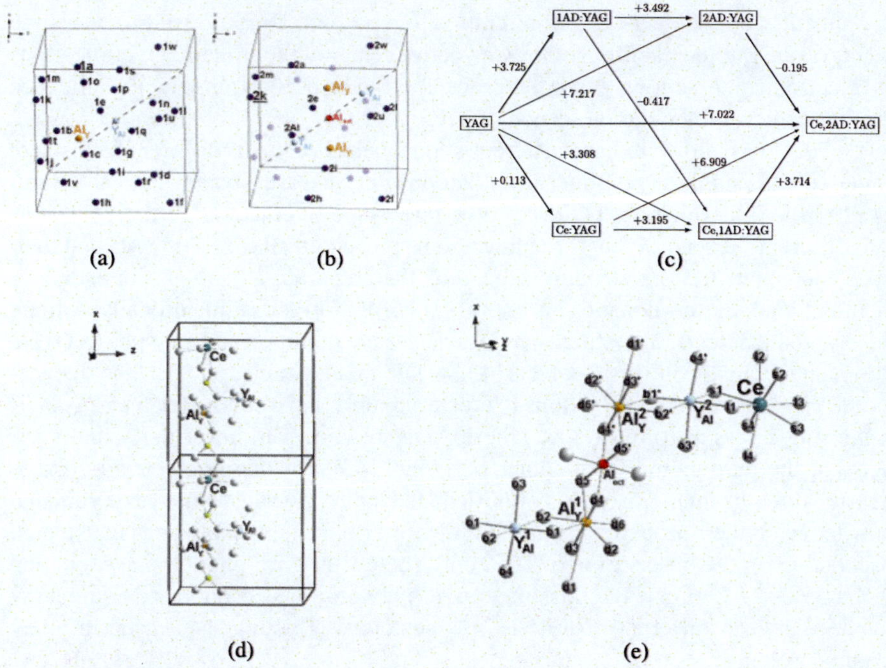

Fig. 8.4 Most stable single (**a**) and double (**b**) Al_Y-Y_{Al} antisite defects of YAG and non-equivalent sites of an additional Ce_Y substitutional defect. The most stable sites of the Ce_Y defects are underlined. **c** Schematic representation of defect formation energies (in eV). **d** Most stable Ce_Y-Al_Y-Y_{Al} defect. Axial periodicity along the ...-Ce_Y-AlO_4-Al_Y-AlO_4-Ce_Y-... tight chains is emphasized. **e** Most stable Ce_Y-Al_Y-Y_{Al}-Al_Y-Y_{Al} defect

of the Ce^{3+} $5d{\rightarrow}4f$ emission in a given garnet, it would be desirable to minimize the concentration of ADs. This result agrees with experimental observations: The $5d{\rightarrow}4f$ emission peak (which is in correspondence with the first $4f{\rightarrow}5d$ absorption) of YAG:Ce^{3+} experiences a small blue shift from single crystalline films (with lower AD concentration) to single crystals (with higher AD concentration) [19]. And the second $4f{\rightarrow}5d$ absorption also experiences a small blue shift from nanocrystals (with significantly lower AD concentration) to single crystals (with higher AD concentration) [40], although the existence of surface defects in nanocrystals can also be playing a role in this case. It can also be remarked that the AD induced blue shift questions the hypothesis of an involvement of these defects in the fact that the $5d{\rightarrow}4f$ YAG:Ce^{3+} emission can be excited below the zero-phonon line [41], which would require a red shift instead.

The analysis of the results showed that the blue shifts are due to a decrease of the ligand-field splitting of the Ce^{3+} $5d^1$ manifold (Fig. 8.3), with negligible electronic effects from the Y and Al site exchanges and irrelevant shifts of the $5d^1$ and $4f^1$ energy centroids. The lowering of the ligand-field strength is a direct consequence of the short-range and long-range structural deformations, i.e. a strongly anisotropic structural expansion around the Ce^{3+} impurities, together with distortions of almost the whole YAG unit cell. In this respect, this is the same effect that blue shifts the Ce^{3+} $5d{\rightarrow}4f$ emission by the presence of Ga_{Al} defects in the doubly doped material $Y_3Al_5O_{12}$:Ce^{3+},Ga^{3+}, except that Ga_{Al} defects produced relevant local distortions around Ce_Y but not significant long-range distortions.

8.2.1.3 Only Cubic and Tetragonal Fields are Able to Red-Shift the Ce^{3+} $5d{\rightarrow}4f$ Emission

The search for new luminescent materials useful for illumination, able to improve the quality of the white light obtained in solid-state lighting devices based on the use of Ce^{3+}-doped Yttrium Aluminum Garnet $Y_3Al_5O_{12}$:Ce^{3+}, naturally focused on modifications of this reference phosphor that could provide better efficiencies and tailored color rendering. In this regard, being able to modify the garnets to get a red shift of the $5d{\rightarrow}4f$ emission became a central issue. Following this goal, the Ce^{3+}-doped garnets $Lu_2CaMg_2Si_3O_{12}$:Ce^{3+} (orange phosphor) [47] and the $Ca_3Sc_2Si_3O_{12}$:Ce^{3+} (green phosphor) [48] were discovered, for instance. However, a clear systematic approach that could make the search more efficient and productive never existed.

In an attempt to get new insight into this issue, a series of multiconfigurational ab initio theoretical works were planned in order to clarify what garnet modifications are able to red shift the Ce^{3+} $5d{\rightarrow}4f$ emission and its corresponding first $4f{\rightarrow}5d$ absorption. The idea was to establish, in a first step, what components of the field created by the garnets on Ce^{3+} are able to red shift these transitions, and, in a second step, what garnet characteristics and modifications enhance these components.

In this Section, we summarize the results of the theoretical study addressed to answer the first question [49], which can be reformulated as: What D_2 fields created by an 8-fold coordination are able to red shift the first $4f{\rightarrow}5d$ transition? Or, more generally, because other states than the lowest of the $4f^1$ and $5d^1$ configurations are also relevant from a broader point of view: How do the $4f^1$ and $5d^1$ energy levels change in 8-fold D_2 fields?

The reason for the specific D_2 local symmetry and 8-fold coordination is because most garnets are represented with the general formula $A_3B'_2B''_3O_{12}$, and Ce^{3+} dopant ions substitute for trivalent A cations that occupy a site with D_2 point group local symmetry and 8-fold oxygen coordination.[3] In principle, additional D_2 field components originated beyond the first 8-fold oxygen coordination shell could also play a role, but this is expected to be of minor importance and the first-shell contributions are assumed to be dominant. Hence to limit the exploration to only the six independent D_2 components that can be generated with an 8-fold coordination.

[3] $A_3B'_2B''_3O_{12}$ garnets have a 160 atom body-centered cubic unit cell (80 atom primitive cell) of the $Ia\bar{3}d$ (230) space group, with 8 formula units. Cations A occupy 24(c) D_2 sites of 8-fold coordination, B′ 16(a) sites of 6-fold coordination, and B″ 24(d) sites of 4-fold coordination, all of them at fixed positions, with the remaining 96 O atoms in (h) sites, which depend on three x, y and z internal parameters [17].

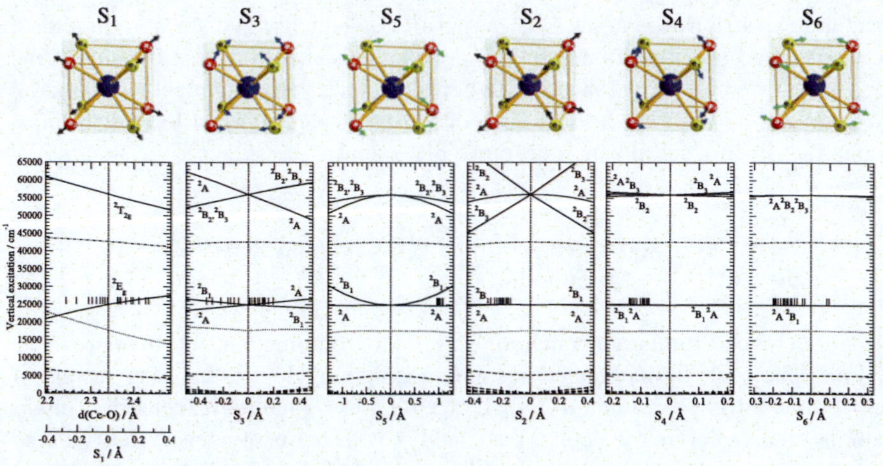

Fig. 8.5 Top: D_2 totally symmetric displacements of the CeO_8 moiety: stretching, bending, and twisting O_h symmetry coordinates S_1 (a_{1g} symmetric stretching-breathing), S_2 ($e_g\epsilon$ asymmetric stretching), S_3 ($e_g\theta$ symmetric bending), S_4 ($e_g\epsilon$ asymmetric bending), S_5 ($e_u\theta$ symmetric twisting), and S_6 ($e_u\epsilon$ asymmetric twisting). Bottom: Energies (relative to the ground state) of the states of the configurations $4f^1$ (dashed lines) and $5d^1$ (full lines) as a function of the totally symmetric D_2 oxygen displacement coordinates, calculated without spin-orbit coupling. The differences between the $5d^1$ and $4f^1$ energy centroids (dot-dashed lines) and between the ligand field stabilization energies of the $5d$ and $4f$ lowest levels (doted lines) are also shown; the lowest $4f \rightarrow 5d$ transition equals the subtraction of these two lines. Experimental values of S_1—S_6 of 21 pure garnets are shown as small vertical lines

Without any lack of generality, it is convenient to look at the six D_2 field components under study as derived from a reference CeO_8 O_h cubic moiety. This choice allows to visualize the D_2 field components from a familiar geometric point of view, as well as to correlate the D_2 $4f^1$ and $5d^1$ energy levels with their respective, also familiar O_h levels. The D_2 field components are represented with the oxygen atoms displacements shown in Fig. 8.5(top). Their symmetry classifications and algebraic details are given in Table 8.2. The S_2 and S_4 orthorombic (dipyramidal) components have equal symmetry, so that other pairs of linear combinations of them are also valid. The present choice corresponds to bond stretching and bond bending vibrational coordinates. The energies of the $4f^1$ and $5d^1$ levels under the effects of the six D_2 fields (stronger fields correspond to smaller Ce-O distances) are shown in Fig. 8.5(bottom).

Before we summarize the results, let us remark that they mean two attainments: On one hand, they clarify the effects of D_2 field components on the energy levels and show what specific fields are able to red shift the $4f \rightarrow 5d$ transition. On the other hand, they give a quantitative measure of the first-neighbor effects on the energy levels and on the $4f \rightarrow 5d$ red shift because the choice of a cubic host embedding potential makes all the energy splittings and changes due to D_2 distortions ascribable to the interactions between Ce and the eight O atoms in its first coordination shell.

Table 8.2 D_2 totally symmetric displacement coordinates S_1–S_6 of a cubic AO_8 moiety. The O_h irreducible representations of the coordinates and the point symmetry groups of the AO_8 moiety resulting from the atomic displacements are given

D_2 symmetrization operator: $\hat{R} = \frac{1}{2}\left(\hat{I} + \hat{C}_{2x} + \hat{C}_{2y} + \hat{C}_{2z}\right)$

Type of distortion			D_2 displacement[a]
Coordinate	O_h irrep	Point group	Coordinate definition[b]
Cubic (hexaoctahedral)			Symmetric bond stretching
S_1	a_{1g}	O_h	$\frac{1}{\sqrt{6}}\hat{R}\left[\left(\sqrt{2}\,\delta x_{a1} + \delta z_{a1}\right) + \left(\sqrt{2}\,\delta y_{e1} + \delta z_{e1}\right)\right]$
Orthorombic (dipyramidal)			Asymmetric bond stretching
S_2	$e_g\epsilon$	D_{2h}	$\frac{1}{\sqrt{6}}\hat{R}\left[\left(\sqrt{2}\,\delta x_{a1} + \delta z_{a1}\right) - \left(\sqrt{2}\,\delta y_{e1} + \delta z_{e1}\right)\right]$
Tetragonal (ditetragonal-dipyramidal)			Symmetric bond bending
S_3	$e_g\theta$	D_{4h}	$\frac{1}{\sqrt{6}}\hat{R}\left[\left(-\delta x_{a1} + \sqrt{2}\,\delta z_{a1}\right) + \left(-\delta y_{e1} + \sqrt{2}\,\delta z_{e1}\right)\right]$
Orthorombic (dipyramidal)			Asymmetric bond bending
S_4	$e_g\epsilon$	D_{2h}	$\frac{1}{\sqrt{6}}\hat{R}\left[\left(-\delta x_{a1} + \sqrt{2}\,\delta z_{a1}\right) - \left(-\delta y_{e1} + \sqrt{2}\,\delta z_{e1}\right)\right]$
Tetragonal (trapezoidal)			Symmetric bond twisting
S_5	$e_u\theta$	D_4	$\frac{1}{\sqrt{2}}\hat{R}\,[-\delta y_{a1} + \delta x_{e1}]$
Tetragonal (scalenoidal)			Asymmetric bond twisting
S_6	$e_u\epsilon$	D_{2d}	$\frac{1}{\sqrt{2}}\hat{R}\,[\delta y_{a1} + \delta x_{e1}]$

[a] With respect to a reference AO_8 cube with the symmetry independent oxygen atoms a_1 and e_1 located at $(\sqrt{2}, 0, 1)d_{\text{A-O,ref}}/\sqrt{3}$ and $(0, \sqrt{2}, 1)d_{\text{A-O,ref}}/\sqrt{3}$ respectively. Labels and cartesian axes correspond to Fig. 8.5. [b] δx_{a1} is the x cartesian displacement of the oxygen atom a_1 from its position in the reference cube: $\delta x_{a1} = x_{a1} - x_{a1,\text{ref}}$; identical definitions stand for the other cartesian displacements

The multiconfigurational ab initio calculations (all-electron SA-CASSCF/MS-CASPT2/RASSI-SO DKH relativistic, with ANO basis sets) were done on a $(CeO_8)^{13-}$ cluster embedded in a cubic AIMP embedding potential. This corresponds to a SrO host with its high pressuere CsCl lattice structure ($Pm\overline{3}m$, no. 221) [50], with a 8-fold cubic oxide coordination of the cationic site, and a lattice constant a=2.87 Å. (A MS-CASPT2 calculation of the Ce-O equilibrium distance in $(CeO_8)^{13-}$ under this cubic embedding potential gives 2.34 Å, which is the Ce-O distance of the reference CeO_8 cube in a hypothetical undistorted Ce-doped YAG [17]).

The results summarized in Fig. 8.5(bottom) span ranges of the S_1-S_6 coordinates that contain their usual values in natural and artificial garnets. These values are shown in Table 8.3 for 21 garnets that will be the subject of further studies of host embedding effects (Sect. 8.2.1.4). They are indicated in Fig. 8.5(bottom) with small vertical lines.

It is clear that only two D_2 components have an important impact on the lowest $4f\rightarrow5d$ transition: the cubic breathing component S_1 and the $e_g\theta$ tetragonal symmetric bond bending component S_3.

Table 8.3 Crystallographic data of $A_3B'_2B''_3O_{12}$ garnets: lattice constant a and special position (h) of the $Ia\bar{3}d$ (230) space group, x_O, y_O, z_O (fractional cell units). Structural data of the AO_8 moieties with D_2 symmetry on (c) sites (Table 8.2). The values of the O_h breathing distortion S_1 from a common reference cube of $d_{A\text{-}O,ref}$=2.34 Å are also given[a]. Length units are Å

		Garnet	Ref.	a	x_O	y_O	z_O		
			$d^a_{A\text{-}O,ref}$	$(S_1)^a$	S_2	S_3	S_4	S_5	S_6
1		$Lu_2CaMg_2Si_3O_{12}$	[47]	11.9750	-0.03510	0.05380	0.15780		
			2.333	(−0.0200)	−0.1522	−0.1619	−0.1139	1.0153	0.0956
2	Pyrope	$Mg_3Al_2Si_3O_{12}$	[51]	11.4566	−0.03290	0.05010	0.15280		
			2.243	(−0.2731)	−0.2191	−0.0768	−0.1215	0.9900	−0.0892
3	Almandine	$Fe_3Al_2Si_3O_{12}$	[52]	11.5250	−0.03401	0.04901	0.15278		
			2.267	(−0.2070)	−0.2334	−0.1126	−0.1279	1.0084	−0.1028
4	Spessartine	$Mn_3Al_2Si_3O_{12}$	[53]	11.6190	−0.03491	0.04791	0.15250		
			2.295	(−0.1278)	−0.2490	−0.1414	−0.1337	1.0281	−0.1243
5	Grossular	$Ca_3Al_2Si_3O_{12}$	[51]	11.8515	−0.03760	0.04540	0.15130		
			2.368	(0.0790)	−0.2752	−0.2203	−0.1378	1.0725	−0.1909
6	Andradite	$Ca_3Fe_2Si_3O_{12}$	[54]	12.0578	−0.03940	0.04870	0.15540		
			2.398	(0.1648)	−0.2102	−0.2895	−0.1260	1.0718	−0.0351
7		$Ca_3Sc_2Si_3O_{12}$	[54]	12.2500	−0.04004	0.05010	0.15887		
			2.427	(0.2454)	−0.1972	−0.3362	−0.1401	1.0894	0.0842
8	LuAG	$Lu_3Al_5O_{12}$	[17]	11.9060	−0.02940	0.05370	0.15090		
			2.303	(−0.1042)	−0.1683	0.0555	−0.0866	0.9766	−0.1044
9	YbAG	$Yb_3Al_5O_{12}$	[17]	11.9310	−0.02960	0.05290	0.15040		
			2.313	(−0.0759)	−0.1819	0.0497	−0.0906	0.9858	−0.1286
10	ErAG	$Er_3Al_5O_{12}$	[55]	11.9928	−0.03039	0.05124	0.14915		
			2.339	(−0.0039)	−0.2047	0.0297	−0.0944	1.0046	−0.1854
11	YAG	$Y_3Al_5O_{12}$	[17]	12.0000	−0.03060	0.05120	0.15000		
			2.339	(−0.0019)	−0.2094	0.0156	−0.1018	1.0099	−0.1613
12	GdAG	$Gd_3Al_5O_{12}$	[17]	12.1130	−0.03110	0.05090	0.14900		
			2.368	(0.0788)	−0.2060	0.0101	−0.0932	1.0180	−0.1957
13	LuGG	$Lu_3Ga_5O_{12}$	[17]	12.1880	−0.02520	0.05700	0.15060		
			2.322	(−0.0520)	−0.1378	0.1961	−0.0727	0.9580	−0.0755
14	YbGG	$Yb_3Ga_5O_{12}$	[17]	12.2040	−0.02590	0.05630	0.15190		
			2.328	(−0.0336)	−0.1561	0.1596	−0.0899	0.9744	−0.0459
15	YGG	$Y_3Ga_5O_{12}$	[56]	12.2730	−0.02740	0.05460	0.14930		
			2.363	(0.0647)	−0.1629	0.1338	−0.0757	0.9876	−0.1440
16	HoGG	$Ho_3Ga_5O_{12}$	[57]	12.2900	−0.02740	0.05520	0.15020		
			2.362	(0.0616)	−0.1550	0.1293	−0.0771	0.9861	−0.1101
17	DyGG	$Dy_3Ga_5O_{12}$	[57]	12.3060	−0.02780	0.05490	0.14990		
			2.369	(0.0823)	−0.1558	0.1194	−0.0752	0.9896	−0.1229
18	TbGG	$Tb_3Ga_5O_{12}$	[58]	12.3474	−0.02820	0.05470	0.14950		
			2.381	(0.1166)	−0.1538	0.1115	−0.0712	0.9933	−0.1377
19	GdGG	$Gd_3Ga_5O_{12}$	[58]	12.3829	−0.02890	0.05420	0.14940		
			2.394	(0.1538)	−0.1578	0.0901	−0.0719	1.0019	−0.1473
20	SmGG	$Sm_3Ga_5O_{12}$	[58]	12.4361	−0.02920	0.05300	0.14900		
			2.412	(0.2031)	−0.1816	0.0777	−0.0818	1.0190	−0.1748
21	NdGG	$Nd_3Ga_5O_{12}$	[58]	12.5051	−0.03000	0.05220	0.14850		
			2.434	(0.2667)	−0.1896	0.0554	−0.0819	1.0321	−0.2009

[a]Two valid alternatives to define D_2 stuctures of the AO_8 moiety are: ($d_{A\text{-}O}$, $S_1 = 0$, S_2, S_3, S_4, S_5, S_6) and ($d_{A\text{-}O}$=$d_{A\text{-}O,ref}$, S_1, S_2, S_3, S_4, S_5, S_6). Note that $S_1 = \sqrt{8}\,(d_{A\text{-}O} - d_{A\text{-}O,ref})$

Symmetric compression of the Ce-O bonds (S_1) lowers the first $4f \rightarrow 5d$ transition and this effect is almost entirely due to the increase in the $5d$ ligand field splitting, with no significant involvement of any change in the energy difference between $4f^1$ and $5d^1$ centroids.

Symmetric bond bending (S_3), which produces a tetragonal (ditetragonal-dipyramidal) D_{4h} distortion of the CeO_8 cube, also lowers the first $4f \rightarrow 5d$ transition. This conclusion supports previous suggestions in the line that tetragonal distortions could be involved in red-shifts of the Ce^{3+} luminescence in garnets [59]. However, other tetragonal distortions (trapezoidal D_4 and scalenoidal D_{2d}, both of them associated with bond twistings) do not shift the first $4f \rightarrow 5d$ transition. In this case, the lowering of the first $5d^1$ level with respect to the $4f^1$ ground state is due in almost equal amounts to an increase of the $5d$ ligand field splittings (the $4f$ ligand field splitting is much smaller) and to a reduction of the centroid energy difference. In the line of what was observed in the analysis of the La^{3+} induced red-shift of the Ce^{3+} $5d \rightarrow 4f$ emission (Sect. 8.2.1.1), the reduction of the centroid energy difference questions the ligand polarizability model for this property [38, 39, 60, 61], according to which a very small increase of the centroid energy difference due to a very small increase of the Ce-O distances should be expected instead.

From a conceptual point of view, it is concluded that the energy of the lowest Ce^{3+} $5d^1$ level in garnets is, basically, the result of a cubic $5d^1$ E_g level under the effect of a strong Jahn-Teller coupling of the type $E_g \times e_g$ [62]. This conclusion stems from the findings that, in spite of the large distortion values explored, no D_2 components other than O_h and D_{4h} have significant effects on the first Ce^{3+} $4f \rightarrow 5d$ transition and the effects on the $4f^1$ ground state are much smaller than in the first $5d^1$ state,

From a practical point of view, the multiconfigurational ab initio study indicates that the search for modifications of Ce^{3+}-doped garnets that could red-shift the $5d \rightarrow 4f$ emission, should be addressed to finding as low breathing S_1 values as possible (i.e. small average Ce-O distances) and as large (in absolute value) symmetric bond bending $e_g\theta$ S_3 values as possible.

8.2.1.4 Unrelaxed Host Effects and Structural Relaxation Effects

In the quest for a recipe for modified Ce^{3+}-doped garnets with red-shifted luminescence with respect to that of YAG:Ce^{3+}, multiconfigurational ab initio calculations were conducted, aimed at establishing the roles of the interactions between Ce^{3+} and the hosting garnets on the $5d \rightarrow 4f$ emission [49, 63, 64].

When a Ce^{3+} ion enters a garnet and substitutes for some cation, it produces a local distortion and the energy of its $5d \rightarrow 4f$ emission is determined by the shifts and splittings experienced by the $4f^1$ and $5d^1$ energy levels of the free-ion as a consequence of its interaction with the distorted host lattice. Such interaction between a Ce^{3+} dopant and a distorted garnet can be decomposed and analyzed in order to find what factors govern the emission energy and, eventually, its red-shift. A natural way of decomposing the influence of the distorted host on the Ce^{3+} energy levels is in terms of two adding effects: (i) the *unrelaxed host* effect due to a hypothetical undistorted substitution, in which Ce^{3+} would substitute for some cation (e.g. Y^{3+} in YAG:Ce^{3+}) but the whole garnet structure would remain intact, i.e. it would be the one of the undoped host; and (ii) the effect of the *structural relaxation* that would follow the undistorted substitution. Quantification of these two effects is hard to realize from the experimental point of view, but it can be achieved in ab initio calculations.

The importance of this quantification lies in the fact that the garnet structure alone would control the emission of the Ce^{3+} dopant if *unrelaxed host* effects are dominant, in which case the choice of host and the syntheses of new garnets oriented towards a red shift of the Ce^{3+} $5d \rightarrow 4f$ emission could follow the line of the analysis of the roles played by the different D_2 field components of garnets (Sect. 8.2.1.3). In this case, the crystallographic structure of the garnet host would be a sufficient control tool. If, on the contrary, *structural relaxation* effects are important, then the detailed knowledge of the local structure established around the Ce^{3+} dopant after the substitution and subsequent reorganization is a key for the understanding of the Ce^{3+} $5d \rightarrow 4f$ emission shift. In this case, the systematics towards the control of the shift would demand not only the knowledge of the garnet structure, but also the local distortion and the rules governing it. This means that additional control tools would be necessary. In this regard, ab initio calculations become instrumental because of their ability to provide high quality detailed local structures, whereas experimental techniques like EXAFS can only provide some features of the local structures, but not all details, particularly those responsible for the different D_2 field components.

A multiconfigurational ab initio study of the *unrelaxed host* effect on the $4f^1$ and $5d^1$ energy levels of Ce^{3+} was completed in [64]. It included 21 Ce^{3+}-doped Si, Al, and Ga garnets of general formula $A_3B'_2B''_3O_{12}$, spanning a broad range of lattice constant values (cf. Table 8.3), where Ce^{3+} substitutes for a trivalent A cation in the D_2 dodecahedral site with 8-fold Oxygen coordination and creates a Ce_A substitutional defect. The *unrelaxed host* effect was further decomposed in two contributions: the short-range, *unrelaxed first-neighbor* effect [49], due to the interactions between Ce^{3+} and its nearest Oxygens when these occupy the positions they have in the AO_8 moieties of the original garnets; and the long-range, *unrelaxed host embedding* effect [64], due to interactions between the $(CeO_8)^{13-}$ cluster and second and more distant neighbors of Ce^{3+}, which are included by means of the electronic embedding potentials of the garnets. Corresponding *structural relaxation* effects were studied in a selection of Ce^{3+}-doped garnets: $Y_3Al_5O_{12}$, $Lu_3Al_5O_{12}$, $Y_3Ga_5O_{12}$, $Lu_3Ga_5O_{12}$ and $Ca_3Sc_2Si_3O_{12}$ [63].

The *unrelaxed host* effects are summarized in Fig. 8.6. The short-range, *unrelaxed first-neighbor* contributions as obtained in [49] from the deformations of the $(CeO_8)^{13-}$ D_2 clusters with respect to a reference cube (all of them under the effects of a cubic, host-independent embedding potential that acts as a confinement potential for the $(CeO_8)^{13-}$ electrons), are represented in panel (a) with solid squares. The long-range, *unrelaxed host embedding* contributions are shown in panel (b) with solid circles. Interestingly, the second and more distant neighbors of Ce^{3+} contribute in all garnets with a red shift of the first $4f \rightarrow 5d$ transition with respect to its value under the effect of the first-neighbor Oxygens only. This role of long-range contributions

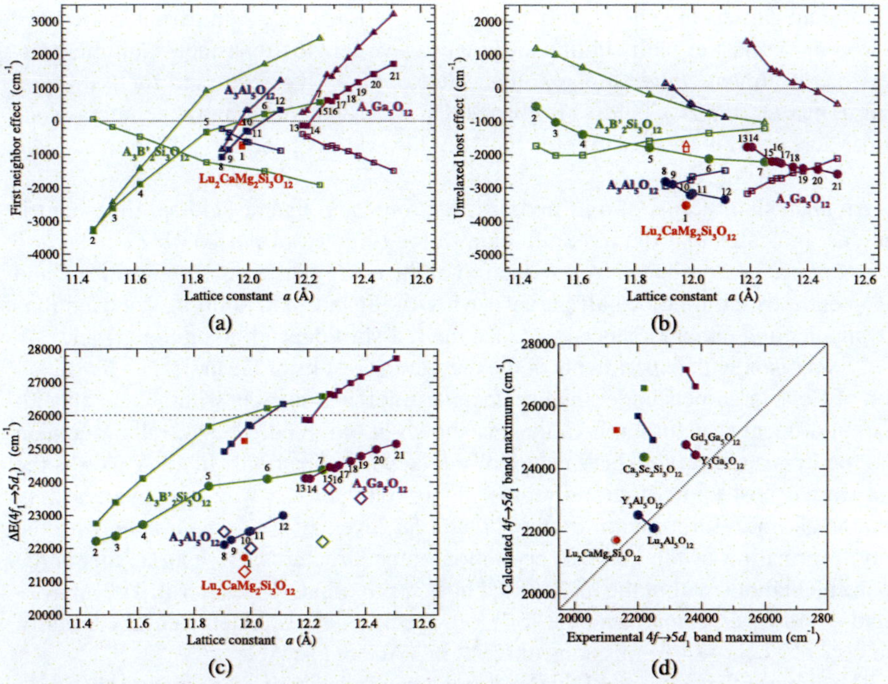

Fig. 8.6 First $4f \to 5d$ vertical (or Frank-Condon) transition of Ce^{3+}-doped garnets calculated in unrelaxed hosts. Red: $Lu_2CaMg_2Si_3O_{12}$; green: $A_3B'_2Si_3O_{12}$; blue: $A_3Al_5O_{12}$; violet: $A_3Ga_5O_{12}$. Identification umbers correspond to Table 8.3. **a** Unrelaxed first-neighbor effect: on the transition (solid squares), on the contribution of the centroids (open squares), and on the contribution of the ligand field splitting (open triangles). **b** Unrelaxed host embedding effect: on the transition (solid circles), on the contribution of the centroids (open squares), and on the contribution of the ligand field splitting (open triangles). **c** Transition energy for $(CeO_8)^{13-}$ clusters with unrelaxed experimental structures under the effect of: a common cubic confinement potential (solid squares) and the embedding potential of the undistorted garnets (solid circles). Horizontal dotted line: Transition corresponding to a reference cubic structure ($d_{\text{Ce-O}}$=2.34 Å) under the effect of the cubic potential. Open diamonds: Experiments. **d** Calculated verses experimental data

in garnets is understood with the conclusions of Sect. 8.2.1.3, i.e.: S_1 (a_{1g}) O_h field components can enhance either red-shifts (via compressions) and blue-shifts (via expansions), but these contributions are basically included by the first-neighbors; S_3 $(e_g\theta)$ D_{4h} tetragonal field components give a red shift; and the remaining field components do not give significant shifts. Altogether one can expect the overall long-range, *unrelaxed host embedding* effect to be a red-shift due to its D_{4h} tetragonal components. Besides, as shown in Fig. 8.6b, the size of this red-shift is important and plays a basic role, as important as the shift produced by the short-range effect, at differentiating the values of the first $4f \to 5d$ transition in different garnet families: $A_3B'_2Si_3O_{12}$ Silicon garnets, $A_3Al_5O_{12}$ Aluminum garnets, and $A_3Ga_5O_{12}$ Gallium garnets. In fact, panel (c) of Fig. 8.6 clearly shows that what makes Ce^{3+}-doped

Aluminum garnets in general, and YAG:Ce^{3+} and LuAG:Ce^{3+} in particular, to have a $5d \rightarrow 4f$ emission with significant longer wave lengths than their Gallium partners, is, in almost equal amounts, their shorter Ce-O distances and their stronger long-range O_h and D_{4h} fields. This panel also shows that the long-range interactions are almost entirely responsible for the very low value of the $5d \rightarrow 4f$ transition in $Lu_2CaMg_2Si_3O_{12}$:Ce^{3+}.

An analysis in terms of $4f^1$ and $5d^1$ centroids and ligand-field splitting contributions, indicates that the red-shifts due to long-range, *unrelaxed host embedding* interactions, are dominated by a reduction of the energy difference between $5d^1$ and $4f^1$ centroids, common to all garnets, whereas ligand-field splitting contributions are smaller and either enhance or reduce the red-shift depending on the garnet.

Comparison with experiments of the theoretical results of the first $4f \rightarrow 5d$ transition of Ce^{3+} in garnets under the *unrelaxed host* effect only, i.e. without any consideration of local structural relaxations, is shown in Fig. 8.6d. The overall agreement and the trends are reasonable in spite of the lack of relaxations. In other words, the undistorted host approximation represents a relatively fast way to provide a first, reliable ab initio information on the $4f$ and $5d$ levels of Ce^{3+} in garnets, with the only information of the garnets' crystallographic structures. As long as this is the goal, calculations within the undistorted host approximation are useful. The approximation, however, cannot account for more specific details, like the energy order of the very close $5d \rightarrow 4f$ emissions of Ce^{3+} in YAG and LuAG.

The effects of *structural relaxation* around the Ce_A substitutional defects in Ce^{3+}-doped $Y_3Al_5O_{12}$, $Lu_3Al_5O_{12}$, $Y_3Ga_5O_{12}$, $Lu_3Ga_5O_{12}$ and $Ca_3Sc_2Si_3O_{12}$ have been studied in [63]. The calculations show that the *unrelaxed host* effect dominates the differences in $4f$-$5d$ transitions between Ce^{3+}-doped garnet families. Local relaxations provide small corrections. However, these subtle corrections, which depend entirely on small displacements of the Oxygens in the first coordination shell of Ce^{3+} that are very hard to determine experimentally, are the ones responsible for the first $4f \rightarrow 5d$ transition to have a slightly lower energy in $Y_3Al_5O_{12}$:Ce^{3+} than in $Lu_3Al_5O_{12}$:Ce^{3+}. In effect, whereas the experimental order is $Y_3Al_5O_{12}$:Ce^{3+} (22000 cm^{-1}) < $Lu_3Al_5O_{12}$:Ce^{3+} (22470 cm^{-1}) [7, 25], the unrelaxed host result is the opposite: $Y_3Al_5O_{12}$:Ce^{3+} (22520 cm^{-1}) > $Lu_3Al_5O_{12}$:Ce^{3+} (22100 cm^{-1}), and it is corrected after including local relaxations around the respective Ce_Y and Ce_{Lu} defects: $Y_3Al_5O_{12}$:Ce^{3+} (23390 cm^{-1}) < $Lu_3Al_5O_{12}$:Ce^{3+} (23420 cm^{-1}).

With the aim of shaping the design of new red phosphors via transformations of YAG:Ce^{3+} oriented by the new theoretical findings, a series of calculations within the undistorted host approximation were performed on a number of transformed garnets doped with Ce^{3+} [65]. The undistorted host approximation was adopted because local structural relaxations are responsible for subtle effects on the lowest $4f \rightarrow 5d$ transition, but the overall trends, which are the goal here, are given by the interaction between the Ce^{3+} dopand and the unrelaxed host. As transformed garnets, a number of Aluminum garnets (which show low $4f \rightarrow 5d$ transitions) resulting from cationic substitutions in $Y_3Al_5O_{12}$ that would enhance the tetragonal field around the Ce^{3+} site were chosen. This is because only O_h and D_{4h} components of the D_2 fields are able to red-shift the first $4f \rightarrow 5d$ transition. A total of 33 hypothetical garnets of the types $M^{IV}_2M^{III}M^{II}_2Al_3O_{12}$:$Ce^{3+}$, $M^{IV}_2M^{I}Al_5O_{12}$:Ce^{3+}, and $M^{IV}M^{III}M^{II}Al_5O_{12}$:$Ce^{3+}$ were studied. In these, trivalent cations (dodecahedral Y^{3+} and octahedral Al^{3+}) in the second coordination shell of Ce^{3+} are substituted by monovalent to tetravalent cations in ways than enhance tetragonal field components. All the transformed garnets were found to have $4f \rightarrow 5d$ transitions red-shifted with respect to that of YAG:Ce^{3+}. The red shifts ranged between –1000 cm^{-1} and –3000 cm^{-1}.

These results shape a driving idea for designing new red phosphors by enhancing the tetragonal field around Ce^{3+} with cationic substitutions made in Ce^{3+}-doped aluminum garnets. We may remark that the Ce^{3+}-doped $MgY_2Al_4SiO_{12}$ garnet, which is a transformation from $Y_3Al_5O_{12}$ via substitution of one dodecahedral Y^{3+} by Mg^{2+} and one tetrahedral Al^{3+} by Si^{4+} and enhances the tetragonal components (besides others, irrelevant for the $4f \rightarrow 5d$ transition), was found to have a clearly redshifted $5d \rightarrow 4f$ transition with respect to YAG:Ce^{3+} [66].

8.2.2 Bi-doped SrB_4O_7

Although divalent bismuth is rarely reported in inorganic chemistry textbooks, Bi^{2+} was considered to be stable and responsible for the red-orange luminescence of bismuth activated SrB_4O_7 in 1994 [67]. In addition to borates, Bi^{2+} active centers were assumed to give their properties to previously reported Bi-based red phosphors [68], like alkaline earth sulfates [69] and phosphates [70], and a line of search for new red phosphors useful in energy efficient solid-state lighting devices was developed around this species [71–74]. The Bi_2^{4+} species was found in crystallographic studies [75], but there is no direct evidence of Bi^{2+} in the red phosphors, in spite of their search with techniques such as EPR and XANES [76], apparently due to its very low concentrations. The existence of this species was mainly supported on optical spectroscopy measurements and the impossibility to assign excitation and emission energies and excited state lifetimes of some Bi-doped materials to any other bismuth species. Quantitative multiconfigurational ab initio calculations could then help in this respect. We will see here that an overall agreement exists between the theoretical electronic structure of Bi^{2+} as a dopant in the host SrB_4O_7 and the optical spectroscopic experiments, which provides a strong support to the existence of Bi^{2+} and the general interpretation of its luminescence [76]. Nevertheless, as it often happens with luminescent materials, the multiconfigurational ab initio calculations also provide some unexpected results: in this case, a new interpretation of the electronic nature of the third excited state.

An emission band and two excitation bands around 17000 cm^{-1} and 21000 cm^{-1} in Bi-doped borates (Fig. 8.7), phosphates, and sulfates have been attributed to the spin-orbit coupling splitting of the $6s^26p^1$–$^2P^o$ term of Bi^{2+} into the $^2P^o_{1/2}$ and $^2P^o_{3/2}$ levels, plus an additional low symmetry splitting of the latter into $^2P^o_{3/2}(1)$ and $^2P^o_{3/2}(2)$ [67, 71, 72]. Transitions between these states are parity forbidden $6p \rightarrow 6p$ transitions in the free-ion. What makes them allowed in the hosts, is the low symmetry

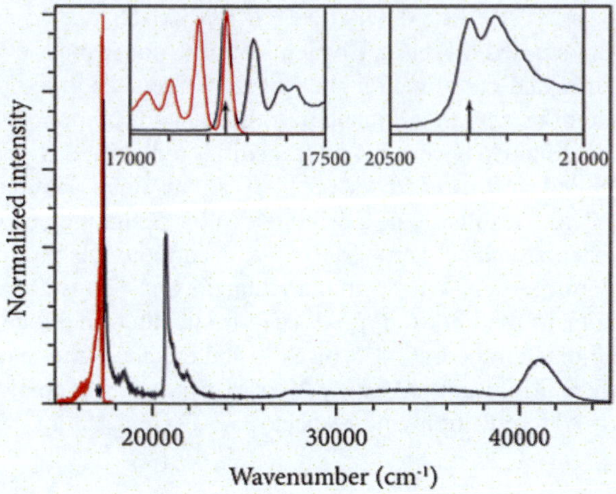

Fig. 8.7 Experimental excitation (for 620 nm emission) and emission (upon 480 nm excitation) spectra of SrB_4O_7:0.02%Bi, measured at 4.2 K [76]. The insets show a more detailed picture in the vicinity of the zero-phonon lines (indicated by arrows) of the two low-energy transitions

at the sites occupied by Bi^{2+}. There is also a third excitation band, more intense, at much higher energy, around 40000 cm^{-1} (Fig. 8.7). It is attributed to a Bi^{2+} parity allowed $6p \rightarrow 7s$ transition: $6s^26p^1\text{–}^2P^o_{1/2} \rightarrow 6s^27s^1\text{–}^2S_{1/2}$.

In SrB_4O_7:Bi^{2+}, Bi substitutes for Sr in a nine-fold oxygen coordination site with C_s point symmetry [77]. Accordingly, SA-CASSCF/MS-CASPT2/RASSI-SO DKH calculations have been performed on a $(BiO_9)^{16-}$ cluster (Fig 8.8) under the effect of the AIMP embedding potential of the SrB_4O_7 host [76]. C_s symmetry has been used in all calculations (and the electronic states are labeled accordingly) except for structures with Bi off-plane, which required C_1 symmetry. They are all-electron calculations with ANO relativistic basis sets for Bismuth $(25s22p16d12f4g)[11s10p9d6f4g]$ and Oxygen $(14s9p4d)[4s3p2d]$ [78]. In the SA-CASSCF calculations that account for mean-field and static correlation effects, the complete-active-space was defined by distributing 3 electrons in 13 active molecular orbitals: nine a' and four a'', which are the ones with main character Bi $6s$, $6p$, $6d$, $7s$, and $7p$. Then, the CAS can be labeled $[9 \times a', 4 \times a'']^3$. The molecular orbital were optimized so as to minimize the energies of the following spin doublet and quartet state averages: The average of 17 states derived from the lowest doublets of free Bi^{2+} ($6s^26p^1\text{–}^2P^o, 6s^27s^1\text{–}^2S, 6s^26d^1\text{–}^2D, 6s^16p^2\text{–}^2P, 6s^16p^2\text{–}^2D$) and the average of the 3 states derived from the term $6s^16p^2\text{–}^4P$ of Bi^{2+}. This means the independent averages of the first 11 $^2A'$, 6 $^2A''$, 2 $^4A'$, and 1 $^4A''$ C_s many-electron states. In the MS-CASPT2 calculations, the dynamic correlation of 85 valence electrons was also considered: that of the $5d$, $6s$, $6p$ electrons of Bi and $2s$, $2p$ electrons of the nine O atoms. In the RASSI-SO calculations, spin-orbit coupling between the first 17 doublets and 3 quartets was included. They are 1-11 $^2A'$, 1-6 $^2A''$, 1-2 $^4A'$, and 1 $^4A''$, which give 23 $E_{1/2}$ Kramers doublets. Some calculations on the corresponding embedded-cluster of Bi^{3+}, $(BiO_9)^{15-}$, used the CAS that results from distributing 2 electrons in 4 active molecular orbitals with main character Bi $6s$ and $6p$; dynamically correlate the same valence electrons; and included spin-orbit cou-

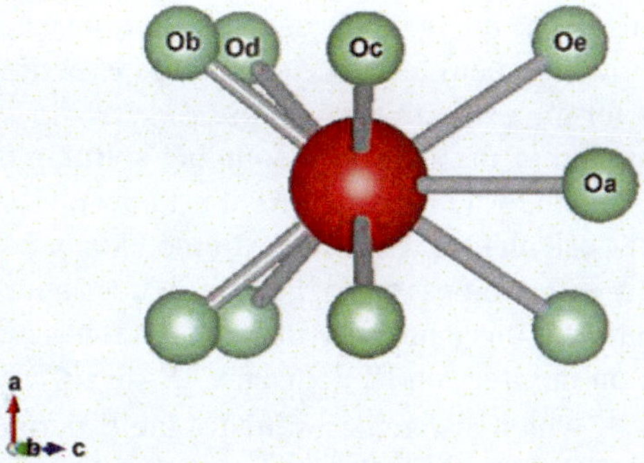

Fig. 8.8 $(BiO_9)^{16-}$ cluster. The a, b, and c axes of the SrB_4O_7 orthorhombic lattice are shown. The Bi-O distances in the unrelaxed host and in the ground state relaxed host (in parenthesis) are, in Angstrom: Bi-O_a 2.73(2.79), Bi-O_b 2.47(2.39), Bi-O_c 2.68(2.60), Bi-O_d 2.79(2.81), and Bi-O_e 2.90(3.02)

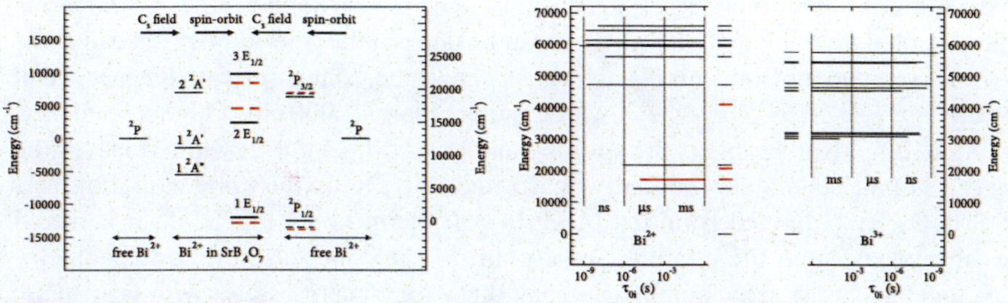

(a) (b)

Fig. 8.9 **a** Quantitative correlation level diagram. Black: ab initio calculations. Red: experiments. **b** Energy levels and radiative lifetimes of the states of Bi^{2+} and Bi^{3+} in SrB_4O_7. Ab initio calculations are shown in black and experimental results in red in both diagrams

pling between the singlets and triplets. The AIMP embedding potential of SrB_4O_7 consists of total-ion quantum-mechanical potentials of Sr^{2+} and B^{3+} cations and O^{2-} anions within a cube of $3 \times 3 \times 3$ unit cells centered on Sr^{2+}, plus 16584 additional point charges at lattice sites that accurately generate the host Ewald potential. Total-ion potentials and point charges are located at experimental lattice sites of SrB_4O_7 [Pmn2_1 orthorhombic; $a = 10.711$ Å, $b = 4.427$ Å, $c = 4.235$ Å] [77].

The results of the calculations and their comparison with experimental data are summarized in Fig 8.9. The pattern of theoretical excited states of Bi^{2+} in SrB_4O_7 resembles well the experimental observations in SrB_4O_7:Bi of two states with similar excitation intensities near 20000 cm^{-1}, followed by a gap up to above 40000 cm^{-1}, where there is a third excitation of higher intensity.

The computed results for the low lying states, derived from the ${}^2P^o$ term of the free-ion Bi^{2+}, are basically described as: three states 1–3 $E_{1/2}$ resulting from a 17980 cm^{-1} ${}^2P_{1/2}$–${}^2P^o_{3/2}$ spin-orbit coupling splitting, plus a 7220 cm^{-1} ${}^2P^o_{3/2}(1)$–${}^2P^o_{3/2}(2)$ C_s low symmetry splitting. This image qualitatively agrees with experiments. A quantitative analysis based on their assignment to Bi^{2+} states gives experimental 18975 cm^{-1} and 3460 cm^{-1} values for the respective splittings.

The 5% underestimation of the spin-orbit coupling splitting is well understood: It has an atomic origin and the way to correct for it is known. However, correcting it in the embedded-cluster calculation would be extremely demanding from the computational point of view and pointless to do, because it does not provide any further information. Let us comment this a bit more in detail. (i) It has an atomic origin. An equivalent calculation on the free-ion Bi^{2+} gives a ${}^2P^{o}_{1/2}$–${}^2P^{o}_{3/2}$ spin-orbit coupling splitting of 18881 cm^{-1}, which also underestimates the experimental 20788 cm^{-1} splitting [79]. This is a well known result of LS-coupling type of calculations (like the present ones) when a unique set of $6p$ orbitals is used for the ${}^2P^{o}_{1/2}$ and ${}^2P^{o}_{3/2}$ states, instead of optimizing the $6p_{1/2}$ and $6p_{3/2}$ spinors with a spin-orbit Hamiltonian in a jj-coupling type of calculation [80]. (ii) The way to correct it is known. An alternative way to use state specific $6p$ orbitals in LS-coupling type of calculations is to include configuration interaction between $6s^2 6p^1 -{}^2 P^{o}$ and $6s^2 7p^1 -{}^2 P^{o}$ terms (the so called double-shell effect). This can be done in the free-ion Bi^{2+} because the symmetry prevents from mixing with the many other states intermediate between $6s^2 6p^1 -{}^2 P^{o}$ and $6s^2 7p^1 -{}^2 P^{o}$, which is more than 110000 cm^{-1} higher in energy. Doing it, effectively corrects the splitting and gives 20345 cm^{-1}, while it leaves the energies of all other states basically unaffected. (iii) Doing the same with the states of SrB_4O_7:Bi^{2+} derived from the $6s^2 7p^1 -{}^2 P^{o}$ term is not practical, or it would be very costly from the computational point of view, because the low symmetry puts together in the same symmetry block these high energy states with very many other states of lower energy and all of them would have to be handled at once. The overestimation of the C_s splitting is consistent with a lack of lattice relaxation in the calculations, because oxygen displacements away from the Bi^{2+} impurity would lower the splitting, and with a possible overestimation of repulsive effects in the SrB_4O_7 embedding potential.

The third excited state $4E_{1/2}$, used to excite the red-orange luminescence, is above 40000 cm^{-1} and has a larger oscillator strength than the first and second excited states. This state was traditionally assumed to have a $6s^2 7s -{}^2 S_{1/2}$ main character [67, 71, 73]. This way, its larger absorption oscillator strength from the ground state could be explained as a consequence of the $6p \rightarrow 7s$ transition being electric dipole allowed, whereas the $6p \rightarrow 6p$ character of the transitions to the two lower states makes them electric dipole forbidden. In the multiconfigurational ab initio calculations, the third excited state has a 63% dominant character of $6s^1 6p^2 -{}^4 P$, with the rest spread over several doublets of $6s^1 6p^2$, $6s^2 6d^1$, and $6s^2 7p^1$ configurations, and no relevant $6s^2 7s^1$ character. In this new interpretation, the third excitation above 40000 cm^{-1} is a configuration interaction mix of $6s \rightarrow 6p$, $6p \rightarrow 6d$, and $6p \rightarrow 7p$ excitations, rather than a $6p \rightarrow 7s$ transition and its larger absorption oscillator strength corresponds to partial $6s \rightarrow 6p$ and $6p \rightarrow 6d$ electric dipole allowed characters involved in the $1E_{1/2} \rightarrow 4E_{1/2}$ absorption.

The computed radiative lifetime of the emitting state of Bi^{2+}, $2E_{1/2}[^2P^o_{3/2}(1)]$, of 3.5 μs, is not far from the experimental values, which range between 10 and 13 μs (Fig 8.9b). The second exited state of Bi^{2+} has a similar one, but it would non-radiatively decay to the the first, which will be the only emitting state, and the low lying states Bi^{3+} lie out of this range of radiative lifetime (its $6p \rightarrow 6s$ emission would have $^3P_0 \rightarrow {}^1S_0$ components in the millisecond regime and $\mu^3P_1 + \nu^1P_1 \rightarrow {}^1S_0$ components in the nanosecond regime, as it is usually found in the luminescence of Bi^{3+} in solid hosts [81]); then, combination of energies and intensities rule out the consideration of Bi^{3+} as possible emitter and leave Bi^{2+} as the only computationally supported possibility.

In summary, there is an overall agreement between multiconfigurational ab initio calculations on SrB_4O_7:Bi^{2+} and experimental optical measurements on SrB_4O_7:Bi, which is within the range of what can be expected from this type of calculations. This agreement means a strong support to the initial hypothesis of Blasse et al. [67] of stable Bi^{2+} species as responsible of the red-orange luminescence of bismuth-doped SrB_4O_7 and provides a new electronic interpretation for its 40000 cm^{-1} excitation.

8.2.3 Pr-Doped $CaTiO_3$ and $CaZrO_3$

The Pr^{3+} ion doped in solid hosts can potentially give two $4f \rightarrow 4f$ emissions. They are respectively originated from the lowest state that results from the crystal-field splitting of the $^3P_0$, $^3P_1$, $^3P_2$, and $^1I_6$ levels of the $4f^2$ configuration of free Pr^{3+}, in one case, and of the $^1D_2$ level, in the other case. The two emission wavelengths are quite constant from host to host, because the splittings are very small, due to the strong inner character of the $4f$ orbitals (they are in fact molecular orbitals, but with a very dominant character of the Pr $4f$ atomic orbitals), which are very efficiently shielded from host fields by the more external $5s^2$ and $5p^6$ electronic shells. The two emissions are traditionally labeled with free Pr^{3+} notation, for simplicity, although, rigorously speaking, the emitting levels depend on the site symmetry of Pr^{3+} in the specific host and on the intensity of the local crystal-field. The more energetic of them is called $^3P_0 \rightarrow {}^3H_4$. Here, $^3H_4$ stands for the set of states derived from the splitting of the free-ion ground level, and $^3P_0$ is used because it is the lowest of the four free-ion levels involved in this emission; the actual emitting state usually has a strong mixed level character. This emission is also referred to as $^3P_0$ emission, greenish-blue emission, and faster emission (its lifetime is in the μs scale). The less energetic emission is referred to as $^1D_2 \rightarrow {}^3H_4$, $^1D_2$, red emission, and slower emission (its lifetime is in the ms scale). The experimental spectra and associated energy level scheme are summarized in Fig. 8.10.

The different colors and time scales of the two emissions opened a menu of application potentialities and set up an important goal in the field of inorganic phosphors: gaining control on the non-radiative decay between the $^3P_0$ and $^1D_2$ levels, which is what ultimately determines the luminescence of the phosphor material activated with Pr^{3+} [83, 84]. E.g., the $^3P_0$, faster emission is the target in the search for Pr^{3+}-doped

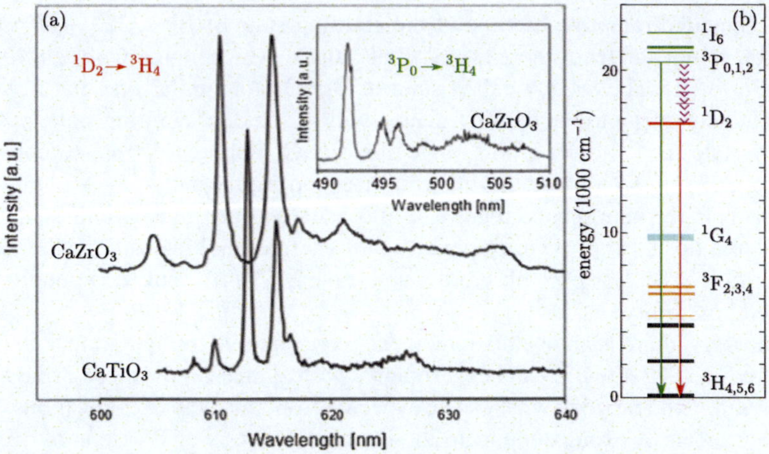

Fig. 8.10 a Experimental emission and upconversion (inset) spectra of $CaTiO_3$:Pr^{3+} and $CaZrO_3$:Pr^{3+}, measured upon 590.3 nm excitation and at 15 K. The inset shows the green $^3P_0 \rightarrow {}^3H_4$ emission which is observed for $CaZrO_3$:Pr^{3+} but not for $CaTiO_3$:Pr^{3+}. Adapted from [82]. **b** Dieke diagram of Pr^{3+} [24], showing the lowest levels of the $4f^2$ configuration of Pr^{3+} and the transitions underlying its red and green emissions. The possibility of non-radiative decay of the $^3P_0$ state via intermediate levels having different configurational character is represented in magenta

materials useful as scintillators for fast decay X-ray computed tomography [85] or for long lasting greenish-blue luminescent materials [86]. Instead, the $^1D_2$, slower emission is the necessary one when Pr-based phosphors for solid-state white illumination, lasers, or other applications are wanted.

Many studies have been devoted to an understanding of the mechanism of the $^3P_0$–$^1D_2$ non-radiative decay that could allow its mastering and, ultimately, the control over the Pr^{3+} luminescence. Combination of experimental observations and empirical modelling led historically to several interpretations: Cross-relaxation within Pr^{3+}–Pr^{3+} ionic pairs; multiphonon relaxation; intersystem crossing through low-lying $4f5d$ levels; and intermediate Pr-to-Metal charge transfer states. Experiments ruled out the cross-relaxation and the multiphonon relaxation models [83, 87]. The intersystem crossing model [83, 88], which was considered by some authors to be unlikely [87], found very large discrepancies with ab initio calculations on Lu_2O_3:Pr^{3+} [89]. The hypothesis of the existence of Pr-to-Metal charge transfer states with an energy intermediate between the $^3P_0$ and $^1D_2$ levels [82], also known as the virtual recharge model [90], was broadly accepted within the luminescence community, mostly because it could provide a reasonable explanation (the only reasonable explanation for a long while) to the fact that $CaTiO_3$:Pr^{3+} exhibits red $^1D_2 \rightarrow {}^3H_4$ luminescence, whereas it is the greenish-blue $^3P_0 \rightarrow {}^3H_4$ emission what dominates the luminescence of the related material $CaZrO_3$:Pr^{3+} [82, 87, 91]. In effect, according to this model, the $^3P_0$–$^1D_2$ non-radiative decay takes place through a mechanism consisting of: (1) an initial electron transfer from Pr^{3+} (in its $^3P_0$ state) to a host cation, which creates an intermediate metal-to-metal charge transfer state (MMCT), followed by (2) an electron transfer from the host cation back to Pr^{4+} leading to a Pr^{3+} in its $^1D_2$ state. Then, considering that the metal-to-metal electron transfer energy must be much smaller in the Pr^{3+}-to-Ti^{4+} case than in the Pr^{3+}-to-Zr^{4+} case, the formation of an intermediate Pr^{4+}–Ti^{3+} MMCT state with an energy between $^3P_0$ and $^1D_2$ in $CaTiO_3$:Pr^{3+} couldn't be ruled out, whereas the formation of an equivalent intermediate Pr^{4+}–Zr^{3+} MMCT state in $CaZrO_3$:Pr^{3+} is very unlikely.[4] According to these hypothetical configuration coordinate energy diagrams for the two luminescent materials, $CaTiO_3$:Pr^{3+} would have all the greenish-blue emission quenched and it would show red luminescence only, whereas both emissions could be simultaneously present in $CaZrO_3$:Pr^{3+}, all of it in agreement with the experiments. Further experimental studies [87, 91, 94–96] fitted within this interpretation, but they could not provide a direct proof of the MMCT nature of the intermediate states in $CaTiO_3$:Pr^{3+}.

A link between the hypothetical intermediate MMCT states and other, relevant experimental data hasn't been established. E.g., the efficiency of the red emission of $CaTiO_3$:Pr^{3+} can be improved with Ba and Sr substitutions for Ca [97]; it can be tuned by using different preparation methods that produce different charge compensations [98]; and it improves upon addition of Lu_2O_3, La_2O_3, and Gd_2O_3 [99]. Similarly, the photoluminescence and afterglow efficiency of $CaTiO_3$:Pr^{3+} was improved by Ca excess [100] and ammonia treatment [84]. None of these treatments is addressed at impacting the Pr-to-Ti charge transfer. All of them produce local defects in the material. And, although these defects might have some minor, indirect effects on the MMCT state energy, it is not clear how such small effects could provide a tool for the tunning of the red emission efficiency.

In the lack of experimental evidence, multiconfigurational ab initio embedded-cluster calculations could provide data that either supported or opposed to the MMCT hypothesis. As we describe below, the calculations performed in [101] couldn't support the MMCT or virtual recharge model. However, they led to a new hypothesis on the nature of the intermediate state never considered before and provided strong support to it. According to this new hypothesis, the intermediate state responsible for the non-radiative decay from $^3P_0$ to $^1D_2$ in $CaTiO_3$:Pr^{3+} is an O-to-Ti ligand-to-metal charge-transfer state (LMCT) of the host.

The $4f^2$ and $4f^15d^1$ ligand-field (LF) states of Pr^{3+} in $CaTiO_3$ have been calculated in two-step spin-orbit coupling SA-CASSCF/MS-CASPT2/RASSI-SO DKH calculations on the $(PrO_{12})^{21-}$ embedded-cluster (Fig 8.11b). Additional equivalent calculations on the $(PrO_{12})^{20-}$ cluster provide ionization data.[5] Results on these two

---

[4] The intermediate state was initially labelled Inter-Valence Charge Transfer state (IVCT) and the model was sometimes referred to as the IVCT model, but this name is conventionally reserved accortiong to IUPAC for the MMCT when the two metals differ only in the oxidation state [92, 93], so that, in this case, MMCT is the right conventional term.

[5] In the SA-CASSCF calculations, a CAS space made of 2 electrons in 13 molecular orbitals (1 electron in 13 MOs for Pr^{4+}) of main character Pr-$4f$, Pr-$5d$, and Pr-$6s$ (1 a_{2u}, 3 t_{1u}, 3 t_{2u}, 2 e_g, 3 t_{2g}, and 1 a_{1g}) was used, together with the following state-averages for the molecular orbital optimizations: For the $4f^2$ configuration of Pr^{3+}, 1 $^3A_{2g}$ and 1 ^3E_g levels; 4 $^3T_{1g}$ and 2 $^3T_{2g}$; 3 $^1A_{1g}$, 1 $^1A_{2g}$, and 3 ^1E_g; and 2 $^1T_{1g}$ and 4 $^1T_{2g}$. For the $4f5d$ and $4f6s$ configurations of Pr^{3+},

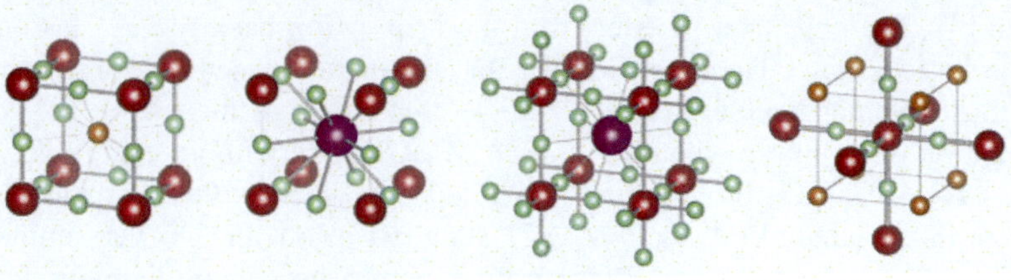

Fig. 8.11 **a** $CaTiO_3$ unit cell. **b** $(PrO_{12})^{21-}$ cluster with neibohring Ti atoms. **c** $(PrO_{12}Ti_8O_{24})^{37-}$ cluster. **d** $(TiO_6Ca_8Ti_6)^{32+}$ cluster. Ca atoms are shown in gold, Ti in red, O in green, and Pr in violet

clusters combined with equivalent calculations on the $(TiO_6)^{8-}$ and $(TiO_6)^{9-}$ clusters give the Pr^{3+}-to-Ti^{4+} MMCT states of Pr^{3+}/Ti^{4+} pairs at the diabatic level [103, 104] (cf. Sect. 7.4). The ground and MMCT states of the Pr^{3+}/Ti^{4+} pair have also been calculated at SA-CASSCF/MS-CASPT2 level in the $(PrO_{12}Ti_8O_{24})^{37-}$ cluster (Fig. 8.11c) with the ground state equilibrium structure, which provided adiabatic corrections that have been added at all geometries.

The energies of the states of the Pr^{3+}-Ti^{4+} and Pr^{4+}-Ti^{3+} configurations of Pr^{3+}/Ti^{4+} pairs in $CaTiO_3$ are shown in Fig. 8.12b. The states of the Pr^{3+}-Ti^{4+} configuration are the LF states of Pr^{3+} combined with the ground state of Ti^{4+}. The states of the Pr^{4+}-Ti^{3+} configuration are the MMCT states. In Fig. 8.12b, they are represented across the MMCT coordination coordinate defined in Fig. 8.12a by the brown line, which is the electron transfer reaction coordinate, i.e. the simultaneous nuclear displacements that accompany an electron transfer from Pr^{3+} to Ti^{4+} (from left to right) and from Ti^{3+} to Pr^{4+} (from right to left).

It is clear in Fig. 8.12b that the Pr^{4+}-Ti^{3+} or MMCT energy levels are too high in energy and cannot play a relevant role in the ${}^3P_0$–${}^1D_2$ non-radiative decay according to these results. For the MMCT states to be intermediate between the ${}^3P_0$ and ${}^1D_2$ levels, they should go down by around 25000 cm^{-1} and this is an amount too large to be ascribable to the intrinsic inaccuracies of the theoretical method. In other words, these multiconfigurational ab initio embedded-cluster calculations do not support the virtual recharge model or MMCT model to explain such decay, but rather question it seriously.

Interestingly, an analysis of the structural dependences of the local electronic states of the doped material $CaTiO_3$:Pr^{3+} indicates that O^{2-}-to-Ti^{4+} LMCT states of the host could play a role in the dopant decay. In effect, as is shown in Fig. 8.12a,

1 ${}^3A_{1u}$, 2 ${}^3A_{2u}$, and 3 ${}^3E_u$ levels; 6 ${}^3T_{1u}$ and 5 ${}^3T_{2u}$; 1 ${}^1A_{1u}$, 2 ${}^1A_{2u}$, and 3 ${}^1E_u$; and 6 ${}^1T_{1u}$ and 5 ${}^1T_{2u}$. For the 4f configuration of Pr^{4+}, 1 ${}^2A_{2u}$ level; and 1 ${}^3T_{1u}$ and 1 ${}^3T_{2u}$ levels. For the 5d and 6s configurations of Pr^{4+}, 1 ${}^2E_g$ and 1 ${}^2A_{1g}$; and 1 ${}^2T_{2g}$. In the MS-CASPT2 calculations all valence electrons were correlated except those with main character Pr-4d and the standard IPEA value of 0.25 au was used [102]. The RASSI-SO calculations included the spin-orbit couplings between all spin singlet and triplet (spin-orbit free) states in the case of Pr^{3+}, and all spin doublets in the case of Pr^{4+}.

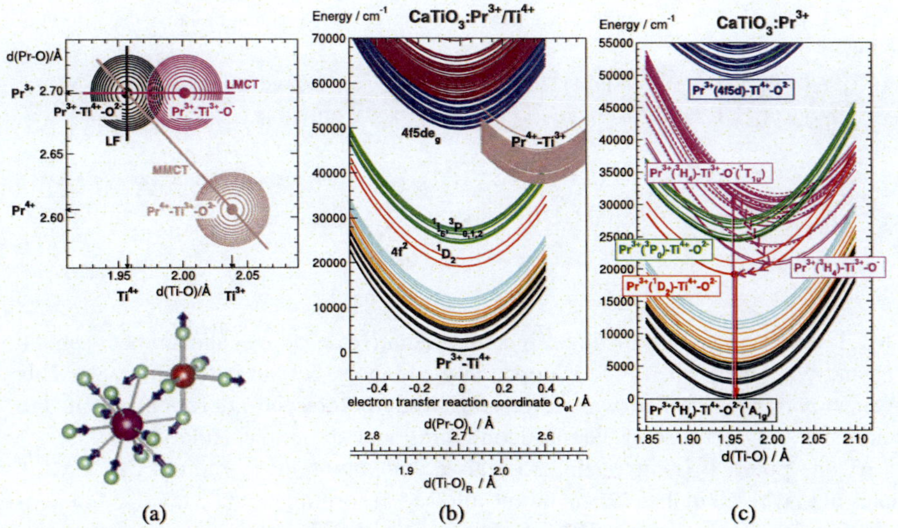

Fig. 8.12 **a** Isoline representation of the potential energy surfaces of three local states of the material $CaTiO_3$:Pr^{3+} as functions of the Pr–O and Ti–O distances. One ligand-field (LF) state of the $4f^2$ configuration of Pr^{3+} in $CaTiO_3$ is shown in black, one Pr^{3+}-to-Ti^{4+} MMCT state in brown, and one O^{2-}-to-Ti^{4+} LMCT state in magenta. Configuration coordinates appropriate for one-dimension energy diagrams are indicated with lines: for LF Pr^{3+} states in black (LF coordinate), for MMCT states of the pair Pr^{3+}/Ti^{4+} in brown (MMCT coordinate), and for O^{2-}-to-Ti^{4+} LMCT states in magenta (LCMT coordinate). Below: MMCT configuration coordinate. Pr is shown in violet, Ti in red, O in green. **b** LF states of Pr^{3+} and Pr^{3+}-to-Ti^{4+} MMCT states across the MMCT configuration coordinate. **c** LF states of Pr^{3+} and O^{2-}-to-Ti^{4+} LMCT states across the LMCT configuration coordinate

the LF states of Pr^{3+} and the LMCT states of $CaTiO_3$ have an offset across the LMCT configuration coordinate (which is just the Ti-O distance), and this offset indicates that the interplay between these sets of states across such coordinate might be physically relevant, specially for the decay mechanisms, and must be explored.

In order to study how the Pr^{3+} $4f^2$ manifold and the O-to-Ti LMCT states of $CaTiO_3$ cross each other across the LMCT coordinate, the LMCT states have been obtained in SA-CASSCF/MS-CASPT2 calculations on the $(TiO_6Ca_8Ti_6)^{32+}$ cluster (Fig. 8.11d) embedded in $CaTiO_3$. They are discussed together with the LMCT states of $CaZrO_3$ in Sect. 7.1 (cf. Fig. 7.1 and Table 7.1), where the high reliability of the results was discussed by comparing them with experimental interband data. The relative energies of the Pr^{3+} $4f^2$ states are basically insensitive to Ti-O distance changes.

The $CaTiO_3$ LMCT states and the Pr^{3+} LF states are shown together in the configuration coordiate energy diagram of Fig. 8.12c. The Figure shows that the lowest states of the LMCT manifold of $CaTiO_3$ lay in the same energy range than the $^3P_0$ and $^1D_2$ levels of Pr^{3+}. On the contrary, the LMCT manifold of $CaZrO_3$ starts at much higher energy, as it is clear in Table 7.1. As a consequence of this, the host

Table 8.4 Electron-hole recombination energy transfer mechanism of non-radiative decay to Pr^{3+} $^1D_2$ in $CaTiO_3$:Pr^{3+}

$Pr^{3+}(^3H_4) + [Ti^{4+}\text{-}O^{2-}] \xrightarrow{+h\nu_{abs}} Pr^{3+}(^3H_4) + [Ti^{3+}\text{-}O^{-}]^*$	(Allowed absorption)
$Pr^{3+}(^3H_4) + [Ti^{3+}\text{-}O^{-}]^* \rightsquigarrow Pr^{3+}(^3H_4) + [Ti^{3+}\text{-}O^{-}]$	(Electronic relaxation, atomic reorganization)
$Pr^{3+}(^3H_4) + [Ti^{3+}\text{-}O^{-}] \rightsquigarrow Pr^{3+}(^1D_2) + [Ti^{4+}\text{-}O^{2-}]$	(e-h recombination energy transfer)
$Pr^{3+}(^1D_2) + [Ti^{4+}\text{-}O^{2-}] \xrightarrow{-h\nu_{em}} Pr^{3+}(^3H_4) + [Ti^{4+}\text{-}O^{2-}]$	(Red emission)

LMCT states can be intermediates in a non-radiative decay mechanism between the $^3P_0$ and $^1D_2$ levels when Pr^{3+} is doped in the titanate, but not in the zirconate. This is an important issue that was never considered before and is brought about into discussion only because of the multiconfigurational ab initio calculations.

At this point, it is interesting to realize and important to remark that LMCT states not involved in the electric dipole allowed absorptions can be involved in non-radiative decays. In $CaTiO_3$:Pr^{3+}, a set of such LMCT states exist starting 8000-9000 cm^{-1} below the optically allowed states. Some of them, with a main one-electron character $t_{1g}(\text{O-}2p_\pi) \rightarrow t_{2g}(\text{Ti}-3d)$, lie between the $^3P_0$ to $^1D_2$ levels of Pr^{3+}.

The lowest LMCT states of $CaTiO_3$ can then provide an efficient non-radiative decay mechanism to selectively populate $^1D_2$ red emission and totally quench $^3P_0$ greenish-blue emission. According to this mechanism (Table 8.4), an allowed O^{2-}-to-Ti^{4+} charge transfer absorption at around 30000 cm^{-1} is followed by structural reorganization around Ti^{3+} (basically an elongation of the Ti–O distances) and electronic relaxation down to the lowest LMCT state. The reorganization and relaxation energies together amount approximately 10000 cm^{-1} and the lowest LMCT state lies below $^3P_0$ and slightly above $^1D_2$. This way, an electron back-transfer from Ti^{3+}-to-O^- takes place and its energy is non-radiatively transferred to the $^1D_2$ level of Pr^{3+}, which yields red emission. Using a different language, more common from a periodic boundary conditions point of view, in this mechanism, after an interband or valence-to-conduction band excitation (or any other excitation that decayed to the conduction band, such as UV or X-ray excitations), and relaxation of the exciton around a Titanium atom, electron and hole recombine and the recombination energy is transferred to Pr^{3+}, which ends up excited in the $^1D_2$ state. From this, the Pr^{3+} $^1D_2 \rightarrow {}^3H_4$ red emission takes place.

In $CaZrO_3$:Pr^{3+}, however, all the O^{2-}-to-Zr^{4+} LMCT states are much higher in energy than the $^3P_0$ and $^1D_2$ levels of Pr^{3+} and they cannot play any role in their selective population. Then, the usual non-radiative decay mechanisms in Pr^{3+} materials lead to partial populations of $^3P_0$ and $^1D_2$, and the luminescence is a superposition of both emissions, greenish-blue and red.

The electron-hole recombination energy transfer decay mechanism to Pr^{3+} $^1D_2$ just described is entirely indicated by the multiconfigurational ab initio calculations. It explains the strong enhancement of the red luminescence in $CaTiO_3$:Pr^{3+} with respect to $CaZrO_3:Pr^{3+}$, but it needs to be confirmed or rejected by further experiments, as it happens with all theoretical statements. Regarding additional experiments presently available, it was remarked above that the efficiency of the red emission of $CaTiO_3:Pr^{3+}$ was found to be tunable with the preparation method, something apparently related with the production of local defects. Local defects should have a very minor effect on the $^1D_2$ energy, but those close to LMCT local states are expected to modulate their energy, because these states are sensitive to electric field changes. Then, if the key of the electron-hole recombination energy transfer decay mechanism is the resonance between the Pr^{3+} $^1D_2$ level and the lowest host LMCT states, it is not strange that local defects have an impact on the red luminescence of $CaTiO_3:Pr^{3+}$. In order to check this computationally, the LMCT states of $CaTiO_3$ have been calculated with a Ca vacancy next to Ti, which is one of the common defects [87], known to produce a shoulder in the diffuse reflectance spectra of $CaTiO_3:Pr^{3+}$ at slightly lower energy than the interband absorption and to play an important role in the enhancement of its red emission [99]. In the calculations, the Ca vacancy lowered the LMCT states 2400 cm^{-1}. This agrees with the shoulder observation and, more importantly, definitely indicates that local defects can impact the lowest LMCT states enough to be able to modulate their resonance with the $^1D_2$ level. Then, according to the mechanism of host-to-impurity energy transfer, the defects can be used to tune the population of the Pr^{3+} $^1D_2$ level and to enhance the red luminescence.

References

1. S. Nakamura, G. Fasol, *The Blue Laser Diode: GaN Based Light Emitters and Lasers* (Springer, Berlin, 1997)
2. E.F. Schubert, J.K. Kim, H. Luo, J.Q. Xi, Rep. Prog. Phys. **69**(12), 3069 (2006)
3. P.F. Smet, A.B. Parmentier, D. Poelman, J. Electrochem. Soc. **158**(6), R37 (2011)
4. P.F. Smet, J.J. Joos, Nat. Mater. **16**(5), 500 (2017)
5. Z. Xia, A. Meijerink, Chem. Soc. Rev. **46**, 275 (2017)
6. J. Gracia, L. Seijo, Z. Barandiarán, D. Curulla, H. Niemansverdriet, W. van Gennip, J. Lumin. **128**, 1248 (2008)
7. G. Blasse, A. Bril, J. Chem. Phys. **47**, 5139 (1967)
8. L. Seijo, J. Chem. Phys. **102**, 8078 (1995)
9. R. Llusar, M. Casarrubios, Z. Barandiarán, L. Seijo, J. Chem. Phys. **105**, 5321 (1996)
10. A.B. Muñoz-García, L. Seijo, Phys. Rev. B **82**, 184118 (2010)
11. D. Koningsberger, R. Prins, *X-Ray Absorption: Principles, Applications, Techniques of EXAFS, SEXAFS and XANES* (Wiley, 1988)
12. B. Ravel, M. Newville, J. Synchrotron Radiat. **12**, 537 (2005)
13. P. Ghigna, S. Pin, C. Ronda, A. Speghini, F. Piccinelli, M. Bettinelli, Opt. Mater. **34**, 19 (2011)
14. A.B. Muñoz-García, J.L. Pascual, Z. Barandiarán, L. Seijo, Phys. Rev. B **82**, 064114 (2010)
15. R.D. Shannon, Acta Crystallogr. A **32**, 751 (1976)
16. L. Seijo, Z. Barandiarán, Int. J. Quantum Chem. **60**, 617 (1996)
17. F. Euler, J.A. Bruce, Acta Crystallogr. **19**, 971 (1965)
18. G.J. Zhao, X.H. Zeng, J. Xu, S.M. Zhou, Y.Z. Zhou, Phys. Status Solidi A **199**, 355 (2003)
19. Y. Zorenko, V. Gorbenko, I. Konstankevych, A. Voloshinovskii, G. Stryganyuk, V. Mikhailin, V. Kolobanov, D. Spassky, J. Lumin. **114**, 85 (2005)
20. V. Bachmann, C. Ronda, A. Meijerink, Chem. Mater. **21**, 2077 (2009)

21. L. Seijo, Z. Barandiarán, J. Chem. Phys. **115**, 5554 (2001)
22. D. Piehler, W.K. Kot, N. Edelstein, J. Chem. Phys. **94**, 942 (1991)
23. N. Edelstein, W.K. Kot, J.C. Krupa, J. Chem. Phys. **96**, 1 (1992)
24. G.H. Dieke, *Spectra and Energy Levels of Rare Earth Ions in Crystals* (Interscience Publishers, New York, 1968)
25. J.M. Ogieglo, A. Zych, K.V. Ivanovskikh, T. Jüstel, C.R. Ronda, A. Meijerink, J. Phys. Chem. A **116**, 8464 (2012)
26. G.F. Herrmann, J.J. Pearson, K.A. Wickersheim, R.A. Buchanan, J. Appl. Phys. **37**, 1312 (1966)
27. L. Seijo, Z. Barandiarán, Phys. Chem. Chem. Phys. **16**, 3830 (2014)
28. Y.S. Lin, R.S. Liu, B.M. Cheng, J. Electrochem. Soc. **152**, J41 (2005)
29. Y.X. Pan, W. Wang, G.K. Liu, S. Skanthakumar, R.A. Rosenberg, X.Z. Guo, K.K. Li, J. Alloys Compd. **488**, 638 (2009)
30. J. Brodrick, J. Disp. Technol. **3**, 91 (2007)
31. A.B. Muñoz-García, L. Seijo, J. Phys. Chem. A **115**, 815 (2011)
32. W.W. Holloway, M. Kestigian, J. Opt. Soc. Am. **59**, 60 (1969)
33. T.Y. Tien, E.F. Gibbons, R.G. DeLosh, P.J. Zacmanidis, D.E. Smith, H.L. Stadler, J. Electrochem. Soc. **120**, 278 (1973)
34. J.M. Robertson, M.W. van Tol, W.H. Smits, J.P.H. Heynen, Philips J. Res. **36**, 15 (1981)
35. Y. Pan, M. Wu, Q. Su, J. Phys. Chem. Solids **65**, 845 (2004)
36. C.K. Jörgensen, Prog. Inorg. Chem. **4**, 73 (1962)
37. S. Sugano, Y. Tanabe, H. Kamimura, *Multiplets of Transition-Metal Ions in Crystals* (Academic, New York, 1970)
38. B.R. Judd, Phys. Rev. Lett. **39**, 242 (1977)
39. C.A. Morrison, J. Chem. Phys. **72**, 1001 (1980)
40. V. Pankratov, L. Grigorjeva, D. Millers, T. Chudoba, Radiat. Meas. **42**, 679 (2007)
41. S.P. Feofilov et al., J. Lumin. (2012). https://doi.org/10.1016/j.jlumin.2012.06.030
42. Y. Zorenko, T. Zorenko, V.V. Gorbenko, T. Voznyak, V. Savchyn, P. Bilski, A. Twardak, Opt. Mater. **34**, 1314 (2012)
43. J. Dong, K. Lu, Phys. Rev. B **43**, 8808 (1991)
44. F.A. Selim, D. Solodovnikov, M.H. Weber, K.G. Lynn, Appl. Phys. Lett. **91**, 104105 (2007)
45. A.B. Muñoz-García, Z. Barandiarán, L. Seijo, J. Mater. Chem. **22**, 19888 (2012)
46. A.B. Muñoz-García, E. Artacho, L. Seijo, Phys. Rev. B **80**, 014105 (2009)
47. A.A. Setlur, W.J. Heward, Y. Gao, A.M. Srivastava, R.G. Chandran, M.V. Shankar, Chem. Mater. **18**, 3314 (2006)
48. Y. Shimomura, T. Honma, M. Shigeiwa, T. Akai, K. Okamoto, N. Kijima, J. Electrochem. Soc. **154**, J35 (2007)
49. L. Seijo, Z. Barandiarán, Opt. Mater. **35**, 1932 (2013)
50. Y. Sato, R. Geanioz, J. Geophys. Res. **86**, 11773 (1981)
51. J. Ganguly, W. Cheng, H.S.C. O'Neill, Am. Mineral. **78**, 583 (1993)
52. T. Armbruster, C.A. Geiger, G.A. Lager, Am. Mineral. **77**, 512 (1992)
53. C.A. Geiger, T. Armbruster, Am. Mineral. **82**, 740 (1997)
54. S. Quartieri, R. Oberti, M. Boiocchi, M.C. Dalconi, F. Boscherini, O. Safonova, A.B. Woodland, Am. Mineral. **91**, 1240 (2006)
55. B. Etschmann, V. Streltsov, N. Ishizawa, E.N. Maslen, Acta Crystallogr. B **57**, 136 (2001)
56. A. Nakatsuka, A. Yoshiasa, S. Takeno, Acta Crystallogr. B **51**, 737 (1995)
57. G. Patzke, R. Wartchow, M. Binnewies, Z. Kristallogr. **214**, 143 (1999)
58. H. Sawada, J. Solid State Chem. **132**, 300 (1997)
59. J.L. Wu, G. Gundiah, A.K. Cheetman, Chem. Phys. Lett. **441**, 250 (2007)
60. M. Bettinelli, R. Moncorgé, J. Lumin. **92**, 287 (2001)
61. P. Dorenbos, J. Lumin. **87–89**, 970 (2000)
62. I.B. Bersuker, *The Jahn-Teller Effect and Vibronic Interactions in Modern Chemistry* (Plenum Press, New York and London, 1984)
63. Q.M. Phung, Z. Barandiarán, L. Seijo, Theor. Chim. Acc. **134**, 37 (2015)

64. L. Seijo, Z. Barandiarán, Phys. Chem. Chem. Phys. **15**, 19221 (2013)
65. L. Seijo, Z. Barandiarán, RSC Adv. **6**, 25741 (2016)
66. Z. Pan, W. Li, Y. Xu, Q. Hu, Y. Zheng, RSC Adv. **6**, 20458 (2016)
67. G. Blasse, A. Meijerink, M. Nomes, J. Zuidema, J. Phys. Chem. Solids **55**, 171 (1994)
68. M.A. Hamstra, H.F. Folkerts, G. Blasse, J. Mater. Chem. **4**, 1349 (1994)
69. L. de Boisbaudran, C.R. Acad, Sci. Paris **103**, 629 (1886)
70. F.A. Kröger, J.T.G. Overbeek, J. Goorissen, J. van den Boomgaard, Trans Electrochem. Soc. **96**, 132 (1949)
71. A.M. Srivastava, J. Lumin. **78**, 239 (1998)
72. M. Peng, L. Wondraczek, Opt. Lett. **35**, 2544 (2010)
73. M. Peng, J. Lei, L. Li, L. Wondraczeki, Q. Zhang, J. Qiu, J. Mater. Chem. C **1**, 5303 (2013)
74. L. Li, J. Cao, B. Viana, S. Xu, M. Peng, Inorg. Chem. **56**, 6499 (2017)
75. B. Wahl, M. Ruck, Z. Anorg, Allg. Chem. **634**, 2873 (2008)
76. M. de Jong, A. Meijerink, R.A. Gordon, Z. Barandiarán, L. Seijo, J. Phys. Chem. C **118**, 9696 (2014)
77. J. Krogh-Moe, Acta Chem. Scand. **18**, 2055 (1964)
78. B.O. Roos, R. Lindh, P.A. Malmqvist, V. Veryazov, P.O. Widmark, J. Phys. Chem. A **109**, 6575 (2005)
79. C.E. Moore, *Atomic Energy Levels as Derived from the Analysis of Optical Spectra* (National Bureau of Standards, Washington, D. C., 1952). Circular 467, Vols 2 and 3
80. T. Fleig, J. Olsen, C.M. Marian, J. Chem. Phys. **114**, 4775 (2001)
81. A. Wolfert, G. Blasse, J. Solid State Chem. **59**, 133 (1985)
82. E. Pinel, P. Boutinaud, R. Mahiou, J. Alloys Compd. **380**, 225 (2004)
83. C. DeMello-Donegá, A. Meijerink, G. Blasse, J. Phys. Chem. Solids **56**, 673 (1995)
84. S. Yoon, E.H. Otal, A.E. Maegli, S.K. Matam, S. Riegg, S.G. Ebbinghaus, J.C. Fallas, H. Hagemann, B. Walfort, S. Pokrant, A. Weidenkaff, Opt. Mater. Express **3**, 248 (2013)
85. G.C. Aumüller, W. Köstler, B.C. Grabmaier, R. Frey, J. Phys. Chem. Solids **55**, 767 (1994)
86. L.L. Noto, M.L. Chithambo, O.M. Ntwaeaborwa, H.C. Swart, J. Alloys Compd. **589**, 88 (2014)
87. P. Boutinaud, E. Pinel, M. Dubois, A.P. Vink, R. Mahiou, J. Lumin. **111**, 69 (2005)
88. P.T. Diallo, P. Boutinaud, R. Mahiou, J.C. Cousseins, Phys. Status Solidi A **160**, 255 (1997)
89. J.L. Pascual, Z. Barandiarán, L. Seijo, Theor. Chim. Acc. **129**, 545 (2011)
90. E.G. Reut, A.I. Ryskin, Phys. Status Solidi A **17**, 47 (1973)
91. P. Boutinaud, R. Mahiou, E. Cavalli, M. Bettinelli, Chem. Phys. Lett. **418**, 185 (2006)
92. J.W. Verhoeven, Pure Appl. Chem. **68**, 2223 (1996)
93. IUPAC (2021) Compendium of Chemical Terminology, 2nd ed. (the "Gold Book"). Compiled by A.D. McNaught, A. Wilkinson, Blackwell Scientific Publications, Oxford (1997). Online version (2019-) created by S.J. Chalk. ISBN 0-9678550-9-8. https://doi.org/10.1351/goldbook.I03125. Accessed on 2021-06-08
94. P. Boutinaud, R. Mahiou, E. Cavalli, M. Bettinelli, J. Lumin. **122–123**, 430 (2007)
95. E. Pinel, P. Boutinaud, R. Mahiou, J. Alloys Compd. **740**, 165 (2004)
96. P. Boutinaud, E. Pinel, M. Oubaha, R. Mahiou, E. Cavalli, M. Bettinelli, Opt. Mater. **28**, 9 (2006)
97. X. Wang, C.N.N. Xu, H. Yamada, Jpn. J. Appl. Phys. **44**, L912 (2005)
98. D. Haranath, A.F. Khan, C. H., J. Phys. D: Appl. Phys. **39**, 4956 (2006)
99. X. Zhang, J. Zhang, X. Zhang, L. Chen, Y. Luo, X. Wang, Chem. Phys. Lett. **434**, 237 (2007)
100. E.H. Otal, A.E. Maegli, N. Vogel-Schäuble, B. Walfort, H. Hagemann, S. Yoon, A. Zeller, A. Weidenkaff, Opt. Mater. Express **2**, 405 (2012)
101. Z. Barandiarán, M. Bettinelli, L. Seijo, J. Phys. Chem. Lett. **8**, 3095 (2017)
102. G. Ghigo, B.O. Roos, P.A. Malmqvist, Chem. Phys. Lett. **396**, 142 (2004)
103. Z. Barandiarán, L. Seijo, J. Chem. Phys. **143**, 144702 (2015)
104. Z. Barandiarán, A. Meijerink, L. Seijo, Phys. Chem. Chem. Phys. **17**, 19874 (2015)

Chapter 9
Fundamental Spectroscopic Studies

Abstract The excited manifolds of luminescent materials activated with lanthanide ions and their associated spectroscopies range in complexity from the simplest case of Ce^{3+}-doped materials, where only one unpaired electron plays the relevant roles, to the much more complex case of Eu^{2+}-doped materials, where many electrons are active in the open-shells and create a variety of problems and opportunities. In this Chapter, a multiconfigurational ab initio approach to fundamental spectroscopic studies of materials with different levels of complexity is discussed.

9.1 Ce^{3+} in Elpasolites, Garnets, and Other Hosts

The single unpaired electron of Ce^{3+}, which occupies the $4f$ shell in the ground state and low-lying excited states of luminescent materials and is excited to split $5d$ shells to yield luminescence, makes this activator specially appropriate for fundamental studies, both experimental and theoretical: Understanding the one-electron case is a necessary condition to deep inside the complexities of many-electron problems. Moreover, when activators occupy high-symmetry sites many spectroscopic complexities inherent to local distortions are removed and only the ones related with the electronic structure are on focus. Since understanding the electronic problem is the first step towards understanding the spectroscopy and controlling its applications, it is not strange that high-symmetry Ce^{3+}-doped materials took a central place in fundamental studies of lanthanide activated phosphors. Ce^{3+}-doped elpasolites belong to this category. This section is focused on them, particularly on the spectroscopy of Cs_2NaYCl_6:Ce^{3+} [1, 2], and $Cs_2LiLuCl_6$:Ce^{3+} [3]. The O_h high-symmetry site of Ce^{3+} in these materials allows to perform high-resolution experimental spectroscopic studies and pinpoint central electronic issues, on one hand, and to give benchmarks for theoretical calculations, on the other. The performance of a theoretical method on these materials indicates its potential in the field, i.e. what can be expected and what cannot in lanthanide-activated luminescent materials in general.

On the other hand, Ce^{3+}-doped garnets, where the activator occupies lower D_2 symmetry sites, have utmost practical importance because they are basic in the development of solid-state-lighting phosphor materials. If multiconfigurational ab initio

Z. Barandiarán et al., *Luminescent Materials*, Springer Series in Materials Science 322,
https://doi.org/10.1007/978-3-030-94984-6_9

methods perform well in Ce^{3+}-doped elpasolites, they might also be useful in Ce^{3+}-doped garnets, where the central electronic problem is enriched with geometrical distortions and more subtle activator-host interactions. Special attention is paid to them in Sects. 8.1 and 8.2.1.

In addition, a number of multiconfigurational ab initio studies of lower symmetry Ce^{3+} active sites in different hosts exist. Among them, the calculations of Ce^{3+} in $LuAlO_3$, YF_3, LaF_3, $KMgF_3$, $LiYF_4$, K_2YF_5, and KY_3F_{10} [4]. E.g., those in $LuAlO_3$:Ce^{3+} [5], as it was the case of Ce^{3+}-doped garnets [6], were shown to predict well the $4f \rightarrow 5d$ absorption onset and the absorption spectrum in general. The study of the Ce^{3+} defect locally compensated with an interstitial fluoride ion in CaF_2:Ce^{3+} [7] revealed not only a good agreement with experimental energies and relative intensities of the $4f \rightarrow 5d$ transitions, but also led to a re-interpretation of its low-temperature absorption spectrum by attributing closely-spaced features to non-equivalent tetragonal and monoclinic defects clusters.

Ce^{3+}-Doped Elpasolites

An early multiconfigurational ab initio calculation on the O_h site of Ce^{3+} in Cs_2NaYCl_6 was performed in [1] in order to confront the method against highest-quality spectroscopy experiments and, eventually, either support or oppose tentative interpretations of emission bands. A more recent calculation on $Cs_2LiLuCl_6$:Ce^{3+} [8] was performed in the framework of the search for an interpretation of the Ce^{3+} anomalous emission and, although it used slightly different conditions following methodological evolutions based on experience, the quality of the results was basically the same. Incidentally, the nature of the Ce^{3+}-anomalous emission, which was understood after [8] as a Ce^{3+}-to-Ce^{4+} inter-valence charge transfer emission, is discussed in Sect. 11.2.2. Table 9.1 summarizes the most spectroscopically relevant results obtained for the $4f^1$, $5d^1$, and ITE states of these materials (cf. Sects. 6.1.1, 6.2.1, and 6.3.1), together with experimental data measured and analyzed in [1–3].

In Cs_2NaYCl_6:Ce^{3+}, the $4f \rightarrow 4f$ transitions are reproduced with a root-mean square deviation of 260 cm^{-1} compared with experiment. This is larger than in empirical crystal-field theory calculations where multiple parameters are optimized to minimize it. To lower it significantly without resorting to empirical parameters, it would most probably be necessary to increase the methodological demands to a very large extent, perhaps even unaffordable and definitely too expensive, specially having in mind that the $4f^N$ manifold is the one where CFT performs without serious limitations. A more detailed analysis of these differences indicates that a slight over-stabilization of the $4fa_{2u}$ molecular orbital is the main responsible for them (the deviation goes down to 80 cm^{-1} for the differences between Γ_{8u} and Γ_{6u} states, which do not depend on $4fa_{2u}$), which implies that these calculations seem to overestimate the effective ligand field.

The $5dt_{2g}$ Γ_{8g} and Γ_{7g} states are 2700 cm^{-1} too low, which can be considered a reasonable result for demanding $4f \rightarrow 5d$ transitions, and the $5de_g$ Γ_{8g} state is in good

Table 9.1 Electronic origins of Ce^{3+} in Cs_2NaYCl_6 and excitation and emission peaks of $Cs_2LiLuCl_6{:}Ce^{3+}$.All energies in cm^{-1}

$Cs_2NaYCl_6{:}Ce^{3+}$				$Cs_2LiLuCl_6{:}Ce^{3+}$		
		exp. [1, 2]	theo. [1]		exp. [3]	theo. [8]
Energy level				Excitation spectrum peaks		
$4f\,^2F_{5/2}$	Γ_{7u}	0	0	$4f\,^2F_{5/2}\,\Gamma_{7u} \rightarrow$		
	Γ_{8u}	599	831	$\rightarrow 5dt_{2g}\,^2T_{2g}\,\Gamma_{8g}, \Gamma_{7g}$	<30000	24500–25700
$4f\,^2F_{7/2}$	Γ_{7u}	2125[a]	2318	$\rightarrow 5dt_{2g}\,^2E_{2g}\,\Gamma_{8g}$ [b]	46190, 47600	45300, 46900
	Γ_{8u}	2689	3019	$\rightarrow \phi_{ITE,a_{1g}}$	57500[c]	59200
	Γ_{6u}	3086	3376			
$5dt_{2g}\,^2T_{2g}$	Γ_{8g}	28193	25510	Emission spectrum peaks		
	Γ_{7g}	29435	26716	$5dt_{2g}\,^2T_{2g}\,\Gamma_{8g} \rightarrow$		
$5de_g\,^2E_g$	Γ_{8g}	47080[a]	47263	$\rightarrow 4f\,^2F_{5/2}, ^2F_{7/2}$	26900, 24700	23300–20200

[a]From $Cs_2NaErCl_6{:}Ce^{3+}$ [1]
[b]Jahn-Teller split levels
[c]Tentatively attributed to free excitations in [3]; interpreted as excitations to impurity-trapped-excitons after [8]

agreement with experiments. Such an agreement is probably the result of an error compensation between overestimations of the $4f \rightarrow 5d$ barycenter and the effective ligand field, but even if the deviation was somehow larger, the result indicates the $5de_g\,\Gamma_{8g}$ state is around 46000–48000 cm^{-1} above the ground state. This is a strong support for the tentative assignment of the 47080 cm^{-1} emission as a $5de_g \rightarrow 4f$ emission [1].

In $Cs_2LiLuCl_6{:}Ce^{3+}$, two $4f \rightarrow 5dt_{2g}\,\Gamma_{8g}$, Γ_{7g} absorption bands are calculated to overlap into a complex band extending in the range 24500–25700 cm^{-1}. Experimentally, the $4f \rightarrow 5dt_{2g}$ excitation gives a broad band below 30000 cm^{-1} with a non-reported maximum. The computed $5dt_{2g}\,\Gamma_{8g} \rightarrow 4f$ emission gives two complex broad bands extending between 20200 cm^{-1} and 23350 cm^{-1}. The first band, due to the three splitting components of $4f\,^2F_{7/2}$, peaks at 20200–21000 cm^{-1} and the second, due to the two splitting components of $4f\,^2F_{5/2}$, peaks at 22750–23350 cm^{-1}. The corresponding experimental emission is made of two bands peaking at 24700 cm^{-1} and 26900 cm^{-1}. The $4f \rightarrow 5de_g$ absorption, which is experimentally found as a Jahn-Teller split band with peaks at 46190 cm^{-1} and 47600 cm^{-1}, is theoretically calculated to peak at 45300 cm^{-1} and 46900 cm^{-1} after consideration of D_{4h} deformation derived from $^2E_g \otimes e_g$ static Jahn-Teller coupling. The $5de_g\,\Gamma_{8g} \rightarrow 4f$ emission is calculated to give two bands with the same structure and composition of the $5dt_{2g}\,\Gamma_{8g} \rightarrow 4f$ emission, peaking at 43000–43700 cm^{-1} and 45500–46000 cm^{-1} respectively, which have not been found experimentally; this was interpreted as due to non-radiative decay to an anomalous state [3], to killer impurities [1] and to Ce^{3+}-to-Ce^{4+} non-radiative inter-valence charge transfer [8]. A weak and broad band in the excitation spectra peaking at 57500 cm^{-1} that was tentatively assigned to free excitons [3] is now attributed to an impurity-trapped-exciton and calculated at 59200 cm^{-1}.

It is interesting to mention that a very broad emission band peaking at 36400 cm^{-1} was found in $Cs_2LiLuCl_6$:Ce^{3+}. It was considered to be an *anomalous* emission of Ce^{3+} and attributed to an impurity-trapped-exciton [3]. However, the multiconfigurational ab initio calculations do not give credibility to such a possibility. Considering all possible sources of deviations from experiments and all the gathered experience that allows to estimate expected deviations within reasonable margins, it is safe to say that no state involving only a Ce^{3+} site in $Cs_2LiLuCl_6$:Ce^{3+}exists that can be responsible for such a low energy and specially broad emission band. This issue is discussed in Sect. 11.2.2, where Ce^{3+}-to-Ce^{4+} inter-valence charge transfer emission is found to perfectly describe the properties of such a singular emission.

9.2 Pr^{3+} in CaF_2

Pr^{3+} owns a special position as an activator of luminescent materials because its electronic structure allows for multiple photonic possibilities, including $5d \rightarrow 4f$ emission and quantum cutting from the high $4f^2\ {}^1S_0$ energy level. Whether one or the other prevails depends on the environment of the ion which selectively affects the $4f5d$ manifolds leaving the $4f^2$ spectrum basically intact. So, whether the environment pushes the $4f5d$ manifolds to cover or not the instrumental ${}^1S_0$ level makes the difference. In consonance with this, the literature on Pr^{3+} phosphor materials is abundant. In particular, many efforts have been put on fluorite-like hosts. Pr^{3+}-doped CaF_2 is a pivotal phosphor in this family [9–15]. Charge compensation enforced by the substitution of Ca^{2+} by Pr^{3+} ions produces a rich set of distinct defects, with single-ion sites of O_h, C_{4v}, and lower symmetries [13], which induced many contradictory results in the literature. In this respect, site-selective laser spectroscopy was able to study the different sites and clarify many of these [9, 12, 14], e.g. that the single-ion C_{4v} site has $5d \rightarrow 4f$ emission whereas a lower symmetry site yields quantum cutting via $4f \rightarrow 4f$ emission from ${}^1S_0$. A third $5d \rightarrow 4f$ emission belongs to Pr^{3+} in an O_h, cubic site, with long-range charge compensations, and this was the subject of theoretical study by means of multiconfigurational ab initio calculations [16], whose spectroscopic conclusions are summarized here.

The SA-CASSCF/MS-CASPT2/RASSI-SO calculations of the excited states of $4f^2$, $4f5de_g$, $4f5dt_{2g}$, and $4f\phi_{ITE}$ configurations discussed in Sects. 6.1, 6.2, and 6.3, together with absorption and emission oscillator strengths of interconfigurational excitations and calculated absorption and emission profiles, are the basis for the spectroscopic discussions presented here.

$4f^2 \rightarrow 4f5d$ *Excitations*

These excitations are discussed using the transition energies from the ground state $4f^2\ 1T_{2g}(^3H_4)$ (Table 6.2) and the theoretical absorption spectrum profile in Fig. 9.1, calculated assuming common breathing mode harmonic vibrational frequencies on the initial and final electronic states of 505 cm^{-1} and a broadening factor of 40 cm^{-1}, which is narrow enough so as to allow to see the totally symmetric vibronic structure. The experiments we are discussing have been performed at less than 10 K and, accordingly, no thermally induced absorptions from the first excited state have been included, because it is 195 cm^{-1} above the ground state. The comparison with the experiments is summarized in Table 9.2. It is to be noted that all sites are observed simultaneously in absorption spectra [10, 11], whereas a single type of site can be selectively measured in excitation spectra [9, 12, 14]. Correspondingly, special attention is paid to the excitation spectra of [12] (cf. its Fig. 7) and [9] (cf. its Fig. 6) because they were explicitly attributed to cubic Pr defects (in CaF_2:Pr^{3+} and CaF_2:Pr^{3+},Na^+ respectively).

Two bands appear in the region 40000–80000 cm^{-1} of the calculated absorption spectrum of the cubic defect. They correspond to $4f^2 \rightarrow 4f5de_g$ and $4f^2 \rightarrow 4f5dt_{2g}$ excitations respectively, the former having an overall intensity slightly larger than the latter. This is not the case of the excitation spectrum of [12] but it agrees with the excitation spectrum of the cubic site of CaF_2:Pr^{3+},Na^+ of [9]. Also, the intensity profiles of absorption and excitation spectra do not necessarily agree as these are intrinsically different measurement techniques. (For excitation spectra, it is thinkable that relaxation steps prior to the probed photon emission affect the shape of the obtained excitation spectrum.)

The $4f5de_g$ band exhibits a sharp onset, three main electronic origins (out of the 23 levels of this configuration), and a smooth tail. These features agree with the experimental band [9, 12], where three dominant origins are found at 44747, 46777,

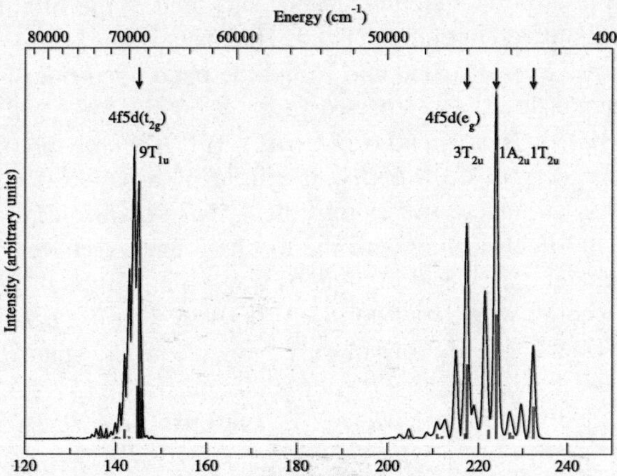

Fig. 9.1 Calculated absorption spectrum profile of the cubic Pr substitutional defect of CaF_2:Pr^{3+}. Oscillator strengths are shown as brown vertical bars. Arrows indicate the most intense electronic origins

Table 9.2 Experimental and theorecital $4f^2 \rightarrow 4f5d$ absorption and excitation transitions, in cm^{-1}

Experiment			Calculation		
Reference	Feature	Energy	Energy	Feature	$1T_{2g}(^3H_4) \rightarrow$
$4f^2 \rightarrow 4f5de_g$					
[9]	Cubic defect A_1 ZPL	44747	43040	ZPL	$\rightarrow 1T_{2u}(1^1T_{2u})$
[9]	Cubic defect A_2 ZPL	46777	44660	ZPL	$\rightarrow 1A_{2u}(1^3T_{2u})$
[9]	Cubic defect A_3 ZPL	49302	46000	ZPL	$\rightarrow 3T_{2u}(1^3T_{2u})$
[9]	Band maximum	45900	44660	Band maximum	$\rightarrow 4f5de_g$
$4f^2 \rightarrow 4f5dt_{2g}$					
[12]	Cubic defect, band onset	60600	67600	ZPL, band onset	$\rightarrow 6E_u(2^1E_u)$
			68930	ZPL, most intense	$\rightarrow 9T_{1u}(3^3T_{2u})$
[9]	Band maximum	64900	69430	Band maximum	$\rightarrow 4f5dt_{2g}$

and 49302 cm^{-1}, the second being the most intense, and labeled A_1, A_2, A_3. The theoretical sharp onset at 43040 cm^{-1}, which is the first dominant origin, is due to the $^1G_4^0$ related $1T_{2u}(1^1T_{2u})$ state and it has the third largest oscillator strength in this band in spite of its dominant spin-singlet character and the ground state being a spin-triplet. It corresponds to peak A_1. The other two main electronic origins are $1A_{2u}(1^3T_{2u})$, at 44660 cm^{-1}, with the largest oscillator strength, and $3T_{2u}(1^3T_{2u})$ at 46000 cm^{-1}. They correspond to peaks A_2 and A_3, respectively. The deviations range between –1700 and –3300 cm^{-1}. This complex multi-origin band extends along 6700 cm^{-1} according to the calculations, between 43000 and 49700 cm^{-1}, and around 6200 cm^{-1} in the experiments, between 44800 and 51000 cm^{-1} (cf. Fig. 10 in [9]).

The $4f5dt_{2g}$ band onset is not so sharp (Note that Fig. 9.1 may be misleading for this purpose because the energy scale is not linear). Its intensity grows gradually from the first two electronic states, with null or very small intensity, to the third and fourth, with significant intensity, and to the fifth, $9T_{1u}(3^3T_{2u})$, with the largest oscillator strength in the band. The onset is marked by the first allowed transition to $6E_u(2^1E_u)$, at 67600 cm^{-1}, and the band maximum is at 68930 cm^{-1}. The experimental onset was reported to be at 60600 cm^{-1} [12] and the band maximum at 63300 cm^{-1} [14] and 64900 cm^{-1} [9]. It seems that the calculations overestimate the energy of the $4f5dt_{2g}$ manifold by around 6000 cm^{-1}, which, together with the small underestimation of the $4f5de_g$ manifold, imply an overestimation of the crystal-field, in line with the too hard energy curves (low distances and high frequencies) discussed in Sects. 6.1 and 6.2. Although a large number of states span an energy window of almost 12000 cm^{-1} in the calculations, between 66500 and 82000 cm^{-1}, only four closely lying electronic states provide most of the intensity of this band: $8T_{2u}(3^3T_{2u})$ at 68600 cm^{-1}, $8T_{1u}(3^3T_{1u})$ at 68770 cm^{-1}, $9T_{1u}(3^3T_{2u})$ at 68930 cm^{-1}, and $3A_{2u}(3^3T_{2u})$ at 69150 cm^{-1}. This corresponds to a relatively narrow band with long tail found experimentally (cf. Fig. 10 in [9]). Unfortunately, electronic origins have not been measured in this band.

$4f5de_g \rightarrow 4f^2$ Emission

The Pr cubic defect of CaF_2:Pr^{3+} was assumed not to be able to yield ${}^1S_0$-based quantum cutting but strong $4f5de_g \rightarrow 4f^2$ emission instead [12]. The multiconfigurational ab initio calculations summarized in Sects. 6.1 and 6.2 showed ${}^1S_0$ to be embedded in the $4f5de_g$ manifold and support this assumption.

The calculated $4f5de_g \rightarrow 4f^2$ emissions are summarized in Fig. 9.2 (also cf. Table 5 in [16]). As for the excitation, site-selective spectroscopy can distinguish the individual emissions from all non-equivalent defects, to a large extent, and pinpoint those from the cubic defect, which are discussed here. They have been reported between 250 nm and 450 nm (cf. Fig. 9a in [14]) and between 220 nm and 300 nm (cf. Fig. 4 in [12]).

The single emission from the lowest state of the $4f5de_g$ manifold, the ${}^1G_4$ related spin-singlet $1T_{2u}({}^1T_{2u})$, is shown in Fig. 9.2top(a). It resembles the experimental emission from ${}^3F_4$ to longer wavelengths [14] and the experimental assignments in terms of free-ion levels agree with the theoretical ones, as well as the intensity order ${}^3F_4$, ${}^1G_4$, ${}^1D_2$, ${}^1I_6$-${}^3P_J$. Here it is remarkable that the most intense emission is to a state with a dominant spin-triplet character in contradiction with a simple prediction from the spin selection rule . This means that not only the percentage parentages are important when spin-orbit coupling breaks the spin quantum number, but also the sizes of the singlet-singlet and triplet-triplet contributions to the oscillator strengths (the emitting state is 80% $S = 0$ and 20% $S = 1$ and the ${}^3F_4$ related $7T_{2g}(2{}^3T_{2g})$ level, with highest oscillator strength, is 30% $S = 0$ and 70% $S = 1$); which illustrates the necessity of spin-orbit coupling calculations. However, the intensity sequence corresponds well with the atomic J quantum number selection rule $|\Delta J| \leq 1$. This is an indication of spin-orbit coupling dominating over crystal-field splitting in the $4f^N$ manifolds of lanthanides in solids, which is well known: both the S and the J selection rules are broken, but the latter holds to a larger extent.

The $1T_{2u}({}^1T_{2u})$ emission, which resembles the experimental emission spectrum from ${}^3F_4$ to longer wavelengths, does not agree, however, with experiment in the region between ${}^3H_4$ and ${}^1G_4$ [12]: the sequence ${}^3H_4$, ${}^3H_6$, ${}^3F_4$, ${}^1G_4$ is correct, but the emissions to ${}^3H_5$ (between ${}^3H_4$ and ${}^3H_6$) and to ${}^3F_2$ (between ${}^3H_6$ and ${}^3F_4$) have far too small intensity. The fact that the disagreement is so specific for some levels suggests the possibility of two electronic origins of the emission. This hypothesis was studied and the emission from the second $4f5de_g$ state $1E_{2u}(1{}^3T_{2u})$, was calculated (Fig. 9.2top(b). The emissions ending in ${}^3H_5$ and ${}^3F_2$ are now much more intense than the others and, interestingly, these are the two "missing" states in the calculated emission from $1T_{2u}$. $1E_{2u}$ is 900 cm^{-1} above $1T_{2u}$ according to the calculations, which is more or less the energy of two phonons, which initally suggests that multiphonon relaxation would be effective at low temperature and the only emitting state would be

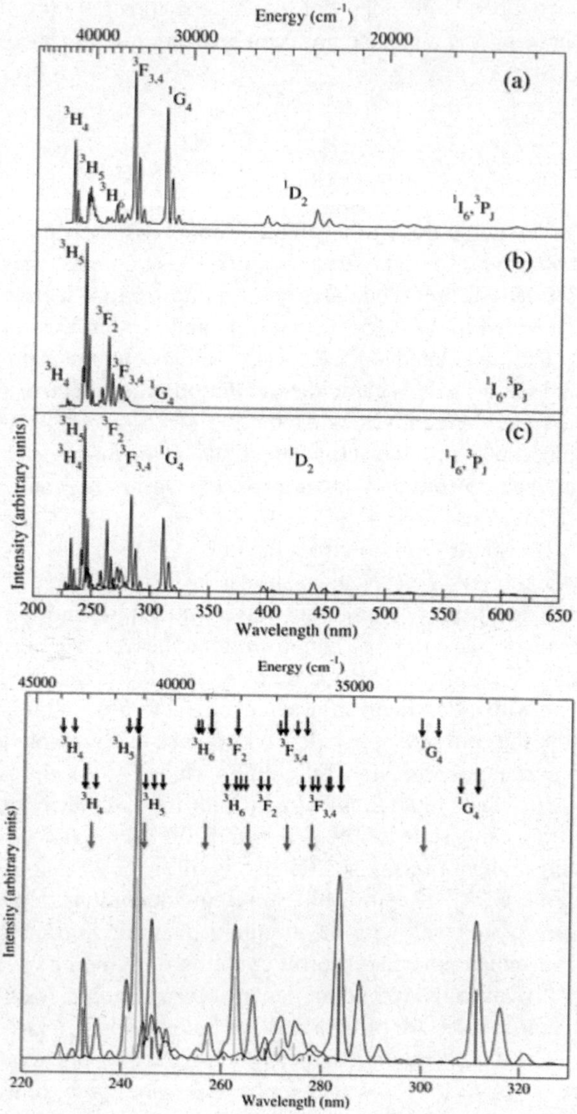

Fig. 9.2 Calculated $4f5de_g \rightarrow 4f^2$ emission spectrum profiles of the cubic Pr substitutional defect of $CaF_2{:}Pr^{3+}$. Top: $1T_{2u}(^1T_{2u})$ emission (**a**), $1E_{2u}(^3T_{1u})$ emission (**b**), and superimposed emissions with arbitrary relative wights 0.6 and 1.0, respectively (**c**). Bottom: Superimposed emissions up to $^1G_4$. Oscillator strengths are shown as vertical bars. Atomic levels responsible for the largest intensities and electronic origins are indicated for $1E_{2u}(^3T_{1u})$ emission (blue) and $1T_{2u}(^1T_{2u})$ emission (black). Green arrows indicate the strongest line within each group of lines of the experimental emission spectrum [12]; they have been shifted $-1500\,\text{cm}^{-1}$ for the sake of comparison with the calculations. Their corresponding lines in the calculated spectrum are indicated with longer arrows

$1T_{2u}$. However, $1T_{2u}$ is 80% spin-singlet and $1E_{2u}$ is 97% spin-triplet, which might help reducing the non-radiative decay rate. If this was the case, it could well be that the observed emission is a superposition of $1T_{2u}(^1T_{2u})$ and $1E_{2u}(1^3T_{2u})$ emissions. Both emissions are superimposed with arbitrary weights 0.6/1 in Fig. 9.2top(c), which shows a remarkable agreement with the experimental emission, with a one-to-one correspondence between theory and experiment. Assuming this complex emission, detailed assignments of the most intense lines have been made (cf. Table 5 in [16]), a regular underestimation of around –1300 cm^{-1} from $^3H_4$ to $^3F_3$ is found, in correspondence with the underestimation of the energy of the emitting level discussed above.

In summary, the experimental $4f5de_g \rightarrow 4f^2$ emission spectrum attributed to the cubic defect of CaF_2:Pr^{3+} can be satisfactorily reproduced (origins and intensities) only when the emissions from the levels $1T_{2u}(^1T_{2u})$ and $1E_{2u}(1^3T_{2u})$ are supposed to occur simultaneously. In that case, all the most intense lines among the $^{2S+1}L_J$ atomic level related groups of lines are identified and assigned. This result suggest the possibility that the experimental emission of the cubic defect of CaF_2:Pr^{3+} is in fact a multiple emission.

9.3 Sm^{3+} and Sm^{2+} in CaF_2

Comparison between theoretical and experimental spectra are shown in Fig. 9.3 and Table 9.3. In the Figure, the experimental $4f^6 \rightarrow 4f^55de_g$ absorption band profiles of Sm^{2+} in CaF_2 at liquid-nitrogen temperature [17] are shown together with theoretical profiles of the SA-RASSCF($4f/5d6s5f$-$4e$)/MS-CASPT2/RASSI-SO calculations. For the latter, the spectroscopic constants and absorption oscillator strengths in Tables 6.4 and 6.12 have been used in Heller's time-dependent semiclassical approach [18, 19], at high and low resolutions (using 5 and 100 cm^{-1} broadening factors respectively) (cf. Sect. 3.3). In the Table, the electronic origins are compared.

Table 9.3 also includes experimental energies and assignments of $4f^5 \rightarrow 4f^5$ electronic transitions of Sm^{3+} in CaF_2 [20, 21], and their comparison with the calculations. These energy levels are important to understand the structure of the $4f^6 \rightarrow 4f^55de_g$ absorptions of Sm^{2+}.

As in other $4f^N$–$4f^N$ and $4f^N$–$4f^{N-1}5d$ transitions [22], the theoretical results of CaF_2:Sm^{3+} $4f^5 \rightarrow 4f^5$ transitions show a systematic overestimation of around 10%. The errors are somehow smaller for the $4f^6 \rightarrow 4f^55de_g$ absorptions of CaF_2:Sm^{2+} however, where underestimations of around 3% are obtained. This result can be associated with the shorter state-average used for the SA-RASSCF of the $4f^55de_g$ manifold in this case, covering only the lowest part of the $4f^55d$ spectrum, which is variationally in favor of the target excited states (as compared with higher excited states not covered here) with the result of slightly underestimated energies.

The interpretation of the fine features of the $4f^6 \rightarrow 4f^55de_g$ absorptions of CaF_2:Sm^{2+} as reflecting the multiplet structure of the $4f^5$ subshell, made by Loh [17], is basically supported by the multiconfigurational ab initio calculations, whose reproduction of almost all details of the experimental bands is remarkable. The calculations provide a detailed analysis of the individual peaks in terms of all valid quantum numbers (O_h irreducible representations and ordinals), the contributions of quasi-valid quantum numbers (spin-orbit-free states, Table 6.12), and the links with the states of the $4f^5$ subshell that couple with the $5de_g$ electron. A similar thing can be said about the features of the $4f^5 \rightarrow 4f^5$ transitions of CaF_2:Sm^{3+}.

Table 9.3 Main features of the $4f^6 \rightarrow 4f^5 5de_g$ absorption spectrum of $CaF_2:Sm^{2+}$ and their correspondence with the lowest multiplets of $CaF_2:Sm^{3+}$. Energies are given in 10^3 cm^{-1}

$CaF_2:Sm^{2+}$						$CaF_2:Sm^{3+}$				
Theory				Experiment [17]		Theory		Experiment [20, 21] [a]		
State[b]	f^{abs}/f_{ref}[c]	Energy[d]	Energy[e]	Energy[d]	Energy[e]		Energy[f]	Energy[f]		
$4f^5(^6H)5de_g$						$4f^5$				
1 T_{1u}	0.12	14.0	0	14.5	0	1 Γ_{8u}	0	0	Z	$^6H_{5/2}$
						1 Γ_{7u}	0.1			
2 T_{1u}	1.00	15.1	1.1	15.47	0.97	1 Γ_{6u}	1.0	1.1	Y	$^6H_{7/2}$
3 T_{1u}	0.13	15.8	1.8	15.8	1.3	2 Γ_{7u}	1.2			
						2 Γ_{8u}	1.4			
4 T_{1u}	0.07	16.5	2.5	16.76	2.26	2 Γ_{6u}	2.3	2.3	X	$^6H_{9/2}$
				16.96	2.46	3 Γ_{8u}	2.5			
						4 Γ_{8u}	2.7			
7 T_{1u}	0.01	17.6	3.6	17.68	3.18	5 Γ_{8u}	3.8	3.6	W	$^6H_{11/2}$
8 T_{1u}	0.01	18.1	4.1	18.68	4.18	6 Γ_{8u}	4.1			
9 T_{1u}	0.02	18.4	4.4			3 Γ_{6u}	4.1			
						3 Γ_{7u}	4.2			
11 T_{1u}	0.01	19.0	5.0	19.71	5.2	7 Γ_{8u}	5.2	5.0	V	$^6H_{13/2}$
12 T_{1u}	0.03	19.3	5.3			4 Γ_{7u}	5.4			
						8 Γ_{8u}	5.7			
						5 Γ_{7u}	5.7			
						4 Γ_{6u}	5.7			
13 T_{1u}	0.14	19.7	5.7	20.12	5.62	9 Γ_{8u}	6.7	5.75	U	$^6F_{1/2}$
						5 Γ_{6u}	6.8			
						6 Γ_{6u}	7.3			
						10 Γ_{8u}	7.3			
						11 Γ_{8u}	7.5			
						12 Γ_{8u}	7.6			
						6 Γ_{7u}	7.6			
$4f^5(^6H+^6F)5de_g$										
18 T_{1u}	0.01	21.8	7.8	21.85	7.35	7 Γ_{7u}	7.9	7	R	$^6F_{5/2}$
21 T_{1u}	0.07	22.1	8.1			13 Γ_{8u}	8.1			
23 T_{1u}	0.18	22.6	8.6	22.47	7.97	8 Γ_{7u}	8.8	7.9	Q	$^6F_{7/2}$
25 T_{1u}	0.04	22.9	8.9			14 Γ_{8u}	8.9			
						7 Γ_{6u}	9.0			
29 T_{1u}	0.25	23.6	9.6	23.64	9.14	8 Γ_{6u}	10.1	9.05	P	$^6F_{9/2}$
30 T_{1u}	0.63	23.8	9.8			15 Γ_{8u}	10.2			
32 T_{1u}	0.02	24.3	10.3			16 Γ_{8u}	10.2			
33 T_{1u}	0.01	24.3	10.3							
36 T_{1u}	0.01	25.2	11.2	24.7	10.2	17 Γ_{8u}	11.6	10.4	O	$^6F_{11/2}$
38 T_{1u}	0.01	25.6	11.6	25.1	10.6	9 Γ_{7u}	11.6			
						18 Γ_{8u}	11.7			
						9 Γ_{6u}	11.7			
42 T_{1u}	0.02	26.6	12.6							

[a] Note the absence of $4f^5$ multiplets $^6H_{15/2}$ (S) and $^6F_{3/2}$ (T). The experimental energies were obtained from ionized gases

[b] Only states with relative absorption oscillator strength $f^{abs}/f_{ref} \geq 0.01$ are listed

[c] f_{ref}=9.724×10^{-3}

[d] Referred to Sm^{2+} $4f^6$ 1 A_{1g}

[e] Referred to Sm^{2+} $4f^5 5de_g$ 1 T_{1u}

[f] Referred to Sm^{3+} $4f^5$ 1 Γ_{8u}

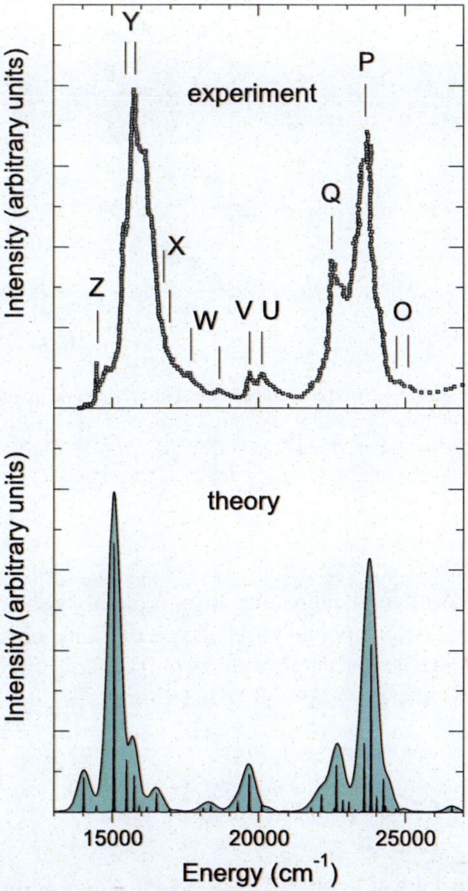

Fig. 9.3 Lowest $4f^6 \rightarrow 4f^5 5de_g$ absorption bands of Sm^{2+} in CaF_2. Top: Experiment [17], Bottom: Theoretical results according to a SA-RASSCF($4f/5d6s5f$-$4e$)/MS-CASPT2/RASSI-SO calculation

9.4 Eu^{2+} and Eu^{3+} in Fluorides and Sulfides

A comprehensive multiconfigurational ab initio study of the $4f^6$ manifold of Eu^{3+} and the $4f^7$, $4f^6 5d$, and $4f^6 \phi_{ITE}$ manifolds of Eu^{2+} in the series of alkaline earth difluorides and sulfides CaF_2, SrF_2, BaF_2, CaS, SrS, and BaS has been performed in [22]. The high quality of the one-electron basis sets used to calculate molecular orbitals of the embedded-clusters and of the many-electron expansions used to calculate wave functions of all the many-electron states involved, together with the acknowledged quality of the relativistic many-electron embedded-cluster Hamiltonians employed, raised the expectation of good theoretical results, The variety of impurity states and solid hosts made the calculations specially appropriate for an overall estimation of accuracies via comparison with experimental data. A comparison of theoretical results and available experimental data on these manifolds and materials is summarized in Fig. 9.4, where a remarkable systematic 10% overestima-

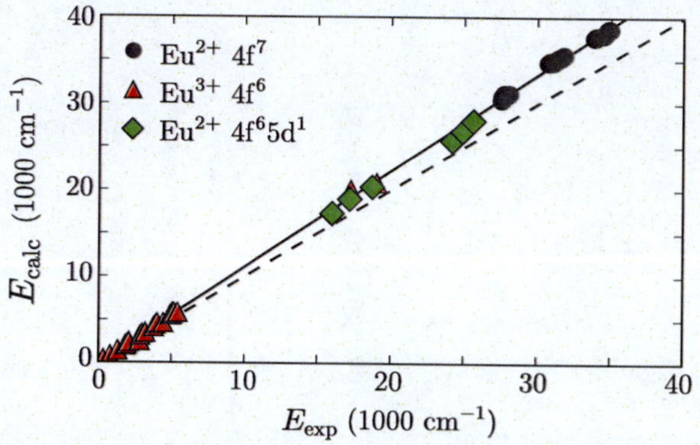

Fig. 9.4 Comparison between calculated and available experimental data [23–29] of excited state energies of Eu^{2+} ($4f^7$ and $4f^65d$ levels) and Eu^{3+} ($4f^6$ levels) in the series of alkaline earth difluorides and sulfides CaF_2, SrF_2, BaF_2, CaS, SrS, and BaS. The straight lines $E_{exp} = E_{calc}$ (dashed line) and $E_{exp} = 0.9E_{calc}$ (solid line) are represented for a reference

tion of the experiments is found. The systematics of the deviations indicates that the raw theoretical results are sufficiently good to discuss and interpret the experiments, to give insight and to make predictions, but a 90% scaling of them provides an even better set of data for these purposes. In line with this, the scaled theoretical results $E^{sc}_{calc} = 0.9E_{calc}$ will be used for discussions in this section.

9.4.1 $4f^6$ Manifold of Eu^{3+}

In spite of its popularity as a red/orange emitter useful in fluorescent lighting [30] and as a sensitive probe of local symmetry [31–33], small crystal-field splittings and low transition strengths make it often difficult to do thorough analyses of the spectrum of Eu^{3+} in solid hosts. An additional complication results from the identification of the signals attributed to cubic defects with long-range charge compensation, which can be formed by a rapid quenching of the crystal during the high-temperature synthesis of fluorides [34]. In the sulfides doped with Ce^{3+}, magnetic resonance studies showed that around 50% of the ions are in a high-symmetry site [35]. In CaF_2:Eu^{3+}, a thorough analysis of combined excitation and emission spectroscopy took advantage of the difference in lifetime of the $^5D_0$ and $^5D_1$ emitting levels and identified the Stark levels of $^7F_{0-6}$, $^5D_0$, and $^5D_1$ due to the O_h defects [29]. Their comparison with the multiconfigurational ab initio calculations are shown in Table 9.4, where the theoretical results with a reduction of systematic errors via 90% scaling show remarkable agreement with the experiments, with deviations of the order of 100 cm^{-1}.

The evolution of the $4f^6$ levels of Eu^{3+} across the series of alkaline earth difluoride and sulfide hosts is summarized in Table 9.5. For simplicity, only the average energies of the Stark components of the $^7F_{0-6}$ and $^5D_{0-3}$ multiplets are shown. As expected,

Table 9.4 Experimental [29] and calculated excitation energies (in cm^{-1}) of the lowest $4f^6$ states of the cubic site of Eu^{3+} in CaF_2

CaF_2:Eu^{3+} cubic site						
$4f^6$		E_{exp} [29]	E_{calc}	$E_{calc} - E_{exp}$	E^{sc}_{calc}	$E^{sc}_{calc} - E_{exp}$
$^7F_0$	1 A_{1g}	0	0	0	0	0
$^7F_1$	1 T_{1g}	337	345	8	311	−26
$^7F_2$	1 T_{2g}	811	803	−8	730	−81
	1 E_g	1342	1406	64	1265	−77
$^7F_3$	2 T_{1g}	1857	1971	114	1774	−83
	2 T_{2g}	1968	2111	143	1900	−68
	1 A_{2g}	2107	2352	245	2117	10
$^7F_4$	2 A_{1g}	2845	2501	−344	2251	−594
	3 T_{1g}	2879	3114	235	2803	−76
	3 T_{2g}	3021	3361	340	3025	4
	2 E_g	3233	3472	239	3125	−108
$^7F_5$	4 T_{2g}	3793	4132	339	3719	−74
	4 T_{1g}	3933	4404	471	3964	31
	3 E_g	4028	4600	572	4140	112
	5 T_{1g}	4438	4602	164	4142	−296
$^7F_6$	3 A_{1g}	4920	5524	604	4972	52
	6 T_{1g}	4946	5580	634	5022	76
	5 T_{2g}	5044	5603	559	5043	−1
	4 E_g	5062	5806	744	5225	163
	6 T_{2g}	5184	5813	629	5232	48
	2 A_{2g}	5411	5827	416	5244	−167
$^5D_0$	4 A_{1g}	17274	20023	2749	18021	747
$^5D_1$	7 T_{1g}	19030	20828	1798	18745	−285

these energies are very insensitive to the chemical environment, although the $^5D_0$ level increases its energy between 50 and 100 cm^{-1} from the sulfide family to the difluoride family and, within each family, in the Ca to Sr and Ba series. The increasing Eu-F and Eu-S bond lengths in the Ca-Sr-Ba series (and associated decreasing force constants) for a given ligand (cf. Table 6.15), result in a small blue shift of the $^7F_0$-$^5D_0$ line; the increasing coordination number and decreasing nephelauxetic effect from the sulfide family to the difluoride family also result in a small blue shift. Since the small energy changes imply wavelength differences never larger than a few nanometers, it indicates that low temperatures and accurately wavelength-calibrated detectors are indispensable for fine spectroscopic studies of the $4f^6$ configuration in different hosts.

Table 9.5 Calculated average energies (in cm^{-1}) of the lowest multiplets of the cubic site of Eu^{3+} in alkaline earth difluorides and sulfides

	$CaF_2:Eu^{3+}$	$SrF_2:Eu^{3+}$	$BaF_2:Eu^{3+}$	$CaS:Eu^{3+}$	$SrS:Eu^{3+}$	$BaS:Eu^{3+}$
$4f^6\,^7F_J$						
$J=0$	0	0	0	0	0	0
1	311	323	327	335	336	338
2	940	949	954	957	961	963
3	1877	1850	1844	1831	1826	1825
4	2887	2849	2836	2803	2797	2796
5	3978	3932	3909	3839	3839	3839
6	5120	5068	5047	4991	4980	4976
$4f^6\,^5D_J$						
$J=0$	18021	18025	18075	17904	17976	18007
1	18745	18752	18804	18629	18703	18735
2	20255	20266	20322	20140	20219	20252
3	22646	22668	22732	22544	22626	22663

9.4.2 $4f^7$ Manifold of Eu^{2+}

All the excited states of the $4f^7$ manifold of Eu^{2+} in alkaline earth difluorides and sulfides, starting with those derived from the ^6P, ^6I and ^6D free-ion terms (Table 6.7), overlap the dense $4f^65d$ manifolds, except for a few of them that lie in the gap between the $4f^6(^7F)5dt_{2g}$ and $4f^6(^7F)5de_g$ manifolds in sulfides (cf. Fig. 6.6). Such overlap poses an important problem for these states to be studied with luminescence spectroscopy [36]. When fortunate gaps in the $4f^65d$ manifolds do not occur, it is still possible to get experimental information on these states in rare cases where Fano antiresonances show up in the excitation spectrum [37], but the most practical approach is two-photon absorption spectroscopy, where $4f \rightarrow 4f$ transitions are allowed [38, 39]. A meticulous analysis of two-photon absorption spectroscopy experiments on $CaF_2:Eu^{2+}$ and $SrF_2:Eu^{2+}$ [23] provided a very detailed and accurate determination of the Eu^{2+} $4f^7$ level energies up to 35000 cm^{-1} in these doped materials; some levels were also identified in $BaF_2:Eu^{2+}$ [24]. In Table 9.6 a comparison is made between calculation and experiments on the average term energies and on the energies of individual states that are derived from the ^6P free-ion term. A satisfactory agreement extends over all reported experimental data not only for the average term energies, but also for the multiplet splittings due to the interaction with the chemical environment and the spin-orbit coupling. The deviations are in the range of 100 cm^{-1} with a maximum deviation of 337 cm^{-1}.

Table 9.6 Experimental (two-photon absorption spectroscopy) and calculated average energies of the lowest terms of the $4f^7$ manifold of Eu^{2+} in CaF_2, SrF_2, and BaF_2 (in cm^{-1})

	CaF_2:Eu^{2+}			SrF_2:Eu^{2+}			BaF_2:Eu^{2+}		
	E^{sc}_{calc}	E_{exp}[23]	diff.[a]	E^{sc}_{calc}	E_{exp}[23]	diff.[a]	E^{sc}_{calc}	E_{exp}[24]	diff.[a]
^8S	0	0	0	0	0	0	0		
^6P	27726	27804	−78	27849	27901	−52	28057		
2 Γ_{8u}	27398	27558	−160	27519	27654	−135	27728	27727	1
2 Γ_{7u}	27408	27564	−156	27527	27658	−132	27734	27729	5
3 Γ_{8u}	27743	27588	155	27866	27672	194	28075	27738	337
2 Γ_{6u}	27948	27959	−11	28075	28066	9	28285		
4 Γ_{8u}	27959	27999	−41	28085	28098	−14	28292		
3 Γ_{7u}	27978			28100			28304		
^6I	31631	31382	249	31767	31491	276	31981		
^6D	34290	34474	−184	34435	34496	−61	34667		
$^6G+{}^6F$	45249			45437			45693		
^6H	50602			50790			51043		

[a] $E^{sc}_{calc} - E_{exp}$

9.4.3 $4f^65d$ Manifold of Eu^{2+}

Spectral Assignments

Calculation of probabilities of electric dipole absorptions in the form of oscillator strengths is made with the many-electron wave functions of the ground $4f^7$ and excited $4f^65d$ states of Eu^{2+} in the hosts. The ones of CaF_2:Eu^{2+} are shown in Table 9.7. These, together with the minimum-to-minimum or adiabatic transition energies (potential energy vertical offsets), Eu-F bond length changes in absorption (potential energy horizontal offsets), and vibrational frequencies (cf. Table 6.14) enable a direct calculation of absorption spectra profiles for the $4f^7 \rightarrow 4f^65d$ transitions, which can be directly compared with experimental results from luminescence and UV-VIS absorption and excitation spectroscopy. The computed absorption spectra are shown along with the available experimental spectra in Fig. 9.5 for the difluorides [24, 28, 40, 41] and Fig. 9.6 for the sulfides [25–27, 42].

The computed and experimental spectra have very similar shapes and energy ranges. Two broad bands and a gap between them are found in all cases, which are in correspondence with the structure of the four $4f^65d$ sub-manifolds discussed in Sect. 6.2.4. The bands are due to transitions from the three almost degenerate states derived from the free-ion level $4f^7(^8S_{7/2})$, 1 Γ_{6u}, 1 Γ_{8u}, and 1 Γ_{7u} (let us call them together the ground level $4f^7(^8S_{7/2})$).

Table 9.7 Absorption oscillator strengths from the $4f^7$ $1\,\Gamma_{6u}$, $1\,\Gamma_{8u}$, $1\,\Gamma_{7u}$ ground levels to the excited states of Eu^{2+} in CaF_2. Calculations are done at $d_{\text{Eu-F}}$=2.400 Å; the reference value is f_{ref}=1.704×10^{-2}. Only states with $f^{\text{abs}}/f_{\text{ref}} \geq 0.01$ are given

$4f^6(^7F_J)5de_g$ main character						Interacting $4f^6(^7F_J)(5dt_{2g}+\phi_{ITE,a_{1g}})$			
High-spin		Mixed-spin		Low-Spin					
State	$f^{\text{abs}}/f_{\text{ref}}$	State	$f^{\text{abs}}/f_{\text{ref}}$	State	$f^{\text{abs}}/f_{\text{ref}}$	State	$f^{\text{abs}}/f_{\text{ref}}$	State	$f^{\text{abs}}/f_{\text{ref}}$
$1\,\Gamma_{8g}$	1.00	$12\,\Gamma_{8g}$	0.24	$8\,\Gamma_{7g}$	0.01	$34\,\Gamma_{8g}$	0.47	$52\,\Gamma_{8g}$	0.98
$1\,\Gamma_{7g}$	0.54	$13\,\Gamma_{8g}$	0.38	$17\,\Gamma_{8g}$	0.17	$35\,\Gamma_{8g}$	0.24	$26\,\Gamma_{6g}$	0.12
$2\,\Gamma_{8g}$	0.09			$10\,\Gamma_{6g}$	0.02	$17\,\Gamma_{6g}$	0.19	$27\,\Gamma_{7g}$	0.61
$2\,\Gamma_{7g}$	0.74			$22\,\Gamma_{8g}$	0.04	$17\,\Gamma_{7g}$	0.58	$53\,\Gamma_{8g}$	0.28
$3\,\Gamma_{8g}$	0.99			$11\,\Gamma_{7g}$	0.03	$36\,\Gamma_{8g}$	0.69	$54\,\Gamma_{8g}$	1.55
$1\,\Gamma_{6g}$	0.38			$11\,\Gamma_{6g}$	0.01	$18\,\Gamma_{7g}$	1.23	$27\,\Gamma_{6g}$	0.01
$4\,\Gamma_{8g}$	2.10			$23\,\Gamma_{8g}$	0.02	$18\,\Gamma_{6g}$	0.49	$55\,\Gamma_{8g}$	3.57
$2\,\Gamma_{6g}$	0.26			$24\,\Gamma_{8g}$	0.01	$37\,\Gamma_{8g}$	0.80	$28\,\Gamma_{7g}$	0.16
$5\,\Gamma_{8g}$	0.25			$12\,\Gamma_{6g}$	0.01	$38\,\Gamma_{8g}$	3.21	$28\,\Gamma_{6g}$	1.05
$3\,\Gamma_{7g}$	0.87			$25\,\Gamma_{8g}$	0.02	$39\,\Gamma_{8g}$	2.11	$56\,\Gamma_{8g}$	5.37
$6\,\Gamma_{8g}$	0.44			$26\,\Gamma_{8g}$	0.01	$40\,\Gamma_{8g}$	4.16	$57\,\Gamma_{8g}$	1.41
$7\,\Gamma_{8g}$	0.72			$27\,\Gamma_{8g}$	0.02	$19\,\Gamma_{7g}$	3.45	$29\,\Gamma_{6g}$	2.94
$3\,\Gamma_{6g}$	0.74			$13\,\Gamma_{6g}$	0.01	$20\,\Gamma_{7g}$	1.59	$58\,\Gamma_{8g}$	3.81
$8\,\Gamma_{8g}$	1.79			$28\,\Gamma_{8g}$	0.01	$19\,\Gamma_{6g}$	0.69	$59\,\Gamma_{8g}$	2.22
$9\,\Gamma_{8g}$	0.48			$14\,\Gamma_{6g}$	0.03	$41\,\Gamma_{8g}$	2.73	$60\,\Gamma_{8g}$	0.80
$4\,\Gamma_{7g}$	0.60			$29\,\Gamma_{8g}$	0.04	$20\,\Gamma_{6g}$	0.01	$30\,\Gamma_{6g}$	0.37
$4\,\Gamma_{6g}$	0.31			$30\,\Gamma_{8g}$	0.04	$21\,\Gamma_{6g}$	0.29	$29\,\Gamma_{7g}$	0.02
$5\,\Gamma_{7g}$	0.99			$31\,\Gamma_{8g}$	0.02	$21\,\Gamma_{7g}$	1.03	$30\,\Gamma_{7g}$	0.03
$10\,\Gamma_{8g}$	0.01			$15\,\Gamma_{6g}$	0.01	$42\,\Gamma_{8g}$	0.15	$31\,\Gamma_{6g}$	0.39
$5\,\Gamma_{6g}$	0.07			$32\,\Gamma_{8g}$	0.01	$43\,\Gamma_{8g}$	0.02	$61\,\Gamma_{8g}$	0.24
$11\,\Gamma_{8g}$	0.98					$22\,\Gamma_{6g}$	0.35	$62\,\Gamma_{8g}$	0.17
$6\,\Gamma_{6g}$	0.57					$22\,\Gamma_{7g}$	0.17	$31\,\Gamma_{7g}$	0.33
$6\,\Gamma_{7g}$	0.01					$44\,\Gamma_{8g}$	1.00	$63\,\Gamma_{8g}$	0.30
$14\,\Gamma_{8g}$	1.37					$45\,\Gamma_{8g}$	3.28	$32\,\Gamma_{6g}$	0.25
$15\,\Gamma_{8g}$	3.19					$23\,\Gamma_{7g}$	0.01	$64\,\Gamma_{8g}$	0.58
$7\,\Gamma_{6g}$	1.71					$23\,\Gamma_{6g}$	4.00	$65\,\Gamma_{8g}$	0.07
$16\,\Gamma_{8g}$	0.87					$46\,\Gamma_{8g}$	0.11	$33\,\Gamma_{6g}$	0.09
$8\,\Gamma_{6g}$	0.14					$47\,\Gamma_{8g}$	2.65	$66\,\Gamma_{8g}$	0.33
$18\,\Gamma_{8g}$	1.27					$24\,\Gamma_{7g}$	1.30	$67\,\Gamma_{8g}$	0.41
$9\,\Gamma_{6g}$	0.05					$48\,\Gamma_{8g}$	0.40	$33\,\Gamma_{7g}$	0.11
$19\,\Gamma_{8g}$	0.16					$49\,\Gamma_{8g}$	0.67	$34\,\Gamma_{6g}$	0.07
$20\,\Gamma_{8g}$	0.02					$24\,\Gamma_{6g}$	0.20	$68\,\Gamma_{8g}$	0.19
$21\,\Gamma_{8g}$	0.01					$50\,\Gamma_{8g}$	1.66	$34\,\Gamma_{7g}$	0.10
						$25\,\Gamma_{7g}$	0.13	$69\,\Gamma_{8g}$	0.37
						$51\,\Gamma_{8g}$	0.20	$35\,\Gamma_{7g}$	0.24
						$25\,\Gamma_{6g}$	0.16	$70\,\Gamma_{8g}$	0.41
						$26\,\Gamma_{7g}$	0.84	$35\,\Gamma_{6g}$	0.20

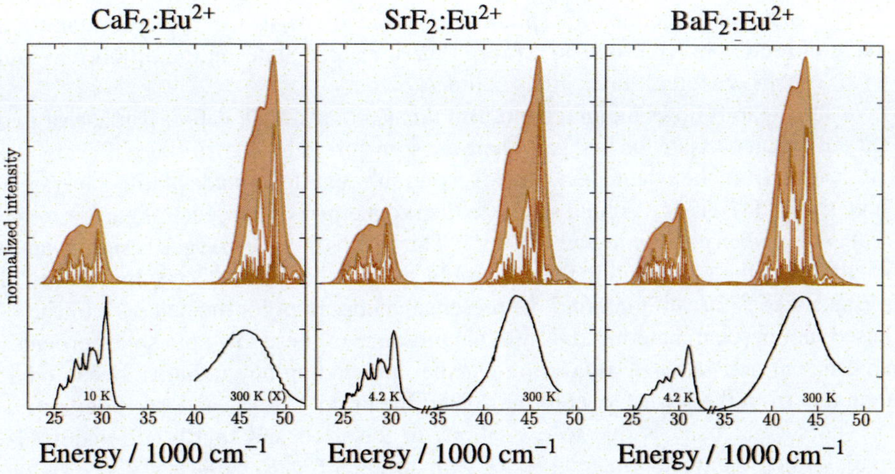

Fig. 9.5 $4f^7 \rightarrow 4f^6 5d$ absorption spectra of Eu^{2+} in CaF_2, SrF_2, and BaF_2. Top: calculated at three resolution levels (broadening parameters of 5, 100, and 300 cm^{-1}) and scaled for improved visibility. Bottom: experimental absorption (or excitation, indicated by 'X') spectra adapted from [28, 41] (CaF_2), [24, 40] (SrF_2, BaF_2)

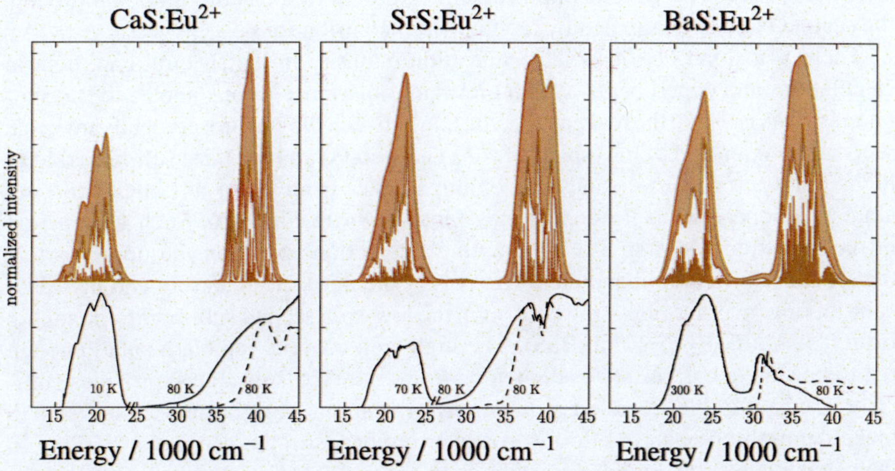

Fig. 9.6 $4f^7 \rightarrow 4f^6 5d$ absorption spectra of Eu^{2+} in CaS, SrS, and BaS. Top: calculated at three resolution levels (broadening parameters of 5, 100, and 300 cm^{-1}). Bottom: experimental excitation spectra adapted from [25, 26] (CaS), [26, 42] (SrS), and [27] (BaS, here the anomalous emission was monitored). The host absorption is also added (dashed lines) to distinguish between intrinsic and Eu^{2+} caused spectral features, adapted from [43]

The first band contains all spin-allowed transitions to the first sub-manifold, made of high-spin (spin-octets) $4f^6(^7F_J)5de_g$ excited states in the difluorides and $4f^6(^7F_J)5dt_{2g}$ excited states in the sulfides.[1]

A large gap follows the first band. But this is not a simple gap: it is made of two very different parts. In the first part, there are spin-forbidden transitions with very low intensities, from the ground level to the second sub-manifold, made of low-spin (spin-sextets) $4f^6(^7F_J)5de_g$ excited states. In the second part of the spectral gap there is a real gap between electronic states of Eu^{2+}. This means that the true gap is significantly smaller than the gap found in the spectra. In spite of the spectroscopic invisibility of the second $4f^65d$ sub-manifold, its presence is important for the behavior of Eu^{2+}-based luminescent materials, because the presence of these low-spin states provides an efficient non-radiative relaxation channel, precluding any radiative decay other than that from the lowest level of the $4f^65d$ manifold.

The second band is due to excitations to the third and fourth sub-manifolds, made of high-spin and low-spin states with strong $4f^6(^7F_J)5dt_{2g} + 4f^6(^7F_J)5de_g$ configurational mixing. It is interesting noticing here that the agreement between calculations and experiments in the second band is common to difluorides and sulfides; however, the contributions of low-spin $4f^65de_g$ spin-orbit-free states to the final spin-orbit coupling states were not taken into account in the calculations on difluorides (Sect. 6.2.4). This implies that, regardless of the configurational mixing, the second band is dominated by the high-spin contributions.

There is a small but observable transition intensity in the spectral gap, both in calculations and experiments, which grows in importance from Ca to Sr to Ba compounds. This is due to the real gap (i.e. the one between the spectroscopically invisible second sub-manifold of low-spin $4f^6(^7F_J)5de_g$ states and the third sub-manifold of high-spin $4f^6(^7F_J)5dt_{2g}$ states) becoming smaller from Ca to Sr compounds and almost disappearing in Ba compounds (not shown in Fig. 6.6). Such a reduction in the gap strengthens spin-orbit coupling interactions between the upper part of the second sub-manifold and the lower part of the third sub-manifold and increases spin-mixing (average high-spin content in the low-spin second sub-manifold ranging from 0.5% in CaF_2 to 2% in BaS). The high-spin content, although small enough not to be able to call the states high-spin, or even mixed-spin states, provides sufficient intensity to the so-called *spin-enabled* transitions [44] towards these states to make them observable. In other words, the tiny bands in the gap are, according to the calculations, *spin-enabled* transitions to some $4f^6(^7F_J)5de_g$ states of the second sub-manifold. Similar *spin-enabled* transitions were assigned in well-resolved low-temperature absorption spectrum of $SrCl_2$:Eu^{2+} single crystals [45, 46].

Interpretation of the spectra of Eu^{2+}-doped sulfides poses an additional difficulty with respect to the fluorides. Their small band gaps make the sulfide hosts to absorb at energies where absorption or photoluminescence spectra are typically measured and, as a result, the second band in the experimental absorption and excitation spectra

[1] For the sake of simplicity, the notation that corresponds to the cubic O_h site of Eu^{2+} in the difluorides will be used in this discussion. For the octahedral O_h site of Eu^{2+} in the sulfides, everything holds, except that $5de_g$ and $5dt_{2g}$ must be interchanged.

of the Eu^{2+}-doped sulfides has to be regarded as the added effect of both Eu^{2+} and host excitations. Only the contributions of Eu^{2+} have been discussed here.

The low-temperature absorption or excitation spectra of the difluorides show a well-resolved fine structure with seven individual peaks, which is not so well identified in the sulfides. Such fine structure is well reproduced by the calculations with small broadening factors. The most commonly accepted explanation of the electronic origin of this fine structure, after a long-standing debate based on empirical and crystal-field theoretical arguments, states, first, that the lowest $4f^65d$ states of Eu^{2+} can be obtained from the multiplets of the $4f^6(^7F_J)$ sub-configuration, which is a reminiscent of the lowest 5000 cm^{-1} of the Dieke diagram for Eu^{3+}, and the lowest crystal-field level of one 5d electron: $5de_g$ in the difluorides, $5dt_{2g}$ in the sulfides; and second, that the coupling between the $4f^6(^7F_J)$ sub-configuration and the 5d electron is of mean-field type, simple and not leading to additional splittings. Then, the fine structure simply reflects the well-known $^7F_{J=0-6}$ seven multiplets of Eu^{3+} [37, 47, 48]. The multiconfigurational ab initio calculations fully support the first statement, as it was discussed above. But not the second one, because the structure of the first sub-manifold of $4f^6(^7F_J)5d_{\rm LOW}$ states is found to be complex, very far from one resulting from a simple, mean-field type of coupling between sub-configurations. Further discussions in Sect. 10.5 give insight on why the fine structure is so simple, apparently corresponding to seven multiplets, when it results from a complex $4f^6(^7F_J) \times 5d_{\rm LOW}$ coupling.

The onset of the $4f^{N-1}5d$ manifold in a lanthanide doped material is known to depend on interactions between the optically active lanthanide ion and the host. Here, there is the opportunity to check the variation across the series of alkaline earth difluorides and sulfides of the zero-phonon lines of the lowest Eu^{2+} $4f{\rightarrow}5d$ transition. They are shown in Table 9.8. The red shift from fluoride to sulfide compounds, which is traditionally attributed to the increase in the chemical bond covalence [49], is reproduced by the multiconfigurational ab initio calculations qualitatively and quantitatively. The changes across the alkaline earth series Ca to Sr to Ba are also pretty well reproduced. And the absolute values of the transitions are computed with deviations of the order of 1000 cm^{-1} or lower. The comparisons evidence the quality of the computed many-electron wave functions and their ability to reproduce physical interactions present in lanthanide-doped luminescent materials, ranging from gross bonding effects to more subtle embedding effects.

Ligand-Field Strength

In Table 9.8, theoretical and experimental quantitative data on ligand-field effects are also compared. Contrary to the case of a single 5d electron under the effect of a ligand-field, the complex interactions between $4f^{N-1}$ subshells and a 5d electron makes quantitative comparisons between theoretical and experimental ligand-field splitting

Table 9.8 Calculated and experimental zero-phonon lines (ZPL) of the lowest $4f \to 5d$ absorption of Eu^{2+} in the cubic site of alkaline earth difluorides ($4f^7(^8S_{7/2})\ 1\,\Gamma_{6u} \to 4f^65de_g\ 1\,\Gamma_{8g}$) and the octahedral site of alkaline earth sulfides ($4f^7(^8S_{7/2})\ 1\,\Gamma_{6u} \to 4f^65dt_{2g}\ 1\,\Gamma_{8g}$). High-spin ligand-field splittings $\varepsilon_{\rm lfs}^{\rm HS}$, and exchange splitting parameters $\varepsilon_{\rm exch}^{e_g}$ for difluorides and $\varepsilon_{\rm exch}^{t_{2g}}$ for sulfides. All numbers in cm^{-1}

	$CaF_2:Eu^{2+}$	$SrF_2:Eu^{2+}$	$BaF_2:Eu^{2+}$	$CaS:Eu^{2+}$	$SrS:Eu^{2+}$	$BaS:Eu^{2+}$
ZPL $4f^7\ 1\,\Gamma_{6u} \to$	$4f^65de_g\ 1\,\Gamma_{8g}$			$4f^65dt_{2g}\ 1\,\Gamma_{8g}$		
Calc.	23150	24145	25240	15580	17080	18320
Exp.[a]	24215	24925	25636	15995	17191	18729
Diff.	−1065	−780	−396	−415	−110	−410
High-spin ligand-field splitting $\varepsilon_{\rm lfs}^{\rm HS}$						
Calc.[b]	18790	16220	14460	22370	20170	16070
Exp.[c]	16480	14150	12670			
Diff.	2310	2070	1790			
Exchange splitting parameter $\varepsilon_{\rm exch}^{\gamma}$						
	$\varepsilon_{\rm exch}^{e_g}$			$\varepsilon_{\rm exch}^{t_{2g}}$		
Calc.[d]	4360	4410	5210	7090	7090	7820

[a] Accurate ZPL experimental observations for $CaF_2:Eu^{2+}$ [28], $SrF_2:Eu^{2+}$ [24], $CaS:Eu^{2+}$ [25] and $SrS:Eu^{2+}$ [26]. Estimated ZPL from band shapes for $BaF_2:Eu^{2+}$ [24] and $BaS:Eu^{2+}$ [27]
[b] Calculated with (9.1). Only many-electron states with ≥90% of spin-octet character are included in the averages
[c] Estimated in the difluorides as the difference of the average values of both excitation bands, $4f \to 5dt_{2g}$ and $4f \to 5de_g$. The upper band is not well isolated in the sulfides
[d] Calculated with (9.2). Only many-electron states with ≥90% of spin-octet and ≥90% of spin-sextet characters are included in the respective HS and LS averages

data difficult for lanthanide ions. In order to facilitate a direct, simple contrast between theory and experiment, it is convenient to define ligand-field splitting parameters that can be calculated and measured.

With this idea in mind, a *high-spin ligand-field splitting* parameter $\varepsilon_{\rm lfs}^{\rm HS}$ for Eu^{2+} in a host can be defined as the positive difference between the averages of the equilibrium energies of the high-spin $4f^6(^7F)5de_g$ and $4f^6(^7F)5dt_{2g}$ states, i.e. the states in the first and third $4f^65d$ sub-manifolds respectively:

$$\varepsilon_{\rm lfs}^{\rm HS} = \left| \langle E_i^{4f^6(^7F)5de_g,\rm HS} \rangle - \langle E_i^{4f^6(^7F)5dt_{2g},\rm HS} \rangle \right| . \qquad (9.1)$$

The value of this parameter is a measure of the ligand-field strength acting on the Eu^{2+} $5d$ shell. It can be calculated from the individual energies of the many-electron states and also measured, from the average energies of absorption and excitation bands, as long as those attributed to the involved sub-manifolds are well separated from other bands and correctly identified. This is the case of Eu^{2+} in the difluorides, but not in the sulfides, where the second absorption band overlaps host excitations and is not well isolated (cf. Table 9.8). In addition, spectral band shapes depend on temperature and doping concentration, bringing additional uncertainties to the empirical values. $\varepsilon_{\rm lfs}^{\rm HS}$ is not to be confused with traditional Crystal-Field Theory parameters like the splitting of one-electron d levels, $10Dq$, even if both of them are measures of ligand-field strength. Furthermore, it is stressed that $\varepsilon_{\rm lfs}^{\rm HS}$ is not a configurational average parameter because only high-spin energy levels are averaged.

Table 9.8 shows that $\varepsilon_{\rm lfs}^{\rm HS}$ indicates the expected drop of ligand-field strength from CaF_2 to SrF_2 to BaF_2, with not only qualitative agreement between theoretical and experimental ligand-field strength variations, but also quantitative. Direct comparison in each luminescent material shows theoretical overestimations of around 2000 cm^{-1}, very similar in the three difluorides. Both a cumulative effect of systematic errors in the calculated energies and the (perhaps also systematic) uncertainties of the empirically obtained parameter may be playing roles in the overestimation. The fact that both the theoretical and the empirical parameters show identical systematic trends is reassuring.

Under identical ligand, coordination number, and crystal structure, the variation of ligand-field strength across a given series, like CaF_2-SrF_2-BaF_2 and CaS-SrS-BaS, is expected to be controlled by the Eu-ligand distance: the shorter the Eu-ligand distance, the stronger the ligand-field strength. Hence the qualitative expectations are confirmed not only in the difluorides, where experimental data are available, but also in the sulfides, where they are missing. Regarding the ligand change, the ligand-field strength increases from difluorides to sulfides (19% from CaF_2 to CaS, 24% from SrF_2 to SrS, and 11% from BaF_2 to BaS) in spite of the longer Eu-ligand distances; it is the result from the balance between the distance effect and opposite contributions from a lower coordination number and a stronger Eu-ligand bond covalency. The multiconfigurational ab initio calculations portray the quantitative results of such a balance, which is hard to achieve a priori by other means.

$4f$-$5d$ Exchange Interactions

Similarly to the definition of a ligand-field splitting parameter that could be calculated and measured, at least in principle, $4f$-$5d$ *exchange splitting* parameters $\varepsilon_{\rm exch}^{e_g}$ (for the difluorides) and $\varepsilon_{\rm exch}^{t_{2g}}$ (for the sulfides) can be defined as the difference between the averages of the equilibrium energies of the low-spin $4f^6(^7F)5de_g$ and high-spin $4f^6(^7F)5de_g$ states for the fluorides, and of the low-spin $4f^6(^7F)5dt_{2g}$ and high-spin $4f^6(^7F)5dt_{2g}$ states for the sulfides, i.e. the difference between the average energies of the second and the first $4f^75d$ sub-manifolds.

$$\varepsilon_{\rm exch}^{\gamma} = \langle E_i^{4f^6(^7F)5d\gamma,{\rm LS}} \rangle - \langle E_i^{4f^6(^7F)5d\gamma,{\rm HS}} \rangle \tag{9.2}$$

with $\gamma = e_g, t_{2g}$. In this case, however, the very low intensity of the transitions to the low-spin states of the second sub-manifold puts severe difficulties to the experimental determination of these exchange splitting parameters. The theoretical results are included in Table 9.8 and the insights deduced from them are discussed in detail in Sect. 10.5.

9.5 Tm^{2+} in Halide Perovskites

The usual behavior of most luminescent materials after an excitation is single emission, i.e. radiative transitions from the lowest state of an excited manifold only, following a rapid non-radiative relaxation from higher to lower excited electronic states. Luminescent materials activated with Tm^{2+} are a notable exception to this because this lanthanide ion is known to show luminescence from up to *three* excited states of the $4f^{12}5d$ manifold [50–52]. Emissions from *two* excited $4f^{N-1}5d$ states had also been reported for other heavy lanthanides ($N>7$), like Er^{3+} and Tm^{3+} doped in fluoride hosts [53] and Yb^{2+} doped in $SrCl_2$ [54, 55] and various perovskites [56], and interpreted as spin-forbidden and spin-allowed emissions originating from high-spin and low-spin states, an interpretation which was questioned on the basis of both empirical Crystal-Field-Theory calculations [55] and multiconfigurational ab initio calculations, which found strong mixing of high-spin and low-spin states and interpreted the two Yb^{2+} emissions as spin-forbidden and spin-enabled transitions [44].

Multiconfigurational ab initio calculations on the multiple emission materials $CsCaCl_3{:}Tm^{2+}$ and $CsCaBr_3{:}Tm^{2+}$, where the lanthanide activator substitutes for Ca^{2+} in an octahedral site of the perovskite, showed that the existence of various energy gaps within a crowded set of states of almost pure $4f^{12}5dt_{2g}$ configurational character with parallel potential energy surfaces, make the multiple emissions possible [57], since a gap large enough below a highly excited state strongly limits non-radiative relaxation and enables emission from it. The calculations of many-electron wave functions and potential energy curves of the $4f^{13}$ and $4f^{12}5d$ manifolds (Sects. 6.1.7 and 6.2.5) allowed for a detailed comparison between theoretical and low temperature experimental absorption and emission spectra and radiative decay times of the emitting states, with remarkably good agreement [57]. The achieved accuracy validates the insights gained in the theoretical study, starting with the support provided to the initial interpretations of the experiments [52] with a few refinements, and continuing with the understanding of the three emission bands in $CsCaBr_3{:}Tm^{2+}$, the faster non-radiative decay in $CsCaCl_3{:}Tm^{2+}$, where only two $4f^{12}5d$ emitting states have been found experimentally, the magnitude and physical origin of the energy gaps that hamper non-radiative relaxation from the emitting states, and the explanation of why the light lanthanides ($N \leq 7$) do not show emission from multiple $4f^{N-1}5d$ states.

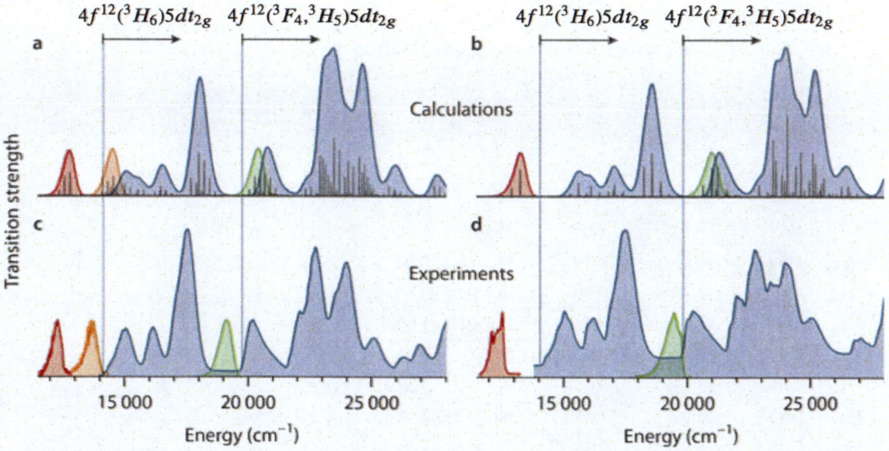

Fig. 9.7 Absorption (in blue) and emission spectra from 1 Γ_{8g} (in red), 3 Γ_{8g} (in orange), and 8 Γ_{6g} (in green), of Tm^{2+} in $CsCaBr_3$ (left panels) and $CsCaCl_3$ (right panels). Top panels: Theoretical spectra with low and high resolution. Bottom panels: Experiment [51]

Also, the availability of several stable excited emitting states made possible highly sensitive excited state excitation (ESE) spectroscopy experiments in these materials in order to expand the experimental information on this complex dopant [58]. They were interpreted on the basis of multiconfigurational ab initio calculations: on one hand, the ESE experiments gave support to the structural and energetic theoretical results; on the other hand, the understanding of widths and energies of narrow and broad ESE bands was supported on the theoretical results, all of it providing a vivid example of the importance of combined experimental and theoretical investigations.

Absorption

Many-electron wave functions of the $4f^{13}$ $1\Gamma_{6u}$ ground state (Sect. 6.1.7) and the $4f^{12}5d$ excited states (Sect. 6.2.5), allow to calculate absorption oscillator strengths necessary for the absorption spectra profiles. The ones for $CsCaCl_3$:Tm^{2+} are given in Table 9.9. These, together with the spectroscopic constants of the states involved (Tables 6.8 and 6.16), give the absorption profiles that can be compared with experiments. This is done in Fig. 9.7 with low and high resolution. More specific details of the most relevant features are summarized in Table 9.10.

There is a remarkable agreement between theory and experiment, with an overall discrepancy of only a few hundred wavenumbers.

Both $CsCaCl_3$:Tm^{2+} and $CsCaBr_3$:Tm^{2+} calculated and experimental absorption spectra start with three peaks with a profile of low–low–high relative intensities between 14000 cm^{-1} and 18500 cm^{-1}. They correspond to transitions to various $4f^{12}(^3H_6)5dt_{2g}$ states which are not the lowest ones with this main character (cf. Table 6.16); the lowest have a strong spin character $S = 3/2$ and too low absorption oscillator strengths from a $4f^{13}$ ground state of pure $S = 1/2$ character (spin-forbidden absorptions). The first group of three peaks is followed by a gap beteen

Table 9.9 Absorption oscillator strengths from the $4f^{13}$ 1Γ_{6u} ground state to the excited states of Tm^{2+} in $CsCaCl_3$. Calculations are done at $d_{\text{Tm-Cl}}$=2.540 Å; the reference value is f_{ref}=7.218×10^{-4}

$4f^{12}5dt_{2g}$ main character						Interacting $4f^{12}5d$, $4f^{12}\phi_{ITE,a_{1g}}$			
$4f^{12}(^3H_6)$		$4f^{12}(^3F_4,^3H_5)$		$4f^{12}(^3H_4,^3F_{3,2})$					
State	$f^{\text{abs}}/f_{\text{ref}}$	State	$f^{\text{abs}}/f_{\text{ref}}$	State	$f^{\text{abs}}/f_{\text{ref}}$	State	$f^{\text{abs}}/f_{\text{ref}}$	State	$f^{\text{abs}}/f_{\text{ref}}$
3 Γ_{8g}	0.21	8 Γ_{6g}	0.62	34 Γ_{8g}	0.04	28 Γ_{6g}	0.01	80 Γ_{8g}	0.01
2 Γ_{6g}	0.01	14 Γ_{8g}	2.49	35 Γ_{8g}	1.56	29 Γ_{6g}	0.10	40 Γ_{6g}	1.45
4 Γ_{8g}	0.27	15 Γ_{8g}	0.10	17 Γ_{6g}	0.06	56 Γ_{8g}	0.01	41 Γ_{6g}	0.09
1 Γ_{7g}	0.02	16 Γ_{8g}	0.40	36 Γ_{8g}	0.05	30 Γ_{6g}	0.01	42 Γ_{6g}	0.22
5 Γ_{8g}	1.00	17 Γ_{8g}	0.17	37 Γ_{8g}	1.43	59 Γ_{8g}	0.14	85 Γ_{8g}	7.09
3 Γ_{6g}	0.46	18 Γ_{8g}	0.08	18 Γ_{6g}	2.21	60 Γ_{8g}	0.01	43 Γ_{6g}	0.51
6 Γ_{8g}	0.06	19 Γ_{8g}	0.01	38 Γ_{8g}	0.74	31 Γ_{6g}	0.26	86 Γ_{8g}	1.81
7 Γ_{8g}	0.98	20 Γ_{8g}	1.25	39 Γ_{8g}	3.32	61 Γ_{8g}	0.04	87 Γ_{8g}	2.78
8 Γ_{8g}	0.44	21 Γ_{8g}	0.04	19 Γ_{6g}	1.00	62 Γ_{8g}	0.57	44 Γ_{6g}	0.35
9 Γ_{8g}	0.73	22 Γ_{8g}	0.14	40 Γ_{8g}	0.40	63 Γ_{8g}	0.14	88 Γ_{8g}	0.42
4 Γ_{6g}	1.17	11 Γ_{6g}	0.17	41 Γ_{8g}	0.36	32 Γ_{6g}	0.96	45 Γ_{6g}	1.30
5 Γ_{6g}	0.01	12 Γ_{6g}	4.93	20 Γ_{6g}	0.94	33 Γ_{6g}	0.25	89 Γ_{8g}	0.98
10 Γ_{8g}	0.04	23 Γ_{8g}	4.24	42 Γ_{8g}	0.18	64 Γ_{8g}	1.09	90 Γ_{8g}	2.48
11 Γ_{8g}	2.33	13 Γ_{6g}	1.24	43 Γ_{8g}	3.64	65 Γ_{8g}	0.21	91 Γ_{8g}	0.01
6 Γ_{6g}	1.36	24 Γ_{8g}	6.65	21 Γ_{6g}	0.80	66 Γ_{8g}	0.04	92 Γ_{8g}	0.62
12 Γ_{8g}	0.02	25 Γ_{8g}	0.29	44 Γ_{8g}	0.04	34 Γ_{6g}	0.87	46 Γ_{6g}	1.81
7 Γ_{6g}	3.51	26 Γ_{8g}	0.13	45 Γ_{8g}	0.34	67 Γ_{8g}	0.27	93 Γ_{8g}	5.46
13 Γ_{8g}	2.68	27 Γ_{8g}	3.83	22 Γ_{6g}	0.22	69 Γ_{8g}	0.03	94 Γ_{8g}	8.51
		14 Γ_{6g}	3.52	46 Γ_{8g}	1.01	70 Γ_{8g}	0.04	95 Γ_{8g}	1.68
		28 Γ_{8g}	0.59	47 Γ_{8g}	1.56	71 Γ_{8g}	8.33	96 Γ_{8g}	0.07
		15 Γ_{6g}	3.86	48 Γ_{8g}	2.51	36 Γ_{6g}	1.74	47 Γ_{6g}	0.01
		15 Γ_{7g}	0.01	23 Γ_{6g}	0.28	73 Γ_{8g}	0.03	97 Γ_{8g}	1.42
		29 Γ_{8g}	0.88	24 Γ_{6g}	0.78	37 Γ_{6g}	1.32	48 Γ_{6g}	2.36
		30 Γ_{8g}	0.28	49 Γ_{8g}	6.99	74 Γ_{8g}	0.11	98 Γ_{8g}	5.94
		31 Γ_{8g}	0.98	50 Γ_{8g}	0.56	76 Γ_{8g}	0.21	49 Γ_{6g}	3.37
		16 Γ_{6g}	0.21	51 Γ_{8g}	2.33	38 Γ_{6g}	0.11	99 Γ_{8g}	0.67
		16 Γ_{7g}	0.05	25 Γ_{6g}	0.17	77 Γ_{8g}	2.18	100 Γ_{8g}	0.03
		32 Γ_{8g}	1.07	52 Γ_{8g}	3.05	78 Γ_{8g}	2.56	101 Γ_{8g}	3.83
		33 Γ_{8g}	0.09	26 Γ_{6g}	0.24	39 Γ_{6g}	0.43	51 Γ_{6g}	6.03
				27 Γ_{6g}	0.17	79 Γ_{8g}	0.33		
				53 Γ_{8g}	0.22				
				54 Γ_{8g}	5.68				

18500 cm^{-1} and 20000 cm^{-1}, where a new group starts corresponding to absorptions to various $4f^{12}(^3F_4,^3H_5)5dt_{2g}$ states. The group between 20000 cm^{-1} and 26000 cm^{-1} is a combination of four main peaks with a profile of low–high–high–low relative intensities, in good correspondence with experiments. The available spectral

Table 9.10 Theoretical and experimental absorption peak energies, in cm^{-1}, of Tm^{2+} in $CsCaBr_3$ and $CsCaCl_3$. Assignments above 28000 cm^{-1} are more uncertain

	$CsCaBr_3$:Tm^{2+}		$CsCaCl_3$:Tm^{2+}	
Transition	Theory	Experiment [51]	Theory	Experiment [51]
$4f^{13}(^2F_{7/2})\ 1\Gamma_{6u} \rightarrow$				
$4f^{12}(^3H_6)5dt_{2g}$ main character				
$5\ \Gamma_{8g}, 7\ \Gamma_{8g}$	15080	15080	15630	15040
$9\ \Gamma_{8g}, 4\ \Gamma_{6g}$	16530	16150	17050	16220
$13\ \Gamma_{8g}, 7\ \Gamma_{6g}$	18050	17560	18550	17530
$4f^{12}(^3F_4)5dt_{2g}$ main character				
$14\ \Gamma_{8g}$	20790	20190	21290	20250
$4f^{12}(^3H_5)5dt_{2g}$ main character				
$12\ \Gamma_{6g}, 23\ \Gamma_{8g}, 24\ \Gamma_{8g}, 27\ \Gamma_{8g}$	23450	22740	23980	22750
$14\ \Gamma_{6g}, 15\ \Gamma_{6g}$	24640	23980	251190	23970
$31\ \Gamma_{8g}$	25960	25090	26460	25080
$4f^{12}(^3H_4, ^3F_{3,2})5dt_{2g}$ main character				
$35\ \Gamma_{8g}$	27630	26940	28130	27090
$39\ \Gamma_{8g}$	29120	[29040][a]	29680	28270
$43\ \Gamma_{8g}$	30330		30790	29940
$49\ \Gamma_{8g}$	31590	30740	32120	[32020][a]
$54\ \Gamma_{8g}$	33720	31990	34190	

[a] Broad unresolved band

assignments identified the first states of a $4f^{13}5de_g$ configurational character right above 26000 cm^{-1} ($CsCaBr_3$:Tm^{2+}) and 29000 cm^{-1} ($CsCaCl_3$:Tm^{2+}) [51]. The multiconfigurational ab initio calculations still find the $4f^{13}5dt_{2g}$ manifold at these energies and the lowest states with a partial $4f^{13}5de_g$ character, horizontally shifted in the configuration coordinate diagram (cf. Fig. 6.7), appear at higher energies, around 35000 cm^{-1}.

Emission

Experimental and calculated emission profiles are also shown in Fig. 9.7. Systematic overestimations between 4% and 8% are found: The three emission bands of $CsCaBr_3$:Tm^{2+} at 12280, 13720, and 19140 cm^{-1} are calculated to peak at 12740, 14500, and 20420 cm^{-1}, respectively; the two of $CsCaCl_3$:Tm^{2+} at 12500 and 19470 cm^{-1} are calculated at 13240 and 21000 cm^{-1}, respectively. Grimm et al. [51] attributed the three emitting states of Tm^{2+} in $CsCaBr_3$ and the two of Tm^{2+} in $CsCaCl_3$ to the $4f^{13}5dt_{2g}$ configuration, and this is what is found in the calculations. The two first emissions of $CsCaBr_3$:Tm^{2+} were considered to be from the lowest high-spin ($S = 3/2$) and low-spin ($S = 1/2$) $4f^{12}(^3H_6)5dt_{2g}$ states; according to the calculations, they are from the $1\ \Gamma_{8g}$ and $3\ \Gamma_{8g}$ states of $4f^{12}(^3H_6)5dt_{2g}$ character, but both of them have mixed spins with a dominant high-spin character (cf.

Table 6.16). The third emission of the bromide was attributed to the lowest state of configurational character $4f^{12}(^3F_4)5dt_{2g}$, where spin-orbit coupling was considered to be considerable and give the emitting state an almost equal weight of $S = 3/2$ and $S = 1/2$ spin characters. The calculations find strong spin-orbit coupling and consequent doublet–quartet mixing already at lower energies, at the high energy side of the $4f^{12}(^3H_6)5dt_{2g}$ states; they attribute the third emission to the 8 Γ_{6g} state, of $4f^{12}(^3F_4,^3H_5)5dt_{2g}$ main configurational character.

The gaps predicted by Grimm et al. [51] to allow the emissions were found and identified in the calculations; they are shown in Table 9.11 for $CsCaBr_3$:Tm^{2+} and $CsCaCl_3$:Tm^{2+}, together with the maximum vibrational quanta. The general assumption that non-radiative decay occurs when energy gaps are smaller than 5 vibrational quanta [59] explains why $CsCaCl_3$:Tm^{2+} only shows two of the emissions of $CsCaBr_3$:Tm^{2+}: the larger breathing mode vibrational frequency of the chloride, which corresponds to lighter breathing Cl ligands, lowers the ratio between gap and vibrational quanta of the $4f^{12}(^3H_6)5dt_{2g}$ 3 Γ_{8g} state enough so as to make the gap unable to stop the non-radiative decay, so quenching the second emission. The gaps in the manifold of excited states thus explain all observed emission bands and the absence of more sufficiently large energy gaps explains why there are not more emission bands in Tm^{2+}-doped $CsCaBr_3$ and $CsCaCl_3$.

Table 9.11 Multiple emissions of Tm^{2+} in $CsCaBr_3$ and $CsCaCl_3$. Calculated maximum vibrational quanta $\bar{\nu}_{a_{1g}}$ are 220 cm^{-1} for $CsCaBr_3$ and 350 cm^{-1} for $CsCaCl_3$

	Theory			
	Gap[a]	Gap/$\bar{\nu}_{a_{1g}}$ [b]	Prediction	Experiment[c]
First emission	$4f^{12}(^3H_6)5dt_{2g}$ 1 Γ_{8g} $S = \frac{3}{2}$			$4f^{12}(^3H_6)5dt_{2g}$ $S = \frac{3}{2}$
$CsCaBr_3$	3822	17.4	Emission	Emission
$CsCaCl_3$	4264	12.2	Emission	Emission
Second emission	$4f^{12}(^3H_6)5dt_{2g}$ 3 Γ_{8g} $S = \frac{3}{2}$			$4f^{12}(^3H_6)5dt_{2g}$ $S = \frac{1}{2}$
$CsCaBr_3$	1460	6.6	Emission	Emission
$CsCaCl_3$	1454	4.2	Quenching	Quenching
Third emission	$4f^{12}(^3F_4,^3H_5)5dt_{2g}$ 8 Γ_{6g} $S = \frac{3}{2}$ mixed			$4f^{12}(^3F_4)5dt_{2g}$ $S = \frac{3}{2}$ mixed
$CsCaBr_3$	2617	11.9	Emission	Emission
$CsCaCl_3$	2597	7.4	Emission	Emission

[a]Energy gap below the emitting state, in cm^{-1}
[b]Number of maximum vibrational quanta contained in the energy gap. Numbers smaller than 5 are expected to tive low temperature luminescence quenching [59]
[c]Observations and interpretations from [50]

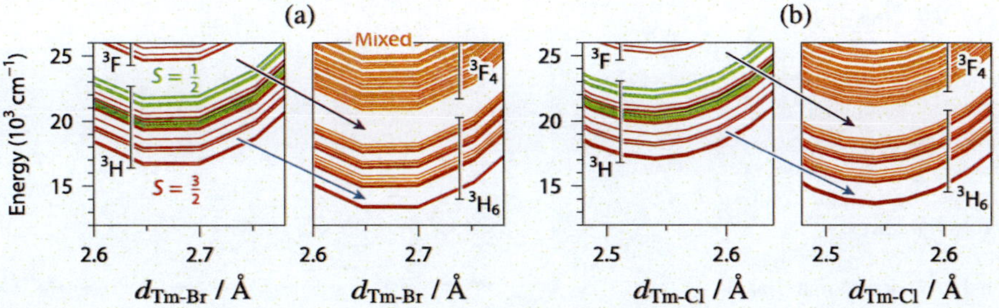

Fig. 9.8 Origin of the energy gaps in the $4f^{12}5dt_{2g}$ emanifold of Tm^{2+} in $CsCaBr_3$ (**a**) and $CsCaCl_3$ (**b**). In both cases, spin-orbit-free MS-RASPT2 calculations are shown in the left panel and spin-orbit coupling RASSI-SO calculations in the right panel. States are colored according to their spin character: red indicates >80% ($S = \frac{3}{2}$); green indicates >80%($S = \frac{1}{2}$); orange indicates none of the spin characters reaches 80%. Blue and purple arrows signal changes in the second and third energy gaps due to spin–orbit coupling. Main free-ion characters of the $4f^{12}$ subshells are indicated

It is interesting to give a closer view to the gaps. The first energy gap in the configuration coordinate diagram, of around 4000 cm^{-1}, separates the highest $4f^{13}$ state from the lowest $4f^{12}5d$ state. This large f–d gap allows for the observation of luminescence from the lowest $4f^{N-1}5d$ state to the $4f^N$ manifold. Accurate calculations of this gap and the corresponding interconfigurational transitions are very demanding both in terms of dynamic and static correlation, and depend on how well the differences between electron correlations in the two configurations are calculated [60]; they require the use of a $5f$ shell or double-shell [61] (Sect. 2.4).

The second gap, of around 1500 cm^{-1}, is better analyzed with the help of Fig. 9.8. Before spin-orbit coupling (left (a) and left (b) panels), there is a small energy gap of around 850 cm^{-1} that separates the three lowest $4f^{12}(^3H)5dt_{2g}$ quartet (S=3/2) states from higher states. It is a manifestation of the crystal-field splitting of the ^3H term of the $4f^{12}$ subshell and its coupling with the $5dt_{2g}$ electron, which is more complex than and not to be confused with the mere superposition of a ^3H pattern on top of a $5dt_{2g}$ pattern. The nearest doublet (S=1/2) states appear 3000 cm^{-1} above them. Spin-orbit coupling interactions mix the lowest quartets basically with other quartets and only very slightly with doublets (the three lowest-energy states show more than 80% S=3/2 character), which results in a slight increase of the gap up to around 1500 cm^{-1}. It is only above the gap that the typical spread of states caused by a spin-orbit coupling mixture with varying spin contributions (S=3/2 and S=1/2), which tends to destroy gaps, is observed.

The third gap, of around 2600 cm^{-1}, has its origin in the separation between the highest $4f^{12}(^3H)5dt_{2g}$ and the lowest $4f^{12}(^3F)5dt_{2g}$ states. It is mostly due to the ^3H–^3F separation in the $4f^{12}$ subshell and in part to the strength of the $4f^{12}\times 5d$ coupling. Spin-orbit coupling does not differ much in the upper and lower $4f^{12}$ terms and it only reduces slightly the gap from 2800 to 2600 cm^{-1}, which becomes now the difference between the highest $4f^{12}(^3H_6)5dt_{2g}$ state, 13 Γ_{8g}, and the lowest

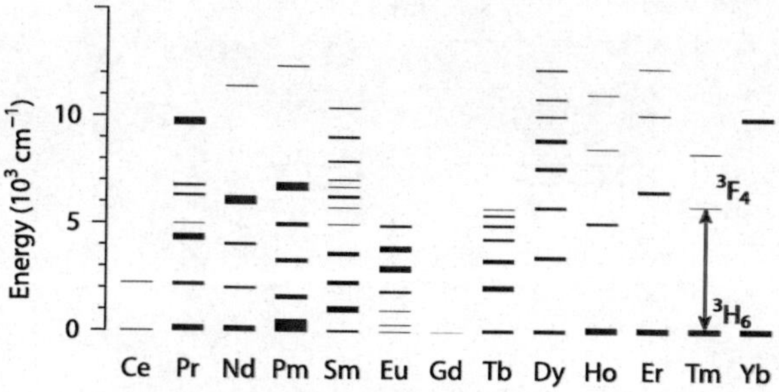

Fig. 9.9 Energy level diagram of the $4f^N$ ground and low excited levels of trivalent lanthanide free-ions [63]. The $^3H_6$–$^3F_4$ energy gap, responsible for the gap that enables emission from the lowest $4f^{12}(^3F_4)5dt_{2g}$ excited state in Tm^{2+}-doped materials is signaled

$4f^{12}(^3F_4)5dt_{2g}$ state, 8 Γ_{6g}. The width of the third gap is thus related to the energy difference between the $^3H_6$ ground level and the first excited $^3F_4$ level of Tm^{3+}.

In order to predict multiple emissions in a lanthanide activated phosphor, the appearance of a sufficiently large energy gap within the crowded $4f^{N-1}5d$ manifold is required, with parallel potential energy surfaces, and many additional factors play their roles, including local symmetry and coordination, overlaps between $4f^N$ and $4f^{N-1}5d$ manifolds, quenching mechanisms through intrvalence charge transfer states, and more. Multiconfigurational ab initio calculations are definitely required. Nevertheless, the increased understanding of the origin of the gaps in the $4f^{12}5d$ manifold of Tm^{2+} can be used for limited, fast predictions. E.g., the second gap is somehow fortunate and hard to predict without sophisticated calculations, but the third gap in Tm^{2+} originates from a large energy gap between the ground and first excited levels in Tm^{3+}: The $^3H_6$–$^3F_4$ energy gap of around 5600 cm^{-1} in Tm^{3+} [62] results in a 2600 cm^{-1} gap in the Tm^{2+} $4f^{12}5d$ manifold, equal to 11.9 vibrational quanta in $CsCaBr_3$ and 7.4 vibrational quanta in $CsCaCl_3$. Inspection to the Dieke diagram [63] (Fig. 9.9) indicates that it is only among the heavy lanthanides (N>7) that suitable candidates can be found, like Dy^{2+}, Ho^{2+}, Er^{2+}, or Yb^{2+}. It explains the absence of luminescence from higher excited states in Eu^{2+} and Sm^{2+}. For Yb^{2+}, with a large energy gap between the $^2F_{7/2}$ and $^2F_{5/2}$ states for the Yb^{3+} core, emission from higher excited $4f^{13}5d$ states can be expected; although multiconfigurational ab initio calculations confirm the presence of a large gap of 4600 cm^{-1} in $CsCaBr_3$:Yb^{2+} [60], no emission from a $4f^{13}(^2F_{5/2})5d$ state seems to have been reported yet.

Finally, calculated 1 Γ_{8g} radiative lifetimes (in vacuum; local field corrections, which would result in shorter lifetimes, require unavailable refractive indexes of the host materials [64]) are 847 μs and 704 μs in $CsCaBr_3$:Tm^{2+} and $CsCaCl_3$:Tm^{2+}, to be compared with the respective experimental results 296 μs and 278 μs [52]. Respective values for the upper emission from 8 Γ_{6g} are 1.54 μs and 1.41 μs theoretical and 1.90 μs and 1.17 μs experimental. The radiative lifetimes of the two states differ by several orders of magnitude, which are well reproduced by the calculations, in correspondence with the different spin characters of the two emitting states: the higher emitting state 8 Γ_{6g}, although of dominant S=3/2 character, has a much higher S=1/2 character than the lower emitting state 1 Γ_{8g}; this makes the emission to the

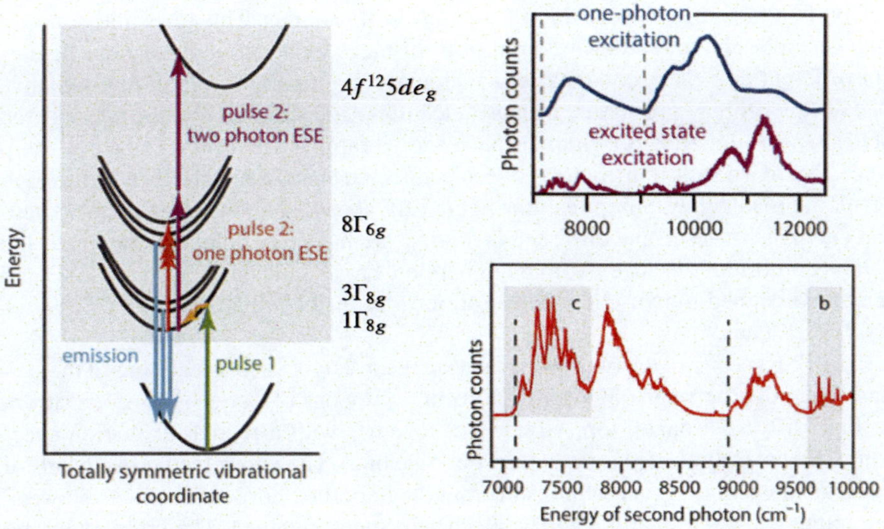

Fig. 9.10 Excited state excitation in $CsCaBr_3$:Tm^{2+} [58]. Left panel: Schematic configuration coordinate diagram of a two-color ESE experiment. A first photon excites to the $4f^{12}(^3H_6)5dt_{2g}$ $1\Gamma_{8g}$ lowest emitting state (green arrow). A second photon excites to higher $4f^{12}5d$ excited states (red arrows) while the emission from the $4f^{12}(^3F_4,^3H_5)5dt_{2g}$ $8\Gamma_{6g}$ state is probed (long blue line). In an ESE spectrum, the intensity of the probed emission is represented vs. the second photon energy. Right panels: Top: Shifted regular excitation spectrum of the $8\Gamma_{6g}$ emission (blue line) and ESE spectrum (violet line). Bottom: More detailed representation of the ESE spectrum 7000–10000 cm^{-1} above $1\Gamma_{8g}$ showing the remarkable sensitivity of this technique to probe electronic states above the upper emitting state $8\Gamma_{6g}$

$4f^{13}$ S=1/2 manifold spin-enabled from 8 Γ_{6g} whereas it is still spin-forbidden from 1 Γ_{8g}.

Excited State Excitation

When emissions from two or more excited states are present, it is possible to do excited state excitation (ESE) experiments, which provide very fine details of the excited states above the higher emitting state. In these, as it is shown in Fig. 9.10 (left panel), after the lower emitting state is populated with a first photon, the intensity of the emission from the higher state is recorded vs. the energy of a second photon that excites from the lower emitting state above the higher emitting state.

Few ESE spectra have been reported on d-d transitions of transition metal ions in solids [65], f-f transitions of lanthanide ions in solids [66, 67], aimed at studying high energy $4f^N$ levels in the vacuum UV or to obtain insight in mismatches for energy transfer in upconversion processes, and f-d transitions of lanthanide ions in solids [68], probing mid-infrared transitions between $4f^{N-1}5d$ states and impurity-trapped-exciton $4f^{N-1}\phi_{ITE}$ states. The multiple $4f^{12}5d$ emitting states of $CsCaBr_3$:Tm^{2+} and $CsCaCl_3$:Tm^{2+} offer the unique possibility of measuring $4f^{N-1}5d \rightarrow 4f^{N-1}5d$ transitions between excited states of a lanthanide ion in solids, i.e. both intraconfigurational $4f$-$4f$ and $5d$-$5d$ transitions. This gives the opportunity to probe the variation in equilibrium distance between activator and ligands for different $4f^{12}5d$ excited states, which plays an important role in non-radiative decay (quenching) processes, because electronic transitions between two excited states with a similar equilibrium distance and vibrational frequency will show sharp line spectra, whereas transitions between contracted and expanded excited states will result in broad bands. Sharp features in the ESE spectra demonstrate that the potential energy surfaces of the different $4f^{12}5dt_{2g}$ are indeed parallel, i.e. they have the same equilibrium distance. At higher energies, broad bands appear in the ESE and they are assigned to two-photon absorptions to mixed configurational $4f^{12}5dt_{2g}$ + $4f^{12}5de_g$ states.

The ESE spectra of the two materials were recorded for varying energies of the second laser pulse between 6897 and 12500 cm^{-1}. The one of $CsCaBr_3$:Tm^{2+} is shown in Fig. 9.10 (right panel, top, violet line) together with a regular excitation spectrum conveniently shifted for comparison (blue line). The unresolved broad bands of the regular excitation spectrum result from the important horizontal offsets between the initial $4f^{13}$ and final $4f^{12}5dt_{2g}$ electronic states involved. The high-resolution sharp lines of the ESE spectrum are an indication of the very small horizontal offsets between different states of the $4f^{12}5dt_{2g}$ configuration; a region is shown in more detail in the bottom part of the right panel. The sharp lines have spectral widths between 5 and 50 cm^{-1} and show irregular patterns that exclude them from being part of vibrational progressions; they are interpreted as zero-phonon-lines of electronic transitions. The ESE spectrum shows onsets of a group of lines at 7100 cm^{-1} followed by a gap and another group of lines starting at 9000 cm^{-1}, all of them attributed to $4f^{12}(^3H_6)5dt_{2g}$ $1\Gamma_{8g} \rightarrow 4f^{12}(^3F_4, ^3H_5)5dt_{2g}$ transitions. According to the multiconfigurational ab initio calculations (cf. Table 6.16 for $CsCaCl_3$:Tm^{2+}), all the states involved have the same bond length within 0.001–0.002 Å and breathing mode vibrational frequency within 4–5 cm^{-1}. It is also interesting to remark that the energy of the $^3H_6 \rightarrow ^3F_4$ transitions of Tm^{3+} of around 5500–6000 cm^{-1} are significantly smaller than the onset at 7100 cm^{-1}, which illustrates that the energy level structure of the $4f^{12}5d$ manifold is far from a simple superposition of $4f^{12}$ and $5d$ splittings, as it is obvious from the calculations.

At energies above 10000 cm^{-1} in the ESE spectra, broad bands with a FWHM of hundreds to 1000 cm^{-1} are found. The broad and structureless nature of these bands indicates that they cannot be assigned by transitions within the $4f^{12}$ subshell of the $4f^{12}5d$ states, but rather to states with potential energy surfaces horizontally shifted with respect to $4f^{12}(^3H_6)5dt_{2g}$ $1\Gamma_{8g}$. However, as shown in Fig. 6.7, horizontally shifted states appear only 20000 cm^{-1} above $1\Gamma_{8g}$. Above that energy, a crowded manifold starts with weakly absorbing impurity-trapped-exciton character $4f^{12}\phi_{ITE}$ and mixed $4f^{12}5dt_{2g}$+ $4f^{12}5de_g$ character, all of them with important horizontal offsets with respect to $1\Gamma_{8g}$. Many transitions from $1\Gamma_{8g}$ to this manifold involve $4f$–$4f$ and $5dt_{2g}$–$5de_g$ excitations. and, whereas both of them are parity forbidden in one-photon absorption, they are allowed in two-photon absorption. This gives us the most plausible interpretation of the broad bands in the ESE spectrum as due to two-photon absorptions to the mixed $4f^{12}5dt_{2g}$+ $4f^{12}5de_g$ manifold.

9.6 Yb^{2+} in $SrCl_2$

The detailed analysis of the electronic structure of the excited states of Yb^{2+}-doped hosts (Sect. 6.2.6) opens the possibility for a theoretical interpretation of their spectroscopy. The potential energy curves and absorption oscillator strengths summarized in Table 6.17 for $SrCl_2$:Yb^{2+} allow to produce theoretical absorption spectral profiles with lower and higher resolution, as shown in Fig. 9.11, which can be directly compared with experiments. In this section, such a comparison will be made following the guidelines established in [69], where the use of a slightly less developed version of the computational choices used in [70] and discussed here, gave some quantitative differences, although maintaining the same qualitative understanding of the spectrum.

Absorption Spectrum

The interpretation of the absorption spectrum of the luminescent material $SrCl_2$:Yb^{2+} emerging from the multiconfigurational ab initio calculations is summarized in Table 9.12, where the theoretical electronic origins and absorption intensities are compared with experiments and previous interpretations.

A pioneering work by McClure and collaborators [71] measured the low temperature absorption spectrum of $SrCl_2$:Yb^{2+} and interpreted it in the framework of Crystal-Field Theory. A strong interaction between the $4f^{13}5de_g$ and $4f^{13}5dt_{2g}$ configurations, induced by a comparable size of $4f^{13}$ spin-orbit coupling and $5d$ crystal-field splitting was reported, which is supported by the ab initio calculations (Sect. 6.2.6). But some of the 12 originally identified peaks were later attributed to three different types of Yb^{2+} sites and Yb^{2+}–Yb^{3+} pairs [54, 72] after new CFT calculations suggesting much stronger cyrstal-field splittings and analyses of photo-

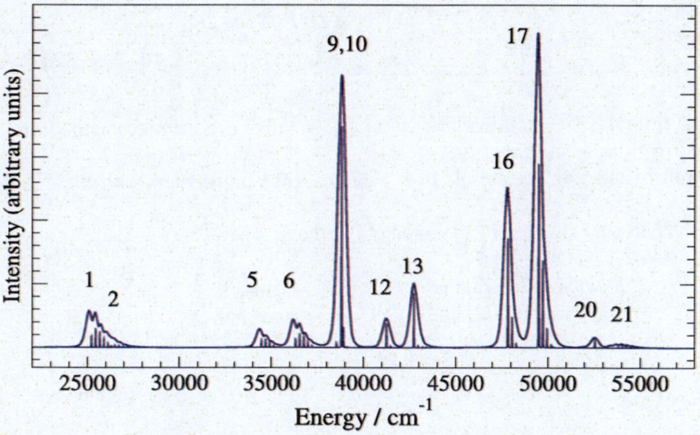

Fig. 9.11 Calculated $4f^{14}\rightarrow 4f^{13}5d$ absorption spectrum of Yb^{2+} in $SrCl_2$ for two different resolutions (broadening parameters: 5 and 35 cm^{-1}). The most intense $1\ A_{1g}\rightarrow$ i T_{1u} electronic origins are indicated (cf. Table 6.17)

Table 9.12 Experimental and theoretical $4f^{14}{\rightarrow}4f^{13}5d$ absorption spectrum of Yb^{2+}-doped $SrCl_2$: absorption energies (in cm^{-1}), relative oscillator strengths and assignments

Experiment[a]					Theory				
Yb^{2+}[b]			Yb^{2+}-Yb^{3+}[b]						
1	1′	1″		f_i/f_1[c]	Peak	Origins	f_i/f_1[d]	Energy	Error
Region I									
$4f^{13}(^2F_{7/2})5de_g$					$4f^{13}(^2F_{7/2})5de_g$				
27049				1.00	#1	$1\ T_{1u}$	1.00	25201	−1848
						$2\ T_{1u}$	0.13	26507	
	28500					$3\ T_{1u}$	0.03	26589	−1911
		29272							
Region II									
					$4f^{13}(^2F_{5/2})5de_g + 4f^{13}(^2F_{7/2})5dt_{2g}$				
			35038[e]	0.13	#2	$5\ T_{1u}$	0.38	34433	−605
$4f^{13}(^2F_{5/2})5de_g$									
37625				0.20	#3	$6\ T_{1u}$	0.69	36288	−1337
	38748			1.1					
		39749		1.1					
$4f^{13}(^2F_{7/2})5dt_{2g}$									
41089				2.4	#4	$9\ T_{1u}$	3.67	38788	−2301
						$10\ T_{1u}$	0.35	38889	
					#5	$12\ T_{1u}$	0.43	41238	
					#6	$13\ T_{1u}$	0.92	42713	
			45749[f]	0.2					
			46649[f]	0.4					
Region III									
					$4f^{13}(^2F_{5/2})5dt_{2g}$				
			48575	2.8	#7	$16\ T_{1u}$	2.40	47820	−755
			50403	1.8	#8	$17\ T_{1u}$	4.91	49471	−932

[a] Band maxima from 4.2 K absorption spectrum [71]. Relative oscillator strengths from [71] with background corrections from [55]

[b] Three types of isolated Yb^{2+} centers 1, 1′, 1″, and several Yb^{2+} centers with Yb^{3+} neighbors are identified [72]

[c] Intensity of the i-th peak divided by the intensity of the first peak

[d] $f_i/f_1 = f(A_{1g}{\rightarrow}i\ T_{1u})/f(A_{1g}{\rightarrow}1\ T_{1u})$

[e] Assigned to $4f^{13}(^2F_{7/2})5de_g$ in [72] and $4f^{13}(^2F_{7/2})6s$ in [55]

[f] Assigned to $4f^{13}(^2F_{5/2})5de_g$ in [72] and $4f^{13}(^2F_{5/2})6s$ in [55]

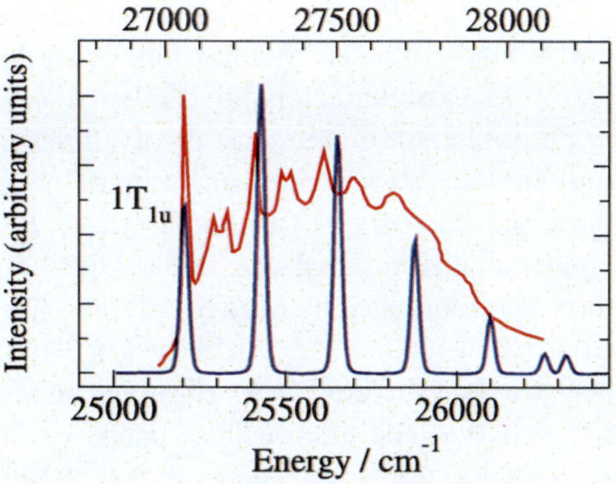

Fig. 9.12 Fine structure of the lowest $4f^{14}{\rightarrow}4f^{13}5d$ absorption band of Yb^{2+}-doped $SrCl_2$. Red line and upper scale: 4.2 K experimental spectrum [73]. Blue line and lower scale: calculated a_{1g} vibrational progression of the $A_{1g}{\rightarrow}1\ T_{1u}$, $2\ T_{1u}$, and $3\ T_{1u}$ absorption bands (cf. Table 6.17)

luminescence spectra showing a highly complex luminescence behavior. New experimental emission spectra at different temperatures and CFT studies [55] reinterpreted the assignments and proposed the co-participation of a low-lying $4f^{13}6s$ configuration. Altogether, after several thorough experimental studies and CFT calculations, the detailed assignment and understanding of the absorption spectrum is not a closed matter, which illustrates the high complexity of the excited manifolds of hosts doped with activators that can coexist in multiple valence states, like Yb^{2+} and Yb^{3+} in this case. Detailed knowledge of the excited states of one species alone, like Yb^{2+} in O_h sites, becomes a key issue for the understanding of the whole picture in these complex luminescent materials. To insist in this complexity, it may also be pointed out that the emission spectrum of Yb^{2+} changes drastically from $SrCl_2$:Yb^{2+} to SrF_2:Yb^{2+}, where it shows a blue emission band similar to the $SrCl_2$:Yb^{2+} emission, plus a broad red band of a different nature; and to CaF_2:Yb^{2+}, where it only shows the broad band; and to BaF_2:Yb^{2+}, where both emissions are quenched. But all this is discussed in Sect. 11.2.1.1.

Table 9.12 shows 4.2 K absorption band maxima and relative oscillator strengths f_i/f_1 from [71], the latter with background corrections obtained in [55] with 10 K data, which could not be determined with great precision, together with theoretical peaks of the $1\ A_{1g}{\rightarrow}i\ T_{1u}$ electric dipole allowed transitions. Three regions are observed in the theoretical absorption spectrum. Region I has only one band with two T_{1u} origins of clear $4f^{13}(^2F_{7/2})5de_g$ configurational character and is dominated by the vibrational progression of the most intense absorption to the $1\ T_{1u}$ level. This is shown in detail in Fig. 9.12, where the theoretical a_{1g} vibrational progression is shown to reproduce reasonably the band shape. The first theoretical peak deviates $-1848\ cm^{-1}$ from experiment and the second peak has not been detected experimentally, perhaps because it is very close to the first and immersed in the vibrational progression. The $3\ T_{1u}$ peak agrees well with the $28500\ cm^{-1}$ peak and no T_{1u} state appears near $29272\ cm^{-1}$. This would agree with the CFT assigmnents of [71], where $2\ T_{1u}$ is not observed and $4\ T_{1u}$ is at higher energies, and it would suggest that, if the interpretation of the third experimental peak could justify the hypothesis of a different type of Yb^{2+} center [72], this is not necessary for the second peak.

The spectral region II extends from 34000 cm^{-1} to 43000 cm^{-1}. Although it corresponds to a set of states with mixed $4f^{13}(^2F_{5/2})5de_g$, $4f^{13}(^2F_{7/2})5dt_{2g}$, and $4f^{13}(^2F_{7/2})6s$ characters (cf. Table 6.17), only the $4f^{13}5de_g$ and $4f^{13}5dt_{2g}$ configurations interact in the states involved in the spectrum; the $4f^{13}6s$ character appears only in states above 43000 cm^{-1}, with low absorption intensity. This region contains the electronic origins from 5 T_{1u} to 13 T_{1u} grouped in 5 bands; the first two have appreciable vibrational structures of the 5 T_{1u} and 6 T_{1u} electronic origins, and the last three are dominated by the zero-phonon lines of the 9 and 10 T_{1u} (third band), 12 T_{1u} (fourth band) and 13 T_{1u} (fifth band). The first three bands agree with experimental peaks within the deviations observed in region I and with not very different intensity patterns. The first of these bands was traditionally hard to interpret and it was attributed to $4f^{13}(^2F_{7/2})5de_g$ of Yb^{2+}-Yb^{3+} pairs in [72] and to $4f^{13}(^2F_{7/2})6s$ in [55]; however, the multiconfigurational ab initio calculations make these assumptions unnecessary because the band energy agrees with a $4f \rightarrow 5d$ transition of a regular Yb^{2+} center and its irregular shape is explained by the mix of two configurational characters with different Yb-Cl bond lengths, $4f^{13}5de_g + 4f^{13}5dt_{2g}$. The two bands peaking at 38748 cm^{-1} and 39749 cm^{-1} and attributed to different Yb^{2+} centers [72] cannot be explained with the ab initio calculations of the regular center. The last two bands in region II don't find a correspondence in the experiments. This is a region where irregular bands attributed to Yb^{2+}-Yb^{3+} pairs appear and this can well be the explanation. For them to correspond to the 45749 cm^{-1} and 46649 cm^{-1} experimental bands, their deviations should jump out of the 2000 cm^{-1} into the 4000 cm^{-1} region. The $4f^{13}(^2F_{5/2})5de_g$ of Yb^{2+}-Yb^{3+} pairs interpretation [72] might well be correct in its Yb^{2+}-Yb^{3+} part, although the pure configurational character of Yb^{2+} would not be justified. The $4f^{13}(^2F_{5/2})6s$ interpretation of these bands [55] is not supported.

Region III of the absorption spectrum involves T_{1u} states of $4f^{13}(^2F_{5/2})5dt_{2g}$ character: 16 T_{1u} and 17 T_{1u}. They agree well with the two experimental bands at 48575 cm^{-1} and 50403 cm^{-1}, which, again, can well be attributed to regular Yb^{2+} centers.

States $4f^{13}(^2F_{5/2})6s$ character are above the explored energies of the absorption spectrum.

Emission Spectrum

The emission spectrum of Yb^{2+}-doped $SrCl_2$ single crystals, studied first at 4.2 K under 334 and 313 nm mercury lines near-ultraviolet excitation [54] and later from 300 K to 10 K under 355 nm excitation [55], was recognized to have a very complex luminescence behavior, with five emission bands peaking at 19000, 23900, 24700, 25400, and 26500 cm^{-1}, and irregular variations of their relative intensities and band structure with temperature, excitation wavelength, and concentration. Three of them were attributed to defects, possibly involving Yb^{2+}-Yb^{3+} pairs, and only two were considered genuine Yb^{2+} bands: The 24700 cm^{-1} band, with a long decay time (few tenths of a millisecond lifetime), was assigned as a vibronic emission from quasi-degenerate 1 E_u and 1 T_{2u} excited states to the 1 A_{1g} ground state; the 26500 cm^{-1} band, with a short decay time (emission lifetime in the 0.7–0.78 μs range), as an electric dipole allowed emission from 1 T_{1u}; and analyses of absorption and emission spectra indicate their electronic origins are 2000 cm^{-1} apart.

The multiconfigurational ab initio results in Table 6.17 give basically the same result for the Yb^{2+} bands: A 1993 cm^{-1} separation was found between 1 T_{1u} and the average of 1 E_u and 1 T_{2u}. And the computed spontaneous lifetime of the electric dipole allowed 1 $T_{1u} \rightarrow 1\ A_{1g}$ emission was 0.36 μs, with a more than reasonable underestimation mainly attributable to missing contributions of vibronic emissions from the close lying 2 T_{2u} and 1 A_{2u} levels. They also support the interpretation of the other three bands: Assignment of the 19000 cm^{-1} and 23900 cm^{-1} bands, of lower energy than the first excited state of Yb^{2+}, to emissions between two Yb^{2+} excited states would require to populate a higher excited metastable level by either excited-state absorption or energy transfer between two excited impurities [50, 74], but the calculated energies indicate that such processes are unlikely to happen at 334 and 313 nm excitation wavelengths because the excited-state absorptions and emissions would basically be two-electron excitations of $u - u$ parity character, hence doubly forbidden transitions. And no state is found 700 cm^{-1} above the first excited state that could justify the 25400 cm^{-1} band.

9.7 Actinide Ions in Hosts

The importance of actinide ions in solids has been recognized long ago and their spectroscopy in condensed phases, specially that of trivalent ions, has been the subject of intense research for a while [75]; then, after a low profile period, the interest for the chemistry and physics of the ground and excited states of actinide ions in condensed phases is experiencing a new bust [76, 77]. As their lanthanide partners do, actinide ions enjoy very rich and complex manifolds of $5f^N, 5f^{N-1}6d, 5f^{N-1}7s$, $5f^{N-1}\phi_{ITE}$ and charge transfer excited states; however, the $5f$ orbitals of actinides are not as internal as the $4f$ orbitals or lanthanides, they are not as well shielded by the more external $6s^2 6p^6$ full shells as the $4f$ are by the $5s^2 5p^6$ shells, and they have much stronger electronic interactions with the external $6d$ orbitals than the $4f$ have with $5d$ [78]. Such stronger interactions increased the difficulties of phenomenological approaches, like crystal-field theory, to handle $5f^N$ and $5f^{N-1}6d$ configurations and limited their applications, but they also make actinide ions more demanding than lanthanide ions from the ab initio methodological point of view. This section summarizes the results of a few multiconfigurational ab initio calculations on excited manifolds of Pa^{4+}, U^{4+}, and U^{3+} ions in six-fold octahedral coordination of chloride hosts.

9.7.1 Pa^{4+} in Cs_2ZrCl_6

The $6d \rightarrow 5f$ fluorescence and $5f \rightarrow 6d$ absorption spectra of Pa^{4+}-doped Cs_2ZrCl_6 have been measured with an exceptionally high-resolution [79, 80]. A multiconfigurational ab initio calculation of the potential energy curves of the manifolds of the $5f^1$ and $6d^1$ configurations along the totally symmetric breathing mode was performed on the octahedral $(PaCl_6)^{2-}$ cluster embedded in Cs_2ZrCl_6 [81] that allowed for direct comparisons between theory and experiment.

Excellent agreements were found with experimental breathing mode vibrational frequencies: A theoretical value of 310 cm^{-1} in all the $5f^1$ states versus 310 cm^{-1} experimental in the two lowest states of this configuration, $5f^1\ {}^2F_{5/2}\ \Gamma_{7u}$ and $5f^1\ {}^2F_{5/2}\ \Gamma_{8u}$. And a theoretical 303 cm^{-1} value versus 302 cm^{-1} experimental in the lowest state of the $6dt_{2g}^1$ configuration, ${}^2T_{2g}\ \Gamma_{8g}$. A lower, theoretical 271 cm^{-1} value was found in the upper $6de_g^1\ {}^2E_g\ \Gamma_{8g}$ state, but experiments have not yet been reported.

Regarding Pa-Cl equilibrium distances in the ground and excited states, they show the same bond length pattern found for lanthanide ions: $d_{e,\text{Pa-Cl}}(6dt_{2g}^1) < d_{e,\text{Pa-Cl}}(5f^1) < d_{e,\text{Pa-Cl}}(6de_g^1)$; as a matter of fact, this was the first case of an f element where such a bond length pattern was found [81]. A narrow bond length range of 2.674–2.681 Å was found in the states of the $5f^1$ configuration, a shorter distance of 2.657 Å in the most stable of the $6d^1$ configurations, $6dt_{2g}^1$, and a longer distance of 2.735 Å in the most unstable $6de_g^1$. Experimental data to make direct comparisons are missing, but an indirect check of the bond length differences can be made by contrasting the intensity profile of the $6dt_{2g}^1\ {}^2T_{2g}\ \Gamma_{8g} \rightarrow 5f^1\ {}^2F_{5/2}\ \Gamma_{7u}$ emission band, which has been measured with very high resolution [79]. In effect, such profile is known to be determined by the absolute value of the bond length offset (difference between the equilibrium distance in the two electronic states involved in the transition) and the vibrational frequencies. Then, since the agreement of the vibrational frequencies is excellent, deviations between the high-resolution experimental emission band profile and the calculated one must be entirely due to disagreements in the absolute values of the bond length offset. Fortunately, the agreement between the calculated and experimental profiles was also remarkable (cf. Fig. 5 of [81]) and the absolute values of the calculated bond length offsets have to be considered to be of a very good quality. Band profiles do not give information on the signs of these offsets.

The energies of the electronic origins with respect to the ground state are in a general good agreement with experiments (Fig. 9.13): $5f^1\ {}^2F_{5/2}\ \Gamma_{8u}$ 1837 versus 2108 cm^{-1}; $5f^1\ {}^2F_{7/2}\ \Gamma_{7u}$ 5709 versus 5350 cm^{-1}, Γ_{8u} 7464 versus 7272 cm^{-1}, Γ_{6u} 8526 versus 8173 cm^{-1}; $6dt_{2g}^1\ {}^2T_{2g}\ \Gamma_{8g}$ 18083 versus 19954 cm^{-1}, Γ_{7g} 21507 versus 23000 cm^{-1}. In contrast with this agreement, $6de_g^1\ {}^2E_g\ \Gamma_{8g}$ is calculated at 49958 cm^{-1}, 10000 cm^{-1} away from the experimental assignment at 40000 cm^{-1}. Such a deviation would imply a calculation of the ligand-field splitting $10Dq$ (energy difference between $6de_g^1\ {}^2E_g$ and $6dt_{2g}^1\ {}^2T_{2g}$ prior to including spin-orbit coupling) with an error of 10000 cm^{-1} in around 20000 cm^{-1}, which is much larger than

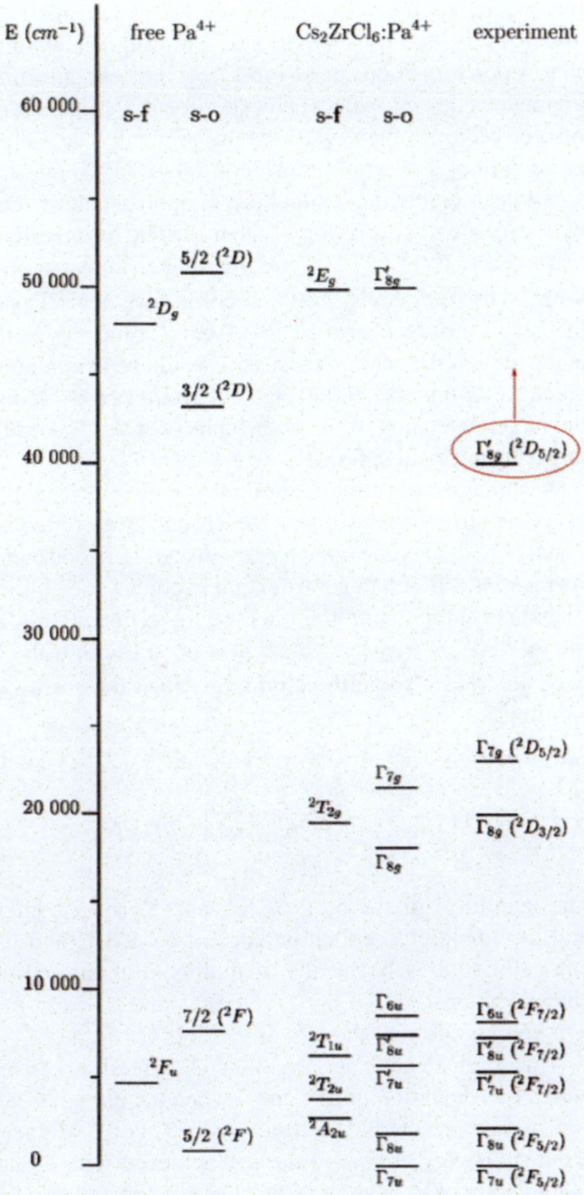

Fig. 9.13 5*f* and 6*d* energy levels of Pa^{4+} *in vacuo* and in Cs_2ZrCl_6, calculated at spin-free (s-f) and spin-orbit (s-o) level [81], compared with the experimentally found levels for $Cs_2ZrCl_6:Pa^{4+}$ [79, 80]. The level which is indicated in red has been re-assigned to a host-related excitation based on the quantum chemical calculations

the accuracies previously achieved for $10Dq$ in transiton metals [82], of around 1000 cm^{-1}. The novelty of a $6d$ splitting, rather than e.g. a $3d$ splitting in the first-series transition metal ions, does not explain this big increment. Instead, the anomalous deviation between experiment and calculation was taken as an indication of a possible problem with the assignment. The value of 40000 cm^{-1} [80] was a correction of a previous consideration of 33000 cm^{-1} [79] and it corresponds to a small peak preceding a small dip at approximately 42000 cm^{-1} which is followed by a very broad absorption of the host. The calculations indicate that the small peak is not due to $6de_g^1\ {}^2E_g\ \Gamma_{8g}$ and this state is in fact at around 50000 cm^{-1}, embedded within a myriad of host states. Accordingly, the CFT parameter $10Dq$ of $Pa^{4+}(6d)$ in Cs_2ZrCl_6 must be of around 30000 cm^{-1}, much larger than the 18600 cm^{-1} assumed at the time of the calculations. This would be in agreement with the expectations of a value larger than the 20000 cm^{-1} of $10Dq$ of $Ce^{3+}(5d)$ in Cs_2NaYCl_6 [83], because the larger extension of the $6d$ orbitals over the $5d$ orbitals would be coherent with a larger ligand-field splitting.

In summary, the multiconfigurational ab initio calculations of Cs_2ZrCl_6:Pa^{4+} gave excellent structural results on the ground and excited states, both in terms of vibrational frequencies and bond length offsets, and good energies of the electronic origins. As insights difficult to get from other sources, they indicate a Pa-Cl bond length contraction accompanying the lowest $5f \rightarrow 6d$ excitation and a new, higher energy for the $6de_g^1\ {}^2E_g\ \Gamma_{8g}$ state, which is coherent with a Pa^{4+} $6d$ ligand-field splitting in hexa-chloride O_h coordination larger than the corresponding Ce^{3+} $5d$ ligand-field splitting.

9.7.2 U^{4+} in Cs_2ZrCl_6 and Cs_2GeF_6

The manifolds of the $5f^2$, $5f6d$, and $5f\phi_{ITE}$ configurations of U^{4+} in octahedral hexa-halide coordination sites of Cs_2ZrCl_6 and Cs_2GeF_6 have also been theoretically studied by means of multiconfigurational ab initio calculations on the embedded-clusters $(UCl_6)^{2-}$, $(UF_6)^{2-}$, and $(UF_6Cs_8)^{6+}$ [84–87], together with the $5f^2$ configuration of the U^{4+} free-ion [84].

The experimental energies of the U^{4+} free-ion $5f^2$ levels [88, 89] are reproduced with a rms deviation of 214 cm^{-1} when the energies of the spin-orbit-free terms ${}^3P$ and ${}^1I$ was corrected −1000 cm^{-1} and 657 cm^{-1} when no correction was added. This accuracy is very similar to the one achieved with Dirac-Coulomb Coupled-Cluster calculations [90], in spite of this being a much more demanding method, presently impractical for many of the luminescent materials discussed in this book.

The local equilibrium structures of the $(UCl_6)^{2-}$ moiety in the manifolds of Cs_2ZrCl_6:U^{4+} are not an exception and the U-Cl bond-lenthgs follow the pattern $d_{e,U\text{-}Cl}(5f6dt_{2g}) < d_{e,U\text{-}Cl}(5f^2) < d_{e,U\text{-}Cl}(5f6de_g)$, with respective values 2.571±0.004 Å, 2.605±0.003 Å, and 2.642±0.004 Å. The same is true for $d_{e,U\text{-}F}$ in Cs_2GeF_6:U^{4+}; in this material, ITE states characterized with shorter bond-lenthgs are found as well, so that the structural pattern is $d_{e,U\text{-}F}(5f\phi_{ITE}) < d_{e,U\text{-}F}(5f6dt_{2g}) < d_{e,U\text{-}F}(5f^2) < d_{e,U\text{-}F}(5f6de_g)$, with respective values 2.099±0.006 Å, 2.155±0.005 Å, 2.174±0.005 Å, and 2.22±0.02 Å. The broader bond length window found in the states of the $5f6de_g$ configuration is a consequence of configurational mixing with $5f\phi_{ITE}$ states at U-F distances longer than 2.2 Å.

As in other cases, breathing-mode vibrational frequencies $\bar{\nu}_{a_{1g}}$ are similar in all manifolds (note that the relationship between bond lengths and vibrational frequencies, the shorter the first the larger the second, only stands for common basic bond interactions and these change from manifold to manifold). The computed values in $Cs_2ZrCl_6{:}U^{4+}$ are: 320±1 cm^{-1} ($5f^2$), 324±1 cm^{-1} ($5f6dt_{2g}$), and 299±3 cm^{-1} ($5f6de_g$), and in $Cs_2GeF_6{:}U^{4+}$ 563±6 cm^{-1} ($5f^2$), 566±8 cm^{-1} ($5f6dt_{2g}$), and 561±72 cm^{-1} ($5f6de_g$), and 631±70 cm^{-1} ($5f\phi_{ITE}$). Configurational mixing of lower $5f\phi_{ITE}$ states with upper $5f6dt_{2g}$ states and of higher $5f\phi_{ITE}$ states with lower $5f6de_g$ states broaden the vibrational frequency windows of the $5f\phi_{ITE}$ and $5f6de_g$ manifolds.

The calculated energies of the states of the $5f^2$ manifold allow to perform a detailed analysis of the $5f\rightarrow5f$ transitions, which compare very satisfactorily with experiments in $Cs_2ZrCl_6{:}U^{4+}$ [91, 92], with rms deviations of 432 cm^{-1} for the states derived from the free-ion levels $^3F_2$ and $^3H_5$; 719 cm^{-1} for the ones derived from $^3F_3$, $^3F_4$, and $^3H_6$; 1459 cm^{-1} for the ones derived from $^1D_2$, $^1G_4$, and $^3P_0$; and 1691 cm^{-1} for the ones derived from $^3P_1$, $^1I_6$, and $^3P_2$. The calculation of the $5f6d$ manifold allows to predict $5f\rightarrow6d$ transitions, which have not been measured experimentally. The energies of the lowest states of the $5f6dt_{2g}$ and $5f6de_g$ manifold are calculated at 31100 cm^{-1} and 67300 cm^{-1} respectively and, although the optical window of Cs_2ZrCl_6 was reported to be 4000–30000 cm^{-1} [93], the UV absorption spectrum of $Cs_2ZrCl_6{:}Pa^{4+}$ at 4.2 K shows the sharp cutoff of the Cs_2ZrCl_6 host crystal at about 37000 cm^{-1} [80], so that the onset of the $5f\rightarrow6dt_{2g}$ transitions, $1\ A_{1g}\rightarrow1\ E_u$ and $1\ A_{1g}\rightarrow1\ T_{2u}$, should be observable. Also, the calculated energies of the $5f6dt_{2g}$ manifold of $Cs_2ZrCl_6{:}U^{4+}$, combined with experimental spectroscopic data on this material, $Cs_2ZrCl_6{:}UO_2^{2+}$, and $Cs_2UO_2Cl_4$, opened the opportunity to discuss a new mechanism to explain the observed green-to-blue up-conversion luminescence of $Cs_2ZrCl_6{:}U^{4+}$ contaminated with UO_2^{2+}.

Combining the calculations on U^{4+}-doped Cs_2ZrCl_6 with new calculations on U^{4+}-doped Cs_2GeF_6, allowed to predict the effect of Cl-to-F chemical substitution, which is an increase of the energy range of this configuration in 6000 cm^{-1} and a non-uniform positive shift of all $5f^2$ levels which makes the order of states in $Cs_2ZrCl_6{:}U^{4+}$ and $Cs_2GeF_6{:}U^{4+}$ different. The comparison between both hosts predicts that the green-to-blue up-conversion luminescence observed in $Cs_2ZrCl_6{:}U^{4+}$ [93, 94] must be quenched in $Cs_2GeF_6{:}U^{4+}$, a result that was used as a high-symmetry model to understand why efficient radiative cascade emissions in the visible do not happen for U^{4+} defects in low-symmetry YF_3 crystals [95]. The $5f\rightarrow6dt_{2g}$ absorptions could be experimentally measured in the fluoride $Cs_2GeF_6{:}U^{4+}$ [86] and they were interpreted with the help of the calculations. The absorption spectrum is a complex superposition of five $5f^2\ 1A_{1g}\rightarrow5f6dt_{2g}\ nT_{1u}$ ($n = 1-5$) electric dipole allowed bands with their respective vibrational structures and the theoretical absorption profile, where the five electronic origins appear at 40400, 42600, 44000, 46200, and 48700 cm^{-1}, in reasonable agreement with experiment, showing an average overestimation of 2500 cm^{-1}.

9.7.3 U^{3+} in Cs_2NaYCl_6

The complex $5f^2 6d$ manifold of U^{3+} in the octahedral hexa-chloride site of the Cs_2NaYCl_6 elpasolite was the subject of an early theoretical multiconfigurational ab initio study in [96]. The absorption spectrum of Cs_2NaYCl_6:U^{3+} had initially been recorded in [97] at 4.2 K. from 4000 cm^{-1} to 25000 cm^{-1} (up to 45000 cm^{-1} in fact, but no significant features appeared above 21000 cm^{-1}) where, besides the identification and CFT analysis of $5f^3 \rightarrow 5f^3$ transitions, a set of very complex intense bands starting at 14000 cm^{-1} were attributed to $5f^3 \rightarrow 5f^2 6d$ excitations without further analysis, which was hard to make with empirical theories only. The multiconfigurational ab initio calculations independently supported this general interpretation, reported transition energies and energy separations between transitions in excellent agreement with the experiments, provided a basic interpretation of the absorption spectrum as due to overlapping vibronic structures built on multiple electronic origins, and explained the absence of intense absorption bands between 25000 cm^{-1} and 45000 cm^{-1}, where a large number of $5f^2 6d$ states should be expected if their onset is at 14000 cm^{-1}, as due to the fact that all $5f^2 6d$ states in that region correspond to two simultaneous electron excitations $5f \rightarrow 5f$ and $5f \rightarrow 5d$, which are strongly electric dipole forbidden.

In the light of the initial ab initio calculations, the energy range of the experimental absorption spectrum was further extended and the assignments were revised [98, 99]. New calculations of absorption band intensities and profiles finally allowed for a detailed interpretation of the $5f \rightarrow 5d$ absorption spectrum of Cs_2NaYCl_6:U^{3+} [100]. In [100], the effects of high pressures (up to 26 kbar) on the $5f \rightarrow 5d$ absorption spectrum of Cs_2NaYCl_6:U^{3+} were also studied, the most relevant feature being a general red shift of around 21 cm^{-1}/kbar. These effects are important to understand the behavior of optically active actinide materials under extreme conditions.

Regarding the $4f^3$ manifold of U^{3+}, the one of the free-ion up to 30000 cm^{-1} was calculated in [101]. In the absence of direct experimental data of the U^{3+} free-ion, it is common to use estimations obtained from the $4f^3$ manifold of U^{3+} in solids [75]; as a matter of fact, these estimations are usually referred to as the experimental energies of U^{3+} in gas phase. After the calculations in [101] and on the basis of a discussion on the available experimental and ab initio data of f-element ions in gas phase and in solids, and on the general agreement with the estimated experimental results and the expected precision of the calculations, it was concluded that the real $4f^3$ energy levels of U^{3+} in the gas phase are higher than the common estimates (and lower than calculations). A recent, improved calculation [102] gives slight quantitative corrections and confirms the underestimations of the data extrapolated from solids.

References

1. P.A. Tanner, C.S.K. Mak, N.M. Edelstein, K.M. Murdoch, G. Liu, J. Huang, L. Seijo, Z. Barandiarán, J. Am. Chem. Soc. **125**, 13225 (2003)
2. R.W. Schwartz, Mol. Phys. **30**, 81 (1975)
3. A. Bessière, P. Dorenbos, C. van Eijk, L. Pidol, K. Krämer, H. Güdel, J. Phys.: Condens. Matter **16**, 1887 (2004)
4. J. Wen, L. Ning, C.K. Duan, Y. Chen, Y. Zhang, M. Yin, J. Phys. Chem. C **116**, 20513 (2012)
5. L. Ning, F. Yang, C. Duan, Y. Zhang, J. Liang, Z. Cui, J. Phys.: Condens. Matter **24**, 05502 (2012)
6. J. Gracia, L. Seijo, Z. Barandiarán, D. Curulla, H. Niemansverdriet, W. van Gennip, J. Lumin. **128**, 1248 (2008)
7. L. Ning, C. Wu, L. Li, L. Lin, C. Duan, Y. Zhang, L. Seijo, J. Phys. Chem. C **116**, 18419 (2012)
8. L. Seijo, Z. Barandiarán, J. Chem. Phys. **141**, 214706/1 (2014)
9. L. van Pieterson, M.F. Reid, R.T. Wegh, S. Soverna, A. Meijerink, Phys. Rev. B **65**, 045113 (2002)
10. E. Loh, Phys. Rev. **158**, 273 (1967)
11. T. Szczurek, M. Schlesinger, in *Rare Earth Spectroscopy*. ed. by J. Legendziewicz, W. Strek (World Scientific, Singapore, 1985), p. 309
12. K.D. Oskam, A.J. Houtepen, A. Meijerink, J. Lumin. **97**, 107 (2002)
13. B.M. Tissue, J.C. Wright, Phys. Rev. B **36**, 9781 (1987)
14. L. van Pieterson, R.P.A. Dullens, P.S. Peijzel, A. Meijerink, J. Chem. Phys. **115**, 9393 (2001)
15. E. Radzhabov, J. Lumin. **129**, 1581 (2009)
16. M. Krośnicki, A. Kedziorski, L. Seijo, Z. Barandiarán, J. Phys. Chem. A **118**, 358 (2014)
17. E. Loh, Phys. Rev. **175**, 533 (1968)
18. E.J. Heller, J. Chem. Phys. **62**, 1544 (1975)
19. E.J. Heller, Acc. Chem. Res. **14**, 368 (1981)
20. G.H. Dieke, H.M. Crosswhite, B. Dunn, J. Opt. Soc. Am. **51**(8), 820 (1961)
21. G.H. Dieke, H.M. Crosswhite, Appl. Opt. **2**(7), 675 (1963)
22. J.J. Joos, P.F. Smet, L. Seijo, Z. Barandiarán, Inorg. Chem. Front. **7**, 871 (2020)
23. M.C. Downer, C.D. Corderomontalvo, H. Crosswhite, Phys. Rev. B **28**(9), 4931 (1983)
24. C. Dujardin, B. Moine, C. Pedrini, J. Lumin. **54**(5), 259 (1993)
25. N. Yamashita, S. Fukumoto, S. Ibuki, H. Ohnishi, Jpn. J. Appl. Phys. **32**, 3135 (1993)
26. N. Yamashita, O. Harada, K. Nakamura, Jpn. J. Appl. Phys. **34**(10), 5539 (1995)
27. P.F. Smet, J.E. Van Haecke, F. Loncke, H. Vrielinck, F. Callens, D. Poelman, Phys. Rev. B **74**(3) (2006)
28. C. Pedrini, M.F. Joubert, D.S. McClure, J. Lumin. **125**(1–2), 230 (2007)
29. S.V. Gastev, J.K. Choi, R.J. Reeves, Phys. Solid State **51**(1), 44 (2009)
30. G. Blasse, B. Grabmaier, in *Luminescent materials* (Springer, 1994), pp. 1–9
31. P.A. Tanner, Chem. Soc. Rev. **42**, 5090 (2013)
32. C. Cascales, J. Fernández, R. Balda, Opt. Express **13**, 2141 (2005)
33. V.A. Morozov, A. Bertha, K.W. Meert, S. Van Rompaey, D. Batuk, G.T. Martinez, S. Van Aert, P.F. Smet, M.V. Raskina, D. Poelman, A.M. Abakumov, J. Hadermann, Chem. Mater. **25**, 4387 (2013)
34. E. Friedman, W. Low, J. Chem. Phys. **33**, 1275 (1960)
35. R. Pandey, S. Sivaraman, J. Phys. Chem. Solids **52**, 211 (1991)
36. G. Blasse, Phys. Status Solidi B **55**, K131 (1973)
37. A. Meijerink, G. Blasse, J. Lumin. **43**, 283 (1989)
38. J.D. Axe, Phys. Rev. A **136**, A42 (1964)
39. N. Bloembergen, J. Lumin. **31**, 23 (1984)
40. B. Moine, C. Pédrini, B. Courtois, J. Lumin. **50**(1), 31 (1991)
41. P.A. Rodnyĭ, A.K. Khadro, A.S. Voloshinovskĭ, G.B. Stryganyuk, Opt. Spectrosc. **103**(4), 568 (2007)

42. J.E. Van Haecke, P.F. Smet, D. Poelman, J. Lumin. **126**(2), 508 (2007)
43. N. Yamashita, Jpn. J. Appl. Phys. **30**(12a), 3335 (1991)
44. G. Sánchez-Sanz, L. Seijo, Z. Barandiarán, J. Chem. Phys. **131**, 024505 (2009)
45. M. Karbowiak, C. Rudowicz, J. Lumin. **199**, 116 (2018)
46. M. Suta, C. Wickleder, J. Lumin. **210**, 210 (2019)
47. M.J. Freiser, S. Methfess, F. Holtzber, J. Appl. Phys. **39**, 900 (1968)
48. S.J. Camardello, P.J. Toscano, M.G. Brik, A.M. Srivastava, J. Lumin. **151**, 256 (2014)
49. P. Dorenbos, ECS J. Solid State Sci. Technol. **2**(2), R3001 (2013)
50. J. Grimm, H.U. Güdel, Chem. Phys. Lett. **404**, 40 (2005)
51. J. Grimm, E. Beurer, H.U. Güdel, Inorg. Chem. **45**, 10905 (2006)
52. J. Grimm, J.F. Suyver, E. Beurer, G. Carver, H.U. Güdel, J. Phys. Chem. B **110**, 2093 (2006)
53. R.T. Wegh, A. Meijerink, Phys. Rev. B **60**, 10820 (1999)
54. H. Witzke, D.S. McClure, B. Mitchell, in *Luminescence of Crystals, Molecules, and Solutions*. ed. by F.E. Williams (Plenum, New York, 1973), p. 598
55. Z. Pan, C. Duan, P.A. Tanner, Phys. Rev. B **77**, 085114 (2008)
56. M. Suta, W. Urland, C. Daul, C. Wickleder, Phys. Chem. Chem. Phys. **18**, 13196 (2016)
57. M. de Jong, A. Meijerink, L. Seijo, Z. Barandiarán, J. Phys. Chem. C **121**, 10095 (2017)
58. M. de Jong, D. Biner, K. Krämer, Z. Barandiarán, L. Seijo, A. Meijerink, J. Phys. Chem. Lett. **7**, 2730 (2016)
59. C.W. Struck, W.H. Fonger, J. Lumin. **10**, 1 (1975)
60. G. Sánchez-Sanz, L. Seijo, Z. Barandiarán, J. Phys. Chem. A **113**, 12591 (2009)
61. Z. Barandiarán, L. Seijo, J. Chem. Phys. **138**, 074102 (2013)
62. NIST Atomic Spectra Database, NIST Standard Reference Database #78
63. G.H. Dieke, *Spectra and Energy Levels of Rare Earth Ions in Crystals* (Interscience Publishers, New York, 1968)
64. B. Henderson, G.F. Imbusch, *Optical Spectroscopy of Inorganic Solids* (Clarendon, Oxford, 1989)
65. U. Oetliker, M.J. Riley, H.U. Güdel, J. Lumin. **63**, 63 (1995)
66. K.W. Krämer, H.U. Güdel, R.N. Schwartz, J. Alloys Compd. **275–277**, 191 (1998)
67. P.S. Peijzel, P. Vermeulen, W.J.M. Schrama, A. Meijerink, M.F. Reid, G.W. Burdick, Phys. Rev. B **71**, 125126 (2005)
68. M.F. Reid, P.S. Senanayake, J.P.R. Wells, G. Berden, A. Meijerink, A.J. Salkeld, C.K. Duan, R.J. Reeves, Phys. Rev. B **84**, 113110 (2011)
69. G. Sánchez-Sanz, L. Seijo, Z. Barandiarán, J. Chem. Phys. **133**, 114509 (2010)
70. Z. Barandiarán, L. Seijo, J. Chem. Phys. **141**, 234704 (2014)
71. T.S. Piper, J.P. Brown, D.S. McClure, J. Chem. Phys. **46**, 1353 (1967)
72. E. Loh, Phys. Rev. B **7**, 1846 (1973)
73. A.A. Kaplyanskii, P.L. Smolyanskii, Opt. Spectrosc. **40**, 301 (1976)
74. E. Beurer, J. Grimm, P. Gerner, H.U. Güdel, J. Am. Chem. Soc. **128**, 3110 (2006)
75. W.T. Carnall, H.M. Crosswhite, *Optical Spectra and Electronic Structure of Actinide Ions in Compounds and in Solution*. (Springer, Dordrecht, 1986), pp. 1235–1277. The Chemistry of the Actinide Elements
76. T. Gouder, R. Eloirdi, *Photoelectron Spectroscopy of Actinide Materials* (John Wiley & Sons, 2019), p. 193. The Heaviest Metals: Science and Technology of the Actinides and Beyond
77. Z.R. Jones, M.Y. Livshits, F.D. White, E. Dalodière, M.G. Ferrier, L.M. Lilley, K.E. Knope, S.A. Kozimor, V. Mocko, B.L. Scott, B.W. Stein, J.N. Wacker, D.H. Woena, Chem. Sci. **44**, (2021). 10.1039/D1SC00233C. http://dx.doi.org/10.1039/D1SC00233C
78. L. Seijo, Z. Barandiarán, E. Harguindey, J. Chem. Phys. **114**, 118 (2001)
79. D. Piehler, W.K. Kot, N. Edelstein, J. Chem. Phys. **94**, 942 (1991)
80. N. Edelstein, W.K. Kot, J.C. Krupa, J. Chem. Phys. **96**, 1 (1992)
81. L. Seijo, Z. Barandiarán, J. Chem. Phys. **115**, 5554 (2001)
82. L. Seijo, Z. Barandiarán, in *Computational Chemistry: Reviews of Current Trends*, vol. 4, ed. by J. Leszczyński (World Scientific, Singapore, 1999), pp. 55–152
83. R.W. Schwartz, P.N. Schatz, Phys. Rev. B **8**, 3229 (1973)

84. Z. Barandiarán, L. Seijo, J. Chem. Phys. **118**, 7439 (2003)
85. B. Ordejón, L. Seijo, Z. Barandiarán, J. Chem. Phys. **123**, 204502 (2005)
86. B. Ordejón, M. Karbowiak, L. Seijo, Z. Barandiarán, J. Chem. Phys. **125**, 074511 (2006)
87. B. Ordejón, L. Seijo, Z. Barandiarán, J. Chem. Phys. **126**, 194712 (2007)
88. J.F. Wyart, V. Kaufman, J. Sugar, Phys. Scr. **22**, 389 (1980)
89. C.H.H. Van Deurzen, K. Rajnak, J.G. Conway, J. Opt. Soc. Am. B **1**, 45 (1984)
90. E. Eliav, U. Kaldor, Y. Ishikawa, Phys. Rev. A **51**, 225 (1995)
91. R.A. Satten, C.L. Schreiber, E.Y. Wong, J. Chem. Phys. **42**, 162 (1965)
92. C.D. Flint, P.A. Tanner, Mol. Phys. **61**, 389 (1987)
93. W. Xu, S. Dai, L.M. Toth, J.R. Peterson, Chem. Phys. **193**, 339 (1995)
94. P.A. Tanner, J. Dexpert-Ghys, Z.W. Pei, J. Lin, Chem. Phys. **215**, 125 (1997)
95. M. Kirm, J.C. Krupa, V.N. Makhov, E. Neodin, G. Zimmerer, J.Y. Gesland, J. Lumin. **104**, 85 (2003)
96. L. Seijo, Z. Barandiarán, J. Chem. Phys. **118**, 5335 (2003)
97. M. Karbowiak, J. Drożdżyński, S. Hubert, E. Simoni, W. Strek, J. Chem. Phys. **108**, 10181 (1998)
98. M. Karbowiak, J. Drozdzynski, J. Phys. Chem. A **108**, 6397 (2004)
99. M. Karbowiak, Chem. Phys. **314**, 189 (2005)
100. F. Ruipérez, Z. Barandiarán, L. Seijo, J. Chem. Phys. **127**, 144712 (2007)
101. F. Ruipérez, B.O. Roos, Z. Barandiarán, L. Seijo, Chem. Phys. Lett. **434**, 1 (2007)
102. S.K. Roy, R. Prasad, Chem. Phys. **535**, 110781 (2020)

Part V
Insights into the Complexity of Luminescent Materials

This last Part of the book is focused on the insightful potential of multiconfigurational *ab initio* calculations in the field of luminescent materials. Such a potential is illustrated by discussing some insights that have already been given by these calculations and that would hardly been obtained by other techniques. They are organized in two Chapters, one dedicated to luminescence relevant properties of individual active centers and another dedicated to phenomena involving two active centers and electron transfers between them.

Chapter 10
Active Centers of Luminescent Materials

Abstract Quantum chemical calculations uniquely allow to reveal how physically observable properties emerge from underlying intricate electronic structures and interactions dictated by the principles of quantum mechanics. As such, they can be used to check traditional models and concepts of luminescence that were developed based on chemical intuition or reasonable guesses and widely accepted by the community. In this chapter, by revisiting several of these, the potential of multiconfigurational ab initio calculations as tools to establish solid conceptual grounds is illustrated. A few misconceptions are unveiled on lanthanide-ligand bond lengths, distribution of defects in a host, $4f^{N-1}6s$ states, and the excited states of Eu-doped luminescent materials, and the importance of a forgotten interaction like Pauli antisymmetry between active center and host is claimed.

10.1 Bond Length Changes upon $4f \rightarrow 5d$ Excitations

When a luminescent center of a solid experiences a change of electronic state, either radiatively or non-radiatively, its local geometrical structure will be prone to a change as well. Even it the transition between states is very fast, much faster than the movement of the atoms, and it happens at a given, frozen, geometry, e.g. in a Frank-Condon electronic transition, the interatomic distances and angles change immediately in order to achieve their equilibrium values in the new electronic state. Such a change in local structure determines the width and shape of the associated optical absorption and emission bands and is crucial to understand non-radiative relaxations between excited states and relaxations to the ground state [1, 2]. If the site shows a high symmetry, the number of geometrical parameters determining the local structure may be as small as one, like the bond length or equilibrium distance between a lanthanide cation and its surrounding ligands in octahedral LnL_6 or cubic LnL_8 sites, $d_{e,\text{Ln-L}}$. This section is focused on the changes experienced by the lanthanide-ligand bond length in a $4f \rightarrow 5d$ transition of a high-symmetry optically active site of a lanthanide-doped luminescent material.

Z. Barandiarán et al., *Luminescent Materials*, Springer Series in Materials Science 322,
https://doi.org/10.1007/978-3-030-94984-6_10

Theoretical Prediction

A finding of an early multiconfigurational ab initio study of Pa^{4+}-doped Cs_2ZrCl_6 [3] (cf. Sect. 9.7) was that a $5f \rightarrow 6dt_{2g}$ absorption of Pa^{4+} in a 6-fold O_h coordination was accompanied by a shortening of the Pa-Cl bond length $d_{e,\text{Pa-Cl}}$ of around −0.02 Å. When the transition was not to the ligand-field stabilized $5d$ orbital, but to the ligand-field destabilized one, $5f \rightarrow 6de_g$, the bond length was elongated +0.06 Å.

The reported bond length shortening in a $5f \rightarrow 6d$ transition of an actinide ion was against many intuitively constructed configurational coordinate models of lanthanide materials that could be found in the literature at that time and still can be found today. In them, all levels of the $4f^{N-1}5d$ manifold of a lanthanide ion in a solid are almost invariably drawn at longer equilibrium distances than their respective $4f^N$ levels, inspired by the fact that $5d$ atomic orbitals have larger orbital radii than $4f$ atomic orbitals and reach further. Obviously, the $5f/6d$ interplay in actinides must be at least qualitatively equivalent to the $4f/5d$ interplay in lanthanides; hence, either the result of the multiconfigurational ab initio calculation on the actinide ion is wrong or the preconceptions behind the configurational coordinate models found in the literature for lanthanide ions are incorrect.

In order to understand the reasons behind the bond length shortening predicted by the multiconfigurational calculations, a detailed constrained space orbital variation quantum chemical analysis (CSOV) was made on the impurity-ligand bond length of Ce^{3+} and Pr^{3+}-doped Cs_2NaYCl_6 and Pa^{4+} and U^{4+}-doped Cs_2ZrCl_6 [4]. In a CSOV analysis [5, 6], the participation of specific orbitals on bonding is controlled on demand by either allowing them to mix freely with all other orbitals or freezing them. Instead of performing a given calculation in a single step, a chosen route made of several steps is followed, each step being defined by taking a set of initial orbitals and signaling a subset of them which are not allowed to mix with the others. Then, a proper choice of steps to make a calculation makes it able to quantify, e.g., the role of ionic bonding, electron transfer, or covalent bonding on any calculated property.

In this study, the impurity-ligand bond lengths of the four materials are calculated at a CASSCF level in several CSOV steps. Taking the $(PrCl_6)^{3-}$ cluster embedded in Cs_2NaYCl_6 as an example, the performed CSOV analysis starts with the calculation of O_h symmetry adapted orbitals of Pr^{3+} and the six Cl^- ions in a preparatory step; they are chosen as starting orbitals of the CASSCF calculations. Core orbitals of Pr and Cl are kept frozen throughout all the calculations by means of the use of AIMP effective core potentials for the Pr [Kr,4*d*] [7] and Cl [Ne] cores [8]. Then, in CSOV step 1, the results of the *ionic model* are calculated by allowing free variational mixing between Pr occupied and virtual (unoccupied) orbitals and between Cl occupied and virtual orbitals, but not allowing any variational mixing between Pr and Cl orbitals, neither occupied nor virtual (there is a residual, non-variational mixing derived from orthogonality requirements). Under these conditions, the molecular orbitals resulting from this step are atomic-like orthogonal orbitals of Pr^{3+} and 6 Cl^-, relaxed and polarized under the effects of one another in the presence of the host, but excluding charge transfer between them and covalency. Hence, this step is a realization of an ionic model. Many-electron wave functions and total energies are calculated for the lowest states of the $4f^2$, $4f5dt_{2g}$, and $4f5de_g$ configurations (this is by choice, because they can be done at any chosen state of any configuration), at several values of the Pr-Cl distance $d_{\text{Pr-Cl}}$. The values of $d_{\text{Pr-Cl}}$ at the minimum energies of the states are their Pr-Cl equilibrium distances or bond lengths $d_{\text{e,Pr-Cl}}$. These are very similar for all states of a configuration and it is their configurational dependence what is interesting here; accordingly, $d_{\text{e,Pr-Cl}}(4f^2)$, $d_{\text{e,Pr-Cl}}(4f5dt_{2g})$, and $d_{\text{e,Pr-Cl}}(4f5de_g)$ are calculated at CSOV step 1, i.e. for the ionic model. They will also be calculated at all steps of the CSOV analysis.

In CSOV step 2, the ionic orbitals resulting from step 1 are used as starting orbitals and free variational mixing is allowed between Cl occupied, Cl virtual, Pr $5d$, and Pr virtual orbitals, but Pr $4f$, $5s$, and $5p$ orbitals are kept frozen. Comparing with step 1, covalent interactions between Cl and Pr $5d$ orbitals are included in the states of the $4f5d$ configuration and Cl-to-Pr charge transfer are taken into account, excluding charge transfer to Pr $4f$. In this step, covalent interactions between the outer Pr shells $5s^2$, $5p^6$ and the Cl ligands are excluded.

In CSOV step 3, also starting with the ionic orbitals of step 1, the same variational mixings as in step 2 are allowed plus the Pr $5s$ and $5p$ occupied orbitals. Only the Pr $4f$ orbitals are kept frozen. Then, comparing step 3 with step 2 gives the role of Pr $5s$ and $5p$ orbitals on bonding.

In the final step of the present analysis, CSOV step 4, no restrictions are imposed and all orbitals are allowed to mix variationally. Then, comparing step 4 with step 3 reveals the role of Cl-to-Pr $4f$ charge transfer, because the efficient shielding of the outer $5s^25p^6$ shell makes Cl-Pr $4f$ covalent interactions negligible.

In Table 10.1, the results of $Cs_2NaYCl_6{:}Pr^{3+}$ are summarized, but equivalent results are obtained in all cases. A $d_{\text{e,Pr-Cl}}(4f5d)$ value calculated as the weighted average of $d_{\text{e,Pr-Cl}}(4f5dt_{2g})$ and $d_{\text{e,Pr-Cl}}(4f5de_g)$ is also shown, which represents the bond length of the $4f5d$ centroid. It is observed that $d_{\text{e,Pr-Cl}}(4f5d)$ is only slightly longer than $d_{\text{e,Pr-Cl}}(4f^2)$ in the ionic model. This is in correspondence with the bond length in the $4f^2$ configuration being determined by the size of the $5s^25p^6$ full outer shell of Pr^{3+} rather than by its $4f^2$ unfilled inner shell; then, when one electron crosses the $5s^25p^6$ shell out to a more external $5d$ orbital, it only gives a small contribution to the bond length and its associated elongation is small, +0.004 Å, very much smaller than the difference between the $5d$ and $4f$ atomic orbital radii.

Still in the ionic model, the differences between the centroid bond length $d_{\text{e,Pr-Cl}}(4f5d)$ and the bond lengths $d_{\text{e,Pr-Cl}}(4f5dt_{2g})$ and $d_{\text{e,Pr-Cl}}(4f5de_g)$ are just the $5d$ ligand-field effects on the bond length: a –0.012 Å shortening in the ligand-field stabilized configuration $4f5dt_{2g}$ and a +0.018 Å elongation in the ligand-field destabilized configuration $4f5de_g$.

The effects of covalency and charge transfer, excluding charge transfer to Pr $4f$, are estimated with the changes experienced from step 1 to step 3. It shortens $d_{\text{e,Pr-Cl}}(4f^2)$ –0.081 Å and it shortens $d_{\text{e,Pr-Cl}}(4f5dt_{2g})$ with additional –0.015 Å. It elongates $d_{\text{e,Pr-Cl}}(4f5de_g)$ with additional +0.009 Å. The contributions of Pr $5s$ and $5p$ to these covalent effects (from step 2 to step 3) are negligible: –0.003 Å shortening of $d_{\text{e,Pr-Cl}}(4f^2)$ and all other bond lengths, which means no additional contributions to $d_{\text{e,Pr-Cl}}(4f5dt_{2g})$ and $d_{\text{e,Pr-Cl}}(4f5de_g)$.

Table 10.1 CSOV analysis of the Pr-Cl bond length $d_{e,Pr\text{-}Cl}$ and its change upon $4f \rightarrow 5d$ excitations in $Cs_2NaYCl_6{:}Pr^{3+}$. All distances in Å

	Step 1[a]	Step 2	Step 3	Step 4	1-to-3[b]	2-to-3[c]	3-to-4[d]
$d_{e,Pr\text{-}Cl}(4f^2)$	2.830	2.752	2.749	2.736	–0.081	–0.003	–0.013
$d_{e,Pr\text{-}Cl}(4f5dt_{2g})$	2.822	2.729	2.726	2.706			
$d_{e,Pr\text{-}Cl}(4f5dt_{2g}) - d_{e,Pr\text{-}Cl}(4f^2)$	–0.008	–0.023	–0.023	–0.030	–0.015	–0.000	–0.007
$d_{e,Pr\text{-}Cl}(4f5d)^e$	2.834	2.751	2.748	2.731			
$d_{e,Pr\text{-}Cl}(4f5d) - d_{e,Pr\text{-}Cl}(4f^2)$	+0.004	–0.001	–0.001	–0.005	–0.005	–0.000	–0.004
$d_{e,Pr\text{-}Cl}(4f5de_g)$	2.852	2.783	2.780	2.766			
$d_{e,Pr\text{-}Cl}(4f5de_g) - d_{e,Pr\text{-}Cl}(4f^2)$	+0.022	+0.031	+0.031	+0.030	+0.009	–0.000	–0.001

[a] Ionic model
[b] Covalency and charge transfer effects, excluding charge transfer to Pr $4f$
[c] Covalent effects of Pr $5s$ and $5p$
[d] Effect of charge transfer to Pr $4f$
[e] Pr-Cl bond length in the hypothetical $4f5d$centroid state

Finally, Cl-to-Pr $4f$ charge transfer (from step 3 to step 4) shortens $d_{e,Pr\text{-}Cl}(4f^2)$ –0.013 Å, the differential effects on the $4f5d$ states being of an additional shortening of $d_{e,Pr\text{-}Cl}(4f5dt_{2g})$ with –0.007 Å and of $d_{e,Pr\text{-}Cl}(4f5de_g)$ with –0.001 Å.

In summary, focusing on the Pr-Cl bond length shortening of –0.030 Å experienced in the $4f \rightarrow 5dt_{2g}$ transition (from 2.736 Å to 2.706 Å), i.e. at the difference $d_{e,Pr\text{-}Cl}(4f5dt_{2g}) - d_{e,Pr\text{-}Cl}(4f^2)$, the CSOV analysis indicates that it results from: (i) +0.004 Å elongation due to the excitation of the electron from a $4f$ to a $5d$ shell and (ii) –0.012 Å shortening due to the ionic $5d$ ligand-field stabilization, all of it adding up to –0.008 Å shortening when only ionic bonding is considered; plus (iii) –0.015 Å shortening due to $5d$ covalent effects and charge transfer other than to $4f$, and (iv) –0.007 Å shortening due to larger charge transfer to the $4f$ shell in the $4f5dt_{2g}$ configuration than in the $4f^2$ configuration.

The numbers differ in Ce^{3+}, Pr^{3+}, Pa^{4+}, and U^{4+}, but the overall image is the same in all cases. The bond length between a lanthanide ion in a $4f^N$ configuration (and, equivalently, of an actinide ion in a $5f^N$ configuration) and its ligands is realized by the interaction between the $5s^25p^6$ outer shell and the ligand's valence electrons. When the ion is excited to a $4f^{N-1}5d$ configuration, covalency and charge transfer are stronger in the excited configuration and they shorten the bond length; the shortening is enhanced by the $5d$ ligand-field stabilization in the $4f^{N-1}5dt_{2g}$ configuration and opposed in the $4f^{N-1}5de_g$ configuration in octahedral sites and the magnitudes of these effects result in a systematic bond length trend: $d_{e,Ln\text{-}L}(4f^{N-1}5dt_{2g}) < d_{e,Ln\text{-}L}(4f^N) < d_{e,Ln\text{-}L}(4f^{N-1}5de_g)$ in these sites.

The initial theoretical finding of the lowest $f \rightarrow d$ excitations of f-element ions in 6-fold coordinated octahedral sites of solids being accompanied by a bond length shortening [3] was later confirmed in 8-fold coordinated cubic sites [9], where the

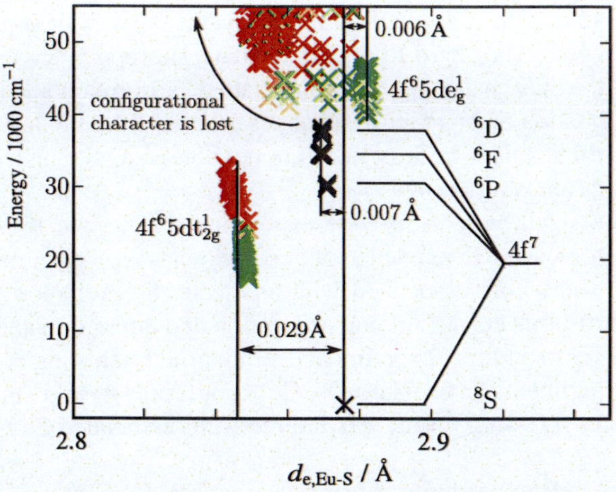

Fig. 10.1 Equilibrium bond lengths for all potential energy curves of CaS:Eu^{2+} [13]. The bond length shortening in $4f^65dt_{2g}$ states and elongation in $4f^65de_g$ states with respect to the $4f^7$ states, are shown. $4f^7$ levels are shown in black crosses. $4f^65d$ and $4f^6\phi_{ITE}$ levels are shown in colored crosses. Colors indicate their spin-multiplicity character (at $d_{\text{Eu-S}}$=2.85 Å): red for $2S + 1 = 6$, green for $2S + 1 = 8$, yelow for mixed character

roles of $5dt_{2g}$ and $5de_g$ are interchanged, and in liquid phase [10], and systematically found in all calculations performed [11, 12] (cf. Sect. 6.2). In high-symmetry sites, the general bond length trend is:

$$d_{\text{e,Ln-L}}(4f^{N-1}5d_{\text{LOW}}) < d_{\text{e,Ln-L}}(4f^N) < d_{\text{e,Ln-L}}(4f^{N-1}5d_{\text{HIGH}}).$$

In activators with excited $4f^{N-1}5d$ manifolds where the $4f^{N-1}5dt_{2g}$ and $4f^{N-1}$ $5de_g$ configurations span separated energy regions, the bond length trend shows up very clearly. Table 6.19 for Yb^{2+} in halide hosts with fluorite structure is a good example of this. It is common, however, to find activator/host combinations where the $4f^{N-1}5d$ states have almost pure $4f^{N-1}5dt_{2g}$ or $4f^{N-1}5de_g$ configurational characters in some energy windows only, whereas in some other energy windows the configurational character is very mixed. When this happens, a variety of excited state bond lengths is found ranging between the limiting bond lengths of the pure configurations. This is shown in Fig. 10.1 for Eu^{2+}-doped CaS.

Experimental Evidence

Given the disagreement between the theoretical prediction of a bond length shortening accompanying the lowest $4f \rightarrow 5d$ transitions of lanthanide ions in luminescent materials and the popular assumption of a bond length elongation, it is interesting to be able to get experimental confirmations. In this respect, direct experimental measurements of bond lengths in excited states of dopants in hosts should be feasible in principle, but they are very hard to do and, to date, no reports have been given in this respect. On the other hand, spectroscopic data on absorption and emission band widths give information on bond length changes occurring during the transitions, but only of their absolute values because they are not able to distinguish between bond length elongations and shortenings.

There is, however, a way to provide experimental evidence to this prediction based on the high-pressure dependence of absorption and emission peaks. In effect, the bond length shortening associated with an electronic excitation was shown to imply the red-shift of its corresponding absorption and emission bands [14], which opened the door for its indirect experimental validation. Interestingly, high-pressure spectroscopic experiments on $Cs_2NaLuCl_6$:Ce^{3+} confirmed the red-shift of its lowest $4f \rightarrow 5d$ transition, $4f \rightarrow 5dt_{2g}$ [15] and, indirectly, its associated Ce-Cl bond length shortening.

A similar red-shift under high-pressure was reported for the $5d \rightarrow 4f$ luminescence of Sm^{2+}-doped SrF_2 in a study of pressure-induced phase transitions in the SrF_2 host, in which the Sm^{2+} luminescence was used as a probe [16]. According to the above discussion, this red-shift is a manifestation of the Sm-F bond length being shorter in the lowest, emitting state of the $4f^5 5de_g$ configuration than in the states of the $4f^6$ configuration. The bond length change and the pressure-dependence of this Sm^{2+} $5d \rightarrow 4f$ luminescence have been confirmed in multiconfigurational ab initio calculations [17]. Nowadays, it is not unusual to find similar experiment results, like those showing the same effect reported for Eu^{2+}-doped fluoride and sulfide hosts [18–20].

10.2 Defects Mutually Attract Each Other and Distribute In-Homogeneously

In luminescent materials made of impurity-doped crystals, it is quite common to assume that the active impurities are homogeneously distributed across the bulk when concentration effects are under consideration or when interactions between dopants and local defects in general could be relevant. This assumption is made, for instance, to discuss interactions between optically active centers and co-dopants, either active ones, which participate in processes of energy transfer and charge transfer with the optical defect, or inactive, which simply modify the excited state manifolds of the active center; or between active centers and defects that are inherent to the host materials, like antisite defects and vacancies, whose concentrations depend on the methods of synthesis and additional physical and chemical treatments. E.g., it is not strange to see the possibility of energy or charge transfer between two dopants discarded when their concentrations are small, just because the distance between them would be very large in a homogeneous distribution.

Confirmation of the validity of this assumption is not simple, neither from the experimental nor from the theoretical point of view, and very often the hypothesis of a uniform distribution of defects over the host is the only practical option, unless bonding or electrostatic interactions between the local defects are known or suspected to play important roles either on their clustering or on their trend to form homogeneous distributions. In the lack of such interactions, when the defects only create local deformations around them, the common practice is to accept that they occupy all the available space with a maximum distance between them.

However, in the course of ab initio calculations on defective luminescent materials discussed in this book, the distribution of defects had to be tackled on several occasions and the image that emerges differs from the homogeneous distribution assumption. In the following, results on rare-earth-doped garnet phosphors are summarized. Similar conclusions have been drawn for other luminescent materials, like Cr^{3+}-doped spinels [21].

In the doubly doped phosphor $Y_3Al_5O_{12}$:Ce^{3+},Ga^{3+} [22], Ce^{3+} substitutes for Y^{3+} and forms dodecahedral Ce_Y defects and Ga^{3+} substitutes for Al^{3+} and forms octahedral Ga_{Al}^{oct} and tetrahedral Ga_{Al}^{tet} defects. The double defect Ce_Y-Ga_{Al}^{oct} is found to be most stable structure at an inter-defect distance d(Ce-Ga) = 5.456 Å. At a shorter distance of 3.436 Å, the double defect energy is increased by 154 meV/defect; this is because the two expansions induced by the two individual substitutions become incompatible. But its energy also increases at longer distances: 17 meV/defect at 6.938 Å and 39 meV/defect at 8.155 Å. Similarly, the double defect Ce_Y-Ga_{Al}^{tet} is found to be most stable at a distance d(Ce-Ga) = 5.655 Å. At shorter distances the defect energy also increases, as in the case of Ga_{Al}^{oct} and for the same reason (223 meV/defect at 3.755 Å and 197 meV/defect at 3.056 Å), the orientations of the defects playing a small role here. And it also increases at longer distances (28 meV/defect at 6.056 Å, 25 meV/defect at 7.104 Å, 28 meV/defect at 8.562 Å, and 26 meV/defect at 9.087 Å). So, there is an optimal, short distance between Ce_Y and Ga_{Al} substitutional defects in YAG:Ce^{3+},Ga^{3+}, which is around 5.5–5.7 Å.

In the doubly doped phosphor $Y_3Al_5O_{12}$:Ce^{3+},La^{3+} [23], as Ce^{3+} does, La^{3+} also substitutes for Y^{3+} and forms dodecahedral La_Y defects. The double defect Ce_Y-La_Y is most stable at the inter-defect distance d(Ce-La) = 3.728 Å. At the slightly longer distance of 5.654 Å, it is almost equaly stable, with only 3 meV/defect higher energy. At longer distances the defect energy increases (48 meV/defect at 6.058 Å, 35 meV/defect at 6.781 Å, 12 meV/defect at 7.117 Å, 20 meV/defect at 8.573 Å, and 21 meV/defect at 10.494 Å). In other words, La_Y defects tend to be as close as possible to Ce_Y defects in $Y_3Al_5O_{12}$:Ce^{3+},La^{3+}.

In undoped $Y_3Al_5O_{12}$ with Y_{Al}-Al_Y antisite defects (AD) [24], the most stable structure of one AD defect (1AD) has the two Y_{Al} and Al_Y individual substitutional defects as near as possible. The formation energy of two AD defects (2AD) is lower than twice the formation energy of 1AD. And the most stable structure of 2AD has the two 1AD defects linked together by an AlO_6 moiety, forming a (Y_{Al}-Al_Y)-AlO_6-(Al_Y-Y_{Al}) chain, with a Y-Al-Y axis and two traversed Y_{Al}-Al_Y vectors showing opposite orientations, as antiparallel as the rigidity of the structure allows. This distribution is understood as due to a relevant role played by multipole electrostatic interactions, according to which two dipole vectors are more stable in an antiparallel than in a parallel orientation. In this case, the role played by mechanical deformations is relevant but not dominant. In summary, the study indicates an ordered distribution of antisite defects with a short distance between them.

In Ce^{3+}-doped YAG with ADs [25], an attractive interaction is found between Ce_Y and Y_{Al}-Al_Y defects. Also, the formation of 1AD is more favorable in the presence of Ce_Y than in its absence, whereas the opposite is true for 2AD, meaning that the presence of Ce_Y tends to lower the concentration of ADs in YAG. The formation energy of the defect Ce_Y-1AD is not minimal at its shortest distance, but it has a clear structural dependence that makes the preferable structure one in which Ce_Y-AlO_4-(Al_Y-Y_{Al})-AlO_4-Ce_Y chains are formed (with the Al_Y-Y_{Al} vector pointing off the Ce-Al-Al-Al-Ce chain). In this case the defects are not as close as possible because of the relevance of multipole electrostatic interactions, but, again, they are at a short distance and orderly distributed, far from homogeneously. In the presence of 2AD defects, the most stable position for Ce_Y is next to a terminal Y_{Al} of the 2AD defect, i.e. forming a (Y_{Al}-Al_Y)-AlO_6-(Al_Y-Y_{Al})-Ce_Y chain. The image provided by this study coincides with that of YAG with 2AD defects: ordered distributions of defects, both in YAG:Ce^{3+},1AD and YAG:Ce^{3+},2AD, with short distances between them.

In the previous cases, several neutral defects with no suspicion of mutual inter-defect bonding nor electrostatic interactions tend to gather together. No apparent reason exists for this phenomenon other than the minimization of the deformation energies, i.e. the fact that a lesser deformation of the perfect host is produced when the two defects are together than when they are far apart. In other words, it is a purely mechanical reason: deformations of the solid attract each other. When the individual deformations of the defects are large, they might be incompatible at very short inter-defect distances, making the most stable distribution not the one with the shortest distance, but the one with the shortest distance that makes the deformations compatible.

In addition, if multipole electrostatic interactions between defects co-exist with mechanical interactions, the combination of both produce most stable distributions with some sort of order, in which the defects may not be as near as possible, but they still are at a short distance from each other.

These results question the hypothesis of a uniform distribution of defects across the bulk, often adopted when the defects are unbound, i.e. when they are not attracted by bonding interactions. Instead, the image that emerges from the calculations is one of in-homogeneous distributions of defects. When multipole electrostatic interactions are relevant, the distribution is ordered; when these interactions do not play a role, the defects gather together in order to minimize the lattice deformations.

10.3 Pauli Antisymmetry Interactions Between Host and Active Center

Optically active centers responsible for the luminescence of doped materials experience Pauli antisymmetry interactions with the solids hosting them. These apparently subtle interactions have a quantum origin and produce steric effects which are commonly ignored in the area of luminescence. However, the steric effects, which are traditionally considered in other fields like organic chemistry, where they are known to play relevant roles in the conformations and reactivities of organic molecules, can also be determinant for some properties of inorganic luminescent materials.

In this section, we discuss the basics of the Pauli antisymmetry interactions and describe two cases in which they have a key role at determining puzzling behaviors of phosphor materials: The case of the red shift experienced by the $5d \rightarrow 4f$ emission of YAG:Ce^{3+} when it is co-doped with La^{3+} and alike ions, and the case of the multiplets formed by the R_1-line emission of Cr^{3+} when it is doped in garnets.

Let us describe what Pauli antisymmetry interactions between an optically active center and its environment are. Firstly, we start with the definition of Pauli antisymmetry, which is the property that all correct N-electron wave functions must fulfill, according to which they must change sign when the positions of any two electrons are interchanged. This is a key property responsible, for instance, for the chemical bond. Then, we can consider the effect of Pauli antisymmetry on the interaction between two electrons of equal spin, one belonging to the active center and one belonging to the environmental host. If both of them have α spin, the first electron (electron 1) occupies a ϕ_a orbital of an optically active center (e.g. made of a lanthanide dopant and its first coordination shell), and the second electron (electron 2) occupies a ϕ_e orbital of the inactive environment (e.g. a host ion or a co-dopant ion), the simplest two-electron wave function describing them would be the product $\phi_a^\alpha(1)\phi_e^\alpha(2)$. However, this two-electron wave function does not fulfill Pauli antisymmetry principle. For it to fulfill it, an antisymmetric product rather than a simple product has to be used. Among all equivalent ways of making an antisymmetric product, all of them with the same final effects, the one that is most illustrative from the point of view of the response of the active center to its Pauli antisymmetry interaction with the host, is: to orthogonalize ϕ_a to its environment and to form a Slater determinant with the new, orthogonal orbitals: $|\phi_{a,ortho}^\alpha \phi_e^\alpha| = \frac{1}{\sqrt{2}}[\phi_{a,ortho}^\alpha(1)\phi_e^\alpha(2) - \phi_e^\alpha(1)\phi_{a,ortho}^\alpha(2)]$. This is illustrated in Fig. 10.2. Such a transformation from a simple product of two non-orthogonal orbitals to an antisymmetric product of two orthogonal orbitals forced by the Pauli antisymmetry principle has several consequences:

(i) *Exchange* interactions between the electrons of the active center and the electrons of the environment are the first consequence of Pauli antisymmetry requirements between active site and host. The central importance of exchange interactions within the active center have always been recognized and they are routinely considered in all theories handling optically active centers, like e.g. crystal field theory. However, exchange interactions between an active center and its environment have long been disregarded [26], based on the assumption that they are small and

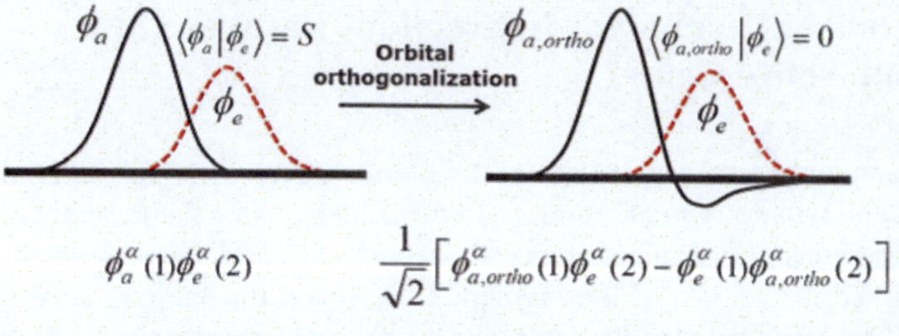

Fig. 10.2 Illustration of the effects on an active orbital ϕ_a of orthogonalization with an environmental orbital ϕ_e: nodes, stronger slope changes, and delocalization. Corresponding wave functions of two electrons of equal spin, not fulfilling (left) and fulfilling (right) Pauli antisymmetry principle

short-range. Interestingly, the AIMP embedding method used throughout this book to represent classical and quantum mechanical interactions between active site and host, includes these interactions [27]. However, they are not the issue here.

(ii) The second consequence is *Pauli repulsion* [28] between active center and environment. This term refers to the increase of the kinetic energy of the active center electrons due to the appearance of nodes and new slope changes in $\phi_{a,ortho}$, as compared with ϕ_a, produced by the orthogonalization. Although the larger total electronic energy of the active center comes from an increment of its kinetic energy, it is called *Pauli repulsion* because it finds its origin in the Pauli antisymmetry principle and it works very similarly to an electrostatic repulsion. The larger the environmental orbital ϕ_e and the shorter the distance between ϕ_a and ϕ_e, the larger the slope changes, the kinetic energy increments, and the Pauli repulsion. In other words, Pauli repulsion is a steric interaction and it is the reason for steric effects.

Pauli repulsion has traditionally been disregarded in phenomenological theories used for optical materials activated with impurity ions, like crystal field theory or ligand field theory [26]. However, it is considered in the embedding method used here, where it is included by means of repulsive operators (one per environmental orbital) of the kind $B_e|\phi_e\rangle\langle\phi_e|$ (1.11). Pauli repulsion (steric) effects from the host onto the active centers have important consequences on their equilibrium structures, up to the point that it is normal to get absurd structures, and even dissociation, when these repulsive effects are disregarded or not properly taken into account [29]. They also mean an important contribution to the ligand field splittings of the active center [29].

The important role played by Pauli repulsion interactions between the La^{3+} co-dopant and the Ce^{3+} optically active center at red-shifting the $5d \rightarrow 4f$ emission in YAG:Ce^{3+},La^{3+}, is discussed in Sect. 10.3.1.

(iii) The third consequence is the *Pauli antisymmetry induced nephelauxetic effect* [30]. This is a more subtle effect, consisting in the reduction of the inter-electronic repulsion *within* the active center as a consequence of the delocalization of ϕ_a and the other active center orbitals, due to their orthogonalization with the environment. Like the well-known Jørgensen's nephelauxetic effect (from the Greek for cloud expanding) [31], which is the lowering of the electronic repulsion within an ion when it enters a solid, this is also a nephelauxetic effect, because it results into more delocalized orbitals. Here, unlike in Jørgensen's effect, the reason for the cloud expanding isn't the covalent interactions between ion and ligands, but the Pauli antisymmetry interactions between the whole active center and the embedding host. Hence its name [30]. In brief we can also call it Pauli's nephelauxetic effect.

In Sect. 10.3.2, the multiplets observed in the R_1-line of Cr^{3+} in garnets with variable composition are shown to be a consequence of the *Pauli antisymmetry induced nephelauxetic effect*. Since this steric effect can be tuned by choosing garnet components, it is a means for a fine control over this famous emission line.

10.3.1 Red Shift of YAG:Ce^{3+} $d-f$ Emission upon Co-doping with La^{3+}

Several factors contribute to the unexpected red-shift of the YAG:Ce^{3+} $5d \rightarrow 4f$ emission upon co-doping with La^{3+} and alike ions, which substitute for dodecahedrally coordinated, smaller Y^{3+} ions and create La_Y substitutional defects and lattice expansions. But the main contribution is the increase of Pauli repulsion effects from the host onto the Ce_Y optically active center when Y^{3+} ions are substituted by larger La^{3+} ions. It is a steric effect, a direct electronic effect resulting from the quantum-mechanical interaction between La^{3+} and the Ce_Y active center. It has nothing to do with the lattice distortions created by the ionic substitutions.

In effect, the analysis of the multiconfigurational ab initio calculations [32] indicates that short-range and long-range lattice expansions contribute 22% of the total red-shift of –220 cm^{-1} and direct La_Y substitutions contribute 78%.

Expectations for a blue-shift were based on the assumption of a dominant contribution of the lowerings in the ligand-field splittings produced by the first-shell expansion (Fig. 8.3). This contribution is indeed a blue-shift (+86 cm^{-1}), but it is not dominant. It is almost entirely compensated by a red-shift of the $5d^1$ energy centroid with respect to the $4f^1$ energy centroid (–77 cm^{-1}), also produced by the first-shell expansion; and long-range expansions also contribute to the red-shift (26%). In other words, lattice distortions induced by the La_Y substitutions do not produce a blue-shift, but a red-shift of the Ce^{3+} $5d \rightarrow 4f$ emission, and, in any case, they only represent 22% of the overall effect of the substitutions. The rest (78%, –171 cm^{-1}) is a direct effect of the substitutions.

Direct interactions between La^{3+} and $(CeO_8)^{13-}$ active sites differ from those between Y^{3+} and $(CeO_8)^{13-}$ not in the electrostatic contributions, but in the electronic ones. The latter are short-range, so that, for them to be relevant, the La_Y defects created after the substitution must be near the active site; otherwise, no significant impact would exist on direct interactions. Here it comes the first issue: The La_Y defects do not distribute randomly but, rather, they are mechanically attracted by Ce_Y

defects and, as a result, both dopants get as close as possible [23] (cf. Sect. 10.2). Then, electronic interactions become relevant. These are not bonding interactions, they are steric interactions derived from Pauli antisymmetry requirements between active site and host: (i) Larger La^{3+} ions increase the Pauli repulsion exerted by the host onto the $(CeO_8)^{13-}$ cluster with respect to smaller Y^{3+} ions. This has an immediate effect on the increase of $5d$ ligand-field splitting, which is quantified as –149 cm^{-1} red-shift of the $5d \rightarrow 4f$ emission, 68% of the total red-shift. (ii) At the same time, larger La^{3+} ions also increase the Pauli antisymmetry induced nephelauxetic effect, which expands the $5d$ orbitals and results in a lowering of the $5d^1$ energy centroid that enhances the red-shift. This is quantified as –22 cm^{-1}, 10% of the total red-shift.

The overall picture is a red-shift of the YAG:Ce^{3+} $5d \rightarrow 4f$ emission upon co-doping with La^{3+} due 22% to lattice distortions and 78% to changes in direct quantum-mechanical interactions with the active site. These interactions play a role because La_Y and Ce_Y defects get together, rather than distribute randomly. They derive from Pauli antisymmetry principle conditions between the active site and its environment, they are twofold, and they are steric. Pauli repulsion is the most important of the two interactions and accounts for 68% of the total red-shift. The Pauli antisymmetry induced nephelauxetic effect enhances the red-shift an additional 10%.

10.3.2 Multiplets of the Cr^{3+} R_1-Line: Controlling Pauli Antisymmetry Strength

The R-line of Cr^{3+} in an octahedral site is the $3dt_2^3$ intraconfigurational emission from the ^2E excited state to the $^4A_2$ ground state. It is a spin-flip transition in which only the spin of one electron changes and it splits into sharp R_1 and R_2 lines by spin-orbit coupling in a trigonal site [33]. It is crystal-field independent and it basically depends on the nephelauxetic effect [31]: More delocalized $3dt_2$ orbitals lower the $3dt_2 - 3dt_2$ repulsion and red-shift the R-lines. Usually, the delocalizations are created by metal-ligand covalent interactions.

The narrow Cr^{3+} R-lines have long been and are used for practical purposes, like measuring high pressures and identifying local defect structures and different luminescent centers in phosphor materials [34, 35], and as benchmarks for theoretical developments [31]. But, in spite of its extensive study and the abundant literature on the $^2E - ^4A_2$ emission, it still rises important questions on basic interactions, like the following one: In Cr^{3+}-doped YAG, substitutions of Y^{3+} with Lu^{3+} and Tb^{3+} produce R_1-line shifts and multiplets in the concentration series $Lu_{3x}Y_{3-3x}Al_5O_{12}$:$Cr^{3+}$ and $Tb_{3z}Y_{3-3z}Al_5O_{12}$:$Cr^{3+}$ (cf. Fig. 10.3). The line shifts might be explained by changes in lattice constant and Cr-O distance, because they affect the Cr-O covalency and the nephelauxetic effect. But, what explains the multiplets? Chemical substitutions of Y^{3+} with Lu^{3+} and Tb^{3+} should not have a direct impact on Cr-O covalency and, through it, on the nephelauxetic effect. Then, what interactions are we missing? The missing interactions must be related to the substitutions because, in spite of

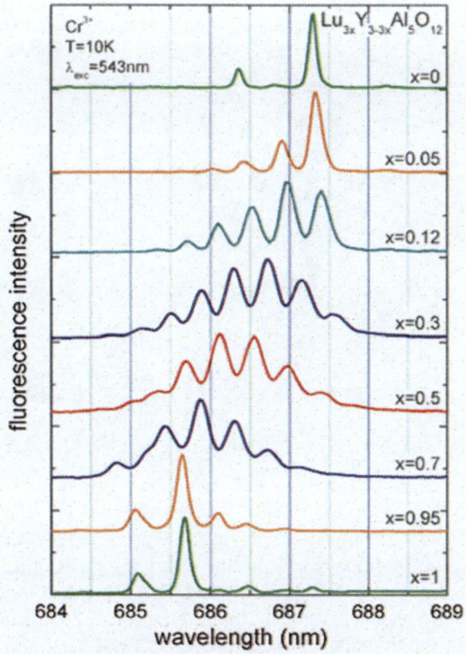

Fig. 10.3 Experimental emission spectra (upon 543 nm excitation) of $Lu_{3x}Y_{3-3x}Al_5O_{12}$:$Cr^{3+}$, measured at 10 K [36]

its incomplete understanding, the multiplet structures have been exploited to obtain critical information on local structures of the disordered materials [36, 37].

Multiconfigurational ab initio calculations [30] have shown that substitutions of Y^{3+} with Lu^{3+} and Tb^{3+} at next-nearest neighbor (NNN) sites produce changes in the strength of the Pauli antisymmetry induced nephelauxetic effect produced by the host on the $(CrO_6)^{9-}$ active site. This being a steric effect, its strength is increased with larger substituents (Tb^{3+}), which red-shift the R-line, and decreased with smaller ones (Lu^{3+}), which blue-shift the R-line, and the change is roughly proportional to the number of NNN substitutions. All of this results in well-structured multiplets of R-lines, each of them assigned to a particular distribution of NNN substituents. In addition, a given line multiplet is shifted as a whole according to the lattice constant: red-shifted for shorter lattice constants (which means shorter Cr-O distances, larger Cr-O covalence, larger Jørgensen's nephelauxetic effect, and smaller $3dt_2$-$3dt_2$ repulsion) and blue-shifted for longer ones. In other words, all the line multiplets of the $Lu_{3x}Y_{3-3x}Al_5O_{12}$:$Cr^{3+}$ and $Tb_{3z}Y_{3-3z}Al_5O_{12}$:$Cr^{3+}$ materials are explained in terms of well understood nephelauxetic effects: Jørgensen's, created by Cr-O covalence and controlled with the lattice constant; and Pauli's, created by antisymmetry conditions between active center and host and controlled with chemical substitutions. A fine tuning of the R_1-line can then be achieved by combining both. We summarize here the results of these calculations.

In $Y_3Al_5O_{12}$:Cr^{3+}, Cr substitutes for Al in a C_{3i} site (point group S_6) with sixfold coordination and creates a $(CrO_6)^{9-}$ active center. The substitutions of Y^{3+} with Lu^{3+}

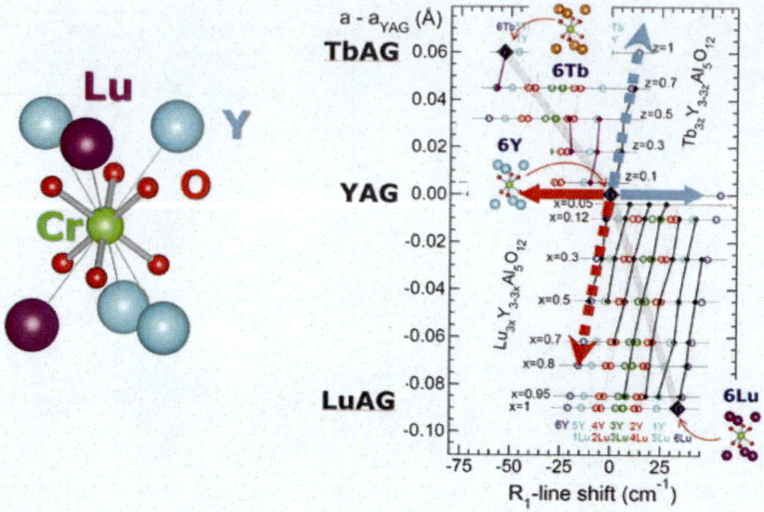

Fig. 10.4 Left: $(CrO_6)^{9-}$ active cluster and the six NNN sites (occupied with 4 original Y^{3+} and 2 substitutional Lu^{3+} in this case). Right: R_1-line multiplets (shifts with respect to YAG:Cr^{3+}) of $Lu_{3x}Y_{3-3x}Al_5O_{12}$:$Cr^{3+}$ and $Tb_{3z}Y_{3-3z}Al_5O_{12}$:$Cr^{3+}$. Full diamonds connected with full lines: experiments, previously interpreted (black) and new assignments (violet); large symbols are used for pure compositions $Tb_3Al_5O_{12}$, $Y_3Al_5O_{12}$, and $Lu_3Al_5O_{12}$, connected with a thick gray line as a guide for the eye. Open circles: ab initio calculations. Colors indicate the NNN occupations associated with the R_1-lines: blue: 6Y, 6Lu, and 6Tb lines; turquoise: 5Y1Lu, 1Y5Lu, 5Y1Tb, and 1Y5Tb lines; red: 4Y2Lu, 2Y4Lu, 4Y2Tb, and 2Y4Tb lines; green: 3Y3Lu and 3Y3Tb lines. Thick arrows indicate red- and blue-shifts due to Jorgensen's nephelauxetic effect (dashed) and Pauli's nephelauxetic effect (full)

and Tb^{3+} that are expected to have an impact on the Cr^{3+} $^2E - ^4A_2$ emission in the concentration series $Lu_{3x}Y_{3-3x}Al_5O_{12}$:$Cr^{3+}$ and $Tb_{3z}Y_{3-3z}Al_5O_{12}$:$Cr^{3+}$ take place in six Cr^{3+} NNN positions (Fig. 10.4(left)). All-electron DKH SA-CASSCF/MS-CASPT2/RASSI-SO calculations were performed on the $(CrO_6)^{9-}$ cluster embedded in $Y_3Al_5O_{12}$, pure and with all possible Lu_Y and Tb_Y substitutions at NNN sites. Gaussian atomic natural orbital relativistic basis sets ANO-RCC of quadruple-zeta with polarization quality were used: $(21s15p10d6f4g2h)[7s5p5d4f3g2h]$ for Chromium [38] and $(14s9p4d3f)[4s3p2d1f]$ for Oxygen [39]. In SA-CASSCF, the active space was made of the 5 molecular orbitals of main character Cr $3d$. The molecular orbitals were optimized by minimizing the respective average energies of ^4A and $^4T_1$ (for spin quartets), and ^2E, $^2T_1$, and $^2T_2$ (for spin doublets). In MS-CASPT2, the O $2s$, $2p$ and Cr $3s$, $3p$, $3d$ electrons were correlated. In RASSI-SO, all calculated spin-orbit-free states were allowed to interact via spin-orbit coupling.

The calculated R_1-line using the structure of the undistorted YAG [$Ia\bar{3}d$ (230) $a = 12.000$ Å, $x_O = -0.03060$, $y_O = 0.05120$, $z_O = 0.1500$] [40], agrees with the experiment [36]: 14329 cm^{-1} versus 14550 cm^{-1}. The line produced by a Cr^{3+} with $6Y^{3+}$ NNN ions (6Y line) has been identified in $Lu_{3x}Y_{3-3x}Al_5O_{12}$:$Cr^{3+}$ and

$Tb_{3z}Y_{3-3z}Al_5O_{12}$:$Cr^{3+}$, and found to red-shift with decreasing lattice constant a (due to Jørgensen's nephelauxetic effect) [36, 37]; a similar red-shift was obtained in the calculations, with a systematic deviation of $0.4(a - a_{YAG})$ cm^{-1}/10^{-3}Å. This is to say: the calculations adequately reproduce Jørgensen's nephelauxetic effect on the Cr^{3+} R_1-line and, when they are corrected for the known, systematic deviation, they reproduce this effect with great accuracy.

As for the Pauli nephelauxetic effect, the calculated shift of the R_1-line due to a Cr^{3+} with 6Lu^{3+} NNN ions (6Lu line) in $Lu_{3x}Y_{3-3x}Al_5O_{12}$:$Cr^{3+}$ ($x = 0.3$ with respect to the 6Y line in the same material, also agrees with experiment: +24 cm^{-1} versus +50 cm^{-1} blue-shift. Equivalent calculations in which the embedding potentials of the NNN Lu^{3+} ions are substituted with Y^{3+} embedding potentials with their Pauli antisymmetry strength scaled, $\lambda_{Pas}\hat{P}_{Y^{3+}}$ (1.11), fully reproduce the calculations when $\lambda_{Pas} = 0.92$ is used and the experiments when $\lambda_{Pas} = 0.82$ is used, which confirms that the blue-shift is due to a reduction in the strength of the Pauli antisymmetry induced nephelauxetic effect. In other words, the real effect of the substitution of a Y^{3+} ion with a smaller Lu^{3+} ion is an 18% reduction of the steric effect or Pauli antisymmetry strength (and this reduction is underestimated by the AIMP embedding method, which predicts it to be of only 8%). Following the same procedure, it is found that the real effect of the substitution of an Y^{3+} ion with a larger Tb^{3+} ion is a 31% increase of the steric, Pauli antisymmetry strength.

Interestingly, chemical substitutions of Y^{3+} produce two opposite effects that have to be combined. E.g. substitution with smaller ions like Lu^{3+} give Cr^{3+} active centers with 1Lu, 2Lu, ..., 6Lu NNN substitutions and increasingly blue-shifted R_1-lines, due to the lower and lower Pauli's nephelauxetic effect and corresponding $3dt_{2g}$ orbital delocalization. This set of R_1-lines constitute a multiplet. But, at the same time, the smaller ions lead to smaller lattice constants, which shorten the Cr-O distances and increase Jørgensen's nephelauxetic effect, with the result of an overall red-shift of the whole multiplet. Something equivalent, but opposite, happens in the case of substitutions of larger ions like Tb^{3+}.

In order to have both effects together, calculations on YAG:Ce^{3+} are performed using different lattice constants and doing all possible substitutions of NNN Y^{3+} ions with Lu^{3+} and Tb^{3+} ions. In these, the substituents are represented by Pauli antisymmetry strength scaled Y^{3+} embedding potentials (λ_{Pas}=0.82 for Lu^{3+}, λ_{Pas}=1.31 for Tb^{3+}) and the empirical correction $-0.4(a - a_{YAG})$ cm^{-1}/10^{-3}Å is applied to the lines. The results are summarized in Fig. 10.4(right).

It is observed that the 13 R_1-lines corresponding to the 13 NNN Lu_Y symmetry independent substitutions (1 for 6Y; 1 for 5Y1Lu; 3 for 4Y2Lu; 3 for 3Y3Lu; 3 for 2Y4Lu; 1 for 1Y5Lu; 1 for 6Lu) gather together in 7 groups, each of them corresponding to a number of substitutions, the spatial distribution of the substitutions being of minor importance. The same is true for Tb_Y substitutions. This indicates that the steric effect is basically not directional here and it explains the fact that the best resolved R_1-line experimental multiplets show 7 peaks. This observation had raised the suggestion that each multiplet peak could correspond to a number of NNN substitutions [36, 37]; the calculations support this assumption and provide its detailed explanation.

Figure 10.4(right) illustrates the interplay between Jørgensen's and Pauli's nephelauxetic effects in order to predict the R_1-line multiplets and their shift across the concentration series. Also, the observation of the multiplets' baricenters shifting in opposite directions across the series of solid solutions [36, 37] is shown here: According to the present interpretation of the multiplets, their baricenters depend on the relative populations of the different distributions of NNN cations; this information is only available in the pure compounds ($z = 1$ $Tb_3Al_5O_{12}$:Cr^{3+}, $z = x = 0$ $Y_3Al_5O_{12}$:Cr^{3+}, and $x = 1$ $Lu_3Al_5O_{12}$:Cr^{3+}), which are highlighted in the Figure with larger full diamonds; the trend is also signaled with a thick gray line showing a kink. As we see, the kink is the result of a different lattice constant change (+0.09 Å from LuAG to YAG and +0.06 Å from YAG to TbAG) together with a different Pauli antisymmetry strength change (+18% from LuAG to YAG and +31% from YAG to TbAG).

The calculations show that chemical substitutions are a means to control the strength of Pauli antisymmetry effects of the host onto the active center, which are basically non-directional. This steric effect grows across the chemical series $Lu^{3+} < Y^{3+} < Tb^{3+}$ in correspondence with the increasing ionic radius, the relative strengths being 82%, 100%, and 131%, and similar behaviors must be expected for other lanthanide ions, depending only on their size. This effect opposes that of the substitutions on the lattice constant and both of them together allow a fine control of the R_1-line energy.

10.4 $4f^{N-1}6s$ States Rise Their Energies in Hosts with Respect to Free-Ions

References to states of lanthanide ions in solids derived from the $4f^{N-1}6s$ configuration of the free-ion are not abundant. This is so in spite of the fact that these states appear at relatively low energies in divalent free-ions, as it is illustrated in Table 10.2 for the $4f^{13}6s$ states of free Yb^{2+}. In the absence of a detailed knowledge of these states' behavior when the free-ions enter solid sites, it seems reasonable to assume their involvement in spectroscopic features of the ions in solids that cannot be understood otherwise and appear at similar energy regions. E.g. absorption peaks of Yb^{2+}-doped $SrCl_2$ difficult to assign and appearing at 35028 cm^{-1} and 45794 cm^{-1} have been attributed to the states derived from the $4f^{13}(^2F_{7/2})6s$ and $4f^{13}(^2F_{5/2})6s$ configurations of Yb^{2+} in the host [43]. Here, not only the closeness of the energies of single states with free-ion data, but also the remarkable agreement between energy differences in the host and in gas phase, suggest the adequacy of the assignment: it is reasonable to expect that the approximate 10000 cm^{-1} difference in the free-ion will remain in the host, because the splitting is due to spin-orbit coupling interactions which are almost entirely intra-atomic interactions. It is clear, in any case, that the 10000 cm^{-1} difference is only a necessary condition but not a sufficient one and that

Table 10.2 Theoretical and experimental energies (in cm^{-1}) of the states of the $4f^{13}6s$ configuration of the Yb^{2+} free-ion above the $4f^{14}\ ^1S_0$ ground state

	Theory [41]		Experiment [42]	
$4f^{13}(^2F_{7/2})6s$	$^3F_4^o$	36800	$^3F_4^o$	34656.13
	53% $^1F_3^o$	37300	54% $^1F_3^o$	34990.66
$4f^{13}(^2F_{5/2})6s$	$^3F_2^o$	47155	$^3F_2^o$	44853.59
	51% $^3F_3^o$	47619	54% $^3F_3^o$	45207.64

Table 10.3 Calculated energies of the states of the $4f^{13}6s$ configuration of Yb^{2+} in $CsCaBr_3$ compared with those of other excited configurations. Energies are given in cm^{-1} and are referred to the $4f^{14}$ 1 A_{1g} ground state

Configurational submanifold	States	Energy range
$4f^{13}(^2F_{7/2})5dt_{2g}$	1 T_{2u} − 2 A_{2u}	23890–29992
$4f^{13}(^2F_{5/2})5dt_{2g}$	3 A_{1u} − 11 T_{1u}	34560–39830
$4f^{13}(^2F_{7/2})5de_g$	8 E_u − 14 T_{2u}	43872–48619
$4f^{13}(^2F_{5/2})5de_g$ + $4f^{13}(^2F_{7/2})6s$	6 A_{1u} − 19 T_{2u}	53859–58932
$4f^{13}(^2F_{5/2})6s$	14 E_u − 7 A_{2u}	66964–67473

the hypothesis rests on the assumption that $4f \rightarrow 6s$ excitation energies are similar in the free-ion and in the ion-in-the-host.

Also, states of the $4f^{N-1}6s$ excited configurations are included in empirical CFT calculations sometimes, which tends to lower rms deviations between theory and experiments. When this happens, it is hard to be sure if the deviations are lowered by the real involvement of these states in the excited manifolds under study or just because there is a larger number of empirical parameters.

In these circumstances, it is insightful to clarify what energy changes experience the $4f^{N-1}6s$ states of an ion when it enters a solid and why. And why these states are so elusive in the absorption spectra of lanthanide-activated luminescent materials.

Multiconfigurational ab initio calculations indicate these states have ion-ligand bond lengths significantly longer than the states of the $4f^{N-1}5d_{HIGH}$ configuration and their energies are significantly larger than in the free-ion (cf. Tables 6.17 and 6.19 and Fig. 6.8(right panels)). In $CsCaBr_3$:Yb^{2+}, where Yb^{2+} substitutes for Ca^{2+} in a 6-fold bromide octahedral coordination, an analysis based on calculations on the Yb^{2+} free-ion, the $(YbBr_6)^{4-}$ cluster *in vacuo*, and the $(YbBr_6)^{4-}$ cluster embedded in $CsCaBr_3$ [41], allowed to pinpoint the separate effects of the Br ligands and the rest of the host.

It was found that the $4f^{13}6s$ states of Yb^{2+} appear around 20000 cm^{-1} above the free-ion energies (cf. Table 10.3); their lower spin-orbit coupling components overlap the higher spin-orbit coupling components of the $4f^{13}5de_g$ states; and their higher spin-orbit coupling components appear approximately 10000 cm^{-1} above, in correspondence with the spin-orbit coupling splitting of the $4f^{13}$ subshell. The elusiveness of $4f^{13}6s$ states seems to have its origin in their lower states overlapping the upper parts of the $4f^{13}5d$ manifold, with much higher absorption intensities, and their upper states being immersed in dense bands of host states.

Furthermore, the important energy increase when the ion enters the solid host is not due to interactions with the Br ligands, because calculations on the $(YbBr_6)^{4-}$ cluster *in vacuo* give $4f \rightarrow 6s$ transition energies only slightly larger than the ones calculated on the Yb^{2+} free-ion. It is the embedding interaction exerted by the $CsCaBr_3$ host on the $(YbBr_6)^{4-}$ cluster what increases the $4f^{13}6s$ energies to the final results.

As a matter of fact, the host embedding effects do not only rise the states' energies, but they are responsible for the very stability of the $4f^{13}6s$ states. In the absence of embedding interactions, their potential energy curves E_{EC} versus $d_{Yb\text{-}Br}$ show a double minimum (cf. Fig. 1 in [41]). Only the shallowest minimum corresponds truly to these states, which are metastable at very long Yb-Br distances, but they collapse into a deeper minimum at short Yb-Br distances, where the electronic character of the states turns into a pre-ionized electronic structure, with the 6*s* electron crossing off the Br_6 ligand cage and remaining in its outer vicinities. When the embedding effects of the $CsCaBr_3$ host are included, the short Yb-Br distances that favored the ionization are penalized and the pre-ionized electronic structure is largely destabilized; the electron remains within the Br_6 ligand cage and retains its 6*s* character at all reasonable Yb-Br distances, but its energy experiences a significant increment. $SrCl_2{:}Yb^{2+}$ is similar to this study case: the combination of a large Cl_8 ligand cage and $SrCl_2$ embedding effects retains the Yb^{2+} 6*s* electron inside the cage and forms a stable high-energy $4f^{13}6s$ manifold with long Yb-Cl bond lengths (cf. Fig. 6.8(right panels)). In $CaF_2{:}Yb^{2+}$, $SrF_2{:}Yb^{2+}$, and $BaF_2{:}Yb^{2+}$, however, the F_8 ligand cages are small and cannot retain inside the 6*s* electron, but the hosts' embedding potentials allow the pre-ionized electronic structures to stabilize in the form of impurity-trapped-excitons (cf. Fig. 6.8).

In summary, lanthanide ions have their $4f^{N-1}6s$ states at significantly larger energies in solids than in gas phase, and this is not as a consequence of their bonding interactions with the ligands, but mostly because of the embedding interactions that the hosts exert on them. Depending on the solid host, the $4f^{N-1}6s$ states can be stable, with long ion-ligand equilibrium distances and energies overlapping the highest $4f^{N-1}5d$ manifolds and host states; or they can turn into impurity-trapped-excitons of $4f^{N-1}\phi_{ITE}$ configuration, with short ion-ligand bond lengths and variable energies.

10.5 The Excited States of Eu-Doped Luminescent Materials

Eu is perhaps the most popular of the activators in lanthanide-doped luminescent materials, both in its Eu^{2+} and Eu^{3+} oxidations states. The complexity of their excited manifolds has been discussed in Sects. 6.1.5, 6.1.6, and 6.2.4, and their associated, rich spectroscopy in Sect. 9.4, especially that of Eu^{2+}. The many-electron states of Eu^{2+} involved in the luminescent phenomena of Eu-doped materials have been shown to be organized in four sub-manifolds, of respective main spin and configurational characters: high-spin ($S = 7/2$) $4f^6(^7F_J)5d_{\mathrm{LOW}}$, low-spin ($S = 5/2$) $4f^6(^7F_J)5d_{\mathrm{LOW}}$, high-spin $4f^6(^7F_J)5d_{\mathrm{HIGH}}$, and low-spin $4f^6(^7F_J)5d_{\mathrm{HIGH}}$, where $5d_{\mathrm{LOW}}$ and $5d_{\mathrm{HIGH}}$ are the ligand-field split $5d$ shells of lower and higher energy respectively (e.g. $5de_g$ and $5dt_{2g}$ in the cubic O_h coordination of Eu^{2+} in CaF_2 and other difluorides, and $5dt_{2g}$ and $5de_g$ in the octahedral O_h coordination of Eu^{2+} in CaS and other sulfides). The first and second sub-manifolds have very pure spin and configurational characters, but the third and fourth sub-manifolds have strong mixed characters, both of spin octets with spin sextets, and of $4f^6(^7F_J)5d_{\mathrm{HIGH}}$ configuration with $4f^6(^7F_J)\phi_{ITE}$ and $4f^6(^5D_J)5d_{\mathrm{LOW}}$ configurations. Any remaining sub-manifolds are higher in energy, immersed in the sea of host-excited levels (the conduction band in the the language of one-electron physical models), and are not detected in luminescence spectroscopies.

In Sect. 9.4, a *high-spin ligand-field splitting* parameter $\varepsilon_{\mathrm{lfs}}^{\mathrm{HS}}$ was defined (9.1). Its definition was chosen so that: (i) the parameter is a measure of ligand-field strength, (ii) it is possible to calculate it from the energies of the many-electron states in the first and third sub-manifolds, and (iii) it is possible to measure it from the first and second intense absorption/excitation bands. These properties enable an insightful, simple and direct contrast between theory and experiment (cf. Table 9.8).

4f 5d Exchange Interactions

In a similar way to the high-spin ligand-field splitting parameter, $4f$-$5d$ *exchange splitting* parameters $\varepsilon_{\mathrm{exch}}^{e_g}$ (for the difluorides) and $\varepsilon_{\mathrm{exch}}^{t_{2g}}$ (for the sulfides) were also defined (9.2). The definitions were chosen so that: (i) the parameter is a measure of the strength of exchange interactions between the $4f$ and $5de_g$ or $5dt_{2g}$ shells, (ii) it is possible to calculate it from the energies of the many-electron states in the first and second sub-manifolds, and (iii) it is possible, in principle, to measure it from the first intense absorption/excitation band and the spin-forbidden band that corresponds to absorptions to the second $4f^65d$ sub-manifold. The very low intensity of the spin-forbidden band prevented the empirical determination of this parameter, though, and only theoretical data is available (cf. Table 9.8). The so defined exchange splitting parameters are informative only when S is a quasi-good quantum number. According to this, only many-electron states with $\geq$90% of spin-octet and $\geq$90% of spin-sextet characters were included in the respective HS and LS averages of $4f^6(^7F_J)5d_{\mathrm{LOW}}$ states in (9.2). When S is not at all a quasi-good quantum number, this parameter loses its meaning. This is the case of the $4f^6(^7F_J)5d_{\mathrm{HIGH}}$ sub-manifold and higher sub-manifolds of Eu^{2+}, as well as the $4f^{N-1}5d$ configurations of heavier lanthanides like Tm^{2+} [44] and Yb^{2+} [45].

The results in Table 9.8 show that the $4f$-$5d$ exchange interactions in the $4f^65d$ configuration of Eu^{2+}, and their corresponding splittings, experience only a slight increase across the series CaF_2-SrF_2-BaF_2 and CaS-SrS-BaS, following the enlarging of the unit cell and the elongation of the Eu-ligand distance, with no significant change in the Eu-ligand bond. However, they experience a very significant increase

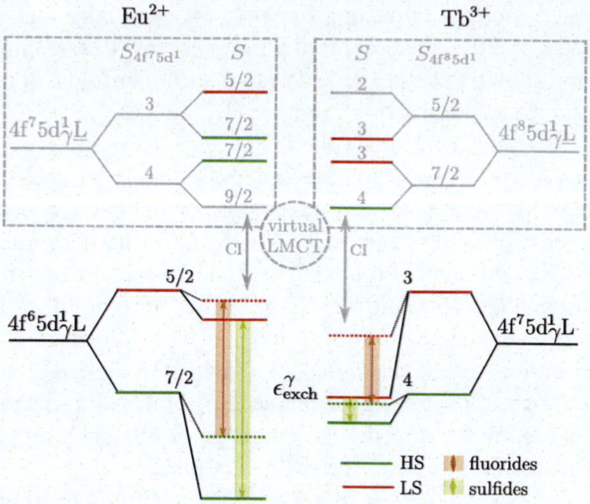

Fig. 10.5 Schematic representation of how configuration interaction (CI) with virtual ligand-to-metal charge transfer (LMCT) states increases or decreases the exchange splitting of the $4f^{N-1}5d$ configurations of Eu^{2+} (N=7) and Tb^{3+} (N=8) respectively. L indicates a full ligand shell (closed-shell); $\underline{L}$ indicates a ligand shell with one hole

with the covalency of the Eu-ligand bond, from the difluorides to the sulfides: from ~4500 cm^{-1} to ~7500 cm^{-1}. Interestingly, this behavior is opposite to the experimentally shown decrease of the $4f$-$5d$ exchange splitting with bond covalency in the $4f^7 5d$ configuration of Tb^{3+} [46, 47] and it deserves some discussion.

The opposite effect of covalency on the exchange splitting of the $4f^6 5d$ configuration of Eu^{2+} and the $4f^7 5d$ configuration of Tb^{3+} is understood by analyzing the role of virtual ligand-to-metal charge transfer (LMCT) excitations (in a theoretical framework and language of valence bond theory (VB) configurations), whose importance grows with the covalency of the lanthanide-ligand bond. This discussion is supported with the scheme in Fig. 10.5.

The idea behind the analysis is the following: (i) virtual LMCT configurations (which are very many) create virtual states with different total spin quantum number S (very many states for each S) and (ii) impurity states of the $4f^{N-1}5d$ configuration with a given total spin quantum number S are stabilized by configuration interaction (CI) with virtual LMCT states of the same S. Then, (iii) impurity states with different S are stabilized differently by virtual LMCT configurations, depending on the numbers of available LMCT virtual states for each S. The different stabilizations of impurity states with different S results on changes of the exchange splittings. The lower the energy of the LMCT virtual configurations, the greater their impact on the exchange splittings. This is why more covalent ligands, with less energetic virtual LMCT excitations, produce larger exchange splittings.

Taking $4f^6 5d$ states of Eu^{2+}-doped sulfides as example, the most important virtual LMCT excitations are S $3p \rightarrow$ Eu $4f$, which produce virtual LMCT configurations of the type $Eu^+(4f^7 5d^1)\times S^-(3p^5)$. Since the hole in the S $3p$ shell is in fact distributed among all the ligands, or several of them, according to some molecular orbital of dominant sulfur character, a notation of the type $4f^7 5d^1 \underline{L}$ describes better the virtual LMCT configuration, in which $\underline{L}$ denotes a hole in a molecular orbital of main ligand character (of main S $3p$ character in this example). The corresponding virtual LMCT configurations of $4f^7 5d$ states of Tb^{3+} in the same host are $4f^8 5d^1 \underline{L}$.

Then, in Eu^{2+}, there will be virtual LMCT states with values of S that result from the coupling $S(4f^7 5d^1) \times S(\underline{L})$, i.e. $(S = 4, 3) \times (S = 1/2)$. (Here, lower spins of the $Eu^{+}(4f^7 5d^1)$ configuration are not being considered because they have higher energy and smaller configuration interaction with the impurity states.) The $4\times1/2$ coupling produces virtual LMCT states of $S = 9/2$ and 7/2; and the $3\times1/2$ coupling produces virtual LMCT states of $S = 7/2$ and 5/2. In summary, the virtual LMCT configurations result in a larger number of virtual LMCT states with $S = 7/2$ than with $S = 5/2$ and, in consequence, the Eu^{2+} $4f^6 5d$ states (which would be $4f^6 5d^1 L$ according to the notation used for the virtual LMCT configurations) with main spin-octet character ($S = 7/2$) will experience a larger stabilization by CI interaction with virtual LMCT states than states with main spin-sextet character ($S = 5/2$), which results in an enhancement of the $4f$-$5d$ exchange splitting. In sulfides, the $4f^7 5d^1 \underline{L}$ virtual LMCT configurations are closer in energy to the $4f^6 5d^1 L$ impurity states than in difluorides; hence their larger exchange splittings.

On the other hand, the opposite effect is found on the high-spin and low-spin states of Tb^{3+}. In this case, the S $3p \rightarrow$ Tb $4f$ virtual excitations originate $4f^8 5d^1 \underline{L}$ virtual LMCT configurations with states resulting from the spin coupling $S(4f^8 5d^1) \times S(\underline{L})$, i.e. $(S = 7/2, 5/2) \times (S = 1/2)$. (As before, $S = 3/2, 1/2$ virtual states of the $4f^8 5d^1$ virtual sub-configuration are very high in energy and are discarded in this analysis.) The $7/2\times1/2$ coupling produces virtual LMCT states of $S = 4$ and 3; and the $5/2\times1/2$ coupling produces virtual LMCT states of $S = 3$ and 2. I.e., in this case there is a larger number of virtual LMCT states with $S = 3$ (low-spin, spin-septets) than with $S = 4$ (high-spin, spin-nonets), which results in a larger stabilization of the low-spin states over the high-spin states, and a corresponding lowering of the $4f$-$5d$ exchange splitting. And, the lowering is, of course, more intense for more covalent ligands.

Origin of the Fine Structure in the Absorption Spectra of Eu^{2+}-Doped Materials

As it was remarked in Sect. 9.4, the low-temperature absorption spectra of the difluorides show a well-resolved fine structure, correctly reproduced by the multiconfigurational ab initio calculations, with seven individual peaks forming a staircase. This fine structure is not so well measured in the sulfides, although the spectra seem to indicate it is also there; the theoretical spectral profiles with low broadening factor confirm this.

The most commonly accepted explanation for the seven individual peaks in the staircase fine structure is based on two assumptions: One is that the lowest $4f^6 5d$ states of Eu^{2+} derive from the multiplets of the $4f^6(^7F_J)$ sub-configuration, which is a reminiscent of the lowest 5000 cm^{-1} of the Dieke diagram for Eu^{3+}, and its coupling with the lowest crystal-field level of one $5d$ electron. This was fully confirmed by the multiconfigurational ab initio calculations. The other is that a decoupled scheme is acceptable, this is to say that the coupling is simple, of mean-field type, and it does not produce any additional splittings [48–50]. But inspection of the multiconfigurational ab initio results do not support this assumption. The many-electron wave functions

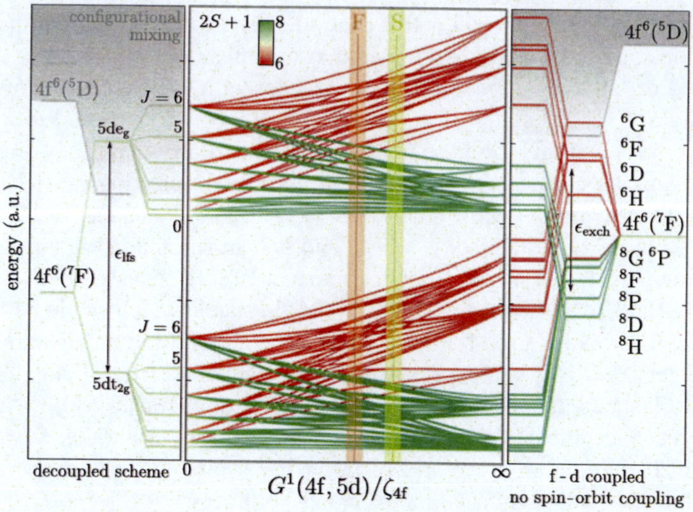

Fig. 10.6 Crystal field theory calculation for the low energy part of the $4f^6 5d^1$ configuration in an octahedral field, obtained in the 490 dimensional $|4f^6(^7F)5d^1\rangle$ basis. The transition from the decoupled scheme (including only $4f$ spin-orbit coupling as additional interaction) to a scheme with only the $4f-5d$ Coulomb interaction is shown. The colors of the curves represent the spin-characters of the eigenstates, ranging from green for pure spin-octets to red for pure spin-sextets. The shaded area indicates the energy range where configurational mixing can be expected. The vertical bands, denoted by 'F' and 'S', denote the parameter ratio where the crystal field calculation resembles best to the ab initio calculations for the fluorides and sulfides, respectively

indicate a rather complex $4f^6(^7F_J) \times 5d$ coupling, much more complex than a simple low-field splitting of the seven $J = 0,1,\ldots6$ atomic levels of Eu^{3+}.

In an attempt to explain why a complex coupling leads to a relatively simple fine structure, and to illustrate how this can happen, a systematic study was made by performing crystal-field theory calculations with a tunable parameter [13]. The parameter was chosen so that it could range between to limits: a fully decoupled scheme, in which spin-orbit coupling interactions in the ^7F_J multiplet dominate the electronic structure, and a coupled scheme with no spin-orbit coupling interactions, in which the electronic structure reflects the dominant character of Coulomb and exchange interactions, and in particular $4f$-$5d$ exchange interactions. In order to do so, the $4f^6(^7F_J)5d$ manifold of Eu^{2+} in an O_h field was calculated as a function of a coupling parameter $G^1(4f,5d)/\zeta_{4f}$ and compared with the multiconfigurational ab initio results. The Slater-Condon integral $G^1(4f,5d)$ represents $4f$-$5d$ exchange interactions and the $4f$ spin-orbit coupling constant ζ_{4f} represents the strength of spin-orbit coupling interactions. The results are shown in Fig. 10.6. Similar analyses inspiring the present one were performed in which all Slater-Condon parameters were scaled in order to study optical and magnetic properties of EuF_2 [51] and a set of $4f^6(^7F)5de_g$ levels [52].

In the left limit, defined by $G^1(4f,5d)/\zeta_{4f}$=0, no $4f$-$5d$ exchange interactions exist, the $4f^6(^7F_J)$ sub-shell is decoupled from the $5dt_{2g}$ and $5de_g$ sub-shells, and the electronic structure simply reflects the spin-orbit coupling splitting of the seven $^7F_{J=0-6}$ levels (of Eu^{3+}) superimposed on the $5dt_{2g}$-$5de_g$ crystal-field splitting of the, additional, $5d$ electron. The lower seven levels in this decoupled scheme (left box in Fig. 10.6) are responsible for the staircase fine structure of the first absorption band in the commonly accepted interpretation. In the center box of the Figure, the energy levels at low values of $G^1(4f,5d)/\zeta_{4f}$ clearly show the $4f$-$5d$ exchange splitting (green lines are used for spin-octets and red lines for spin-sextets). It is clear in the Figure that the decoupled scheme is fundamentally incorrect because, although it correctly predicts an absorption band with staircase structure of seven peaks, it totally fails at representing the exchange splitting.

In the right limit, defined by $G^1(4f,5d)/\zeta_{4f}=\infty$, no spin-orbit coupling interactions exist and the spin-orbit-free $4f$-$5d$ coupling $4f^6(^7F)\times(^2T_{2g}+^2E_g)$ producing spin-octets and spin-sextets controls the electronic structure, as shown in the right box in Fig. 10.6. Obviously, this limiting case is far from reality, as far as the decoupled scheme is, but it cannot induce to confusion because nothing in it resembles experimental features.

Energy levels at intermediate couplings are shown in the center box in Fig. 10.6, where best fittings to the multiconfigurational ab initio results on difluorides and sulfides are indicated with respective orange and yellow vertical strips. These areas correspond to the existence of two simultaneous couplings: an intense spin-orbit coupling in the $4f^6(^7F_J)$ subshell and an intense $4f^6\times5d$ coupling with important $4f$-$5d$ exchange interactions. These couplings are superimposed on a large $5dt^1_{2g}(^2T_{2g})$-$5de^1_g(^2E_g)$ ligand-field splitting. As a result of all couplings and interactions, the organization of the $4f^6(^7F_J)5d$ energy levels in four sub-manifolds discussed in Sect. 6.2.4 is very clear in Fig. 10.6: A first sub-manifold of high-spin $4f^6(^7F_J)5d_{\mathrm{LOW}}$ states and a second sub-manifold of low-spin $4f^6(^7F_J)5d_{\mathrm{LOW}}$ states, respectively represented by the lower sets of green and red lines crossing the orange and yellow bands; and a third sub-manifold of high-spin $4f^6(^7F_J)5d_{\mathrm{HIGH}}$ states and a fourth sub-manifold of low-spin $4f^6(^7F_J)5d_{\mathrm{HIGH}}$ states, respectively represented by the upper sets of green and red lines crossing the orange and yellow bands. The fact that the $4f$-$5d$ coupling is intense and the exchange interactions are large, is what separates the first sub-manifold from the second; lower exchange interactions would make these two sub-manifolds to overlap, as they correspond to a left shift on the Figure. Large ligand-field strength is what separates the two lower sub-manifolds from the two upper ones and opens a gap in the set of states.

References

1. C.W. Struck, W.H. Fonger, J. Lumin. **10**, 1 (1975)
2. N.F. Mott, Proc. R. Soc. London, Ser. A **167**, 384 (1938)
3. L. Seijo, Z. Barandiarán, J. Chem. Phys. **115**, 5554 (2001)

4. Z. Barandiarán, L. Seijo, J. Chem. Phys. **119**, 3785 (2003)
5. P.S. Bagus, K. Hermann, C.W. Bauschlicher, J. Chem. Phys. **80**, 4378 (1984)
6. C.W. Bauschlicher, P.S. Bagus, J. Chem. Phys. **81**, 5889 (1984)
7. L. Seijo, Z. Barandiarán, E. Harguindey, J. Chem. Phys. **114**, 118 (2001)
8. Z. Barandiarán, L. Seijo, Can. J. Chem. **70**, 409 (1992)
9. Z. Barandiarán, L. Seijo, Theor. Chem. Acc. **116**, 505 (2006)
10. Z. Barandiarán, N.M. Edelstein, B. Ordejón, F. Ruipérez, L. Seijo, J. Solid State Chem. **178**, 464 (2005)
11. Z. Barandiarán, L. Seijo, in Computational Methods, in *Lanthanide and Actinide Chemistry*. ed. by M. Dolg (John Wiley & Sons, New York, 2015), pp. 217–240
12. L. Seijo, Z. Barandiarán, *Ab Initio Calculations on Excited States of Lanthanide Containing Materials* (Elsevier, 2016), *Handbook on the Physics and Chemistry of Rare Earths*, vol. 50, chap. 285, pp. 65–89
13. J.J. Joos, P.F. Smet, L. Seijo, Z. Barandiarán, Inorg. Chem. Front. **7**, 871 (2020)
14. F. Ruipérez, L. Seijo, Z. Barandiarán, J. Chem. Phys. **122**, 234507 (2005)
15. R. Valiente, F. Rodríguez, J. González, H.U. Güdel, R. Martín-Rodríguez, L. Nataf, M.N. Sanz-Ortiz, K. Krämer, Chem. Phys. Lett. **481**, 149 (2009)
16. C.S. Yoo, H.B. Radousky, N.C. Holmes, N.M. Edelstein, Phys. Rev. B **44**, 830 (1991)
17. J.L. Pascual, Z. Barandiarán, L. Seijo, Phys. Rev. B **76**, 104109 (2007)
18. F.H. Su, W. Chen, K. Ding, G.H. Li, J. Phys. Chem. A **112**(21), 4772 (2008)
19. S. Mahlik, K. Wiśniewski, M. Grinberg, R.S. Meltzer, J. Phys.: Condens. Matter **21**(24) (2009)
20. C. Chen, K.L. Teo, T.C. Chong, Y.H. Wu, T.S. Low, R. Carey, D.M. Newman, I. Viney, J.P. Wu, J. Bickerton, J. Appl. Phys. **93**(5), 2559 (2003)
21. A.D. Vos, K. Lejaeghere, D.E.P. Vanpoucke, J.J. Joos, P.F. Smet, K. Hemelsoet, Inorg. Chem. **55**(5), 2402 (2016)
22. A.B. Muñoz-García, L. Seijo, Phys. Rev. B **82**, 184118 (2010)
23. A.B. Muñoz-García, L. Seijo, J. Phys. Chem. A **115**, 815 (2011)
24. A.B. Muñoz-García, E. Artacho, L. Seijo, Phys. Rev. B **80**, 014105 (2009)
25. A.B. Muñoz-García, Z. Barandiarán, L. Seijo, J. Mater. Chem. **22**, 19888 (2012)
26. A.B.P. Lever, *Inorganic Electronic Spectroscopy, Studies in Physical and Theoretical Chemistry*, vol. 33 (Elsevier, Amsterdam, 1984)
27. Z. Barandiarán, L. Seijo, J. Chem. Phys. **89**, 5739 (1988)
28. C.W. Wilson Jr., W.A. Goddard III., J. Phys. Chem. **56**, 5913 (1972)
29. J.L. Pascual, J. Schamps, Z. Barandiarán, L. Seijo, Phys. Rev. B **74**, 104105 (2006)
30. L. Seijo, S.P. Feofilov, Z. Barandiarán, J. Phys. Chem. Lett. **10**, 3176 (2019)
31. C.K. Jörgensen, Prog. Inorg. Chem. **4**, 73 (1962)
32. A.B. Muñoz-García, J.L. Pascual, Z. Barandiarán, L. Seijo, Phys. Rev. B **82**, 064114 (2010)
33. B. Henderson, G.F. Imbusch, *Optical Spectroscopy of Inorganic Solids* (Clarendon, Oxford, 1989)
34. P.H. Horn, Y.M. Gupta, Appl. Phys. Lett. **49**, 856 (1986)
35. A. Bessière, S.K. Sharma, N. Basavaraju, K.R. Priolkar, L. Binet, B. Viana, A.J.J. Bos, T. Maldiney, C. Richard, D. Scherman, D. Gourier, Chem. Mater. **26**, 1365 (2014)
36. S. Feofilov, A. Kulinkin, K. Ovanesyan, A. Petrosyan, C. Dujardin, Phys. Chem. Chem. Phys. **16**, 22583 (2014)
37. S.P. Feofilov, A.B. Kulinkin, K.L. Ovanesyan, A.G. Petrosyan, Solid State Commun. **226**, 39 (2016)
38. B.O. Roos, R. Lindh, P.A. Malmqvist, V. Veryazov, P.O. Widmark, J. Phys. Chem. A **109**, 6575 (2005)
39. B.O. Roos, R. Lindh, P.A. Malmqvist, V. Veryazov, P.O. Widmark, J. Phys. Chem. A **108**, 2851 (2004)
40. F. Euler, J.A. Bruce, Acta Crystallogr. **19**, 971 (1965)
41. G. Sánchez-Sanz, L. Seijo, Z. Barandiarán, Spec. Letters **43**, 393 (2010)
42. B.W. Bryant, J. Opt. Soc. Am. **55**, 771 (1965)
43. Z. Pan, C. Duan, P.A. Tanner, Phys. Rev. B **77**, 085114 (2008)

44. M. de Jong, A. Meijerink, L. Seijo, Z. Barandiarán, J. Phys. Chem. C **121**, 10095 (2017)
45. G. Sánchez-Sanz, L. Seijo, Z. Barandiarán, J. Chem. Phys. **131**, 024505 (2009)
46. J.S. Shi, S.Y. Zhang, J. Phys.: Condens. Matter **15**, 4101 (2003)
47. P. Dorenbos, J. Phys.: Condens. Matter **15**, 6249 (2003)
48. A. Meijerink, G. Blasse, J. Lumin. **43**, 283 (1989)
49. M.J. Freiser, S. Methfess, F. Holtzber, J. Appl. Phys. **39**, 900 (1968)
50. S.J. Camardello, P.J. Toscano, M.G. Brik, A.M. Srivastava, J. Lumin. **151**, 256 (2014)
51. A. Yanase, T. Kasuya, Progress of Theoretical Physics **46**, 388 (1970)
52. H.A. Weakliem, Phys. Rev. B **6**, 2743 (1972)

Chapter 11
Electron Transfer and Luminescence

Abstract In Chap. 7, the diabatic approximation was shown to offer an elegant recipe to obtain qualitative and in many cases quantitative information on electron transfer processes involving multiple impurity centers. Here, this machinery is put in action to reveal novel insights into how electron transfer impacts the properties of luminescent materials. In some cases, it will just lead to charge-transfer quenching. In more interesting cases, it gives a means for color control via selective non-radiative decay, it makes novel emission bands to emerge, or it enables energy storage. It is shown how multiconfigurational ab initio quantum chemistry can aid in the design of novel functional materials based on electron transfer phenomena, both by providing specific information on given materials and by unveiling systematic charge transfer behaviors.

11.1 Role of LMCT States in the Color Control of Pr Luminescence

In Sect. 8.2.3, in the context of the search for new efficient red phosphors useful in solid-state lighting, the focus was put on non-radiative decay mechanisms between the Stark splittings of the $^3P_0$ and $^1D_2$ states of Pr^{3+} doped in hosts, in particular in $CaTiO_3$ and $CaZrO_3$. Here we focus on the new insights revealed by the multiconfigurational ab initio study and hard to deduce from experiments alone.

It was discussed how, after different proposals of mechanisms responsible for the $^3P_0$–$^1D_2$ decay, based on experimental observations and empirical modeling, like cross-relaxation within Pr^{3+}–Pr^{3+} ionic pairs, multiphonon relaxation, and intersystem crossing through low-lying $4f5d$ levels, the virtual recharge model [1] was adopted as the accepted explanation. In a series of experimental works, first, it was clarified that the states intermediate between Pr^{3+} $^3P_0$ and $^1D_2$ are charge transfer states, and second, it was postulated that they are Pr-to-Metal charge transfer states, i.e. MMCT states [2–4].

In the absence of direct calculations of MMCT transitions and empirical models for them which could be as reliable as Crystal Field Theory is for *d*-*d* transitions of transition metal ions and *f*-*f* transitions of lanthanide ions in solids, the search for

Z. Barandiarán et al., *Luminescent Materials*, Springer Series in Materials Science 322,
https://doi.org/10.1007/978-3-030-94984-6_11

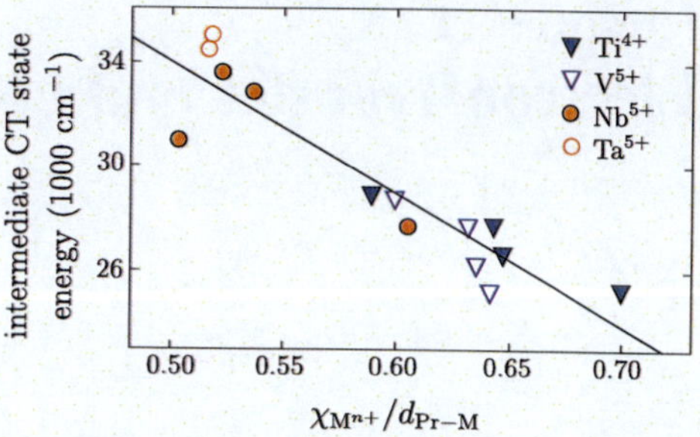

Fig. 11.1 Variation of the energies of the intermediate charge transfer (CT) state with the ratio between the optical electronegativities of the $3d^0$, $4d^0$ and $5d^0$ M^{n+} cations and the shortest Pr-M distance. Adapted from [3]

empirical correlations became a key tool. Along this line, it was found that in many hosts doped with Pr^{3+}, the ${}^1D_2$ emission could be directly excited with an absorption to a state located at the lower energy side of the host absorption and identified as a charge transfer state. And, when this happened, a linear correlation was found between the energy of this intermediate charge transfer state able to transfer energy to Pr^{3+} ${}^1D_2$ and the ratio between the optical electronegativities of the $3d^0$ and $4d^0$ M^{n+} cations $\chi_{M^{n+}}$ and the shortest Pr-M distance $d_{\text{Pr-M}}$ [3] (cf. Fig. 11.1),

$$E_{\text{CTS}} = A - B\frac{\chi_{M^{n+}}}{d_{\text{Pr-M}}}. \tag{11.1}$$

This linear correlation was a strong support to the charge transfer hypothesis and the existence of Pr-to-M MMCT states intermediate between ${}^1D_2$ and ${}^3P_0$ became the standard interpretation for the enhancement of the ${}^1D_2$ red emissions in materials that presented it, and the simultaneous quenching of the ${}^3P_0$ emission, like the archetypal $CaTiO_3$:Pr^{3+}.

Fortunately, in the last years, multiconfigurational ab initio calculations of MMCT transitions became feasible and reliable and, as discussed in Sect. 8.2.3, the energies of Pr-to-Ti and Pr-to-Zr MMCT absorptions in $CaTiO_3$:Pr^{3+} and $CaZrO_3$:Pr^{3+} were calculated with the aim of further supporting the MMCT interpretation with an independent source of information. However, the MMCT transitions were found to be much larger than the Pr^{3+} ${}^3H_4 \rightarrow {}^3P_0$ absorption [5]. This result implies that MMCT states cannot play a role in the ${}^3P_0$–${}^1D_2$ decay mechanism and it seriously questions the accepted interpretation.

But ab initio calculations are general and, contrary to empirical models, they are not formulated for specific types of materials and properties; so, the same methods that ruled out the MMCT hypothesis were used in the search for a solution of a different kind. In this process, they were also applied to LMCT states of the host materials. The theoretical studies showed that the calculated optically allowed O-to-Ti LMCT absorptions of $CaTiO_3$ (host interband absorptions) agreed very well with the spectroscopic measurements, which find them at significantly higher energy than the Pr^{3+} ${}^3H_4 \rightarrow {}^3P_0$ absorption. But they also showed that there are optically forbidden O-to-Ti LMCT states well below the allowed absorptions, and these can play a key role in the ${}^3P_0$–${}^1D_2$ decay mechanism because they have an energy intermediate between them (cf. Sect. 8.2.3).

In other words, when it was possible to perform multiconfigurational ab initio calculations of sufficient quality of ligand-field, MMCT, and LMCT states, the ab initio theoretical approach, as a tool independent from experiments but aimed at complementing them, strongly supported the long ago formulated hypothesis of CT states intermediate between ${}^3P_0$ and ${}^1D_2$ as responsible for the non-radiative decay between them; seriously questioned the hypothesis of these being Pr-to-Metal MMCT states; and suggested and strongly supported a new hypothesis of these intermediate states being LMCT states of the host.

The calculations also showed that the lowest LMCT states of the host can be down-shifted by the presence of defects. Since defects also break the absorption selection rule, all of it together can explain the existence of an absorbing CT state located at the lower energy side of the host absorption when Pr^{3+} is doped into the host.

Let us finish by commenting on an interesting issue: If the intermediate state is a host LMCT state rather than a Pr-to-M MMCT state, why does the empirical relationship (11.1) hold? At this point, the Jørgensen's empirical correlation for the energies of LMCT transitions must be recalled [6]

$$E_{\text{LMCT}} = A - B\frac{\chi_{\text{M}^{n+}}}{d_{\text{L-M}}}, \tag{11.2}$$

where $\chi_{\text{M}^{n+}}$ is, as above, the optical electronegativity of M^{n+} and $d_{\text{L-M}}$ the ligand-metal distance, which should hold for the intermediate state if it is a LMCT state. Besides, the Pr–M distances used in (11.1) were not experimental distances, which are very hard to measure in doped materials, but distances between host sites occupied by Pr and M host sites, and these distances scale as the O–M distances. In other words, (11.1) and (11.2) are equivalent, except for the value of the slope B, and it is normal that both of them hold if the intermediate CT state is a LMCT state.

Then, if an intermediate LMCT state of the host enhances a fast ${}^3P_0$–${}^1D_2$ decay by an electron-hole recombination mechanism (cf. Table 8.4) and subsequent red ${}^1D_2$ luminescence, and the energy of the LMCT state can be modulated by the presence of defects, here there is an obvious means for an efficient control over the color control of Pr^{3+} luminescence.

11.2 IVCT States of Mixed-Valence Lanthanide-Activated Phosphors

In Chap. 7, inter-valence charge transfer (IVCT) states were introduced, naturally emerging from the mixed-valence nature of transition metal or lanthanide ions. Although IVCT states of mixed-valence transition metal compounds are long known and their interest is widely recognized [7, 8], those of lanthanide compounds received little attention before they were the subject of multiconfigurational ab initio calculations. Their occurrence in scientific literature seems to be limited to reports on thermally induced electron hopping within mixed-valence pairs in Eu_3S_4 [9] and photoinduced inter-valence electron transfer in mixed-valence europium chloride crystals $Na_5Eu_7Cl_{22}$ and KEu_2Cl_6 [10, 11].

In lanthanide-doped materials, where the ions are diluted in a crystalline host, the only identification of an IVCT absorption is a tentative assignment of a broad band in the reflection spectrum of Ce-doped $LaPO_4$ as a Ce^{3+}/Ce^{4+} IVCT absorption [12].

However, lanthanides are multivalent and it is important to know the roles that IVCT states play or can eventually play in the luminescence of mixed-valence lanthanide-doped phosphor materials. That is the focus of this section. The so-called anomalous luminescence of several lanthanide-doped compounds are reviewed and re-interpreted as IVCT luminescence; direct evidence of IVCT in Eu luminescent materials is given; and the systematic behavior of the IVCT absorption onset across the lanthanide series is unveiled.

11.2.1 Role of IVCT States in the Complex Interplay Between Regular and Anomalous Emission of Yb^{2+} in Fluorite-Type Hosts

The $4f \rightarrow 4f$ and $5d \rightarrow 4f$ luminescences of lanthanide ions in solids used in solid state lighting, lasers, fiber amplifiers, or medical imaging, among other applications, are often compromised by quenching or by replacement of the expected luminescence by another of a different and unexpected nature, usually referred to as an anomalous emission. When this is the case, detailed understanding of the new emission is important to control the design of luminescent materials in order to get efficient target luminescences and, eventually, to take advantage of the new luminescence.

An archetype anomalous emission is that of Yb^{2+} in Yb-doped materials. It was detected for the first time in the 50's in CaF_2:Yb and SrF_2:Yb [13] and intensively studied afterward [14–16]. It is characterized by a very broad emission band, with a peak very much red-shifted with respect to the expected Yb^{2+} $5d \rightarrow 4f$ emission and a very large Stokes shift when it is excited by the Yb^{2+} UV $4f^{14} \rightarrow 4f^{13}5d$ absorption.

In addition to its surprising properties, the Yb anomalous emission exhibits a very complex, puzzling interplay with the regular emission in the family of isostructural hosts with fluorite structure: CaF_2, SrF_2, BaF_2, and $SrCl_2$: When Yb^{2+} is doped in CaF_2, the emission is anomalous [13, 15, 16, 18]. In SrF_2, both the regular and the anomalous emissions coexist (cf. Fig. 11.2) [14, 16, 17, 19]. And going on in the series, in BaF_2 all emissions are quenched up to 1.5μm [15, 16]. In $SrCl_2$, the change of ligand with respect to SrF_2 results in the fourth case: the anomalous emission disappears and the emission is regular, although with some unexplained features and a complex temperature dependence [20, 21].

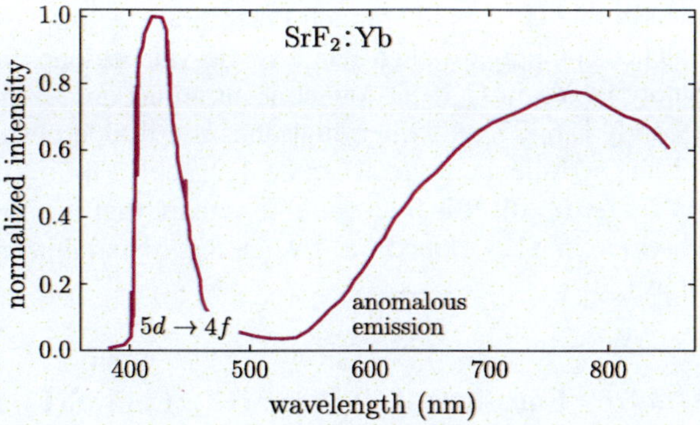

Fig. 11.2 Emission spectrum of SrF_2:Yb under 355 nm excitation, showing the simultaneous occurrence of the regular $5d{\rightarrow}4f$ and anomalous emissions, measured at 4.4 K. Adapted from [17]

The impurity-trapped exciton (ITE) model was proposed to explain the Yb anomalous emission [14, 15]. According to it, the emission is an Yb^{2+} $4f^{13}\phi_{ITE}{\rightarrow}4f^{14}$ transition with its origin in an ITE excited state of Yb^{2+} in which a bound electron-hole pair exists, with the hole localized at the Yb site and the electron delocalized outside the YbF_8 volume, over the nearest cation neighbors. An Yb-Ligand bond length much shorter in the ITE than in the ground state would explain the very large band width of the emission. Its large red shift with respect to the regular $5d{\rightarrow}4f$ emission would depend on the energy of the ITE state, much harder to predict in the lack of a detailed knowledge of the ITE state. And the complex interplay with the regular emission remains an unsolved puzzle. Nevertheless, this model was accepted as the standard explanation for over three decades, not only for the anomalous emission of Yb-doped materials but also for those of compounds doped with other lanthanides like Ce and Eu [22, 23].

However, attempts to find and understand ITE states of Yb^{2+} in materials with multiconfigurational ab initio calculations and systematize their energies [24], fail to confirm the ITE hypothesis because ITE states, which were indeed found, had their energies too high so as to be involved in the Yb anomalous luminescence (cf. Fig. 6.8 and Sects. 6.2.6 and 6.3.4).

This theoretical finding was experimentally proven by combined X-ray absorption near edge structure (XANES) and UV spectroscopy studies on CaF_2:Yb [25], where it was possible to determine fractions of Yb^{2+} and Yb^{3+} at low Yb concentrations, and thus the net concentrations of Yb^{2+}. They provided direct experimental evidence that the ITE model cannot explain the anomalous luminescence in CaF_2:Yb by showing that the anomalous luminescence intensity is not proportional to the Yb^{2+} concentration and that only a small fraction of Yb^{2+} ions is involved in the anomalous emission.

As it is discussed next, a new interpretation of the Yb anomalous emission as an IVCT emission, i.e. a radiative electron transfer between Yb^{2+} and Yb^{3+} ions in Yb^{2+}/Yb^{3+} mixed-valence pairs, emerged from multiconfigurational ab initio calculations on the electronic structure of Yb^{2+}/Yb^{3+} pairs in the above hosts, which found it to be able to explain perfectly well the complex features of the Yb anomalous emission [24]. This interpretation requires the luminescent material to be a mixed-valence compound rather than a single-valence one. In this respect, the multivalent nature of Yb in CaF_2 at low concentrations has been experimentally proven only recently [26, 27], although high intrinsic difficulties to prevent the coexistence of +2 and +3 valence states when dealing with Yb-doped materials have always been recognized [20]. It is interesting to remark that the Yb anomalous emission is also observed in Yb^{3+}-doped CaF_2 when this material is brought under reducing conditions [28].

11.2.1.1 Emissions of Yb^{2+} in $SrCl_2$, CaF_2, SrF_2, and BaF_2: Regular, Anomalous, Both, and None

A first step in a multiconfigurational ab initio study on the excited manifold of Yb^{2+}/Yb^{3+} mixed-valence pairs aimed at exploring the hypothesis of their involvement in the anomalous emission of Yb^{2+}-doped materials, is the calculation of their diabatic IVCT configuration coordinate energy diagrams. These are expected to provide the fundamental characteristics of the electronic structure of the mixed-valence pairs. Further steps that could provide more quantitative details imply the calculation of much more computationally demanding adiabatic energy diagrams. But the latter would only change the basic descriptions of the former if the electronic couplings between Yb^{2+} and Yb^{3+} active sites are very strong, something that is not expected at all for a pair of cations at distances long enough so as to have negligible mutual bonding interactions.

Following the above arguments, the diabatic IVCT diagrams of Yb^{2+}/Yb^{3+} pairs in CaF_2, SrF_2, BaF_2, and $SrCl_2$ have been calculated in [24] according to the descriptions of Sect. 7.4, using as ingredients the configuration coordinate energy diagrams of the individual Yb^{2+} and Yb^{3+} cubic active sites in the same hosts (cf. Table 6.9, Fig. 6.8). The most relevant part of these diagrams are shown in Fig. 11.3.

In Fig. 11.3, left and right black parabolas represent states of the $Yb^{2+}(1A_{1g})$-$Yb^{3+}(^2F_{7/2})$ and $Yb^{3+}(^2F_{7/2})$-$Yb^{2+}(1A_{1g})$ equivalent configurations of the Yb^{2+}/Yb^{3+} pair, respectively, where Yb^{2+} is in its ground state and Yb^{3+} in the lowest states of the $4f^{13}$ electronic configuration. (In the following and for simplicity, only one of two equivalent configurations of the pairs will be explicitly mentioned, in the knowledge that the other one also exists.) Red curves represent $Yb^{2+}(1A_{1g})$-$Yb^{3+}(^2F_{5/2})$ states, where Yb^{3+} has been excited to the highest $4f^{13}$ states, and blue curves represent $Yb^{2+}(4f^{13}5de_g)$-$Yb^{3+}(^2F_{7/2})$ states, where Yb^{2+} has been excited to the $4f^{13}5de_g$ configuration. In the orange curves, both Yb^{2+} and Yb^{3+} have been excited and they represent $Yb^{2+}(4f^{13}5de_g)$-$Yb^{3+}(^2F_{5/2})$ states. Note than only the lowest, relevant states of the manifolds are represented in the Figure.

The configuration coordinate diagrams of the four materials look very similar. However, the apparently small differences between them are responsible for important differences in their luminescent behavior, as it will be discussed. The key region is around Q_{et}=–0.3 Å (and its equivalent Q_{et}=+0.3 Å) and 24000–26000 cm^{-1}. In this area, two important $Yb^{2+}5de_g \rightarrow Yb^{3+}4f$ IVCT non-radiative decays are competing: After accepting an electron, Yb^{3+} converts into Yb^{2+} in its ground state in both of them; however, the Yb^{3+} resulting from Yb^{2+} after donation of its $5de_g$ electron ends up in its ground $^2F_{7/2}$ configuration at the blue-black crossing and in its excited $^2F_{5/2}$ configuration at the blue-red crossing.

The availability of these two decays after an $Yb^{2+}4f \rightarrow 5de_g$ excitation (process I, black arrow) makes the differences between hosts. The different possibilities are highlighted in the bottom right panel of the Figure and labelled cases A, B, C, and D.

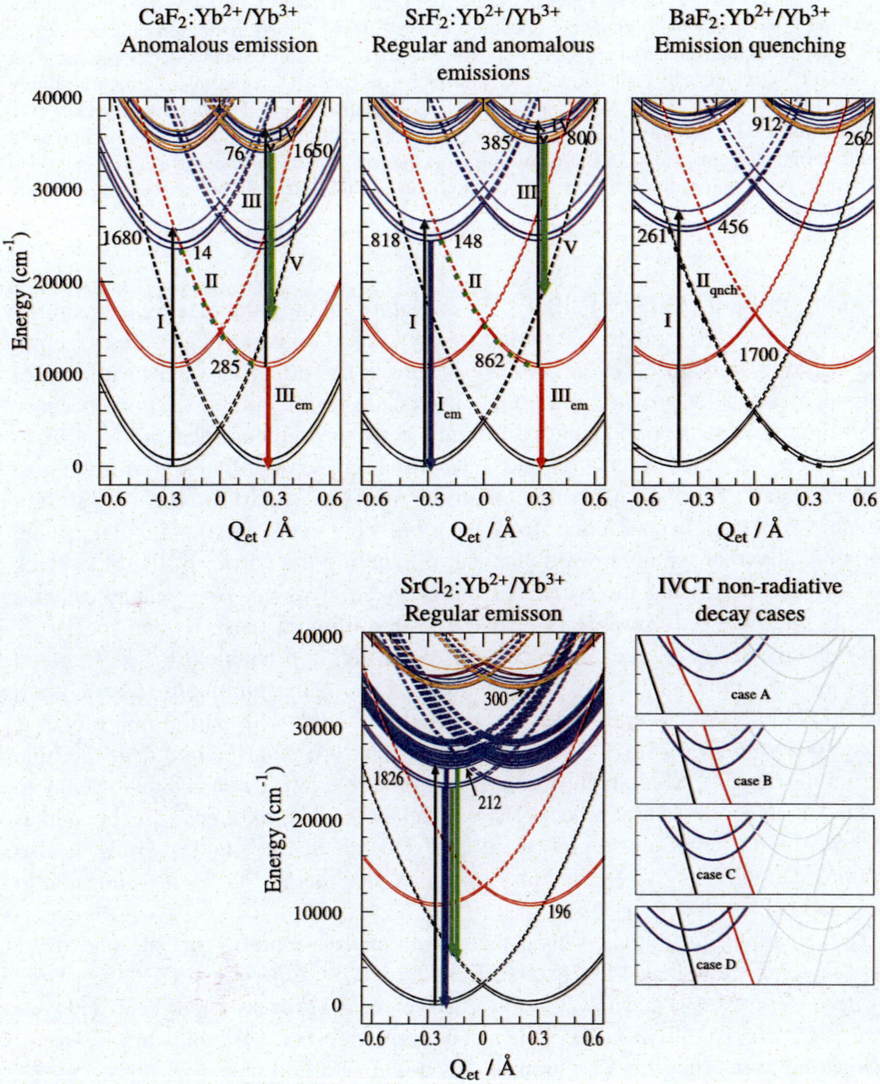

Fig. 11.3 (Continued)

◂**Fig. 11.3** (Continued) Diabatic IVCT configuration coordinate diagrams of Yb^{2+}/Yb^{3+} mixed-valence pairs in CaF_2, SrF_2, BaF_2, and $SrCl_2$ and possible IVCT non-radiative decay cases. The following processes are indicated: I: $Yb^{2+}4f\rightarrow 5de_g$ excitation. I_{em}: $Yb^{2+}5de_g\rightarrow 4f$ emission (regular emission). II and II_{qnch}: $Yb^{2+}5de_g\rightarrow Yb^{3+}4f$ non-radiative IVCT plus structural relaxation from $Yb^{2+}(4f^{13}5de_g)$-$Yb^{3+}(^2F_{7/2})$ states to $Yb^{3+}(^2F_{5/2})$-$Yb^{2+}(1A_{1g})$ and $Yb^{3+}(^2F_{7/2})$-$Yb^{2+}(1A_{1g})$ states, respectively. III: $Yb^{2+}4f\rightarrow 5de_g$ excitation combined with an excited $Yb^{3+}(^2F_{5/2})$. III_{em}: $Yb^{3+}4f\rightarrow 4f$ emission $^2F_{5/2}\rightarrow {}^2F_{7/2}$. IV: non-radiative decay made of a $^2F_{5/2}\rightarrow {}^2F_{7/2}$ de-excitation of Yb^{3+} and a $^2F_{7/2}\rightarrow {}^2F_{5/2}$ excitation of the $4f^{13}$ inner shell of $Yb^{2+}(4f^{13}5de_g)$. V: $Yb^{2+}5de_g\rightarrow Yb^{3+}4f$ IVCT emission (anomalous emission). Bottom right panel: Possible IVCT non-radiative decays from the lowest $Yb^{2+}(4f^{13}5de_g)$-$Yb^{3+}(^2F_{7/2})$ states (blue lines) to $Yb^{3+}(^2F_{7/2})$-$Yb^{2+}(1A_{1g})$ states (black lines) and $Yb^{3+}(^2F_{5/2})$-$Yb^{2+}(1A_{1g})$ states (red lines). After excitation I, the following options are opened: In case A ($SrCl_2$) there is no non-radiative decay, which yields regular emission (process I_{em}). In case B (CaF_2) non-radiative decay produces an excited Yb^{3+}, which can yield slow Yb^{3+} emission (process III_{em}) and, also, create a doubly excited $Yb^{2+}(4f^{13}5de_g)$-$Yb^{3+}(^2F_{5/2})$ pair (process III); after decay IV, the latter yields IVCT emission (process V). In case C (SrF_2) the decays are only partial and both regular emission as in A and anomalous emission as in B are possible, together with partial quenching as in D. In case D (BaF_2) the IVCT non-radiative decay goes down to the ground states of Yb^{3+} and Yb^{2+} with total emission quenching

Across the series CaF_2-SrF_2-BaF_2, the curvatures of the parabolas are very similar because the breathing mode vibrational frequencies of donor $(YbF_8)^{6-}$ and acceptor $(YbF_8)^{5-}$ vary very little. On the contrary, the horizontal offset between the minima of equivalent parabolas grows continuously from CaF_2 to SrF_2 to BaF_2. This is because it is proportional to the bond length difference between donor and acceptor, $d_e(Yb^{2+}$-$F^-)$–$d_e(Yb^{3+}$-$F^-)$ (cf. Sects. 7.4 and 7.5), which grows significantly with the host lattice constant. Then, the main effect of going from CaF_2 to SrF_2 to BaF_2 is a shift of the Yb^{3+}-Yb^{2+} (right) parabolas away from the Yb^{2+}-Yb^{3+} (left) parabolas, which shifts the above crossings towards the right, lowering the barrier of the blue-black crossing and increasing the barrier of the blue-red crossing. As a result, whereas in CaF_2 the blue-red crossing is near the blue minimum (case B) and the IVCT non-radiative decay to an excited $Yb^{3+}(^2F_{5/2})$ is highly probable (process II, green dots), in SrF_2 none of the crossings is very close to the minimum (case C) and $Yb^{2+}(4f^{13}5de_g)$ metastable states are formed, although with partial non-radiative decay to $Yb^{3+}(^2F_{5/2})$. In BaF_2 it is the blue-black crossing the one near the blue minimum (case D), which results in a highly probable IVCT non-radiative decay to an Yb^{3+} in its ground state (process II_{qnch}, black dots). This explains why the regular $Yb^{2+}5de_g\rightarrow 4f$ emission is present in SrF_2 but not in CaF_2 and in BaF_2. It also explains that the $Yb^{2+}4f\rightarrow 5de_g$ absorption excites the $Yb^{3+}4f\rightarrow 4f$ emission in CaF_2 and SrF_2.

The remaining question is: Why is there an anomalous emission in CaF_2 and SrF_2? The diagrams show that, after excitation of an Yb^{3+} to $^2F_{5/2}$ via $Yb^{2+}4f\rightarrow 5de_g$ absorption (process I) and IVCT non-radiative decay (process II), an $Yb^{3+}(^2F_{5/2})$-$Yb^{2+}(4f^{13}5de_g)$ doubly-excited state can be formed (process III, think black arrow). (It must be noted that the Yb^{2+} ion in this doubly-excited pair does not necessarily have to be the same ion that participated in processes I and II; it can be the

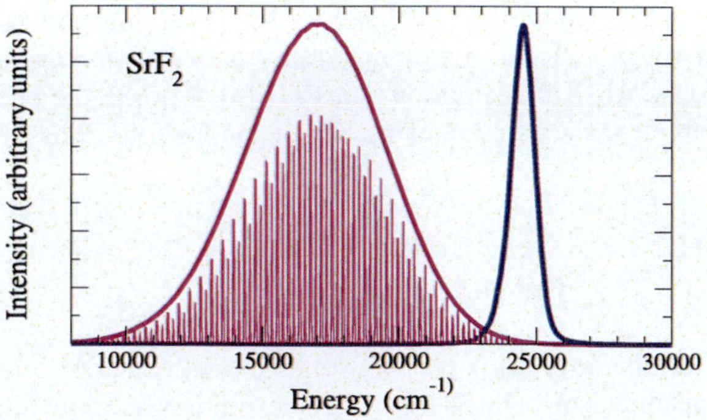

Fig. 11.4 Calculated band profiles of the $Yb^{2+}5de_g \rightarrow 4f$ regular emission (blue) and $Yb^{2+}5de_g \rightarrow Yb^{3+}4f$ IVCT emission (magenta) in the SrF_2 host

same one if it experiences a $4f \rightarrow 5de_g$ excitation with a second photon, but it can also be another Yb^{2+} in the sample that has been excited and is not far away from the excited Yb^{3+} ion.) The doubly-excited pair state $Yb^{3+}(^2F_{5/2})$-$Yb^{2+}(4f^{13}5de_g)$ is in an energy area shared by a number of states characterized by having one electron of Yb^{2+} excited to the $5de_g$ molecular orbital and one of the two $4f^{13}$ shells (the one of Yb^{3+} and the inner shell of Yb^{2+}) also excited to $^2F_{5/2}$. Then, a non-radiative decay populates the lowest of these states (process IV), which is $Yb^{3+}(^2F_{7/2}\ 1\Gamma_{7u})$-$Yb^{2+}(4f^{13}(^2F_{5/2})5de_g\ 2A_{1u})$ which is metastable and a potential emitter. From this state, an $Yb^{2+}5de_g \rightarrow Yb^{3+}4f$ IVCT emission to the lowest states $Yb^{2+}(4f^{14})$-$Yb^{3+}(^2F_{7/2})$ takes place (process V, green arrow). This is the anomalous emission.

This large horizontal offset between the initial and final states of the IVCT emission results in very broad emission bands with FWHM about 5800 cm^{-1}, in agreement with experiments, and much broader than the regular $Yb^{2+}5de_g \rightarrow 4f$ emission, as it is illustrated in the simulations of SrF_2:Yb^{2+}/Yb^{3+} in Fig. 11.4.

In $SrCl_2$, the main effect of the chemical substitution from SrF_2 is an important lowering of the vibrational force constant which translates into significantly lower curvatures. This shifts the blue-black and blue-red crossing very much to the left and far from the blue minimum (case A), so allowing for the regular emission and preventing from the Yb^{3+} excitation that could lead to the IVCT emission. This is why $SrCl_2$ only presents regular emission. Yet, the presence of Yb^{2+}/Yb^{3+} mixed-valence pairs still has its fingerprint because IVCT emissions (green lines) on top of the regular emission can provide additional features.

In summary, the puzzling multiple emission options of regular ($SrCl_2$), anomalous (CaF_2), both (SrF_2), and none (BaF_2), in Yb-doped hosts with fluorite structure can be explained if Yb^{2+}/Yb^{3+} mixed-valence pairs play a key role and the anomalous emission is the $Yb^{2+}5de_g \rightarrow Yb^{3+}4f$ IVCT emission, because they can be associated with the multiple IVCT non-radiative decay possibilities after $Yb^{2+}4f \rightarrow 5de_g$ excitation.

It is also remarkable that the IVCT diagrams in Fig. 11.3 also explain the fine features of 10 K two-frequency transient photoluminescence enhancement experiments on CaF_2:Yb^{2+} single crystals [18] that probe the energy levels lying immediately above the state responsible for the anomalous emission [24].

11.2.1.2 X-ray Excitation Triggers Yb^{2+} Anomalous Emission in CaF_2 But Not in SrF_2

Another intriguing property of the so called anomalous emission of Yb^{2+} was found in studies on scintillators. High-resolution X-ray-excited optical luminescence spectra of Yb-doped CaF_2 and SrF_2 crystals, where samples are excited with 8949 eV radiation at 80 K, show that ionizing radiation excites the anomalous emission of ytterbium in the first host but not in the second, in spite of their strong similarities [27]. The different behavior of the two similar materials is independent of the Yb concentration and it is also observed in luminescence experiments where the excitation is into the host free-exciton levels [29].

If the anomalous emission of Yb^{2+} is an IVCT emission, the IVCT diagrams of Yb^{2+}/Yb^{3+} pairs in CaF_2 and SrF_2 of Fig. 11.3 are applicable here. And, if electron capture by Ca^{2+} and Sr^{2+} after the thermalization step in X-ray-excited luminescence (and also in free-exciton-excited luminescence) gives Ca^{+} and Sr^{+}, MMCT configuration coordinate energy diagrams of Yb^{2+}/Ca^{2+} and Yb^{2+}/Sr^{2+} pairs should also be applicable. These have been calculated in [30] and used to explain photoconductivity experiments in these materials. Both MMCT and IVCT diagrams are shown in Fig. 11.5. As it is discussed next, the MMCT diagram shows that, after capture, electron transfer from the host to the impurity ($Yb^{3+}+Ca^{+}\rightarrow Yb^{2+}+Ca^{2+}$ and $Yb^{3+}+Sr^{+}\rightarrow Yb^{2+}+Sr^{2+}$) leaves the latter in the IVCT emitting state in the CaF_2 host and in the $5d\rightarrow 4f$ emitting state in the SrF_2 host.

According to the diagrams, the host-to-dopant electron back-transfer can be as indicated by the dashed red arrows: After the ionizing excitation and thermalization, a host electron relaxes to the lowest states of Yb^{3+}-Ca^{+} and Yb^{3+}-Sr^{+}. These are $Yb^{3+}(^2F_{7/2})$-$Ca^{+}(3de_g)$ and $Yb^{3+}(^2F_{7/2})$-$Sr^{+}(4de_g)$ respectively (cyan energy curves). At their respective minima, when the Yb^{3+}-Ca^{+}, and Yb^{3+}-Sr^{+} pairs are structurally relaxed, they are electronically unstable and respective non-radiative electron back-transfers (from cyan curves to blue curves) give structurally stressed Yb^{2+}-Ca^{2+} and Yb^{2+}-Sr^{2+} pairs. After geometry relaxation (dashed red arrows) they produce metastable states. What is relevant here is that, whereas the metastable state is $Yb^{2+}(4f^{13}(^2F_{5/2})5de_g\ 2A_{1u})$-$Ca^{2+}(^1A_{1g})$ in CaF_2, it is $Yb^{2+}(4f^{13}(^2F_{7/2})5de_g\ 1E_u)$-$Sr^{2+}(^1A_{1g})$ in SrF_2. This means that the host ion is in its ground state and the dopant in the $4f^{13}5de_g$ configuration in both cases, but Yb^{2+} has its inner $4f^{13}$ shell excited into $^2F_{5/2}$ only in CaF_2 (there is a 5700 cm^{-1} energy gap below that hinders further multiphonon non-radiative decay). And the double excitation in the $4f^{13}$ shell and in the $5de_g$ molecular orbital of Yb^{2+} is the configurational character of the IVCT emitting state of an Yb^{2+}/Yb^{3+} pair in which the Yb^{3+} ion is in its ground $^2F_{7/2}$ configuration (Sect. 11.2.1.1). In consequence, the

$Yb^{2+}5de_g \rightarrow Yb^{3+}4f$ IVCT emission takes place in CaF_2 (green arrow in the bottom left panel). In SrF_2, instead, the metastable state of Yb^{2+} reached after thermalization and electron back-transfer is the lowest of the $4f^{13}5de_g$ electronic configuration, which yields regular $Yb^{2+}5d \rightarrow 4f$ emission (blue arrow in the bottom right panel).

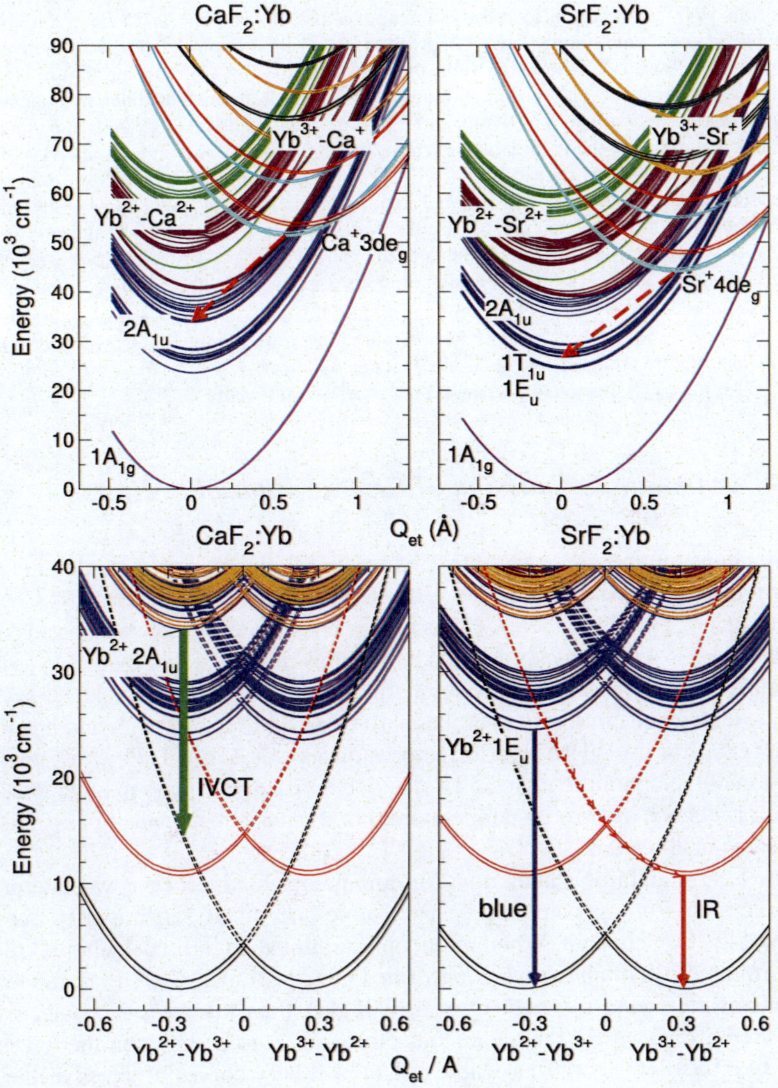

Fig. 11.5 (Continued)

◂**Fig. 11.5** (Continued) Diabatic MMCT configuration coordinate diagrams of Yb^{2+}/Ca^{2+} and Yb^{2+}/Sr^{2+} pairs (above) and IVCT configuration coordinate diagrams of Yb^{2+}/Yb^{3+} pairs (below) in Ytterbium-doped CaF_2 (left) and SrF_2 (right). Dashed red arrows in the MMCT diagrams indicate Ca^+ $3de_g \rightarrow Yb^{3+}5de_g$ and Sr^+ $4de_g \rightarrow Yb^{3+}5de_g$ non-radiative electron back-transfer from Yb^{3+}-Ca^+ and Yb^{3+}-Sr^+ configurations of the pairs, followed by structural relaxation to $Yb^{2+}(2A_{1u})$-Ca^{2+} and $Yb^{2+}(1E_u)$-Sr^{2+} states respectively. $2A_{1u}$ and $1E_u$ are the lowest states of the $4f^{13}(^2F_{5/2})5de_g$ and $4f^{13}(^2F_{7/2})5de_g$ configurations of Yb^{2+}, respectively. $2A_{1u}$ is the origin of the IVCT emission (also called anomalous emission) of Yb^{2+} and $1E_u$ is the origin of the regular $5d \rightarrow 4f$ emission of Yb^{2+}. A green arrow in the bottom-left diagram indicates the $Yb^{2+}5de_g \rightarrow Yb^{3+}4f$ IVCT emission from the $Yb^{2+}(2A_{1u})$-$Yb^{3+}(1\Gamma_{7u})$ state to $Yb^{3+}(^2F_{7/2})$-$Yb^{2+}(1A_{1g})$ states in the CaF_2 host. A blue arrow in the bottom-right diagram indicates the $5de_g \rightarrow 4f$ regular emission of Yb^{2+} from its $1E_u$ excited state to its $1A_{1g}$ ground state in the SrF_2 host. In this diagram, a dashed red arrow indicates $Yb^{2+}5de_g \rightarrow Yb^{3+}4f$ non-radiative IVCT from the $Yb^{2+}(1E_u)$-$Yb^{3+}(1\Gamma_{7u})$ state to $Yb^{3+}(^2F_{5/2})$-$Yb^{2+}(1A_{1g})$ states of a Yb^{2+}/Yb^{3+} pair in SrF_2 followed by structural relaxation; and a full red arrow indicates the subsequent $4f \rightarrow 4f$ IR emission of Yb^{3+} from $^2F_{5/2}$ to $^2F_{7/2}$ states. MMCT color codes: Violet, blue, green, and marroon are used in Yb^{2+}-Ca^{2+} and Yb^{2+}-Sr^{2+} configurations of the pairs with Yb^{2+} $4f^{14}$, $4f^{13}5de_g$, $4f^{13}5dt_{2g}$, and interacting $4f^{13}5d$-$4f^{13}\phi_{ITE,a_{1g}}$ configurations, respectively. Cyan, red, orange, and black are used in Yb^{3+}-Ca^+ and Yb^{3+}-Sr^+ configurations of the pairs with Ca^+ $3de_g$, $4s$, $3dt_{2g}$, and $4p$ configurations, and Sr^+ $4de_g$, $5s$, $4dt_{2g}$, and $5p$ configurations, respectively. IVCT color codes: Black, red, blue, and orange are used for Yb^{2+}-Yb^{3+} states $4f^{14}$-$^2F_{7/2}$, $4f^{14}$-$^2F_{5/2}$, $4f^{13}5de_g$-$^2F_{7/2}$, and $4f^{13}5de_g$-$^2F_{5/2}$, respectively, and their Yb^{3+}-Yb^{2+} counterparts

11.2.2 Anomalous Emission of Ce^{3+} in Elpasolite Hosts

An anomalous luminescence was reported in scintillating materials made of elpasolites and related hosts doped with Ce^{3+}, like $Cs_2LiLuCl_6$:Ce^{3+}, Cs_2NaYCl_6:Ce^{3+}, or Cs_3LuCl_6:Ce^{3+} [31–33], where Ce^{3+} occupies O_h sites of 6-fold coordination. Like the usual $5de_g \rightarrow 4f$ luminescence, the anomalous one is excited by the $4f \rightarrow 5de_g$ absorption, but its emission band peaks at much lower energy (more than 9000 cm^{-1} below), which results into a huge Stokes shift of around 10000 cm^{-1}, and it is much broader (4800 cm^{-1} FWHM at room temperature). In $Cs_2LiLuCl_6$:Ce^{3+}, its intensity increases with temperature up to 250 K and decreases above it and in other hosts the emission is quenched at room temperature, but the basic mechanism is assumed to be the same in all cases.

In the lack of ab initio calculations, the anomalous luminescence was interpreted as an emission from an anomalous state, more specifically an impurity-trapped-exciton of the Ce^{3+} dopant. When multiconfigurational ab initio calculations aimed at confirming this hypothesis were performed on $Cs_2LiLuCl_6$:Ce^{3+} [34], the lowest ITE states were found too high in energy, higher than the $5de_g\,^2E_g$ state, which makes them irrelevant for the anomalous emission. This is shown in the left panel of Fig. 11.6 (also cf. Sect. 9.1), where potential energy curves of states of the $4f$, $5dt_{2g}$, and $5de_g$ configurations of the Ce^{3+} active center are shown together with its lowest ITE state (magenta line).

As it is discussed next, the ab initio calculations provided a new interpretation of the anomalous emission as an IVCT $Ce^{3+}5de_g \rightarrow Ce^{4+}4f$ emission. This interpre-

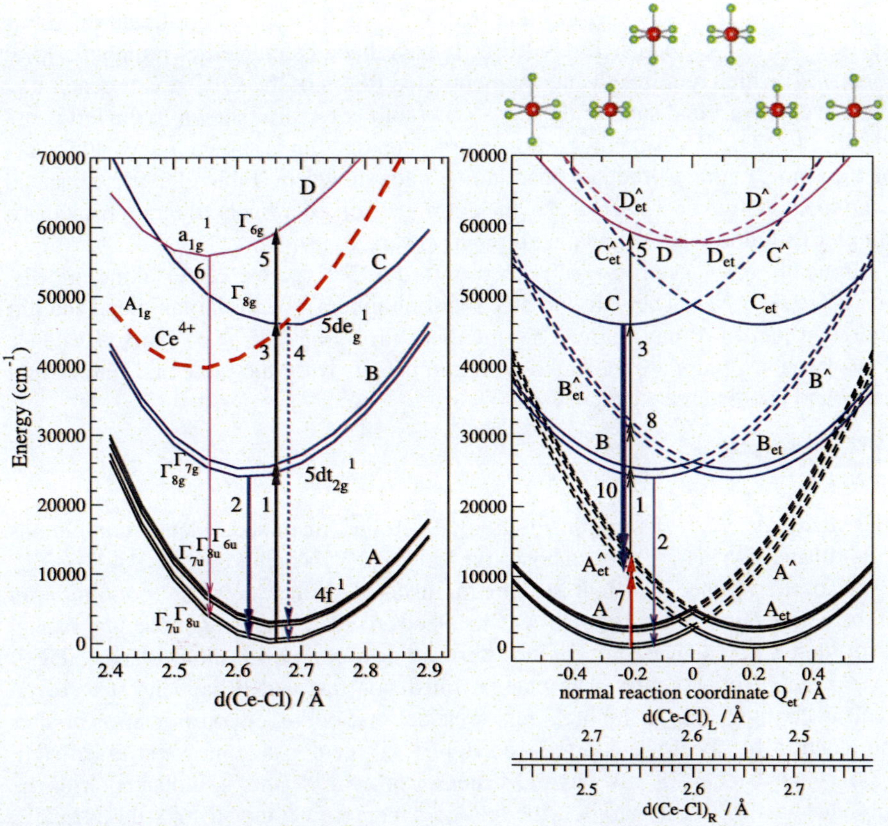

Fig. 11.6 Left panel: RASSI-SO potential energy curves of the donor and acceptor active centers $(CeCl_6)^{3-}$ and $(CeCl_6)^{2-}$, respectively, of Ce-doped $Cs_2LiLuCl_6$. States of the $4f^1$(black), $5d^1$ (blue), and $\phi^1_{ITE,a_{1g}}$ (magenta) configurations of the Ce^{3+} donor active site are shown together with the ground state of the $4f^0$ configuration of the Ce^{4+} acceptor (dashed red). Right panel: Diabatic IVCT configuration coordinate energy diagram of a Ce^{3+}/Ce^{4+} D/A pair in $Cs_2LiLuCl_6$. Qualitative representations of the left and right $CeCl_6$ moieties' structures that correspond to three areas of the diagram are shown at the top: left brach ($Q_{et} < 0$), right branch ($Q_{et} > 0$), and activated complex ($Q_{et} = 0$). The arrow numbers in both panels correspond to the processes indicated in Table 11.1

tation requires the presence of Ce^{4+} impurities together with the Ce^{3+} ones in the samples, but the great difficulties, if not the impossibility, to get rid of Ce^{4+} when experiments are being made with Ce^{3+} is a well-known fact. Moreover, when the states of Ce^{3+}/Ce^{4+} donor/acceptor active pairs are considered, they also other features of these materials can be explained, like unassigned low-temperature excitation bands or the anomalous behavior of the 351 nm emission of $Cs_2NaYF/_6$:Ce under various excitation wavelengths [35, 36].

The potential energy surfaces and the IVCT configuration coordinate diagrams of a Ce^{3+}/Ce^{4+} D/A pair in $Cs_2LiLuCl_6$ have been calculated at a diabatic level (Sect. 7.4) which requires the above mentioned PES's of the Ce^{3+} donor center and the one of the ground state of the Ce^{4+} acceptor center, also shown in the left panel of Fig. 11.6. The IVCT configuration coordinate diagram is shown in the right panel of the same figure. Although quantitative data shown in Table 11.1 are extracted from the potential energy surfaces, the configuration coordinate diagram provides a simpler and more understandable, though powerful view.

Optically relevant processes related with the Ce^{3+} active center alone are discussed in Sect. 9.1. They are indicated with numbers 1–6 in the table and the left and right panels of the figure. The processes involving Ce^{3+}/Ce^{4+} mixed-valence donor/acceptor pairs are indicated with numbers 7–10 in the table and figure right panel and are discussed next.

IVCT absorptions

$Ce^{3+}4f \rightarrow Ce^{4+}4f$

The standard IVCT absorption, traditionally found in mixed-valence compounds [38–40] and also in Ce^{3+}-doped hosts like $LaPO_4$ [12], is in this case $Ce^{3+}4f \rightarrow Ce^{4+}_{sts}$ $4f$ absorption (process 7). In it, starting for instance from a Ce^{3+}-Ce^{4+} configuration of the pair at equilibrium (branch A), one electron occupying a $4f$ molecular orbital of the left Ce^{3+} active center is transferred to an empty $4f$ molecular orbital of a right Ce^{4+} active center. The charge transfer is fast and it happens at a frozen nuclear configuration of both centers (vertical or Frank-Condon transition). After the electron has been transferred, the right Ce^{4+} center becomes a stressed right Ce^{3+} center because it still has the structure of a Ce^{4+} (this is indicated with the subindex "sts") and, in parallel, the left Ce^{3+} center becomes a stressed left Ce^{4+} center. In other words, the IVCT absorption leads to a structurally stressed Ce^{4+}-Ce^{3+} configuration of the pair (branch $A^{\wedge}_{et}$). After the absorption, both centers relax immediately down to their respective equilibrium structures, with a large contraction of the Ce-Cl distance around the new Ce^{4+} and a large elongation of the Ce-Cl distance around the new Ce^{3+}, and reach the Ce^{4+}-Ce^{3+} configuration of the pair at equilibrium (branch A_{et}), with the same energy than the starting point (branch A). During this relaxation, a large reorganization energy is released, which coincides with the IVCT absorption energy. Obviously, an IVCT absorption like this is never followed by an emission. The calculation predicts a very broad IVCT absorption band peaking at about 10000–13300 cm^{-1}, which has never been experimentally reported.

$Ce^{3+}4f \rightarrow Ce^{4+}5dt_{2g}$

A second and a third additional IVCT absorptions of higher energy correspond to electron transfers to excited molecular orbitals of the right Ce^{4+} active center, $5dt_{2g}$ and $5de_g$. The second IVCT is $Ce^{3+}4f \rightarrow Ce^{4+}_{sts}$ $5dt_{2g}$ (process 8). It is predicted as a very broad band peaking at 31000–32200 cm^{-1}, 6500 cm^{-1} above the regular Ce^{3+} absorption $4f \rightarrow 5dt_{2g}$ (process 1). This excess energy is the reorganization energy released after the IVCT absorption, in a structural relaxation from a stressed

Table 11.1 Excitations and emissions of Ce-doped $Cs_2LiLuCl_6$

Experiment [33]			Theory [34]	
Feature	cm^{-1}	Assignment	Process	cm^{-1}
Emission peaks with $4f \rightarrow 5de_g$ excitation				
E1	24700, 26900	[33, 34] Ce^{3+} $5dt_{2g} \rightarrow 4f(^2F_{7/2}, ^2F_{5/2})$	2	20200–23300
E2	36400 (34100–38900)[a]	[33] Anomalous state $\rightarrow 4f$ [34] IVCT $Ce^{3+}5de_g \rightarrow Ce^{4+}_{sts}\ 4f$	10	32000–35300
Excitation peaks of the $5dt_{2g} \rightarrow 4f$ emission				
A1	<30300	[33, 34] Ce^{3+} $4f \rightarrow 5dt_{2g}$	1	24500, 25700
A2[b]	46190, 47600	[33, 34] Ce^{3+} $4f \rightarrow 5de_g$ (JT split)	3	46100 (45300, 46900)[c]
A3	33000	[33] Unassigned [34] IVCT $Ce^{3+}4f \rightarrow Ce^{4+}_{sts}\ 5dt_{2g}$	8	31010, 32170
A4	35700	[33] Low symmetry Ce^{3+}		
A5	37900, 41200	[33] Low symmetry Ce^{3+}		
A6	50500	[33] Host self-trapped exciton		
A7	54900	[33] Impurity-trapped exciton [34] IVCT $Ce^{3+}4f \rightarrow Ce^{4+}_{sts}\ 5de_g$	9	56240
A8	56800	[33] Free excitons [34] Ce^{3+} $4f \rightarrow a_{1g,ITE}$	5	59190
Excitation peaks of the anomalous emission				
A2	46190,47600	[33, 34] Ce^{3+} $4f \rightarrow 5de_g$ (JT split)	3	46100 (45300, 46900) [c]
A9	54300	[34] IVCT $Ce^{3+}4f \rightarrow Ce^{4+}_{sts}\ 5de_g$	9	56240
A10	57500	[34] Ce^{3+} $4f \rightarrow a_{1g,ITE}$	5	59190
Unobserved transitions				
		[34] $Ce^{3+}5de_g \rightarrow 4f$ emission	4	43000–46030
		[34] ITE emission $Ce^{3+}\phi_{ITE,a_{1g}} \rightarrow 4f$	6	53000–56200
		[34] IVCT emission $Ce^{3+}5dt_{2g} \rightarrow Ce^{4+}_{sts}4f$		13500–16800
		[34] IVCT absorption $Ce^{3+}4f \rightarrow Ce^{4+}_{sts}4f$	7	10000–13300

[a] Full width at half maximum at 290 K

[b] Absent at low temperature

[c] 1600 cm^{-1} room temperature Jahn-Teller (JT) splitting calculated with $\Delta E_{\text{peaks}} = 2\sqrt{2E_{JT}kT}$ [37]

to a relaxed Ce^{4+}-$Ce^{3+}(5dt_{2g}\,{}^2T_{2g})$ excited configuration of the pair (from branch $B_{et}^{\wedge}$ to branch B_{et}). The theoretical result is expected to have the same degree of underestimation found for the regular absorption, which indicates that the very broad bands of the excitation spectrum peaking at 33000 cm^{-1} (feature A3, unassigned) and 35700 cm^{-1} (feature A4, assigned to low symmetry Ce^{3+} sites), can be attributed to this second IVCT absorption.

$Ce^{3+}4f \rightarrow Ce^{4+}5de_g$

The third IVCT absorption is $Ce^{3+}4f \rightarrow Ce_{sts}^{4+}\, 5de_g$ (process 9). It is predicted as a very broad band peaking at 56240 cm^{-1}, 10200 cm^{-1} above the regular Ce^{3+} absorption $4f \rightarrow 5de_g$ (process 3). As before, this excess energy is a reorganization energy released after the IVCT absorption, from a stressed to a relaxed Ce^{4+}-$Ce^{3+}(5de_g\,{}^2E_g)$ excited configuration of the pair (from branch $C_{et}^{\wedge}$ to branch C_{et}). This means it has a reorganization energy similar to the standard IVCT absorption and higher than the second. This time, the theoretical result is expected to agree fairly well with experiments, like the regular absorption, which suggests that the experimental peak in the excitation spectrum at 54900 cm^{-1} (feature A7) can be attributed to the third IVCT absorption.

IVCT emissions

$Ce^{3+}5de_g \rightarrow Ce^{4+}4f$

When Ce^{3+} is in its $5de_g\,{}^2E_g$ excited state, an IVCT emission $Ce^{3+}5de_g \rightarrow Ce_{sts}^{4+}\, 4f$ is possible in a Ce^{3+}/Ce^{4+} pair (process 10). It is predicted to be a very broad band (of 5000 cm^{-1} FWHM) peaking at 34500 cm^{-1}, around 10000 cm^{-1} below the regular Ce^{3+} emission $5de_g \rightarrow 4f$ (process 4). This energy difference is the reorganization energy of the standard IVCT absorption discussed above, which is the difference between the $Ce^{3+}5de_g$ electron landing in the $4f$ molecular orbital of the same Ce^{3+} center or in the $4f$ molecular orbital of a Ce^{4+} center. Besides, the 10000 cm^{-1} must also be very similar to the Stokes shift between the regular Ce^{3+} $4f \rightarrow 5de_g$ absorption and the IVCT $Ce^{3+}5de_g \rightarrow Ce_{sts}^{4+}\, 4f$ emission. All these properties agree with the so called anomalous emission (feature E2), which has a Stokes shift of around 10000 cm^{-1} with respect to its $4f \rightarrow 5de_g$ excitation, peaks around 9000 cm^{-1} below the regular $5de_g \rightarrow 4f$ emission, and is much broader (4800 cm^{-1} FWHM at room temperature).

These agreements between calculations and experiments in such an unconventional emission suggest the so called anomalous emission of Ce^{3+} is an IVCT emission that takes place in Ce^{3+}/Ce^{4+} donor/acceptor mixed-valence pairs, the $Ce^{3+}5de_g \rightarrow Ce_{sts}^{4+}\, 4f$ emission.

Further supporting arguments exist in addition to the agreements in energy, band width, and Stokes shift.

If the anomalous emission is an IVCT emission, its band width should basically coincide with that of the IVCT absorption, which is related with the IVCT absorption maximum according to FWHM $= (\bar{\nu}_{max}\, 7.71\, T)^{1/2}$ except at low T [41]. And, as discussed above, the IVCT absorption maximum coincides with the Stokes shift of the IVCT emission. Since a value of $\bar{\nu}_{max}$=10500 cm^{-1} gives FWHM = 4850 cm^{-1} at T = 290 K, all these data are in agreement with the IVCT emission interpretation.

The principal excitation of the anomalous emission is the Ce^{3+} $4f \rightarrow 5de_g$ absorption, but additional excitation peaks exist, and, as shown in Table 11.1, all of them are consistent with Ce^{3+} $5de_g{}^2E_g$ being the emitting state rather than an anomalous state and with the existence of excited states of Ce^{3+}/Ce^{4+} pairs.

The presence of excited states of Ce^{3+}/Ce^{4+} pairs also explains the features of the excitation spectrum of the $5dt_{2g} \rightarrow 4f$ emission. Some of them remained unexplained and had to be vaguely attributed to low symmetry Ce^{3+} sites, ITEs and free excitons.

The correlation of the decay times of the anomalous emission excited with the $4f \rightarrow 5de_g$ absorption and the $5dt_{2g} \rightarrow 4f$ emission, showed an anomalous emission quenching by energy transfer to the $5dt_{2g}{}^2T_{2g}$ states of Ce^{3+}, whith an activation energy of 2900 cm^{-1}. And, in parallel, the $4f \rightarrow 5de_g$ absorption was found to excite the $5dt_{2g} \rightarrow 4f$ emission. Again, all these features are consistent with Ce^{3+} $5de_g{}^2E_g$ being the emitting state and the configuration coordinate energy diagram of Ce^{3+}/Ce^{4+} pairs, which shows an IVCT non-radiative decay mechanism between the C branch ($5de_g{}^2E_g$ emitting level) and the $B^{\wedge}_{et}$ branch ($5dt_{2g}{}^2T_{2g}$ emitting level), with a diabatic barrier of 5100 cm^{-1}, which must be lower at the adiabatic level.

$Ce^{3+}5dt_{2g} \rightarrow Ce^{4+}4f$

An IVCT emission $Ce^{3+}5dt_{2g} \rightarrow Ce^{4+}_{sts}$ $4f$ is also possible in a Ce^{3+}/Ce^{4+} pair (from branch B to branch $A^{\wedge}_{et}$), which is expected to appear at an energy lower than the regular $5dt_{2g} \rightarrow 4f$ emission by an amount very similar the 10000 cm^{-1} if the first IVCT absorption, i.e. at about 14000–16000 cm^{-1}, but it has not been reported so far.

11.2.3 Anomalous Red and Infrared Emission of Ce^{3+} in SrS

Cerium-doped SrS is a luminescent material with a blue-green emission of high brightness peaking at around 480 nm wavelength which is known to red-shift 20–40 nm for high Ce contents, something attributed to formation of Ce^{3+} pairs and clusters [42]. Similar red-shifts are found in powder, thin film, and nanocrystalline SrS:Ce, and the excitation spectra of the red-shifted emission and the regular Ce^{3+} $5d \rightarrow 4f$ emission agree. However, sintered ceramics SrS:Ce, prepared with the goal to reduce porosity and thereby increase chemical stability, exhibit a new, very broad and strongly red-shifted luminescence peaking at around 620 nm in addition to the regular one [43] (cf. Fig. 11.7). The sintered ceramics samples have a darker body color due to the higher sintering temperatures and higher Ce concentrations, and the experiments indicate the presence of two Ce sites with different local structure.

A previous interpretation of the anomalous emission of Ce^{3+} in elpasolites as an IVCT emission involving Ce^{3+}/Ce^{4+} pairs (Sect. 11.2.2), together with the fact that the almost black color of the samples with higher Ce concentrations are typical of mixed-valence compounds and due to a broad IVCT absorption which necessarily

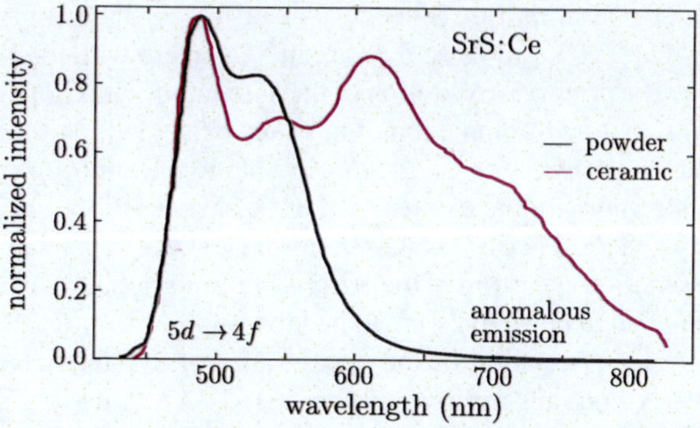

Fig. 11.7 Emission spectrum of SrS:Ce under 425 nm excitation, showing the simultaneous occurrence of the regular $5d \rightarrow 4f$ and anomalous emissions for the sintered ceramic sample, measured at room temperature. Adapted from [43]

decays to the ground state with no emission [44], suggest to consider the hypothesis of the involvement of Ce^{3+}/Ce^{4+} pairs in the present, anomalous red and infrared emission. This time, instead of performing demanding multiconfigurational ab initio calculations, the hypothesis is explored by verifying whether it is possible to make a semi-quantitative, empirical IVCT configuration coordinate diagram according to the model in Sect. 7.5, which could explain the experimental observations.

Empirical data was used to construct the IVCT configuration coordinate diagram in Fig. 11.8. The differences between Ce^{3+}-S and Ce^{4+}-S distances were estimated to be 90% of the 6-fold coordination ionic radii difference between Ce^{3+} and Ce^{4+}). For the vibrational frequencies of Ce^{3+} and Ce^{4+} defects, values between 200 cm^{-1} and 250 cm^{-1} were taken. Excited state energies of the Ce^{3+} centers were extracted from the regular emission. Finally, a small vertical offset between both ground states of 3000 cm^{-1} was assumed, originating from the observation that two non-equivalent Ce sites are involved.

In the figure, the lower black lines correspond to the $^2F_{5/2}$ and $^2F_{7/2}$ states of Ce^{3+} in the Ce_1^{3+}-Ce_2^{4+} and Ce_1^{4+}-Ce_2^{3+} configurations of a Ce^{3+}/Ce^{4+} mixed-valence pair, where subindex 1 makes reference to the highest symmetry site and subindex 2 to the lowest symmetry site. I.e. $Ce_1^{3+}(^2F_{5/2})$-Ce_2^{4+} and $Ce_1^{3+}(^2F_{7/2})$-Ce_2^{4+} with their minima in the left side, and Ce_1^{4+}-$Ce_2^{3+}(^2F_{5/2})$ and Ce_1^{4+}-$Ce_2^{3+}(^2F_{7/2})$ with their minima in the right side. Equivalently, blue lines represent $Ce_1^{3+}(5dt_{2g})$-Ce_2^{4+} and Ce_1^{4+}-$Ce_2^{3+}(5dt_{2g})$ states, resulting from low symmetry splittings of $Ce^{3+}5dt_{2g}\,^2T_{2g}$, which is larger in site 2 than in site 1. And green lines represent split $Ce_1^{3+}(5de_g)$-Ce_2^{4+} and Ce_1^{4+}-$Ce_2^{3+}(5de_g)$ states.

Vertical arrows that almost connect two minima represent regular Ce^{3+} absorptions and emissions. Those connecting the minimum of a curve with a large slope branch of another curve are IVCT absorptions and emissions, where radiative electron transfers from Ce^{3+} to Ce^{4+} take place. Dotted arrows are used for radiative transitions of assumed lower intensity: It is assumed that the structures of sites 1 and 2 make the electron transfer $Ce_1^{3+}4f \rightarrow Ce_2^{4+}5dt_{2g}$ much more probable than the electron transfer $Ce_2^{3+}4f \rightarrow Ce_1^{4+}5dt_{2g}$. Broken lines represent non-radiative decays.

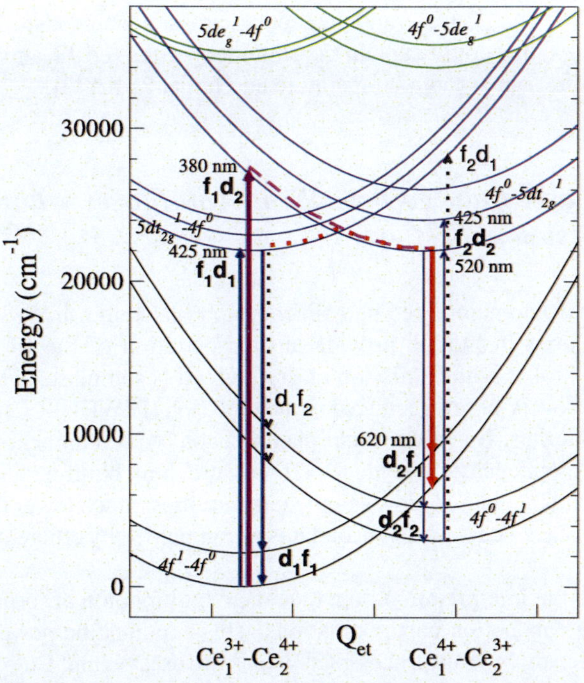

Fig. 11.8 Empiricial IVCT configuration coordinate diagram of a Ce^{3+}/Ce^{4+} pair in SrS consistent with the experimental observations. The two Ce ions occupy two symmetry inequivalent sites labelled 1 and 2 respectively. $f_1 d_1$ stands for a $Ce_1^{3+} 4f \rightarrow 5dt_{2g}$ excitation; $f_1 d_2$ for a $Ce_1^{3+} 4f \rightarrow Ce_2^{4+} 5dt_{2g}$ excitation; $d_2 f_1$ for a $Ce_2^{3+} 5dt_{2g} \rightarrow Ce_1^{4+} 4f$ emission; and so on. The violet arrow, magenta dashed line, and red arrows represent the sequence: vertical or Frank-Condon $Ce_1^{3+} 4f \rightarrow Ce_2^{4+} 5dt_{2g}$ IVCT absorption, followed by a strong structural relaxation in both sites and, finally, by a $Ce_2^{3+} 5dt_{2g} \rightarrow Ce_1^{4+} 4f$ IVCT emission

Under the above assumptions, the empirical IVCT configuration coordinate diagram can explain the experimental observations. The most important ones are the following, but finer features are also explained in detail [43]:

(1) The 425 nm absorption excites only the regular emission (480 and 540 nm) in powders but it excites both the regular and the anomalous emission (605 and 700 nm) in ceramics. The IVCT diagram explains the excitation of both emissions (violet up-arrow exciting blue and red down-arrows), but this diagram only applies to ceramics, where the Ce^{4+} centers are created by the high sintering temperatures.

(2) The 380 nm absorption only excites the anomalous emission. This is explained if the $Ce_1^{3+} 4f \rightarrow Ce_2^{4+} 5dt_{2g}$ IVCT absorption decays mostly to the Ce_1^{4+}-$Ce_2^{3+}(5dt_{2g})$ configuration of the pair, which gives the broad, intense, double peaked, red-IR $Ce_2^{3+} 5dt_{2g} \rightarrow Ce_1^{4+} 4f$ emission. Branching to $Ce_1^{3+}(5dt_{2g})$-Ce_2^{4+} during the non-radiative decay takes some time and it is hindered by the fast decay. As a consequence, only the red-IR emission would be excited. Thermal equilibrium would transfer some

population to $Ce_1^{3+}(5dt_{2g})$-Ce_2^{4+} and the regular emission could also show weakly at high temperatures: this could explain the shoulders of the red-IR emission found in the 425–550 nm region at room temperature and not at 25 K [43].

11.2.4 IR Laser Induced Broadband Anti-Stokes White Emission of Sr_2CeO_4

As in the case of anomalous red and IR emission of Cerium in SrS:Ce, a similar procedure was used in [45] to provide an interpretation of the IR laser-induced broadband anti-Stokes white emission of Sr_2CeO_4. It is summarized here.

Since the discovery of laser induced white emission (LIWE) [46], it was found in a number of materials. It is a threshold phenomenon, with nonlinear power dependence of emission intensity, dependence on pressure, long build-up and decay times (a few milliseconds), relatively low sample temperatures, much lower than those corresponding to a black body emission, and it is accompanied by efficient photocurrent generation.

According to the interpretation, in a first stage multiphoton absorption promotes valence band to conduction band excitations. This explains the power dependence of the LIWE intensity. In this process, Ce^{3+} ions are created and Ce^{3+}/Ce^{4+} mixed-valence pairs are formed. The number of such pairs increases with the excitation power and with the exposure time of the sample to CW laser irradiation. The states of these pairs may well be represented with the IVCT configuration coordinate diagram of Fig. 11.9, which, as before, was made with reasonable values of bond length differences, vibrational frequencies, and Ce^{3+} state energy differences. This time, small adiabatic electronic couplings were used between diabatic states (ranging between

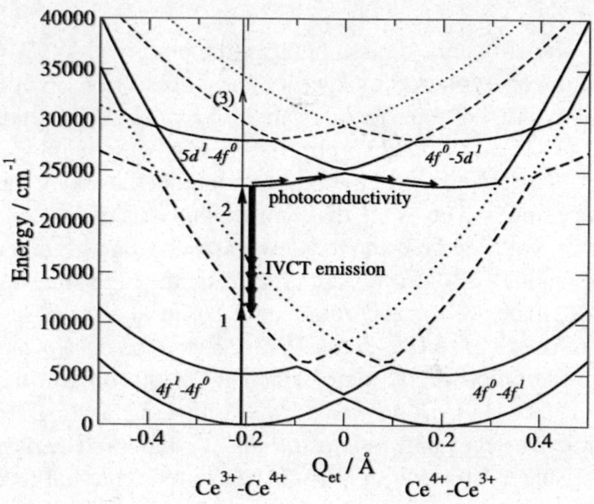

Fig. 11.9 IVCT configuration coordinate diagram of a Ce^{3+}/Ce^{4+} pair in Sr_2CeO_4 consistent with the experimental observations

100 and 200 cm^{-1}) and an adiabatic diagram is shown in fact. It illustrates basic effects of the electronic couplings of producing avoided crossings and lowering thermal electron transfer barriers. The multiphoton absorption is able to excite pairs to $Ce^{3+}(5d)$-Ce^{4+} states and, from them, efficient photocurrent and very broad photoemission are possible. A hoping mechanism of photoconductivity in excited pairs is indicated: $Ce^{3+}(5d)$-Ce^{4+} $\rightarrow$ Ce^{4+}-$Ce^{3+}(5d)$ and the $Ce^{3+}(5d)$-Ce^{4+} $\rightarrow$ Ce^{4+}-$Ce^{3+}(4f)$ IVCT white emission is emphasized.

11.2.5 Direct Evidence of IVCT States of Eu-Doped Luminescent Materials

In previous sections, the existence of IVCT states of mixed-valence lanthanide pairs and their involvement in absorption, emission and non-radiative decay processes was shown to be able to explain very complex and puzzling luminescence phenomena of lanthanide-activated phosphor materials that could not be understood otherwise. Direct evidence for the existence and optical involvement of such states, in particular of IVCT states of Eu-doped luminescent materials, was given in [47]. It is summarized here.

By means of a combined theoretical/experimental study, it was shown that IVCT states of lanthanide-doped materials absorb light so as to give significant absorption spectra fingerprints, which opens the door to their effective participation in other mechanisms that may range from limiting the efficiency or regular emissions to producing new forms of luminescence.

In [47], the IVCT levels of Eu^{2+}/Eu^{3+} pairs in the series of CaF_2, SrF_2, and BaF_2 hosts with fluorite structure were calculated with multiconfigurational ab initio methods. Theoretical $Eu^{2+}4f \rightarrow Eu^{3+}4f$ IVCT absorption bands were compared with Eu^{2+} $4f^7 \rightarrow 4f^6(^7F_J)5de_g$ absorptions and shown to give rise to additional features in the low-energy edges of the Eu^{2+} $4f \rightarrow 5de_g$ band that present a systematic trend. Then, diffuse reflectance spectroscopy experiments on in-house synthesized polycrystalline powders of Eu-doped CaF_2, SrF_2, and BaF_2 confirmed the low energy features and their systematics.

The theoretical results are summarized in the configuration coordinate energy diagrams shown in the bottom panel of Fig. 11.10. The multiconfigurational ab initio calculations assumed a non-local charge compensation of the Eu^{3+} center, which results in symmetric configuration coordinate diagrams. The IVCT states reached in a vertical, Frank-Condon absorption transition (green arrows) are structurally very stressed, hence the corresponding absorption bands are very broad; their calculated peak energies span wide intervals: 15100–20800 cm^{-1} (CaF_2); 18200–23900 cm^{-1} (SrF_2); and 22000–27700 cm^{-1} (BaF_2). The superpositions of all the IVCT transitions lead to the very broad IVCT absorption envelopes plotted in green in the middle panel of Fig. 11.10, where the low energy edge profiles of the Eu^{2+} $4f^7 \rightarrow 4f^6(^7F_J)5de_g$ absorptions are also shown, in brown. A host-driven shift of the IVCT band towards

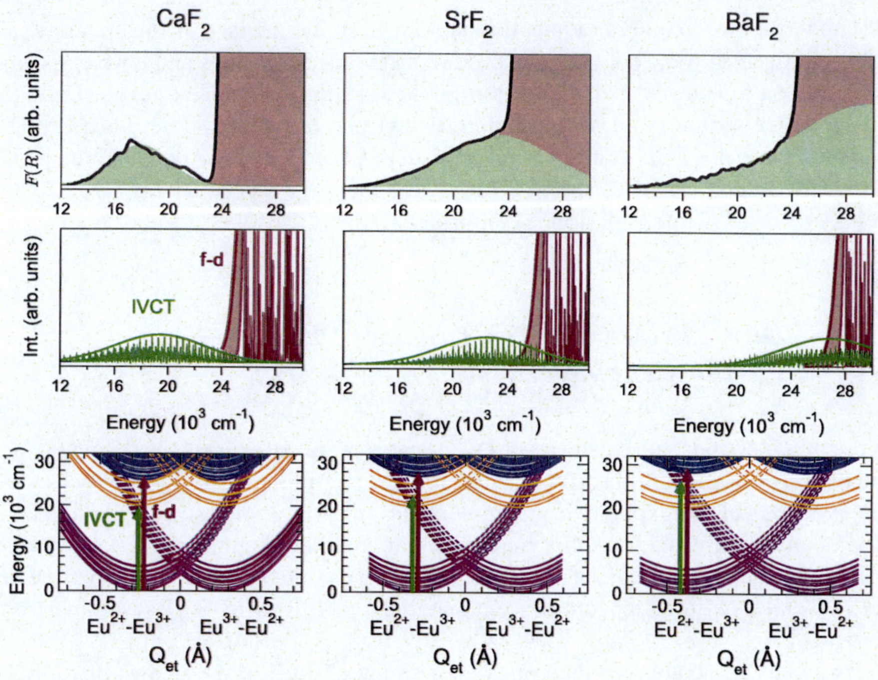

Fig. 11.10 IVCT absorption of Eu-doped CaF_2, SrF_2, and BaF_2. Top: Experimental absorption spectra (black lines) from Kubelka-Munk-transformed diffuse reflectance spectra. Fits are shown in green. Middle: Theoretical absorption spectra profiles. $Eu^{2+}4f \rightarrow Eu^{3+}4f$ IVCT absorption bands are shown in green. The low-energy edges of Eu^{2+} $4f^7 \rightarrow 4f^6(^7F_J)5de_g$ absorptions are shown in brown. Bottom: Multiconfigurational ab initio diabatic Eu^{2+}/Eu^{3+} IVCT configurational coordinate energy diagrams. $Eu^{2+}(^8S_{7/2})$/$Eu^{3+}(^7F_J)$ states are shown in violet, $Eu^{2+}(^8S_{7/2})$/$Eu^{3+}(^5D_J)$ states in orange, and $Eu^{2+}4f^6(^7F_J)5de_g$/$Eu^{3+}(^7F_J)$ states in blue. The $Eu^{2+}4f \rightarrow Eu^{3+}4f$ IVCT absorption is indicated with a green arrow and the $Eu^{2+}4f \rightarrow 5de_g$ aborption with a black arrow

higher energies is evident. It is caused by the increase of the mismatch between the Eu^{2+}-F^- and Eu^{3+}-F^- bond lengths across the CaF_2 - SrF_2 - BaF_2 series (cf. Sect. 7.4), which increases the differences between the minima of the two configurations of the pair, Eu^{2+}-Eu^{3+} and Eu^{3+}-Eu^{2+}, and shifts the right parabolas away from the left parabolas. It is larger than the shift of the Eu^{2+} $4f \rightarrow 5de_g$ band, which brings the IVCT absorption bands closer to the bottom of the Eu^{2+} $4f \rightarrow 5de_g$ band, so that the IVCT and $4f \rightarrow 5d$ bands are well separated in CaF_2, they get closer in SrF_2, and they overlap significantly in BaF_2. This is related with the IVCT states getting closer to the bottom of the $4f^65de_g$ manifold, which also enable non-radiative decay after a $4f \rightarrow 5d$ absorption through electron transfer crossings, up to the point that Eu^{3+} is predicted to quench Eu^{2+} $4f^65de_g \rightarrow 4f^7$ luminescence in BaF_2:Eu^{2+}. Such non-radiative decay channel explains that the blue Eu^{2+} $5d \rightarrow 4f$ emission observed in CaF_2 and SrF_2, is quenched in BaF_2 [48].

Experimental absorption spectra of Eu-doped CaF_2, SrF_2, and BaF_2 were obtained from Kubelka-Munk transformation of diffuse reflectance spectra, measured on polycrystalline powder samples with a doping concentration of 1% Eu. These are shown in the top panel of Fig. 11.10. As expected, the spectra are dominated by the well known $4f^7 \rightarrow 4f^6 5d$ band of Eu^{2+} [48–50], but a careful inspection of the lower-energy part of the reflection spectra identifies an additional weak absorption band partly overlapping the main band and extending up to 12500 cm^{-1} on the low-energy side. These identifications via diffuse reflectance spectroscopy require carefully obtained white baselines from undoped fluorides that underwent the same thermal treatment as the doped samples, ensuring that only absorption features due to the presence of Eu are measured. The low-energy absorption bands are, as the theoretical IVCT absorptions, broad bands that show a blueshift from CaF_2 to SrF_2 to BaF_2, closer approaching the onset of the Eu^{2+} $4f \rightarrow 5d$ absorptions. Figure 11.10 shows strikingly similar theoretical and experimental spectra. The IVCT absorption band was fitted with a Voigt spectral profile in order to allow for a rough estimation of its position and its width, as indicated with the green filled area in Fig. 11.10(top). The correspondence between computed and experimental band positions is remarkable, of the order of 2000 cm^{-1}, indicating further the correctness of the theoretical IVCT prediction. The fact that these absorptions are very similar with IVCT absorptions identified in mixed-valence Eu chloride compounds, where Eu^{2+} and Eu^{3+} occupy fixed crystallographic positions [10, 11], reinforces the theoretical assignment.

Once there is evidence of the involvement of IVCT states of Eu^{2+}/Eu^{3+} pairs in photon absorptions of Eu-doped phosphors, there is no reason to believe they will not be involved in other radiative and non-radiative decay processes of the same materials. It is obvious from the diagrams that these states can act as emission quenchers and it should be remarked that their impact on phosphor performance can be substantial even when few mixed-valence pairs are present. This is so because, even at low doping concentrations, fast thermal IVCT makes the distribution of Eu^{2+}/Eu^{3+} pairs across the sample very dynamic, which makes the IVCT non-radiative decay path created by a few Eu^{3+} ions virtually accessible to all Eu^{2+} ions. This can explain, e.g., that oxidation of only a small fraction of Eu^{2+} in the blue phosphor $BaMgAl_{10}O_{17}$:Eu has a much more severe effect on the luminescence efficiency than decreasing Eu^{2+} content [51]. Non-radiative IVCT decay can hence be a microscopic explanation for concentration quenching.

11.2.6 Invariance of IVCT Absorption Onset Across the Lanthanide Series

It is common to have traces of Eu^{3+} in Eu^{2+}-doped materials and traces of Eu^{2+} in Eu^{3+}-doped materials even after strong efforts to get rid of them. The same is true for Yb^{2+} and Yb^{3+}. As a matter of fact, the multivalent nature of these elements sets complex scenarios that challenge widespread single-valent-ion models [52] (cf.

Sect. 11.4). But the interest of the interplay between divalent and trivalent ions in a doped phosphor extends in fact to the whole lanthanide series because, although with a few exceptions the trivalent species is the most stable and only a few divalent doped lanthanide ions are stable in regular conditions, the entire series of rare earths can exist in the divalent state under suitable treatments, like X-ray and gamma irradiation, e.g. in CaF_2 [53]. After detailed theoretical studies on IVCT manifolds of Eu^{2+}/Eu^{3+} (Sect. 11.2.5) and Yb^{2+}/Yb^{3+} (Sect. 11.2.1), those of divalent/trivalent Ln^{2+}/Ln^{3+} pairs in Ln-doped CaF_2 for the entire lanthanide series (Ln=Ce, Pr, Nd, Pm, Sm, Eu, Gd, Tb, Dy, Ho, Er, Tm, Yb) have also been addressed in multiconfigurational ab initio calculations [54] and the results are summarized here.

The finding is that the IVCT states of all Ln^{2+}/Ln^{3+} lanthanide pairs in CaF_2 start at basically the same energy. This result is compatible with a simple model of IVCT states [55] (Sect. 7.5) according to which, for given ligands and coordination number, the maximum of the absorption to the first IVCT state (i.e. the onset of the full IVCT absorption band) depends only on two simple structural properties of the electron donor and acceptor active centers that show very small variation across the series: the difference between their lanthanide-ligand bond lengths, $\Delta d_{DA} = d(\text{Ln}^{2+}\text{-L}) - d(\text{Ln}^{3+}\text{-L})$, and the average of their breathing mode vibrational frequencies.

This implies the invariance of the IVCT absorption onset across the lanthanide series in a given host. The onset sets an energy threshold for partial or total luminescence quenching when mixed-valence lanthanide pairs are formed in a doped host. Such an important host property can be predicted with relatively simple structural calculations of two oxidation states of a single lanthanide element in the host.

The multiconfigurational ab initio theoretical IVCT configuration coordinate energy diagrams of Ln^{2+}/Ln^{3+} mixed-valence pairs in CaF_2 are shown in Fig. 11.11. In it, the full IVCT absorption band is also shown for the Sm^{2+}/Sm^{3+} pair, together with the contribution of the absorption to the lowest IVCT state, which defines the IVCT absorption onset (in blue in the rightmost second row panel), the quantity which is invariant across the series.

According to the simple predictive model of IVCT states [55], the IVCT onset $E^{\text{abs}}_{\text{IVCT}}$ can be expressed in terms of the coordination number of donor and acceptor n, the ligand mass μ (in amu), the average of the breathing mode vibrational frequencies of donor and acceptor centers $\bar{\nu}_{ave}$ (in cm^{-1}), and the difference between the lanthanide-ligand distances in the donor and acceptor centers Δd_{DA} (in Å), as

$$E^{\text{abs}}_{\text{IVCT}} = 0.02966 \cdot n \cdot \mu \cdot \bar{\nu}^2_{ave} \cdot (\Delta d_{DA})^2 . \tag{11.3}$$

The validity of this model is shown in Table 11.2, which illustrates not only the invariance across the series of the IVCT absorption onset, but also of the structural properties that occur in Eq. 11.3. Application of these results to predictions of IVCT absorption onsets in other hosts, which is to say their energy thresholds for partial or total luminescence quenching due to the formation of mixed-valence lanthanide pairs are illustrated in Table 11.3.

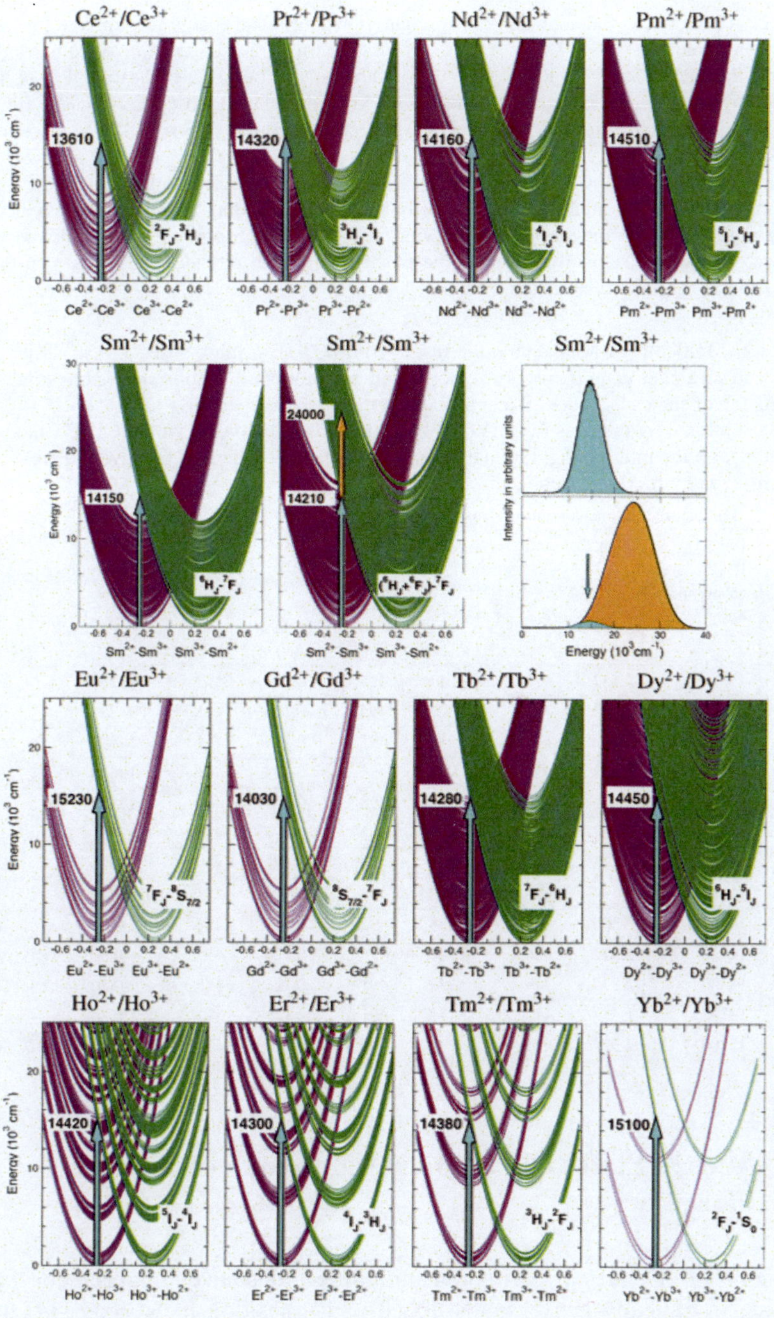

Fig. 11.11 (Continued)

◂**Fig. 11.11** (Continued) Calculated IVCT configuration coordinate energy diagrams of Ln^{2+}/Ln^{3+} mixed-valence pairs in CaF_2. Abscissa axes represent electron transfer reaction coordinates Q_{et}, in Å. Only indicated $Ln^{2+}(4f^{N+1})/Ln^{3+}(4f^{N})$ levels are shown. The onsets of the $Ln^{2+}4f \rightarrow Ln^{3+}4f$ IVCT absorptions, i.e. the maxima of the lowest individual IVCT absorption bands, are given with turquoise vertical arrows. In the second row, the diagrams of Sm^{2+}/Sm^{3+} are given for $Sm^{2+}(4f^{6}\,{}^{7}F_J)/Sm^{3+}(4f^{5}\,{}^{6}H_J)$ states (left panel) and $Sm^{2+}(4f^{6}\,{}^{7}F_J)/Sm^{3+}(4f^{5}\,{}^{6}H_J,{}^{6}F_J)$ states (center panel). In the right panel, the IVCT onset, i.e. the first individual IVCT absorption band, is shown in turquoise; above, as calculated with the left diagram and below, as calculated with the with the center diagram, together with the full IVCT absorption band, which is shown in orange. The turquoise arrow indicates the position of the so defined IVCT onset on the full $Sm^{2+}4f \rightarrow Sm^{3+}4f$ IVCT absorption band

Table 11.2 Multiconfigurational ab initio results for IVCT absorption onsets of Ln^{2+}/Ln^{3+} pairs in CaF_2 and several structural properties: donor and acceptor bond lengths d_{Ln-F}, breathing mode vibrational frequencies $\bar{\nu}_{a_{1g}}$, donor-aceptor bond length differences $\Delta d_{DA}=d_{Ln^{2+}-F}-d_{Ln^{3+}-F}$, and average vibrational frequencies $\bar{\nu}_{ave}$. IVCT absorption onsets predicted with the IVCT model and differences with the multiconfigurational ab initio calculations. Energies are given in cm^{-1}, bond lengths in Å, and vibrational frequencies in cm^{-1}

	Multiconfigurational ab initio calculations							Model	
	Ln^{2+}/Ln^{3+} IVCT absorption onset	Ln^{2+} donor		Ln^{3+} acceptor				IVCT absorption onset (Eq. 11.3)	
		d_{Ln-F}	$\bar{\nu}_{a_{1g}}$	d_{Ln-F}	$\bar{\nu}_{a_{1g}}$	Δd_{DA}	$\bar{\nu}_{ave}$		diff.
Ce	13610	2.432	420	2.311	494	0.121	457	13790	180
Pr	14320	2.422	426	2.300	495	0.122	461	14230	−90
Nd	14160	2.412	427	2.289	495	0.123	461	14500	340
Pm	14510	2.401	426	2.278	495	0.123	461	14460	−50
Sm	14150	2.387	422	2.266	498	0.121	460	13970	−180
Eu	15230	2.387	421	2.260	495	0.127	458	15250	20
Gd	14030	2.375	423	2.252	494	0.123	459	14340	310
Tb	14280	2.368	422	2.244	494	0.124	458	14540	260
Dy	14450	2.360	420	2.235	492	0.125	456	14650	200
Ho	14420	2.352	420	2.226	495	0.126	458	14980	560
Er	14300	2.345	419	2.219	491	0.126	455	14820	520
Tm	14380	2.337	417	2.212	491	0.125	454	14520	140
Yb	15100	2.329	421	2.200	491	0.129	456	15600	500

11.3 MMCT States of Co-doped Lanthanide-Activated Phosphors

Although MMCT states involving transition metal ions have long been considered important in optically active solids doped with transition metal ions [44], those involving lanthanide ions haven't been the subject of systematic research. Sometimes they have been, however, tentatively considered to play interesting roles in luminescent processes without further proof, like Pr-to-Ti MMCT, which was assumed to quench the green-blue luminescence and enhance the red emission in Pr^{3+}-doped

Table 11.3 Theoretical lanthanide IVCT absorption onsets (or energy thresholds for partial or total luminescence quenching due to the formation of mixed-valence lanthanide pairs) of several hosts.

Host	Mixed-valence pair	IVCT absorption onset	Calculated Ln	Reference
CaF_2	Ln^{2+}/Ln^{3+}	14500	All	[24, 47, 54]
SrF_2	Ln^{2+}/Ln^{3+}	18000	Eu,Yb	[24, 47]
BaF_2	Ln^{2+}/Ln^{3+}	22500	Eu,Yb	[24, 47]
$SrCl_2$	Ln^{2+}/Ln^{3+}	8100	Yb	[24]
CaS	Ln^{2+}/Ln^{3+}	10400	Eu	[54]
SrS	Ln^{2+}/Ln^{3+}	12800	Eu	[54]
BaS	Ln^{2+}/Ln^{3+}	18100	Eu	[54]
SrS	Ln^{3+}/Ln^{4+}	10000	Ce	[43] (empirical)
$Cs_2LiLuCl_6$	Ln^{3+}/Ln^{4+}	10000	Ce	[34]
$CaTiO_3$	Ln^{3+}/Ln^{4+}	12000	Pr	[5]
$CaZrO_3$	Ln^{3+}/Ln^{4+}	12800	Pr	[5]

compounds [3] before systematic multiconfigurational ab initio calculations showed it was due to host LMCT states (Sect. 11.1), or an absorption in YPO_4:Ce^{3+},Sm^{3+} tentatively assigned to Ce-to-Sm MMCT [56].

Here the focus is put in multiconfigurational ab initio calculations on MMCT between lanthanide ions. Section 11.3.1 covers the experimentally proven Eu-to-Tb MMCT IR luminescence in the doubly doped phosphor CaS:Eu^{2+},Tb^{3+} and its multiconfigurational ab initio calculation, which also explains why the similar material SrS:Eu^{2+},Tb^{3+} does not show an equivalent emission. Section 11.3.2 is dedicated to MMCT from Eu^{2+} to all trivalent lanthanides in CaF_2.

11.3.1 Broadband Infrared LEDs Based on Europium-to-Terbium Charge Transfer Luminescence

Efficient broadband IR luminescence, cooperatively originating from Eu^{2+} and Tb^{3+} dopants in CaS has been reported in [57]. The singular IR emission (cf. Fig. 11.12), which is useful in a wide range of emerging applications that exploit near-IR spectroscopy, overlaps with the red Eu^{2+} emission, ranges up to 1200 nm with a FWHM of 195 nm, and is efficiently excited with visible light. And a near-IR LED for versatile spectroscopic use was fabricated with this emission. But, besides its practical values, the emission has been experimentally proven to be a metal-to-metal charge transfer luminescence, using data from luminescence spectroscopy, microscopy and X-ray spectroscopy. This Eu-to-Tb radiative electron transfer is, in fact, the first

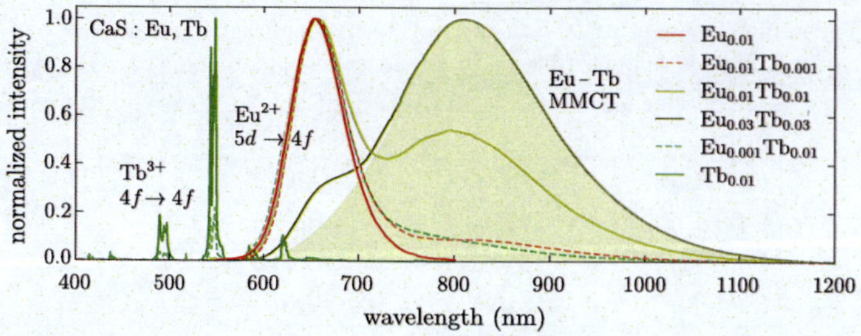

Fig. 11.12 Emission spectra of CaS:Eu,Tb for various doping concentrations under 285 nm excitation, showing the simultaneous occurrence of the regular Eu^{2+} $5d \rightarrow 4f$ and Tb^{3+} $4f \rightarrow 4f$ emissions and the broadband MMCT emission. Measured at room temperature. Adapted from [57]

experimentally proven MMCT emission. Interestingly, the emission is absent in the similar co-doped material SrS:Eu^{2+},Tb^{3+}.

Multiconfigurational ab initio calculations attribute the IR emission of Eu^{2+},Tb^{3+}-co-doped CaS to radiative decay of a metastable MMCT state of a Eu^{2+}/Tb^{3+} pair, i.e. to emission from an excited Eu^{3+}-Tb^{2+} configuration to the ground state of the Eu^{2+}-Tb^{3+} configuration [57]. Equivalent calculations on SrS:Eu^{2+},Tb^{3+} show that the MMCT emission is quenched because the excited Eu^{3+}-Tb^{2+} configuration, rather than being metastable, decays non-radiatively to Eu^{2+}-Tb^{3+}. The interpretative success of the calculations allow them to anticipate how to fine-tune the characteristics of the MMCT luminescence.

The multiconfigurational ab initio results for the excited states of Eu^{2+}/Tb^{3+} pairs in CaS and SrS are summarized in Fig. 11.13.

Regarding the individual ions, the electronic manifold of Eu^{2+} consists in a single ground state level $4f^7(^8S_{7/2})$ separated from a very dense $4f^65d^1$ manifold by about 16000–17000 cm^{-1} [57]. Eu^{3+} and Tb^{3+} have conjugate ground state configurations with $4f^6(^7F_0)$ and $4f^8(^7F_6)$ multiplets, respectively, and their higher-lying $4f^6(^5D_0)$ and $4f^8(^5D_4)$ states are the main $4f$-$4f$ emitting levels. The high energies of the $4f^{N-1}5d^1$ manifolds make them unnecessary for the study of the IR luminescence. In Tb^{2+}, the $4f^85d^1$ and $4f^9$ configurations are close in energy and, whereas the ground state belongs to the former in CaS ($1\ \Gamma_{7g}$), it belongs to the latter in SrS ($1\ \Gamma_{7u}$). Comparisons between theoretical and experimental levels of Eu^{2+} and Eu^{3+} were excellent, with quantitative agreements of less than 300 cm^{-1} for excitation energies and at most a few % for equilibrium bond lengths and vibrational frequencies [58].

The configuration coordinate diagrams show that the lowest states of the Eu^{3+}-Tb^{2+} configurations of the Eu^{2+}/Tb^{3+} pairs lie around 20000 cm^{-1} above the ground state and that these MMCT states have an impact on the decay dynamics after an $4f^7 \rightarrow 4f^65d$ excitation of the Eu^{2+} ion. In effect, the excited electron that occupies an Eu^{2+} $5d$ molecular orbital can be non-radiatively transferred to a Tb^{3+} ion and create a transient Eu^{3+}-Tb^{2+} configuration. Shortly after, an electron is transferred

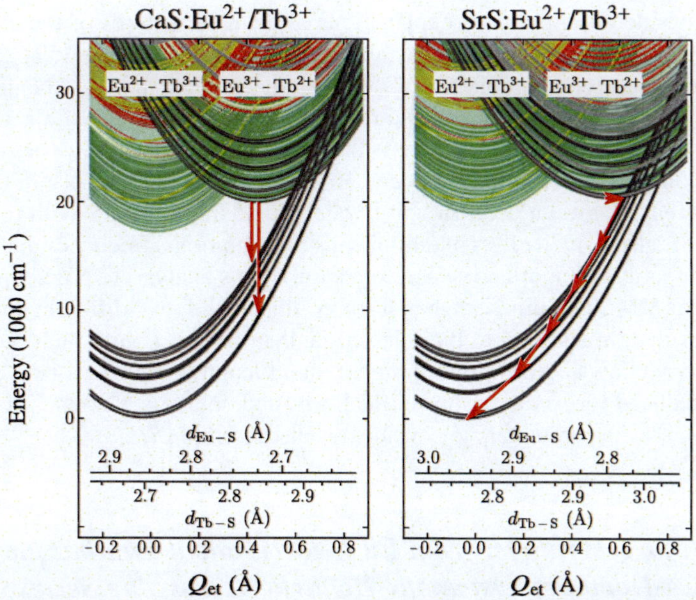

Fig. 11.13 MMCT configuration coordinate energy diagram of an Eu^{2+}/Tb^{3+} pair in CaS and SrS. For the Eu^{2+}-Tb^{3+} configuration of the pair, $4f^7(^8S)$-$4f^8(^7F)$ (black), $4f^65dt_{2g}$-$4f^8(1A_{1g})$ (green for high-spin, red for low-spin), and $4f^7(^8S)$-$4f^8(^5D,^5L,^5G)$ (gold) states are shown. For the Eu^{3+}-Tb^{2+} configuration of the pair, $4f^6(^7F)$-$4f^85d(1\Gamma_{7g})$ (black), $4f^6(1A_{1g})$-$4f^85dt_{2g}$ (green for high-spin, red for low-spin states), and $4f^6(1A_{1g})$-$4f^9$ (grey) states are shown. A green background represents a quasi-continuum formed by all other levels of the pair where both ions are in excited states

back from Tb^{2+} to Eu^{3+} recreating the original Eu^{2+}-Tb^{3+} configuration of the pair, with Eu^{2+} in its $4f^7(^8S_{7/2})$ electronic configuration and Tb^{3+} in its $4f^8(^7F_J)$ electronic configuration.

But there is an important difference in the electron back-transfer between the CaS and the SrS hosts: In CaS, the minimum of the lowest MMCT state is metastable and it decays radiatively to the structurally stressed ground configuration of Eu^{2+}-Tb^{3+} (vertical red arrows in Fig. 11.13 (left panel)), so yielding the IR emission of the material. The large horizontal offset between MMCT and ground states is the reason for the band to be very broad. In SrS, however, the minimum of the lowest MMCT state is crossed by the branches of the stressed Eu^{2+}-Tb^{3+} states, enabling efficient non-radiative decay due to fast bottom crossover (red arrows in Fig. 11.13 (right panel)) that quenches any radiative decay. This is in full agreement with the different experimental behaviors of Eu^{2+}, Tb^{3+} co-doped CaS and SrS and it provides an explanation.

It is interesting to remark that the presence of the MMCT manifold does not only allow the existence of a new IR emission, but it also has an effect on the excitation and their consideration provides explanations of detailed features of the excitation spectra of the IR emission, but also of the regular Eu^{2+} red emission [57].

As it was discussed, contrary to CaS, SrS is not a suitable host to obtain a luminescent MMCT state. Taking CaS as a reference, the right shift of the Eu^{3+}-Tb^{2+} parabolas in SrS that makes them to cross the Eu^{2+}-Tb^{3+} stressed states at their minimum and enable the non-radiative decay, is mainly due to an increase in the bond length difference between divalent and trivalent lanthanides in SrS with respect to CaS. (There is also a small effect of diminishing the vibrational frequency, but this has a minor comparative importance.) An expansion of this analysis to predict presence or absence of MMCT luminescence in hosts co-doped with two different lanthanide ions implies that, in addition to the selected lanthanide pair (whith their respective ionization potentials and electron affinities), bond length differences between oxidized and reduced forms of the ions in the hosts, and, in a lesser extent, vibrational frequencies, are the parameters to consider, as discussed in [57].

11.3.2 Charge Transfer from Eu^{2+} to Trivalent Lanthanide Co-Dopants: Systematic Behavior Across the Series

Multiconfigurational ab initio calculations on direct metal-to-metal charge transfer between the Eu^{2+} luminescence activator and a trivalent lanthanide co-dopant Ln^{3+} ($Ln^{3+} = Ce^{3+}, Pr^{3+}, Nd^{3+}, Pm^{3+}, Sm^{3+}, Eu^{3+}, Gd^{3+}, Tb^{3+}, Dy^{3+}, Ho^{3+}, Er^{3+}, Tm^{3+}$ and Yb^{3+}) in CaF_2 revealed how the structures and energies of the electronic manifolds of the co-dopants, their vibrational force constants and their structural properties in the host, determine five different luminescent behaviors of CaF_2:Eu^{2+},Ln^{3+} co-doped materials after MMCT absorption: formation of stable traps, MMCT emission, emission quenching, Ln^{3+} emission, and Eu^{2+} emission [59].

Partial configuration coordinate energy diagrams of Eu^{2+}/Ln^{3+} with the states of mixed-valence pairs in CaF_2 including the relevant terms of the divalent and trivalent ions (only the states derived from a single-SLJ set of free-ion multiplets are included) are shown in Fig. 11.14.

In Eu^{2+}, Ln^{3+} co-doped materials it is obviously interesting to be able to predict cases of potential metastability of the Eu^{3+}-Ln^{2+} configuration of the pair, i.e. the lowest MMCT state, which could either (i) become an electron trap and induce the possibility of energy storage by getting involved in trapping/detrapping phenomena, or (ii) decay radiatively yielding Ln^{2+}-to-Eu^{3+} back-transfer MMCT emission (e.g. the MMCT IR luminescence of CaS:Eu^{2+},Tb^{3+} (Sect. 11.3.1)). These cases would compete with non-radiative decay of the MMCT state, which may result in luminescence quenching or in new luminescences from metastable excited states of Eu^{2+} or Ln^{3+}.

Five possibilities schematically represented in Fig. 11.15 have been found. They are the following:

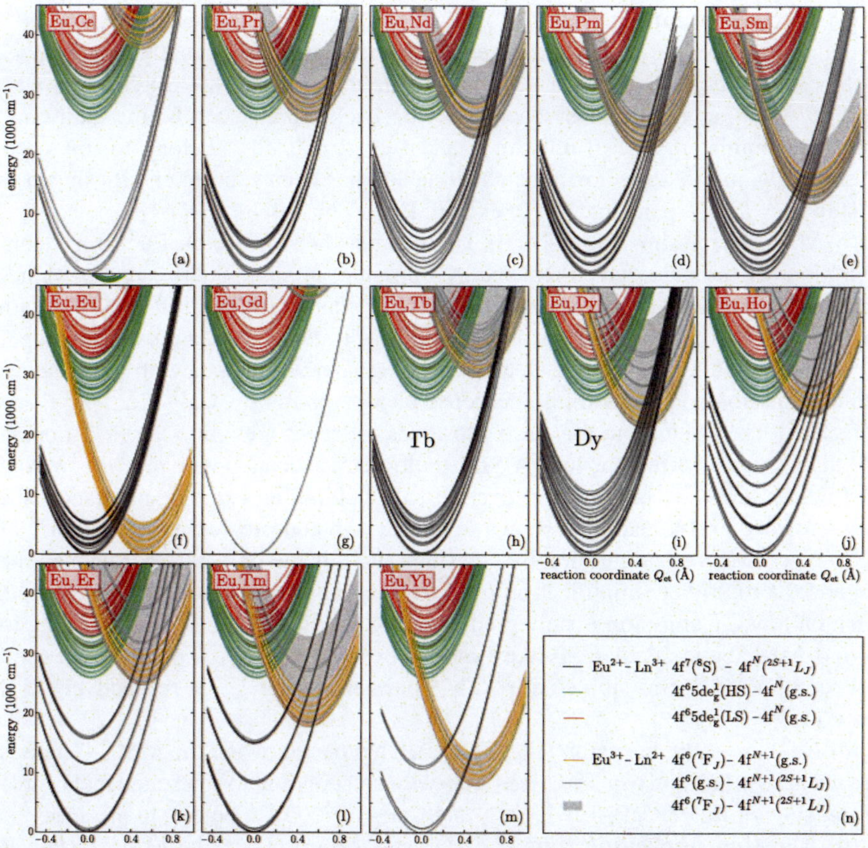

Fig. 11.14 Panels **a–m** MMCT configuration coordinate energy diagrams of Eu^{2+}/Ln^{3+} pairs in CaF_2. Abscissa axes are Q_{et}, in Å. Ordinate axes are energies relative to the ground state, in $1000\ cm^{-1}$. Panel **n** Legend for the other panels. Colors indicate dominant electron configuration or dominant total spin

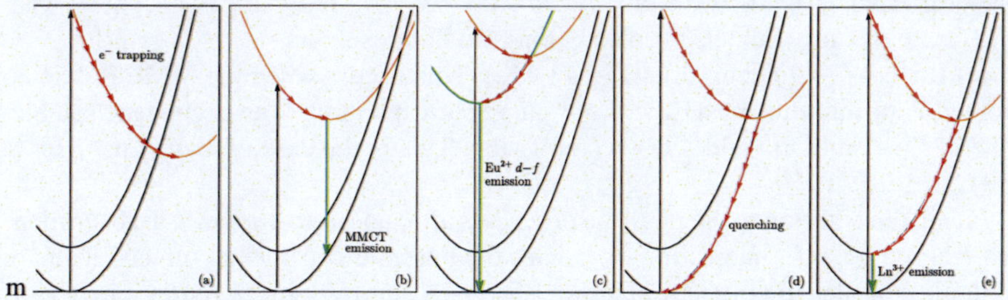

Fig. 11.15 Potential phenomena triggered by MMCT absorption: **a** electron trapping, **b** MMCT emission, **c** Eu^{2+} $d-f$ emission, **d** quenching, **e** Ln^{3+} $f-f$ emission. Abscissa axes are electron transfer reaction coordinates Q_{et} and ordinate axes are energies

(a) **Electron trapping** (Fig. 11.15a). This occurs when metastable Eu^{3+}-Ln^{2+} configurations of the pairs have their energy minimum at significantly larger values of the electron transfer configuration coordinate than their rightmost crossing with the Eu^{2+}-Ln^{3+} configurations. Then, electron back-transfer can be thermally and optically stimulated. CaF_2:Eu^{2+},Sm^{3+} and CaF_2:Eu^{2+},Yb^{3+} belong to this case. Interestingly, in the CaF_2 host, this phenomenon has been experimentally shown to exist in Eu^{2+}/Sm^{3+} pairs and not to exist in Eu^{2+}/Tm^{3+} pairs [60, 61].

(b) **MMCT emission** (Fig. 11.15b). This occurs when metastable Eu^{3+}-Ln^{2+} configurations of the pairs have their energy minimum at significantly smaller values of the electron transfer configuration coordinate than their leftmost crossing with the Eu^{2+}-Ln^{3+} configurations, provided the MMCT state does not cross with higher Eu^{2+}-Ln^{3+} excited states. Then, electron back-transfer can happen with spontaneous photon emission and a broad luminescence band. Apparently, CaF_2:Eu^{2+},Pr^{3+} and CaF_2:Eu^{2+},Tb^{3+} belong to this case. However, a deeper view into the Pr^{3+} dopant $4f^2$ manifold shows that the lowest SLJ multiplets associated with the ^3H, ^3F, and ^1G free-ion terms do interact and overlap [5], yielding an excited state landscape more crowded and spread than shown in Fig. 11.14b and precluding the stability of the lowest MMCT configuration by crossings (Pr^{3+} is the only case here where the single-SLJ multiplet simplification used for the diagrams is a drawback.) Hence, potential MMCT emission is only predicted in CaF_2:Eu^{2+},Tb^{3+}. It is interesting to remind that diabatic calculations like these and experiments have shown presence and absence of MMCT emission of Eu^{2+}/Tb^{3+} pairs in CaS and SrS, respectively [57] (Sect. 11.3.1).

(c) **Eu^{2+}** $5d{\rightarrow}4f$ emission (Fig. 11.15c). This occurs when the MMCT state is crossed at its minimum by a Eu^{2+}-Ln^{3+} curve corresponding to an excited metastable $4f^65d$ state of Eu^{2+}. CaF_2:Eu^{2+},Ce^{3+} and CaF_2:Eu^{2+},Gd^{3+} belong to this case.

(d) **Emission quenching** (Fig. 11.15d). This occurs when the MMCT state is crossed at its minimum by the ground state curve, like in Eu^{2+}/Tm^{3+}, and when it is crossed by low-lying states that relax further down to the ground state, like CaF_2:Eu^{2+},Pr^{3+}, CaF_2:Eu^{2+},Nd^{3+}, CaF_2:Eu^{2+},Pm^{3+}, and CaF_2:Eu^{2+},Dy^{3+}. Then, emission quenching is predicted for the co-doped materials CaF_2:Eu^{2+},Pr^{3+}, CaF_2:Eu^{2+},Nd^{3+}, CaF_2:Eu^{2+},Pm^{3+}, CaF_2:Eu^{2+},Dy^{3+}, and CaF_2:Eu^{2+},Tm^{3+}. The IVCT case of CaF_2:Eu^{2+},Eu^{3+} also belongs to this case.

(e) **Ln^{3+}** $4f{\rightarrow}4f$ emission (Fig. 11.15e). This occurs when the MMCT state is crossed at its minimum by a Eu^{2+}-Ln^{3+} curve corresponding to an excited metastable state of Ln^{3+} able to yield $4f{\rightarrow}4f$ emission. This is the case of CaF_2:Eu^{2+},Ho^{3+} and CaF_2:Eu^{2+},Er^{3+}.

Finally, in addition to the possible processes after an Eu-to-Ln MMCT absorption just considered, it is interesting to remark the possible detrimental effects of Ln^{3+} co-doping on the important CaF_2:Eu^{2+} $5d{\rightarrow}4f$ luminescence. As the energy of the lowest MMCT state goes down towards Eu^{2+},Sm^{3+} in the light lanthanides and towards Eu^{2+},Tm^{3+} and Eu^{2+},Yb^{3+} in the heavy lanthanides, the stressed branches of the lowest MMCT states may serve as non-radiative channels to directly or thermally depopulate the Eu^{2+} $4f^65d$ emitting state and quench the Eu^{2+} $5d{\rightarrow}4f$ emission.

11.4 Role of Compensator-to-Dopant Charge Transfer in the Complex Luminescence of the CaF_2:Yb Multivalent System

The multivalent nature of lanthanide optical activators sets a complex scenario that challenges widespread single-valent-ion models. The archetypal complexity inherent to the 2+/3+ mixed-valence nature of Yb in CaF_2 was faced in a combined experimental and theoretical effort [52], using XANES, extended X-ray absorption fine structure (EXAFS), and UV-vis absorption and emission spectroscopy, plus multiconfigurational ab initio calculations.

The facts are summarized as follows: CaF_2:Yb samples with low Yb concentration (0.01%) and +2 and +3 valence fractions $f(2+) = 0.3$, $f(3+) = 0.7$, exhibit anomalous emission (cf. Sect. 11.2.1) at 200 K. After long X-ray exposure at 200 K they show maximal, huge Yb^{3+} reduction to $f(2+) = 0.7$, $f(3+) = 0.3$. Although the reduction doubles the number of Yb^{2+} ions in the samples, the anomalous emission completely disappears. The reduction is stable at low temperatures, but, upon warming to 300 K, it is reverted to the original state ($f(2+) = 0.3$, $f(3+) = 0.7$); however, the re-oxidation does not recover the anomalous emission at 300 K nor at low temperature. Surprisingly, after annealing at 900 K, some anomalous emission is again observed even below 150 K, even though annealing does not change the Yb^{2+} concentration or the +2 and +3 fractions.

The complexity of the reversible conversion between +2 and +3 valences of Yb is understood considering that: (1) Yb^{3+} sites require a charge compensating defect such as interstitial fluorine, F_i^- (the samples grown in an O-free environment with a slight excess of F do not show any O^{2-}); (2) under X-ray exposure, electrons are ejected from them and metastable interstitial F_i^0 neutral atoms, much more mobile than F_i^-, are created; and (3) multiconfigurational ab initio calculations on compensator-to-dopant F_i^--to-Yb^{3+} electron transfer reveal the existence of an important, critical dopant-compensator distance: when F_i^- and Yb^{3+} are closer than the critical distance, spontaneous re-oxidation happens; when they are at a longer distance, re-oxidation is not spontaneous. This is the central message of the calculations and it would hold if the diabatic approximation were overcome.

The results of the diabatic configuration coordinate energy diagrams of F_i^-/Yb^{3+} pairs obtained in multiconfigurational ab initio calculations [52] are summarized in Fig. 11.16.

The distance-dependence of the (vertical) energy difference between the lowest states of the F_i^-/Yb^{3+} and F_i^0/Yb^{2+} configurations originates directly from the change in ionic charges upon the transfer, modeled here by a Coulomb potential (given by the last contribution in Eq. 7.13). The figure shows that the dopant-compensator distance determines the stability of the X-ray induced reduction: Locally compensated trivalent ytterbium Yb^{3+}_{LC} will re-oxidize spontaneously after X-ray induced reduction (Fig. 11.16b), but non-locally compensated trivalent ytterbium Yb^{3+}_{NLC} will remain reduced (Fig. 11.16d).

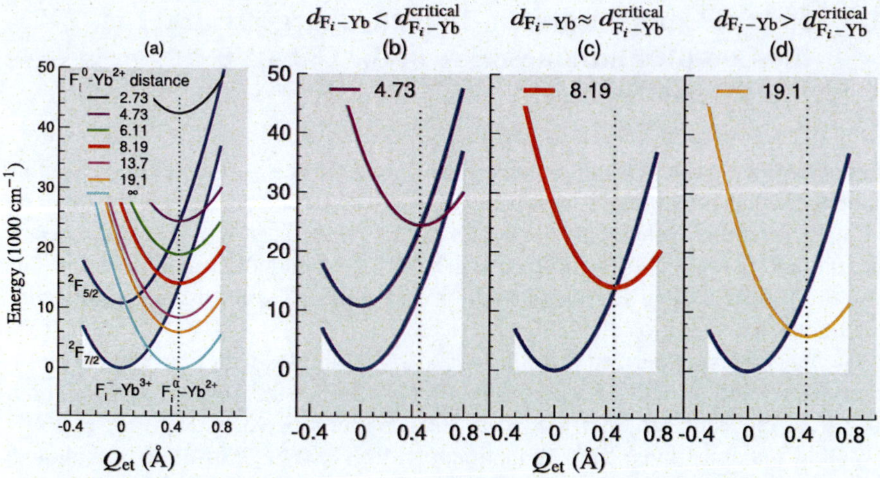

Fig. 11.16 Configuration coordinate energy diagram of F_i^-/Yb^{3+} compensator/dopant pairs in CaF_2 as a function of the F_i-Yb distance d_{F_i-Yb}. In **a**, diagrams at several d_{F_i-Yb} values are plotted together. Typical diagrams at d_{F_i-Yb} shorter than, similar to, and longer than the critical F_i-Yb distance $d^{critical}_{F_i-Yb}$ are shown in **b**, **c**, and **d**, respectively. The respective F_i-Yb distances chosen are 4.73 Å, 8.19 Å, and 19.19 Å. In **b**, the hole-electron pair F_i^0-Yb^{2+} spontaneously recombines. In **d**, the hole-electron pair F_i^0-Yb^{2+} is metastable

$$F_i^- + Yb_{LC}^{3+} \overset{\text{x-ray}}{\rightleftharpoons} F_i^0 + Yb^{2+} \;,\; d_{F_i-Yb} \leq d^{critical}_{F_{i-Yb}} \tag{11.4}$$

$$F_i^- + Yb_{NLC}^{3+} \overset{\text{x-ray}}{\longrightarrow} F_i^0 + Yb^{2+} \;,\; d_{F_i-Yb} > d^{critical}_{F_i-Yb}\,. \tag{11.5}$$

This conclusion agrees with extensive experimental data that suggest that only those trivalent lanthanide ions in cubic sites are reduced to the divalent state in the CaF_2 host [62]. Putting it together with considerations on the mobility of F_i^0 ions as it is done in [52] allows to explain that there is a temperature at which X-ray induced reduction is maximal, that both X-ray exposure and warming after X-ray exposure enhance local compensation and quench Yb anomalous emission (which is an Yb^{2+}/Yb^{3+}_{NLC} IVCT emission), and that annealing at high temperatures increases non-local compensation and recovers Yb anomalous emission.

References

1. E.G. Reut, A.I. Ryskin, Phys. Status Solidi A **17**, 47 (1973)
2. E. Pinel, P. Boutinaud, R. Mahiou, J. Alloys Compd. **380**, 225 (2004)
3. P. Boutinaud, R. Mahiou, E. Cavalli, M. Bettinelli, J. Lumin. **122–123**, 430 (2007)

4. P. Boutinaud, E. Pinel, M. Dubois, A.P. Vink, R. Mahiou, J. Lumin. **111**, 69 (2005)
5. Z. Barandiarán, M. Bettinelli, L. Seijo, J. Phys. Chem. Lett. **8**, 3095 (2017)
6. C.K. Jrgensen, Prog. Inorg. Chem. **12**, 101 (1970)
7. R.A. Marcus, Annu. Rev. Phys. Chem. **15**, 155 (1964)
8. G.C. Allen, N.S. Hush, Prog. Inorg. Chem. **8**, 357 (1967)
9. O. Berkooz, M. Malamud, S. Shtrikman, ssc **6**, 185 (1968)
10. C. Wickleder, Z. Naturforsch. **57b**, 901 (2002)
11. C. Wickleder, Z. Anorg, Allg. Chem. **628**, 1815 (2002)
12. W. van Schaik, S. Lizzo, W. Smit, G. Blasse, J. Electrochem. Soc. **140**, 216 (1993)
13. P.P. Feofilov, Opt. Spektrosk. **1**, 992 (1956)
14. D.S. McClure, C. Pédrini, Phys. Rev. B **32**, 8465 (1985)
15. B. Moine, B. Courtois, C. Pédrini, J. Phys. France **50**, 2105 (1989)
16. E.G. Reut, Opt. Spectrosk. **40**, 99 (1976)
17. B. Moine, C. Pédrini, D.S. McClure, H. Bill, J. Lumin. **40 & 41**, 299 (1988)
18. M.F. Reid, P.S. Senanayake, J.P.R. Wells, G. Berden, A. Meijerink, A.J. Salkeld, C.K. Duan, R.J. Reeves, Phys. Rev. B **84**, 113110 (2011)
19. P.S. Senanayake, J.P.R. Wells, M.F. Reid, G. Berden, A. Meijerink, R.J. Reeves, J. Lumin. **133**, 81 (2013)
20. H. Witzke, D.S. McClure, B. Mitchell, in *Luminescence of Crystals, Molecules, and Solutions*, ed. by F.E. Williams (Plenum, New York, 1973), p. 598
21. Z. Pan, C. Duan, P.A. Tanner, Phys. Rev. B **77**, 085114 (2008)
22. P. Dorenbos, J. Phys.: Condens. Matter **15**, 2645 (2003)
23. M. Grinberg, J. Lumin. **131**, 433 (2011)
24. Z. Barandiarán, L. Seijo, J. Chem. Phys. **141**, 234704 (2014)
25. C. MacKeen, F. Bridges, M. Kozina, A. Mehta, M.F. Reid, J.P.R. Wells, Z. Barandiarán, J. Phys. Chem. Lett. **8**, 3313 (2017)
26. R.B. Hughes-Currie, K.V. Ivanovskikh, J.P.R. Wells, M.F. Reid, R.A. Gordon, J. Phys.: Condens. Matter **28**, 135502 (2016)
27. R.B. Hughes-Currie, K.V. Ivanovskikh, J.P.R. Wells, M.F. Reid, R.A. Gordon, L. Seijo, Z. Barandiarán, J. Phys. Chem. Lett. **8**, 1175 (2017)
28. S.M. Kaczmarek, T. Tsuboi, M. Ito, G. Boulon, G. Leniec, J. Phys.: Condens. Matter **17**, 3771 (2005)
29. R.B. Hughes-Currie, A.J. Salkeld, K.V. Ivanovskikh, M.F. Reid, J.P.R. Wells, R.J. Reeves, J. Lumin. **158**, 197 (2015)
30. Z. Barandiarán, L. Seijo, J. Chem. Phys. **143**, 144702 (2015)
31. A. Bessière, P. Dorenbos, C. van Eijk, L. Pidol, K. Krämer, H. Güdel, J. Phys.: Condens. Matter **16**, 1887 (2004)
32. P. Dorenbos, E. van Loef, C. van Eijk, K. Krämer, H. Güdel, Phys. Rev. B **68**, 125108/1 (2003)
33. A. Bessière, P. Dorenbos, C. van Eijk, K. Krämer, H. Güdel, A. Galtayries, J. Lumin. **117**, 187 (2006)
34. L. Seijo, Z. Barandiarán, J. Chem. Phys. **141**, 214706/1 (2014)
35. C.K. Duan, P.A. Tanner, A. Meijerink, V. Babin, J. Phys.: Condens. Matter **21**, 395501 (2009)
36. C.K. Duan, P.A. Tanner, V. Makhov, N. Khaidukov, J. Phys. Chem. A **115**, 8870 (2011)
37. I.B. Bersuker, *The Jahn-Teller Effect and Vibronic Interactions in Modern Chemistry* (Plenum Press, New York and London, 1984)
38. G.C. Allen, N.S. Hush, Prog. Inorg. Chem. **8**, 357 (2007)
39. M. Robin, P. Day, Adv. Inorg. Chem. Radiochem. **10**, 247 (1968)
40. A. Werner, Z. Anorg, Chem. **12**, 46 (1896)
41. N.S. Hush, Prog. Inorg. Chem. **8**, 391 (1967)
42. P.F. Smet, L. Moreels, Z. Hens, D. Poelman, Materials **3**, 2834 (2010)
43. D. Kulesza, J. Cybińska, L. Seijo, Z. Barandiarán, E. Zych, J. Phys. Chem. C **119**, 27649 (2015)
44. G. Blasse, Struct. Bond. **76**, 153 (1991)
45. W. Strek, R. Tomala, L. Marciniak, M. Lukaszewicz, B. Cichy, M. Stefanski, D. Hreniak, A. Kedziorski, M. Krosnicki, L. Seijo, Phys. Chem. Chem. Phys. **18**, 27921 (2016)

46. J. Wang, P.A. Tanner, J. Amer. Chem. Soc. **132**, 947 (2010)
47. J. Joos, L. Seijo, Z. Barandiarán, J. Phys. Chem. Lett. **10**, 1581 (2019)
48. A.A. Kaplyanskii, P.P. Feofilov, Opt. Spectrosc. **13**, 129 (1962)
49. E. Loh, Phys. Rev. **175**, 533 (1968)
50. B. Moine, C. Pédrini, B. Courtois, J. Lumin. **50**(1), 31 (1991)
51. L. Amidani, K. Korthout, J.J. Joos, M. van der Linden, H.F. Sijbom, A. Meijerink, D. Poelman, P. F., P. Glatzel, Chem. Mater. **29**(23), 10122 (2017)
52. C. MacKeen, F. Bridges, L. Seijo, Z. Barandiarán, M. Kozina, A. Mehta, M.F. Reid, J.P.R. Wells, J. Phys. Chem. C **121**, 28435 (2017)
53. D.S. McClure, Z. Kiss, J. Chem. Phys. **39**, 3251 (1963)
54. I. Neefjes, J.J. Joos, Z. Barandiarán, L. Seijo, J. Phys. Chem. C **124**, 2619 (2020)
55. Z. Barandiarán, A. Meijerink, L. Seijo, Phys. Chem. Chem. Phys. **17**, 19874 (2015)
56. N.R.J. Poolton, A.J.J. Bos, G.O. Jones, P. Dorenbos, J. Phys.: Condens. Matter **22**, 185403 (2010)
57. J.J. Joos, D.V. der Heggen, L.I.D.J. Martin, L. Amidani, P.F. Smet, Z. Barandiarán, L. Seijo, Nat. Commun. **11**, 3647 (2020)
58. J.J. Joos, P.F. Smet, L. Seijo, Z. Barandiarán, Inorg. Chem. Front. **7**, 871 (2020)
59. J.J. Joos, I. Neefjes, L. Seijo, Z. Barandiarán, J. Chem. Phys. **154**, 064704 (2021)
60. P.P. Feofilov, Opt. Spektrosk. **12**, 531 (1962)
61. B. Welber, J. Chem. Phys. **42**, 4262 (1965)
62. J.L. Merz, P.S. Pershan, Phys. Rev. **162**, 217 (1967)